Resilient Energy Systems

TOPICS IN SAFETY, RISK, RELIABILITY AND QUALITY

Volume 19

Aims and Scope. Fundamental questions which are being asked these days of all products, processes and services with ever increasing frequency are:

What is the risk?
How safe is it?
How reliable is it?
How good is the quality?
How much does it cost?

This is particularly true as the government, industry, public, customers and society become increasingly informed and articulate.

In practice none of the three topics can be considered in isolation as they all interact and interrelate in very complex and subtle ways and require a range of disciplines for their description and application; they encompass the social, engineering and physical sciences and quantitative disciplines including mathematics, probability theory and statistics.

The major objective of the series is to provide series of authoritative texts suitable for academic taught courses, reference purposes, postgraduate and other research and practitioners generally working or strongly associated with areas such as:

Safety Assessment and Management
Emergency Planning
Risk Management
Reliability Analysis and Assessment
Vulnerability Assessment and Management
Quality Assurance and Management

Special emphasis is placed on texts with regard to readability, relevance, clarity, applicability, rigour and generally sound quantitative content.

For further volumes:
http://www.springer.com/series/6653

Ion Bostan • Adrian Gheorghe • Valeriu Dulgheru
Ion Sobor • Viorel Bostan • Anatolie Sochirean

Resilient Energy Systems

Renewables: Wind, Solar, Hydro

Springer

Ion Bostan
Technical University of Moldova
Stefan Cel Mare Boulevard 168
2004 Chişinău
Republic of Moldova

Valeriu Dulgheru
Mechanical Engineering
Technical University of Moldova
Stefan Cel Mare Boulevard 168
2004 Chişinău
Republic of Moldova

Viorel Bostan
Technical University of Moldova
Stefan Cel Mare Boulevard 168
2004 Chişinău
Republic of Moldova

Adrian Gheorghe
Engineering Management
and Systems Engineering
Old Dominion University
Norfolk, VA, USA

Ion Sobor
Technical University of Moldova
Stefan Cel Mare Boulevard 168
2004 Chişinău
Republic of Moldova

Anatolie Sochirean
Technical University of Moldova
Stefan Cel Mare Boulevard 168
2004 Chişinău
Republic of Moldova

ISSN 1566-0443
ISBN 978-94-007-4188-1 ISBN 978-94-007-4189-8 (eBook)
DOI 10.1007/978-94-007-4189-8
Springer Dordrecht Heidelberg New York London

Library of Congress Control Number: 2012938960

Printed on acid-free paper

Springer is part of Springer Science+Business Media (www.springer.com)

Contents

Chapter 1
Introduction

Can you imagine life without television, cars or computers, without being able to prepare your food every day, without lighting in the house, without heating during the cold seasons of the year, etc.? But all this is the result of creative activity of scientists and inventors, especially during the last 200 years. All this may disappear during the first half of the present century, following the drastic depletion of natural reserves of fossil fuels. Increased energy consumption leads to a continuous increase in the volume of extracting fossil fuels, which provides more than 85% of energy use today. Currently, the annual energy consumption is equivalent to more than 11 billion tons of conventional fuel or 459 EJ (459 10^{18} J), of which only 15.4% is of non-fossil origin. As the world population increases, and the level of energy endowment of the economy grows, simultaneously, this figure is steadily increasing, which fact will have serious consequences. Most acceptable fuels, economically, – oil and natural gas – are supposed to be about exhausted in 30–50 years.

Today, most of the energy needed for daily consumption is produced by burning fossil fuels – coal, oil and natural gas. Several million years, plants and animals decomposing led to the formation of fossil fuels, which, however, were consumed during about 200 years, practically. Millions of years, Earth's atmosphere formed a whole plant system, and during a 200 years period, but, particularly in the last 100 years, the environment was seriously jeopardized and the world is facing an ecological disaster.

In 1960, 3,000 TWh of electricity were produced and consumed. In 1970 it increased up to 6,000 TWh. 150,000 TWh were consumed in 2000. Even, if it is possible to reduce electricity consumption in industrialized countries (U.S., Germany, Japan, etc.) by half, and at the same time to increase consumption per capita, by only 25% of global electricity, in India, China etc. – third world countries, the overall demand would double from the today's one. What energy sources are able to meet these requirements? Increasing power generation by burning traditional fossil fuels, would further endanger the ecological system. The expectation of power engineering professionals is based on finding new solutions and processes that

I. Bostan et al., *Resilient Energy Systems*, Topics in Safety, Risk, Reliability and Quality 19, DOI 10.1007/978-94-007-4189-8_1,

would meet the energy needs of the mankind in the coming decades or centuries. At the forefront, nuclear energy solutions have been related to, but after the power failures (the U.S. Three Miles Island, Chernobyl in Ukraine, and Fukushima in Japan), the need to develop alternative energy solutions, environmentally friendly, has become an imperative.

The concept of energy efficiency (or energy optimization) became, at present, one of the main concerns of the mankind in the whole world. With the first oil crisis of the early 70s, human society began to realize, more than ever the need for a sustainable strategy, by increasing the efficiency of energy use and implementing energy efficiency programs by taking into account the depletion of fossil fuel reserves on Earth. Today, we speak of a global energy policy and a concerted strategy to reduce harmful emissions into the atmosphere, based on concrete economic and technical solutions for rational use of fossil fuel reserves (which still have the main share of energy production) and valorization of renewable energy resources on a large scale, the so-called "clean" energy or non-conventional energy, as an alternative to the current system of fuel reserves on Earth. Renewable energies (solar, wind, hydro, etc.) are environmentally friendly but today they are not able to meet these ever-growing needs.

These two serious issues -the energy crisis and environmental impact- are global problems of humanity, which settlement falls on the shoulders of engineers. Because the world is so dependent on energy, because most of Earth's population uses fossil fuels to meet energy needs, which causes a high degree of environmental pollution, it is strictly necessary to seek sustainable and environmentally friendly energy systems. Energy sources producing the least possible pollution will eventually be found. Since all traditional energy sources pollute the environment, renewable energy is practically to a large extent, devoid of this negative effect of environmental pollution.

Diversification of energy sources becomes an economic and environmental imperative. These alternative energies are called renewable energy. What are these alternative sources of energy? The best known renewable and acceptable energy sources are solar energy (direct, photovoltaic and thermal), wind (as a derivative of solar energy), hydraulic (using potential and kinetic energy of water), geothermal, bioenergy, etc.

Renewable energy techniques can be used, both, as a centralized and/or largely decentralized energy source. Decentralized sources are particularly advantageous, especially for rural and isolated consumers. According to UN information, about two billion people lack access to electricity, while about 40 countries have no national electricity networks. The cost of the electricity transmission network is bigger in proportion of 4:1 or more to the cost of power plants. From this point of view, promoting decentralized energy sources is advantageous, as key programs of rural electrification and poverty reduction in rural areas.

Disadvantages of decentralized energy systems are, inter alia, the operational instability and inability of electricity storage and redistribution, distribution networks themselves having the role of electricity storage, too.

The current status of renewable energy impact includes social, political, economic, environmental and technological aspects.

Special attention is paid to energy potential, history of development and production of renewable energy conversion systems: solar, wind, hydro, sea waves. Today the European Parliament declared a clear signal how to promote renewable energies in the EU until 2020 in order to achieve 25% share of primary energy. At the same time, the European Council on Renewable Energy (ECRE) has been established for this purpose. "*Parliament's vote today is an historic opportunity for the Commission to test citizens' demands for renewable energy. Together with the Parliament, leaders in building and securing legislative proposals for all three sectors must be: for electricity, heating and biofuel. The Commission should focus its attention on eliminating gaps in EU legislation for renewable energy - heating and cooling*", said Oliver Schafer, policy director of ECRE. Leaders in research and professionals in various fields of renewable energies have met at EUREC Agency (European *Renewable Energy* Research Centres *Agency)* to redefine the position of renewable energy conversion technologies, in the context of meeting global energy needs, and recommend specific directions for development of technology for each branch based on that analysis.

Aspiring to the future, Freeman Dyson of the University of Oxford, UK, has justified that technological exchanges fundamentally alter our ethical and social arrangements and that three new technologies that are growing rapidly – renewable energy, genetic engineering and global communication, have the potential to create a more uniform distribution of global health today.

Developing countries that have low or inadequate resources of e.g. oil and coal have cleared large areas of forests, are today in a situation of using non-conventional energy resources, such as solar, hydraulic, wind, or combined with conventional fuels for higher efficiency.

Increased costs associated with fuel procurement, transportation and maintenance of engines, coupled with difficulties in quantifying the environmental costs make renewable energy an attractive alternative to combustion engines fuel-based generators.

The efforts of researchers are increasingly targeted to revitalization of existing technologies to reduce energy consumption and waste production, and also to use unconventional energy sources, wherever possible. Desire to have more efficient production processes in terms of energy consumption has occurred, especially after the energy crisis of 1970, which led to rapid price increase.

During about 200 years, mankind has created a great energy complex, difficult to imagine, providing basic services: lighting, heating, refrigeration, transport, technological processes, etc. Modern standards of welfare, education and health cannot be maintained without energy. However, it was recognized that the emergence of modern energy is guilty of many environmental problems. It is necessary to find a compromise between the growing demand for energy services and the critical need to protect the environment. In the opinion of the authors of this book, the solution is to return humanity to renewable energy sources. In this way, it will naturally repair the chain, broken 200 years ago. The nineteenth century was the century of steam, the twentieth century – of the electricity, and the twenty-first century will be the century of renewable energy or will not be at all.

Recognizing the importance of developing technologies of renewable energy exploration, authors dedicate this book to young researchers and inventors, whose creativity will depend on whether the twenty-first century will be the century of renewable energy. It is youth with creative thinking mostly mission to realize fully the contents of this manual.

Given the first approach to the subject of such an extensive and high complexity area, authors are grateful to the readers for the understanding shown to any errors; remain indebted for the communication of impressions and suggestions.

Finally, the authors wish to express their gratitude for the help offered by colleagues from the Departments of "Theory of Mechanisms and Machine Parts", "Electro mechanics" and "Construction and Mechanical Structures" of the Technical University of Moldova, financial support from the Swiss Science Foundation, doctoral student Ersin Ancel from Old Dominion University, Norfolk, VA for editorial support. Discussions and comments offered by the Scientific Steering Committee of the current Integrated Risk Governance (IRG) – Project are highly acknowledged.

Chapter 2
Use of Renewable Energy: World, European and National Perspectives

2.1 Recent Consumption of Primary Energy at World and at National Levels

2.1.1 World Consumption of Primary Energy in the World

Energy conversion, consumption and conservation: The Law of Conservation of Energy is one of fundamental laws of physics: in physical processes energy cannot be destroyed or decreased, it can be converted (transformed) from one form of energy to another [1]. Table 2.1 lists several processes of energy conversion from one form to another [2].

In this context some questions arise: What is primary energy? What is energy consumption? When energy is consumed the following things occur: the conversion of chemical energy stored in coal, oil, natural gas, wood or of the energy stored in the atomic kernel, or the conversion of water kinetic and gravity energy, or wind kinetic energy, or the conversion of solar radiation energy into heating and lighting of our houses, or into electrical energy to make machine-tools operate or into kinetic energy to move the vehicles. In other words, energy consumption is equivalent to energy conversion. Energy conservation means the production of more goods and services by converting a small quantity of primary energy into heating, electricity, lighting (electromagnetic energy), etc.

Primary energy is all energy contained in the original sources. Nowadays, the main original sources are fossil fuels (coal, oil and natural gas), biofuels – wood, wooden wastes, agricultural waste, manure, etc. Other sources can be added, like hydraulic and geothermal energy, and other sources of renewable energy such as solar and wind energy, and nuclear energy as well. Consumers are interested in the satisfaction of their energy needs: they need thermal energy for heating and cooking, electrical energy for lighting, transportation and production of goods, etc.

The quantity of supplied energy, the form of useful energy needed, the amount of energy losses and the price to pay for the used energy are important things for consumers. Figures 2.1 and 2.2 show the three notions of energy (primary, supplied

I. Bostan et al., *Resilient Energy Systems*, Topics in Safety, Risk, Reliability and Quality 19, DOI 10.1007/978-94-007-4189-8_2,

Table 2.1 Conversion of various forms of energy

Forms of energy	To chemical	To thermal	To electrical	To electro-magnetic	To mechanical
From chemical	Plants, food stuffs	Burning, anaerobic digestion	Battery, fuel cell	Candle and votive candle radiation, phosphorescence	Human and animal muscles
From thermal	Pyrolytic gasification	Heat pump, heat exchanger	Thermo-couple	Fire	Gas turbine, steam turbine, memory materials
From electrical	Battery, electrifier	Boiling plate, toaster, iron	Converter, frequency converter	Fluorescent lamp, luminescent diode	Electrical motor, electromagnet
From electro-magnetic	Photosynthesis	Solar collector	Photovoltaic cell	Laser	Solar radiation pressure
From mecha-nical	Crystallization (formation of solid crystalline out of liquid)	Friction brake	Generator	Firestone	Water wheel, windmill, pendulum, flywheel

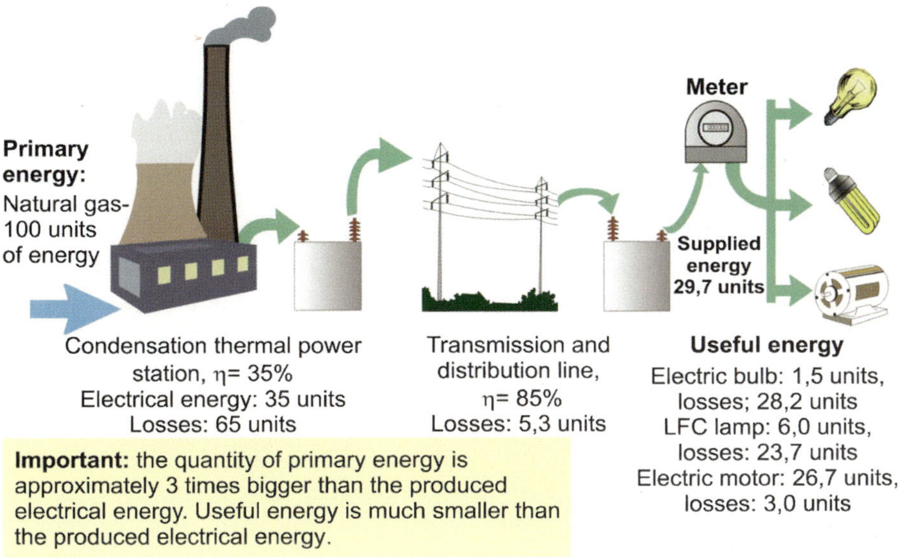

Fig. 2.1 Primary, supplied and useful energy: the case of energy from fossil sources

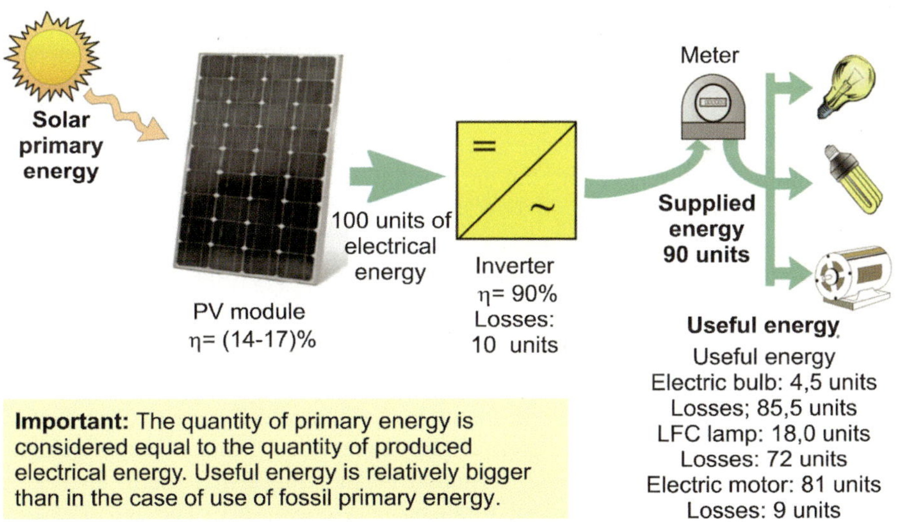

Fig. 2.2 Primary energy, delivered and useful: the case of renewable energy

and useful), for two systems of electrical energy production – from natural gas at a condensation thermal power station (Fig. 2.1) and from conversion of solar energy into electrical energy using photovoltaic modules (PV). In both cases three devices are used: the electric bulb, the compact fluorescent lamp (CFL) and the electrical motor. The efficiency of devices is equal to 5%, 20% and 90%, respectively.

For systems based on fossil energy sources the biggest losses occur at the production of electrical energy (about 66%), followed by losses in the electrical networks at transportation and distribution, and losses in the devices. The last can be quite significant, if device efficiency is small. For example, for an electric bulb the global efficiency of energy conversion equals to 1.5% only, in other words, from 100 units of primary energy only 1.5 units are applied usefully, and 98.5 units bring about atmosphere thermal and gas pollution with the greenhouse effect. Global efficiency of a CFL lamp is 6%, and for an electric motor – 26.7%.

National [3] and international statistics published data about the production and consumption of various forms of energy sources. Data is presented both in natural measurement units and in conventional measurement units. The concept of *"consumption of primary energy sources"* is applied in economic estimations, in the determination of energy efficiency and in the specific energy consumption, etc. The above concept contains all forms of the consumed energy – fuel, electrical or thermal energy obtained from the fossil sources, from the nuclear or hydraulic energy, from the energy of various forms of biomass, or from geothermal, solar and wind energy, etc. It is important to know generally accepted agreements concerning the presentation of statistical data related to the consumption of primary energy resources. UNO, International Energy Agency (IEA) and some countries recommend the following methodology for primary energy resources calculation [4, 5]:

- Primary energy used for the production of electrical energy at thermal power stations (TPS: energy of fossil fuels is transformed into steam thermal energy, then into mechanical energy and, finally, into electrical energy) is defined as the result of multiplying the quantity of the produced electrical energy by 3 or division by 0.33, that in fact is the average efficiency of the thermal power station;
- Contribution of primary energy to the production of hydraulic or wind electrical energy or from other renewable sources is considered equivalent with the quantity of electrical energy. In other words, in the case of electrical energy production from renewable sources, the efficiency of the conversion process is not taken into consideration.

Obviously, the value of the conversion process efficiency is very important in this case too (Fig. 2.2). The higher the efficiency, the lower the unit cost of the produced electrical energy. As well, the overall efficiency of energy conversion is defined as the relationship of the useful energy to the solar primary energy incident on the surface of the PV module. It should be mentioned that useful energy is relatively bigger than in the case of fossil primary energy use. The agreement to use the above accepted concept of primary energy aims at outlining the fact that for the production of one unit of electrical energy from a renewable source, the same unit of primary energy will be spent, which circulates in the environment and does not change the balance of nature. The methods of energy (fossil or renewable) conversion are characterized by the efficiency factor E. The higher efficiency E, less primary energy will be spent at input for the production of one energy unit at output.

Table 2.2 World reserves of oil and gas, January 2005

	Crude oil (billion of barrels)		Natural gas (millions of cubic km)	
Countries/regions	*Oil and Gas Journal*	*World Oil*	*Oil and Gas Journal*	*World Oil*
North America	214, 771	40, 874	7,477.7	7,752.4
Central and South America	100, 595	75, 973	7,096.9	7,837.3
Europe	17, 613	18, 125	5,497.1	5,419.3
Eurasia	77, 832	89, 898	55,314.4	79,978.9
Middle East	729, 341	708, 289	71,448.3	72,030.9
Africa	100, 784	112, 410	13,498.8	14,185.8
Asia and Oceania	36, 246	36, 244	10,875.8	12,036.4
World Total	**1,277,182**	**1,081,813**	**171,940**	**192,720**

Efficiency factor ***E*** is determined by the mathematical relation:

$$E = \frac{E_{useful}}{E_{primary}} 100\%. \tag{2.1}$$

The biggest part of primary energy sources on Earth belongs to fossil fuels based on carbon. World reserves of fossil fuels (oil, coal, gas) estimate at approximately [6] (BBEP – billions of barrels equivalent petroleum; MB – millions of barrels; MBEP – millions of barrels equivalent petroleum):

- Oil: from 1,050,691 to 1,277,702 millions of barrels (from 167 to 203 km^3);
- Gas: from 171,040 to 192,720 km^3 ($6.8 \cdot 0.182 = 1.239$ BBEP);
- Coal: 1,081,279 million tons ($1{,}081{,}279 \cdot 0.9 \cdot 4.879 = 4.786$ BBEP),

Daily consumption of fuels in 2002 (7.9 is the conversion ratio of tons equivalent petroleum in barrels equivalent petroleum BEP):

- Oil: $(10{,}230 \cdot 0{,}349) \cdot 7.9/365 = 77$ MB/day;
- Gas: $(10{,}230 \cdot 0{,}212) \cdot 7.9/365 = 47$ MBEP/day;
- Coal: $(10{,}230 \cdot 0{,}235) \cdot 7.9/365 = 52$ MBEP/day.

According to the most optimistic forecast done by *Oil and Gas Journal, World Oil* [7] these reserves of fossil fuels will last for:

- **32 years –** oil;
- **72 years –** gas;
- **252 years** – coal.

World reserves of oil and gas, distributed among the biggest producers, are given in Table 2.2 [6]. To note the fact that these estimates were done according to constant daily fuel consumption for year 2002. Natural gas is the most environment friendly fossil fuel and might be a partial solution to the climate warming and atmosphere pollution issues (Table 2.3).

Table 2.3 Atmosphere pollution by fossil fuels burning (kg emissions per TJ of consumed energy)

Emissions	Natural gas	Oil	Coal
Nitrogen oxides	43	142	359
Sulphur dioxides	0.3	430	731
Solid particles	2	36	1,333

Source: US Environmental Protection Agency; American Gas Association (AGA)

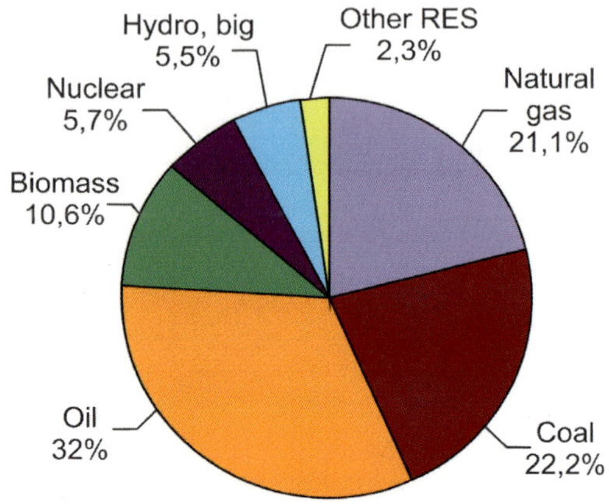

Fig. 2.3 World consumption of primary energy: 2002, 451 EJ

Natural gas could be largely used in future for the production of hydrogen from methane (CH_4), necessary for the functioning of fuel cells that will substitute gradually internal combustion engines.

There are other estimations too. According to the estimations of ARAMCO, presented at a workshop organized by OPEC in 2005, until now it was extracted about one trillion barrels or approximately 18% of the proven reserves of 5.7 billion barrels of oil. These quantities are estimated to maintain the recent level of consumption for a 140 years' time. "*Still, this should not impede the development of alternative sources of oil, because energy demands supplementary energy sources*" mentioned the reporter.

Oil, coal and gas made up 78.7% of primary energy production in 2002 (Fig. 2.3). World consumption of primary energy of all forms increased in the former century more than 10 times and, in 2002, it reached about 451EJ (451×10^{18}J) or 10,800 mln. t.e.p[1]. The share of renewable energy sources (RES) equals (15.4%). Figure 2.4 shows the structure of primary energy consumption from renewable sources: traditional biomass supplies 71.5%, hydraulic energy – 13.0%, other renewable sources – 15.5%. The concept of "*other renewable sources*" includes geothermal energy, wind and solar energy, micro- and mini-hydro energy, new forms of energy from biomass – liquid and solid biofuels, biogas.

[1]www.bp.com/centres/energy2002/index.asp and www.undp.org/seed/eap/activities/wea

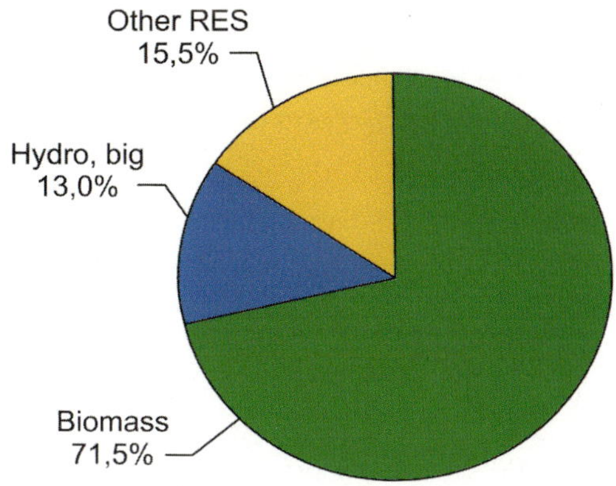

Fig. 2.4 Structure of RES consumption in the World: 2002, 69,4 EJ

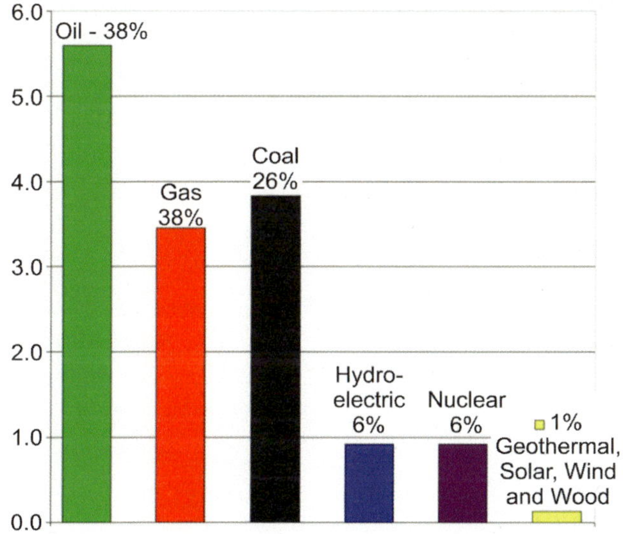

Fig. 2.5 World energy needs supply in TW, 2004. USA Department of Energy

According to the 2004 estimations of the US Department of Energy, the consumption of fossil fuels is increasing continuously; as these fuels continue to be the main energy sources used in the world (see Fig. 2.5). The coming of such nations as China and India on the world market of fuel consumption confirms the hypothesis of future growth of daily consumption of fossil fuels. These countries are very big consumers as their population makes about half of the Earth's population.

The same information, for EU countries and Romania, is presented in Fig. 2.6. Renewable energy has reached 13.8% of the world primary energy of 9,958 Mtep in 2000 [8].

The consumption of renewable energy has increased by 2% in the last 30 years. At the same time, the term *"other categories"* refers to new renewable sources

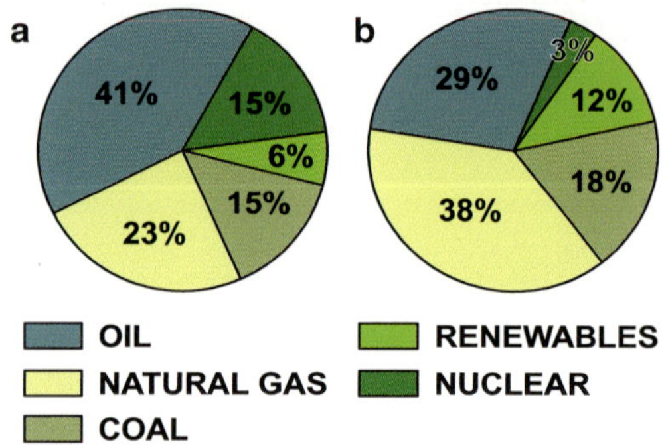

Fig. 2.6 Primary energy consumption in EU member countries (**a**) and Romania (**b**)

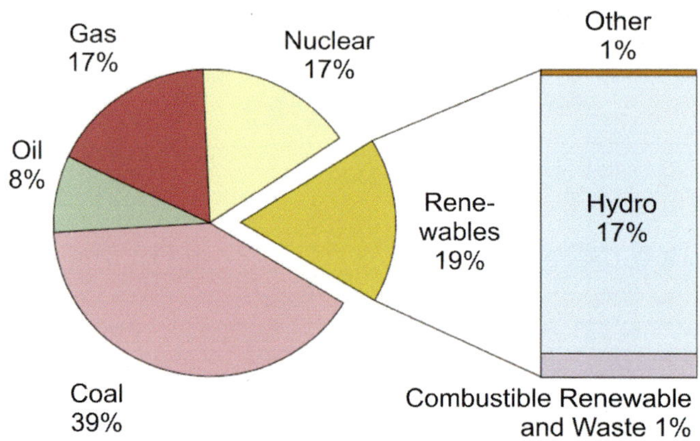

Fig. 2.7 Share of renewable energy in global production of electrical energy

and includes geothermal, solar, wind, and other energy that has grown by 9%. In comparison with the lowest consumption in 1971, the wind energy increased by 52%, followed by solar energy – by 32%. About 58% of the renewable energy is consumed by the domestic, commercial and public sectors.

Renewable energy sources have reached the volume of 19% in 2000 (Fig. 2.7) and are on the second place in the world production of electrical energy (after coal – 39%), followed by nuclear energy (17%), natural gas (17%) and oil (8%). The biggest share of renewable electrical energy belongs to hydraulic energy (92%), produced by large hydro stations constructed in dams. For the time being, the world production of electrical energy is environment unfriendly, and this is an issue for researchers to consider as well as the issue of energy resources exhaust.

In the long run, the world production of fossil fuels will reach a maximum growth, followed by a continuous reduction until it will reach the lowest level. Afterwards, the extraction will become disadvantageous economically or impossible from the engineering point of view. Countries, importing energy sources, will found

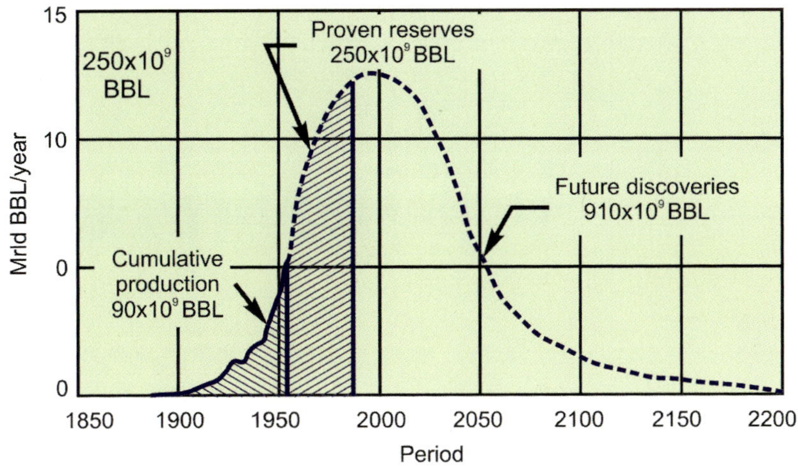

Fig. 2.8 Curve of oil production, suggested by M. K. Hubbert in 1956

themselves in difficult situation. The curve in Fig. 2.8 shows an example of oil production decline forecast, based on the Hubbert peak theory of maximum [9, 10].

The highest rate on the curve was estimated on the basis of production and exploration rates. Hubbert peak theory is based on the fundamental observation that oil reserves are limited. Hubbert has developed a method for modeling the knowledge about oil reserves and extraction rate. At first, viewed with skepticism, nowadays this method is used widely by oil producing companies to forecast the future of the oil industry. Hubbert envisaged correctly the maximum rate of the world oil production half a century before the event occurred. Still, it's a topic to discuss which fossil fuel will be considered as *the fuel of the future*. In the opinion of some experts none of the traditional fossil fuels can be considered as *the fuel of the future*, because of their reserves depletion (except coal) in the near future. Although coal reserves are relatively big, the impact of coal burning on the environment is so big that the use of coal in the current form has no future.

Further on some graphs of oil production in several oil producing countries are presented. The standard Hubbert curve is taken as the starting point (Fig. 2.9a). Coordinates $\boldsymbol{x}$ and $\boldsymbol{y}$ replace the scales of time and oil production. The curve is drawn on the basis of equation:

$$x = \frac{e^{-t}}{(1 + e^{-t})^2} = \frac{1}{2 + 2\cosh t}. \tag{2.2}$$

Figure 2.9b shows the Hubbert curve for Norway. Summary analysis of the curve demonstrates its similarity with the standard Hubbert curve. The curves in Fig. 2.10 show the production and import of oil in the USA in the years 1920–2005. The analysis of the curve shows that, in the USA, the maximum production rate was in 1970.

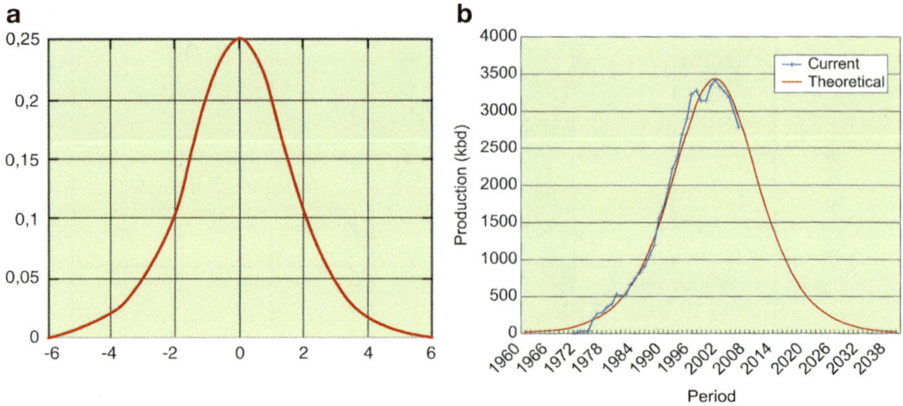

Fig. 2.9 Hubbert curve: (**a**) standard; (**b**) for Norway

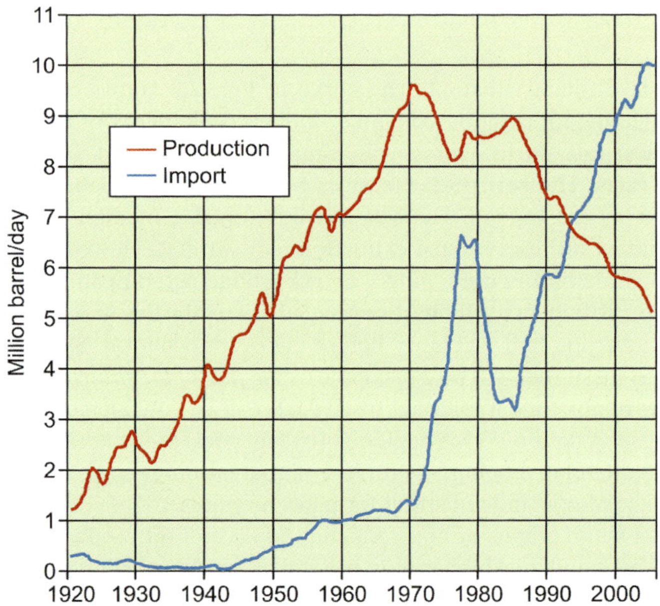

Fig. 2.10 Curve of oil production and import, SUA, 1920–2005

As result of the investigations Hubbert came to the conclusion that, after the discovery of fuel reserves (oil, coal and natural gas), at the beginning, the production grows provisionally exponential, more efficient facilities being installed. After reaching the highest rate, the decline of production follows, described by a provisionally exponential curve.

Having the data of oil production from the past times, Hubbert curve can be drawn by approximating data and it can be used to estimate future production.

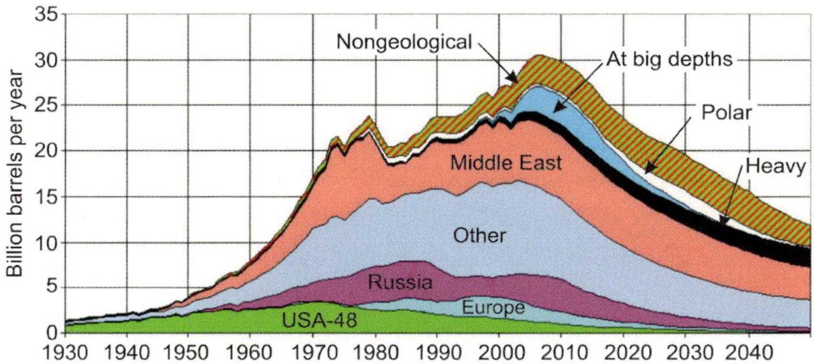

Fig. 2.11 Estimation of maximum share of oil and gas production, 2004 (conventional and unconventional)

In particular, data of maximum rates of oil production or the total quantity of the produced oil can be estimated likewise. Cavallo [9] defined Hubbert curve used to forecast the maximum rate of oil production in the USA:

$$Q(t) = \frac{Q_{\max}}{\left(1 + ae^{bt}\right)^2}. \tag{2.3}$$

where $Q_{\max}$ are the total available reserves; $Q(t)$ – cumulative production, a and b are constant. Maximum annual production is defined by the relation:

$$t_{\max} = \left(\frac{1}{b}\right) \ln \left(\frac{1}{a}\right). \tag{2.4}$$

Other methods were applied to define the maximum rate of oil production and they give almost the same results, as the whole. According to ASPO (Association for the Study of Peak Oil and Gas) (Fig. 2.11), both conventional and unconventional production will reach the maximum rate in the year 2007 [11]. About 30% of unconventional fuel for cars (Diesel fuel, in particular) is produced from coal.

Similar modeling was undertaken for other forms of fossil fuels (coal, gas). World energy consumption (its history and future) for the years 1970–2025 is shown in Fig. 2.12. Rough analysis of the graphs shows that the world stakes on oil, gas and oil use as main primary energy sources, further on. But the exploitation of resources becomes more difficult: the exploration of Russian regions, rich in hydrocarbons, expands to the Polar Circle, and the exploration of the world offshore resources – to the new depths.

Geographical distribution of hydrocarbon reserves continues to be unchanged with a big concentration of oil in the Middle East (65%) and OPEC countries, and a more or less distribution between the Russia (37%) and the Middle East (35%) for gas. The newly discovered world reserves of about 500 mln billions of barrels of oil

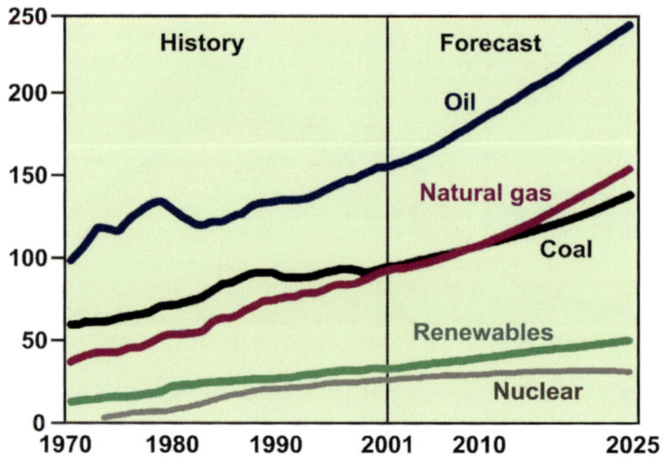

Fig. 2.12 World energy consumption, 1970–2025

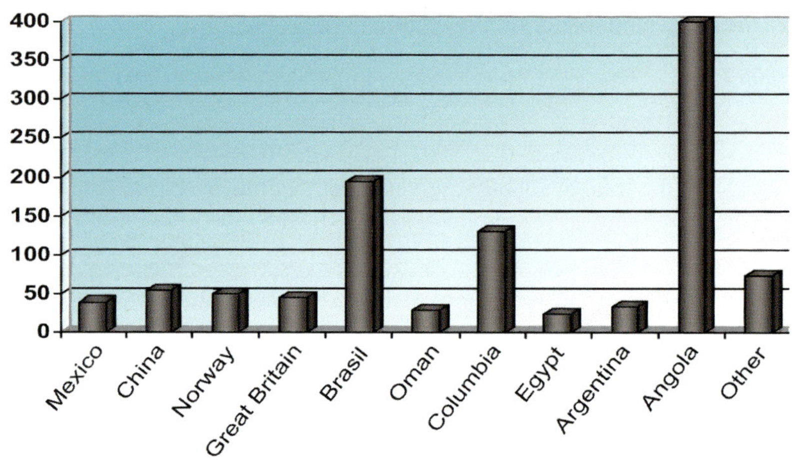

Fig. 2.13 Brasil and Angola are in the top of countries with new developed reserves of oil, 1990–1999

are concentrated in 29 oil deposits areas, including 7 coastal areas at drilling depths over 1,000 m and 29 gas deposits areas, 8 of which located in the Middle Asia and 5 in Australia. From the total of 29 gas areas 5 are located at depths over 1,000 m. Fifty percent of newly discovered gas deposits are concentrated in 10 countries, on the one hand, in Iran and Saudi Arabia, on the other hand, in Brazil and Angola (Fig. 2.13).

The new oil deposits are explored at bigger and bigger depths. The depth of coastal exploration increased from 300 m in 1978 to 1,900 m (Fig. 2.14) in 1999.

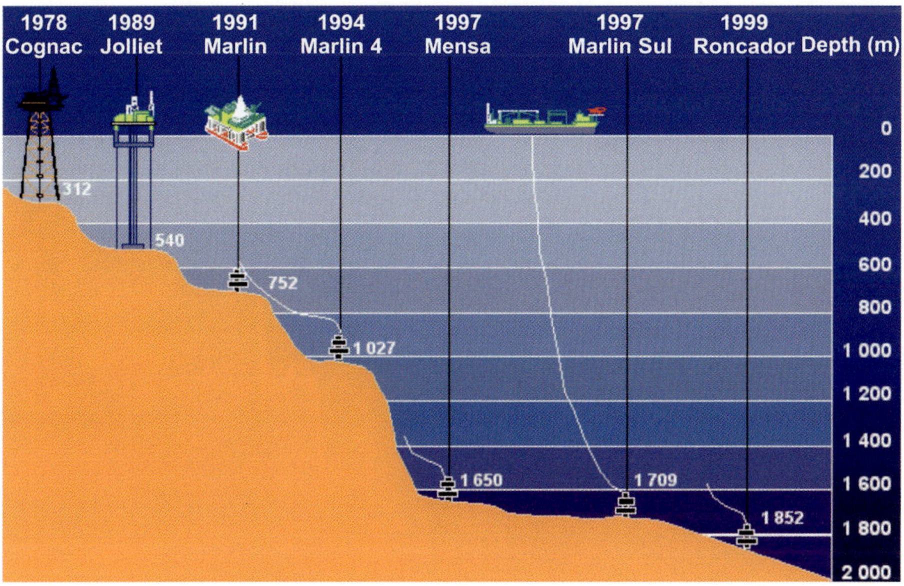

Fig. 2.14 Depth records at exploring offshore oil

During these 20 years the depth of continental shelf oil production continued to be limited by technological aspects. Nowadays, the potential of oil resources at depths have not been defined yet. But the issues of technological exploitation of these resources are significant: high pressure, low temperature, considerable depths, constant presence of ocean currents, etc. The use of conventional coastal solutions for big water depths is inconceivable, and new technologies are demanded. The majority of companies producing offshore oil have a drilling experience of about 3,000 m. The achievement of these objectives depends on the technological challenges for the next 5–10 years. An option of a future technology is shown in Fig. 2.15. We hope that the exhaustion of oil and natural gas reserves, the increase of difficulties related to their exploitation that will lead inevitably to price growth and to the understanding the threat of the ecological disaster towards which the mankind is straightening, will change the balance in favor of renewable energy sources friendly to the environment. The world energy crisis leads to the growth of prices as result of partial dissatisfaction of current needs. The prices on oil on the world market have increased constantly with some fluctuations, which can be seen in Fig. 2.16.

In the last 3 years (2002–2006) the price for one barrel of oil increased 2.5 times. The same trend is observed for the natural gas, in particular, after Russia's attempt to monopolize this market. The tendency to raise continuously the price on oil and natural gas is impending.

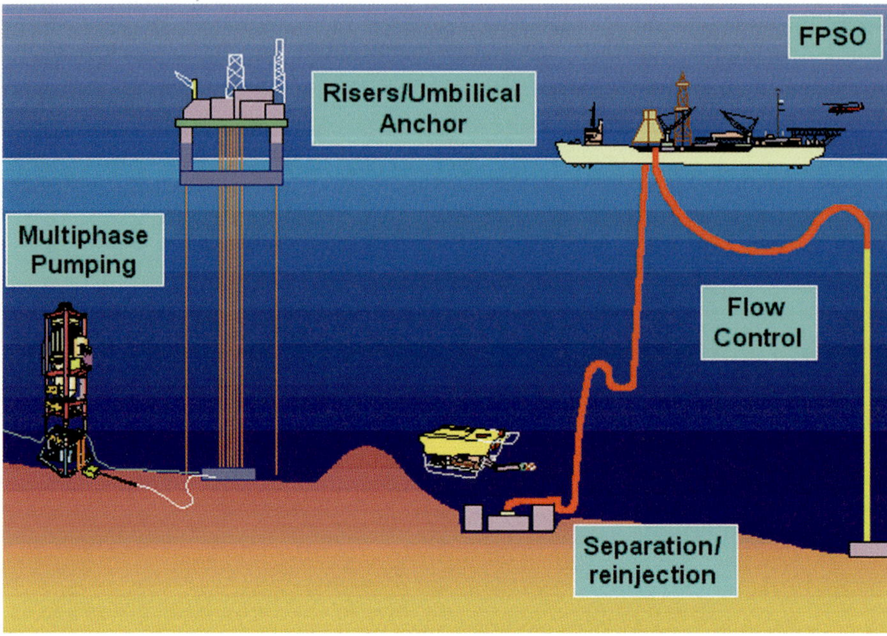

Fig. 2.15 Key-technologies for oil drilling from deep deposits

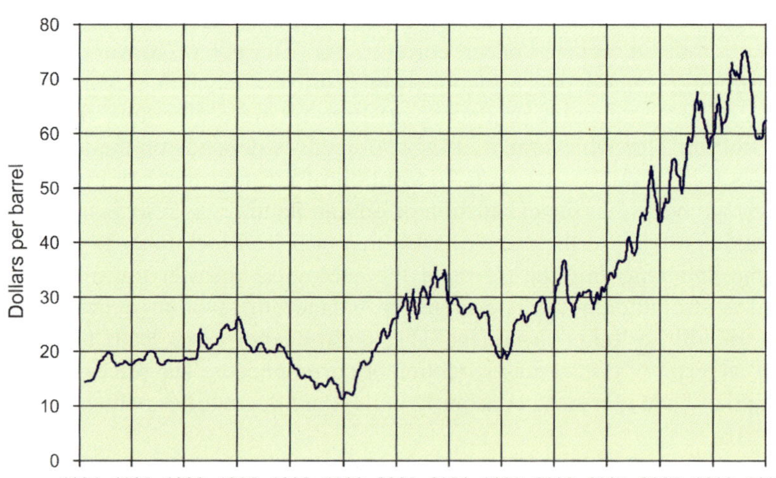

Fig. 2.16 Oil prices in 1994–2007

2.2 Energy and the Environment

2.2.1 *Factors that Change the Climate*

Climate changes are due to the variations of the Earth's atmosphere and to the various processes that take place in some regions of Terra, namely the oceans and ice calottes. These changes can be caused by the internal processes of the Planet, external factors or, more recently, by human activity. The external factors that can modify the climate are often called forcers of climate and include such processes as solar radiation variations and shifts of the planet orbit, and the concentration of gases producing "*greenhouse effect*".

2.2.1.1 Variations of Solar Radiation

Solar variations are fluctuations of the Sun's energy. Figure 2.17 portrays the last three solar cycles as solar radiation unit, number of black spots, solar explosions and 10.7 cm radio waves. Solar radiation received by the external surface of the planet does not vary much (solar constant) from the approximate value 1,366 W/m^2. The measured variation of the last two solar cycles is about 0.1% or approx. 1.3 W/m^2 [12, 13].

Only a small fraction with the ultraviolet wave length is varying to 5%. Radio emissions of the Sun define solar activity and can be measured from the Earth, as well. They are not affected seriously by the atmosphere. Lastly, solar explosions are considered a form of solar activity that might influence the life on Earth by damaging the electrical equipment of satellites, in particular.

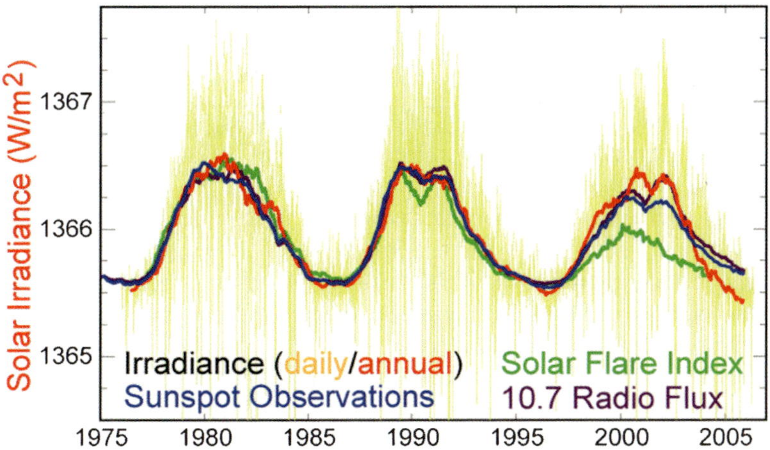

Fig. 2.17 The last three cycles of solar radiation variation

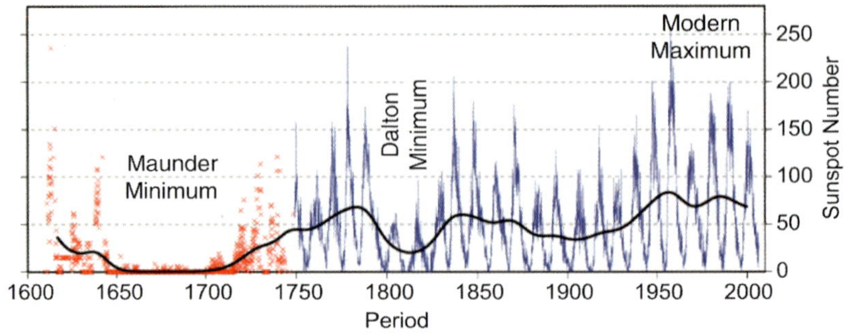

Fig. 2.18 Observations of the black spots during the last 400 years

From the researchers' point of view these variations have a considerable impact on the temperature of the planet's atmosphere. Measurements of solar radiation have been taken only for the last three cycles and are based on the observations of a big number of satellites [14]. By correlating the measurements of solar radiation and other forms of solar activity it is possible to estimate the solar activity in the past. The black spots on the Sun (areas marked by relatively low temperatures and very intensive magnetic activity) and the bright spots (generated by solar particles with dimensions from 1,000 km to 10,000 km that have a rather short lifetime–tens of minutes) are generators of solar radiation fluctuations. And this factor is one of the reasons of climate change. The observations of the last 400 years have demonstrated certain grow of the black spots on the Sun that can also be a reason of climate change (Fig. 2.18). The magnetic solar cycles last 11 years and the maximum value of the year 1950 is partially responsible for the global heating. Due to high solar activity in 1950, residual heating is considered responsible for 16–36% of current heating (1950–1999) [15].

2.2.1.2 "*Greenhouse Effect*" Concept

What is the so-called "*greenhouse effect*"? How does it appear and what is its impact on the environment? What are the current possibilities of decreasing the negative impact of the greenhouse effect on the climate? These are important issues that the mankind is challenging. It is well known that the Sun produces all energy that reaches the planet's surface – about 1,366 W/m^2. At the same time, geometrical factors and reflection properties of the surfaces are limiting the radiation absorbed by typical location of about 235 W/m^2. If this is total received heating, then the Earth surface could have a temperature of -18°C.

On the other hand, Planet's atmosphere recycles a part of radiation reflected by the Earth's surface and provides additional 324 W/m^2, and this is the source of a temperature on the surface of about $+14$°C [16]. The heating captured by atmosphere more than 75% can be attributed to the action of the greenhouse gases

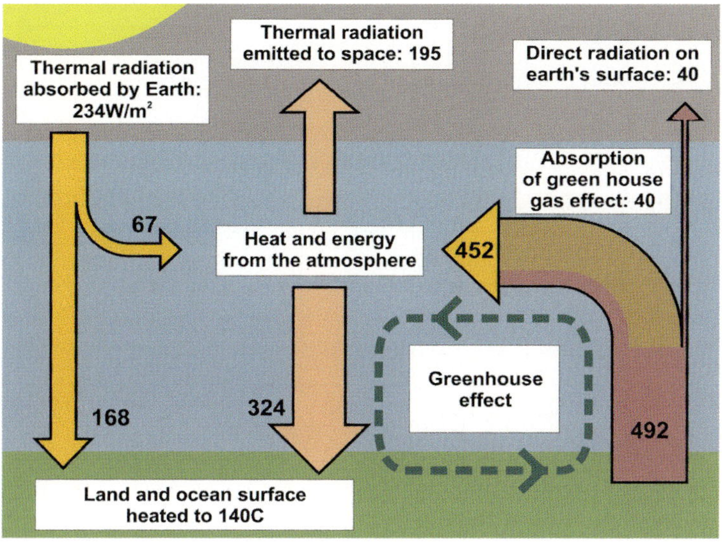

Fig. 2.19 Simplified presentation of greenhouse effect

that absorb the thermal radiation emitted by the planet's surface. In its turn, the atmosphere transfers the received energy into the space (38%) and back to the planet's surface (62%), that is transferred to any direction depending on the thermal properties and structure of the atmosphere. This process that recycles the energy into atmosphere to heat the planet's surface is known as the "greenhouse effect" and represents an essential element in the generation of climate on Terra. Under steady conditions, when the total quantity of energy is balanced exactly by the quantity of produced heat in the space, this allows the planet to maintain a constant average temperature. At the same time, recent measurements suggest that, nowadays, the planet absorbs 0.85 ± 0.15 W/m^2 more than it emits into the space. This growth associated with the global heating is considered to be caused by the recent increase of the greenhouse gas concentration. Figure 2.19 shows, in a simple way, energy flows between the space, atmosphere and the planet's surface and how these flows combine to generate heating close to the surface and produce greenhouse effect [17].

Energy exchanges are expressed in Watt per square meter (W/m^2) and have been obtained by Kiehl and Trenberth (1997). According to recent research gas concentration with the "*greenhouse*" effect in the atmosphere is the primary reason of global heating. Greenhouse gas is important for the understanding of the planet's climate history. According to these studies, greenhouse effect has a decisive role in the temperature regulation of the planet. During the last 600 million years, the concentration of carbon dioxide varied from 5,000 ppmv to less than 200 ppmv (parts per million volumes) due to the impact of geological processes and biological innovations, in particular. During the modern period, the increase of the carbon dioxide level has been considered the main reason for global heating since 1950.

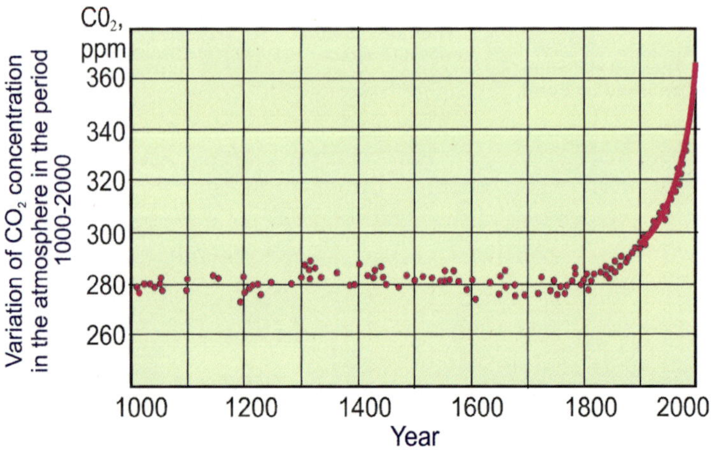

Fig. 2.20 Variation of carbon dioxide concentration

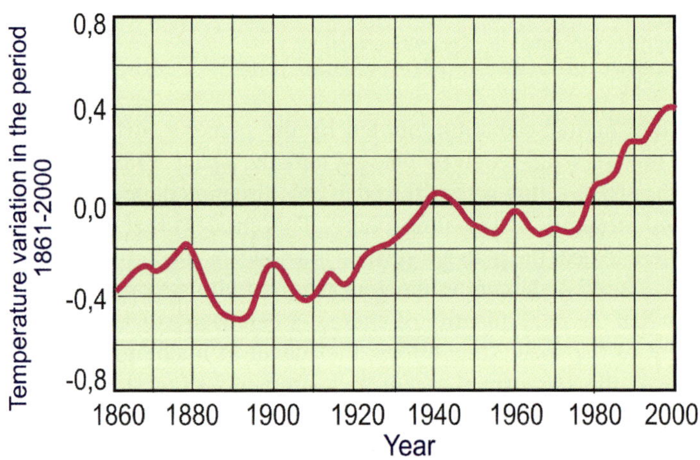

Fig. 2.21 Variation of Earth surface temperature

The main components of fossil fuels are carbon and hydrogen that produce carbon dioxide and water steam at burning. More and more CO_2 is emitted into the atmosphere. This surplus of greenhouse gas (GHG) triggers the growth of the Earth's surface temperature globally. According to the study carried out in 2001 by IPCC (*Intergovernmental Panel on Climate Change*) [18], a certain correlation has been stated between particles concentration of GHG in the atmosphere and the average temperature on the Earth's surface (see Fig. 2.20) [15].

During about 800 years, CO_2 concentration varied around 280 ppmv. Starting with the nineteenth century, CO_2 concentration grew so that at the end of twentieth century it reached 370 ppmv, simultaneously the average temperature of the Earth's surface increased by 0.6°C (Fig. 2.21) [19].

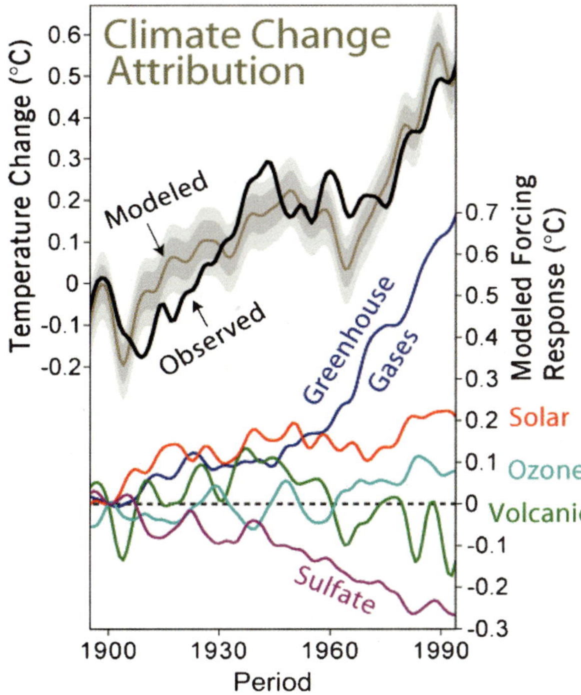

Fig. 2.22 Effects contribution to climate change

If this process doesn't stop, CO_2 concentration will reach about 700 ppmv by the end of the twenty-first century and the temperature will increase by 1.4–5.8°C. Probably, the heating of the Earth's atmosphere will cause the intensification of extreme climatic phenomena, including heavy rains, more frequent tropical cyclones, and longer droughts and deserts expansion as heavier consequences. Occasionally or not, during the summer of 1998, the USA meteorological stations registered the highest air temperature for the last 123 years since the beginning of meteorological measurements. According to the data provided by HYDROMETEO State Department, the same phenomenon was registered for the Republic of Moldova – since 1999 the highest temperatures for the last 112 years are: 1999 – 38.7°C; 2000 – 40.0°C; 2001 – 38.0°C; 2002 – 40.0°C; 2007 – 42.0°C.

Diagrams in Fig. 2.22, drawn on the basis of research carried out by Meehl and others (2004) [20], show the capacity of the global climate change to reconstitute the historical temperature record and the degree at which total temperature variation can be decomposed into elements generated by various factors, as: greenhouse gases, sulphate emissions as a result of human activity, solar variableness, changes in the stratosphere and troposphere, and volcanic emissions (including natural sulphates). Greenhouse gas exercises a major influence on the growth of global average temperature. The result of computer simulation of the climate heating curve for the period 1900–1990 coincides fairly with the results of observations (see Fig. 2.23). Table 2.4 lists temperature changes in the years 1940, 1970 and 1994,

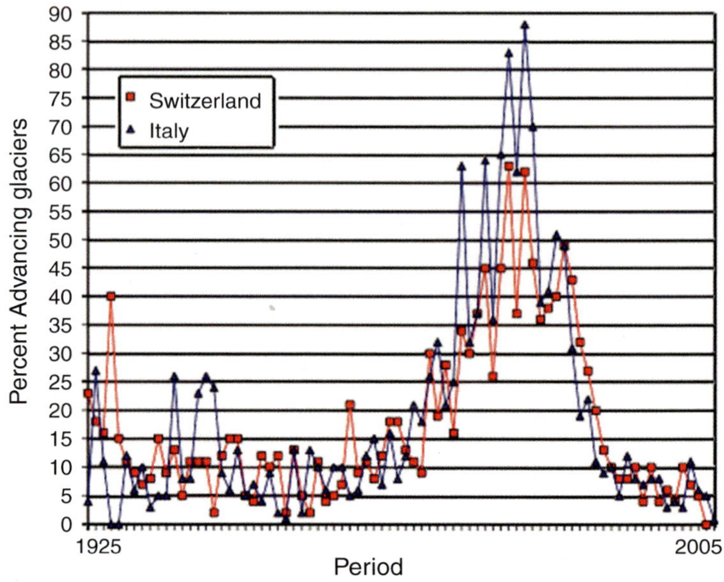

Fig. 2.23 Glaciers expansion in the Alps

Table 2.4 Temperature changes in 1940, 1970 and 1994 compared to 1900

	Temperature change compared to 1900		
Factors	1940	1970	1994
Greenhouse gas	0.10	0.38	0.69
Sulphate emissions	−0.04	−0.19	−0.27
Solar activity	0.18	0.10	0.21
Volcanic activity	0.11	−0.04	−0.14
Ozone	−0.06	0.05	0.08
Simulated	**0.19**	**0.17**	**0.53**
Observed	**0.26**	**0.21**	**0.52**

compared to 1900. Global climate heating by 0.42°C between 1900 and 1994 is due to an increase of 0.69°C generated by the greenhouse gas, and partially reduced by 0.27°C cooling due to sulphate emissions generated by human activities.

2.2.1.3 Greenhouse Gas and Climate Change

Climate represents a day-to-day state of the atmosphere and, generally, is a dynamic, non-linear and chaotic system. Particularly, climate change can be a self-perpetuating process in time, as various environment aspects react in different ways to inevitable fluctuations.

Climate includes general temperature, degree of precipitations, other variable factors that can be measured, and also other modifications of the planet's

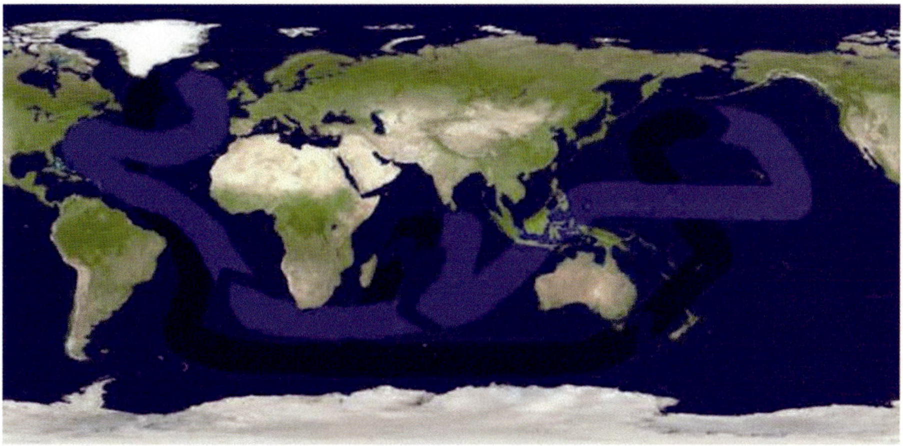

Fig. 2.24 Current diagram of thermal flows circulation

environment that can have an impact on climate. Glaciers are known as most sensible indicators of climate changes. Figure 2.23 portrays the evolution of glaciers in the Alps. Increase of the planet's temperature leads to glaciers melting, which, in its turn, raises the sea level and reduces the possibilities of sufficient regeneration of glaciers for maintaining "*winter–summer*" seasons.

Often, sea level growth has a decisive role in observing climate reaction in most regions. Climate change can result in changes of the ocean atmosphere system, as well.

At large scale, such oceanic processes like thermal flows circulation are critical for the redistribution of heating, and when modified, influence the climate dramatically (Fig. 2.24).

The closed dark color arrow shows the motion of deep water flows and the light color arrow shows the motion of surface flows.

Figure 2.25 shows surface temperature anomalies in 1995–2004, compared to 1940–1980 period. An increase of temperatures on the planet is observed. This phenomenon leads, primarily, to glacier melting and rise of the sea level. For the last 120 years the sea level increased by about 20 cm and this process is on (Fig. 2.26).

Let's analyze human factors that influence climate change. One of the basic factors that leads to a significant modification of climate, in which human being is involved directly, is the growth of carbon dioxide concentration in the atmosphere, leading to the greenhouse effect. Diagram in Fig. 2.27 shows the increase of CO_2 concentration in the atmosphere in the last 400,000 years. The biggest changes can be observed at the boundary of glacial/interglacial cycles (caused by the Earth's orbit shift), that corresponds to current glaciers age. Since Industrial Revolution (around 1,800), fossil fuels burning caused a dramatic CO_2 increase in the atmosphere, reaching unexpected concentrations for the last 400,000 years.

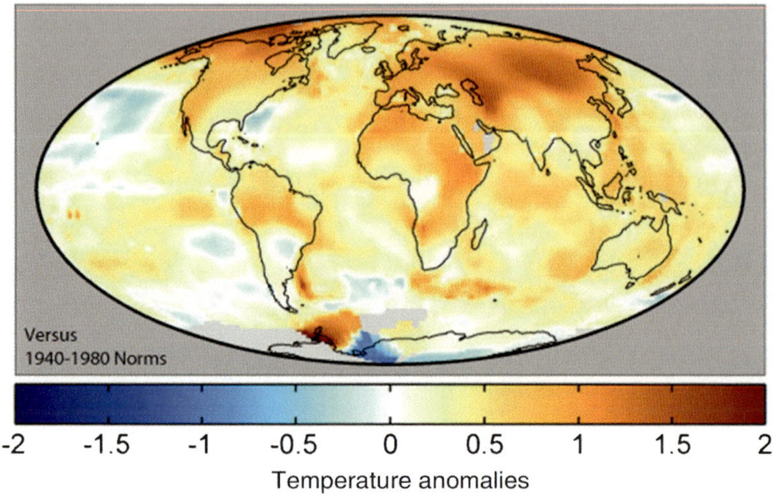

Fig. 2.25 Surface temperature anomalies in 1995–2004 compared to average temperatures in 1940–1980

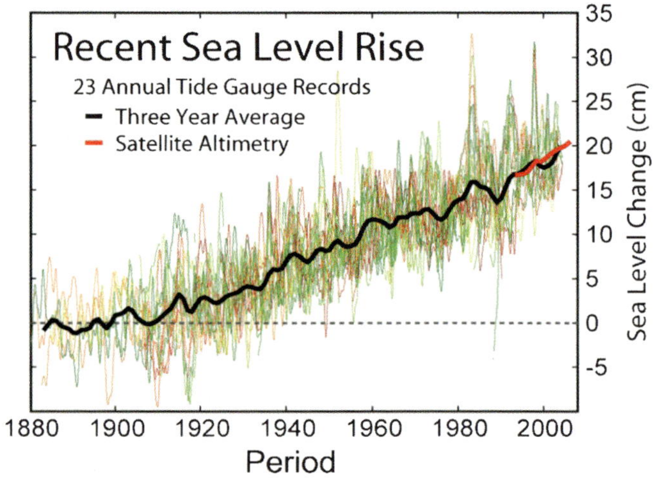

Fig. 2.26 Recent rise of sea level

This increase is considered as primary reason of global warming. Figure 2.28 outlines global carbon emissions growth at logarithmic scale. More detailed information regarding the share of carbon emissions for various fossil fuels is depicted from the linear diagram in Fig. 2.29. Due to coal burning, a certain decrease of carbon emissions can be observed, which, however, is followed by a constant growth of carbon emissions from oil and natural gas burning.

This fact is explained by an enormous growth in the number of transportation means – auto vehicles, with a share of 14% taking the third position after thermal

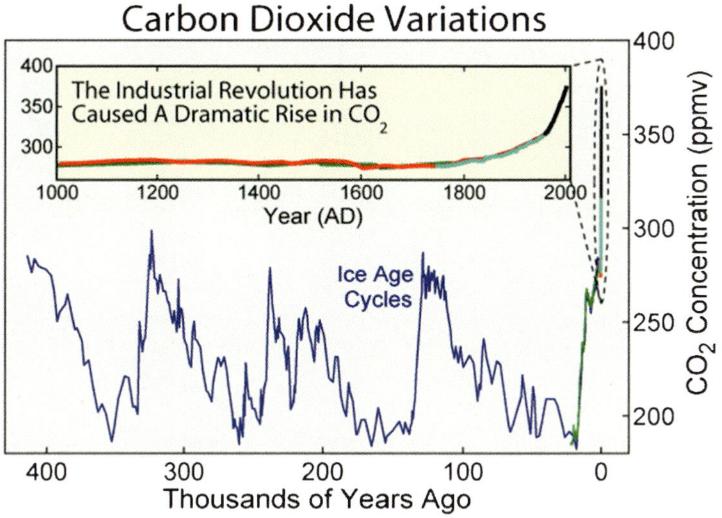

Fig. 2.27 Atmospheric CO_2 concentration variation

Fig. 2.28 Fossil carbon global emissions

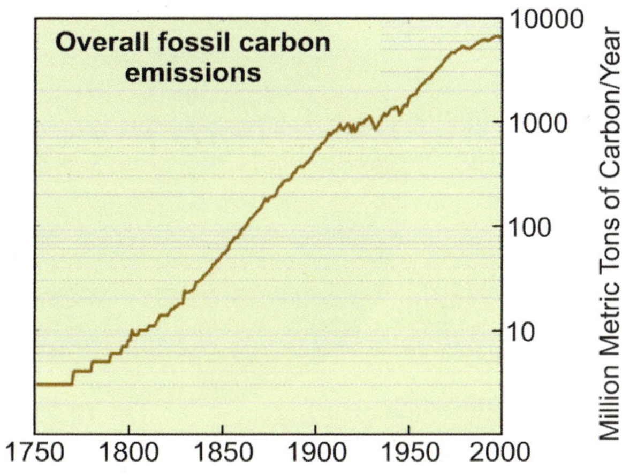

power plants (21.3%) and industrial processes (16.8%) concerning CO_2 emissions into atmosphere (Fig. 2.30), despite the fact that internal combustion motors have become more advanced in view of noxious substances emissions from exhaust gas.

The following diagrams present the total sum of greenhouse gas resulting from human activity for the last 100 years. These emissions contain 72% carbon dioxide, 18% methane and 9% nitrogen oxides. Due to the industrial revolution of 1850 and, particularly, due to the fast rate of recent human consumption of fossil fuels, CO_2 concentration in the atmosphere has increased from ~280 to 370 ppmv, nowadays. This increase is estimated to reach the rate of 560 ppmv till the end of the twenty-first century. Obviously, the world nations should engage into economical,

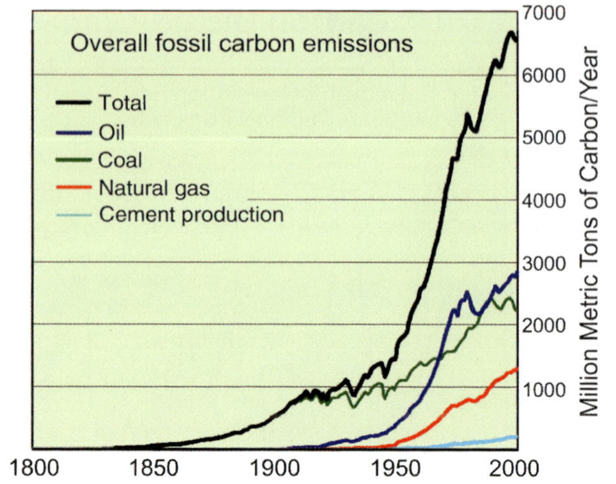

Fig. 2.29 Fossil carbon global emissions from various sources

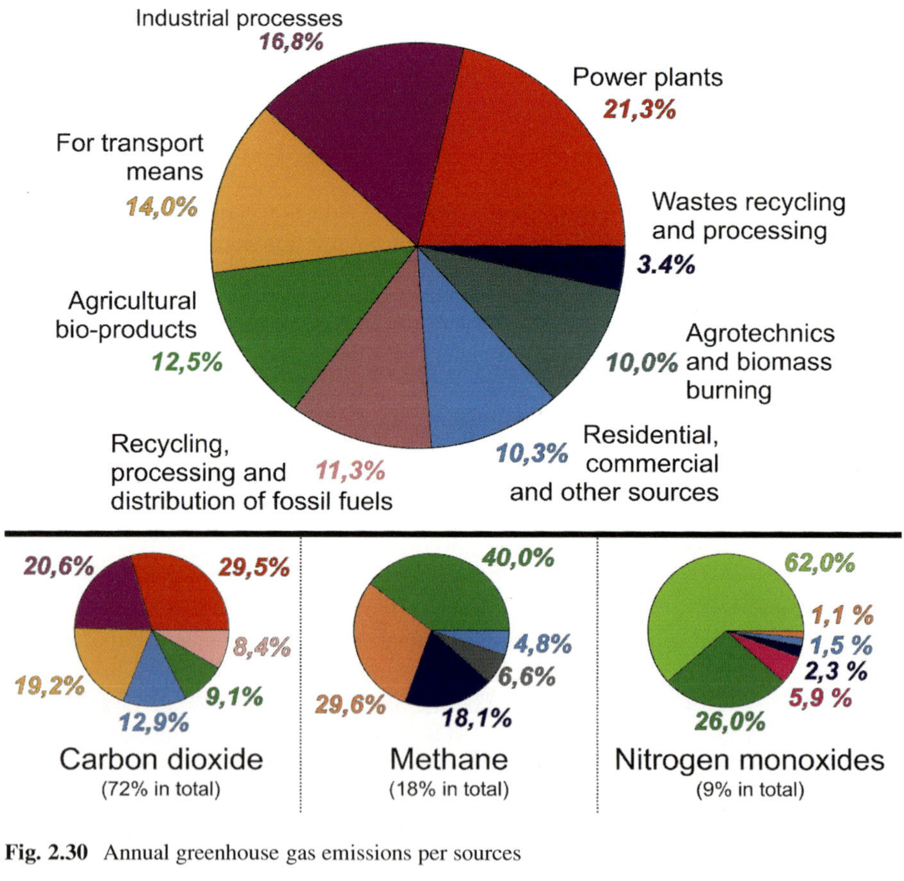

Fig. 2.30 Annual greenhouse gas emissions per sources

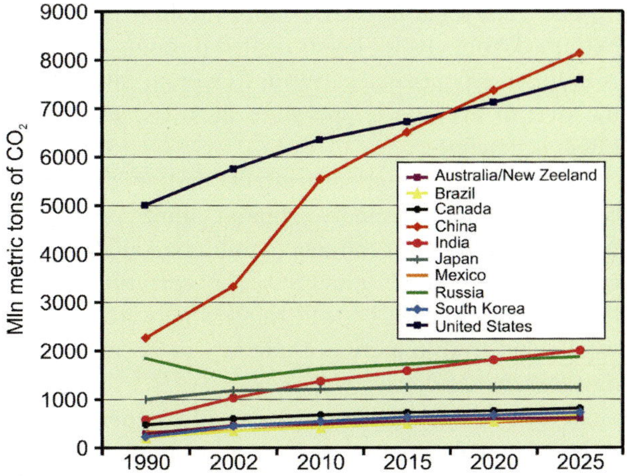

Fig. 2.31 Global carbon dioxide emission, 1990–2025

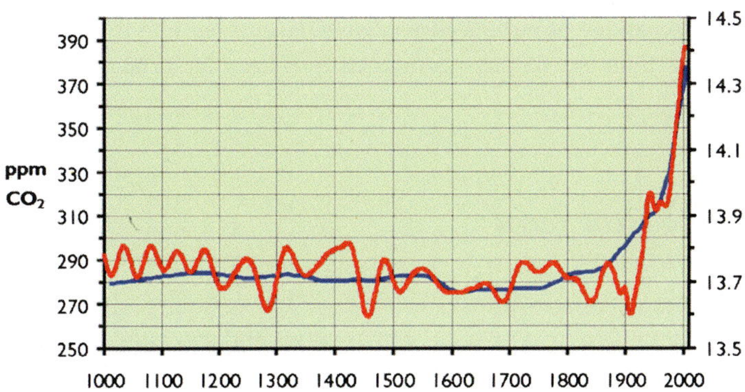

Fig. 2.32 Growth of global CO_2 emissions and of planet's average temperature in the last 1,000 years

ecological and social activities in order to reduce CO_2 emissions. However, the forecasts for 1990–2025 period, related to global carbon emissions are not optimistic (Fig. 2.31). Regretfully, USA is the highest polluting country, but, at the same time – the biggest producer of green energy. Most disturbing is the dangerous increase of CO_2 emissions in China. In just 20 years, (1990–2010), the volume of CO_2 emissions increased about 2.5 times in China, so that in 2020 – China will outrun USA, and in 2025 China will generate to the atmosphere four times more CO_2 than in 1990. This volume is equivalent to CO_2 emissions from all countries except USA. The same situation is in India (see Fig. 2.31).

The concentration of carbon dioxide in the atmosphere influenced temperature increase in the last 1,000 years (Fig. 2.32) [15]. CO_2 concentration (blue line on Y-coordinate) is shown in parts per million, and the temperatures (red line, Y-coordinate) – in Celsius degrees.

The data regarding CO_2 concentration is based on historical records of CO_2 extracted from glacier layers drilled through at Law Dome in Antarctica [15]. There is a direct dependence between CO_2 concentration increase in the atmosphere and the overall average temperature increase. In this case, the so-called greenhouse effect is generated.

Besides the greenhouse effect, carbon dioxide (and sulphur dioxide, nitrous oxides) has an extremely negative impact on the environment – air and water pollution with major influence on human and vegetal health. Recent investigations on the increase of CO_2 and SO_2 concentrations in the atmosphere demonstrate that these are the reason of acid rains which are extremely dangerous for the vegetal world, and lead to a critical increase of ocean water acidity, a factor that seriously destabilizes sea ecosystems.

Another negative effect generated by fossil fuels burning is the penetration of radioactive substances into the atmosphere, such as uranium and ionium found in fossil fuels. In 2000 about 12,000 t of ionium and 5,000 t of uranium have been emitted into atmosphere by coal burning. It was stated that in 1982 in the USA, after coal burning, the volume of radioactive substances emitted into the atmosphere was 155 times higher than the damage produced by Three Mile Island nuclear power plant in 1979.

2.2.2 How to Heal Terra

Conventional energy sources based on oil, coal and natural gas, are very efficient means of economic progress, which, meanwhile, endanger the environment and human health. These traditional energy sources based on fossil fuels significantly increase the pressure on the environment, and dangerously influence the climate. Greenhouse effects leading to global warming of the planet's climate, more frequent acid rains as a result of increased atmospheric concentration of gas emissions from fossil fuels burning (CO_2, SO_2, NO_2, etc.), so-called natural calamities (although some of them no longer natural due to human activities) with disastrous consequences: hurricanes, tornadoes, floods, draughts, etc., are all signs of planet's serious illness. Figure 2.33 portrays two natural images: the left one – without the presence of greenhouse gas; the right one – the same image observed through the fog caused by pollutant gas and other substances resulting from human activity. Thus, the organism named Terra is alive and sick. What measures should be undertaken to save this organism, or at least to stop the evolution of disease? Obviously, any effort to maintain CO_2 concentration in the atmosphere under the rate of 550 ppmv cannot rely on economics based on oil and coal. One of the most feasible measures could be the reduction of greenhouse gas emissions into the atmosphere. The main source of atmosphere pollution with greenhouse gas and other noxious substances in the most developed countries is the production of electricity by fossil fuels burning. As electricity consumption will constantly grow one of the solutions would be the use of alternative energy sources. Diversification

Fig. 2.33 Fragment of landscape observed through a clean lense (on the *top*) and via the fog (gas and other pollutant substances)

of energy is very important from the ecological point of view. These alternative forms of energy are called renewable energy. The most known renewable energy sources are: solar (direct, photovoltaic and thermal) energy, wind (derivative of solar energy), hydraulic (use of water potential and kinetic energy), geothermal, bioenergy, etc. Renewable energy has no polluting effect on the environment. Still, the development of renewable energy conversion systems demands financial assistance at present. Therefore, the development of renewable energy needs support from the governments. Regretfully, in certain European countries, big companies lobby the production, distribution and consumption of fossil fuels, limiting the use of renewable energy sources. Table 2.5, developed by Greenpeace, shows direct annual subsidies paid by the governments of most European countries for the production of traditional forms of energy.

Just a few countries, like Austria, Finland, Greece, Luxemburg, Netherlands, Sweden, Switzerland, have provided higher subsidies for renewable energy than for fossil fuels and nuclear energy. The country with the highest subsidies for renewable energy (459.3 M$) is France.

Table 2.5 Direct annual subsidies paid by EU and European governments for energy production, mln USD

Country	Fossil fuel	Nuclear energy	Renewable energy
EU	520.7	428.3	131.3
Austria	4.7	1.4	35.7
Belgium	61.6	146.8	5.5
Denmark	368.2	2.8	108.8
Finland	68.7	8.9	129.0
France	280.5	563.3	459.3
Germany	6,890.4	314.6	149.3
Greece	1.3	0	5.2
Ireland	32.4	0	5.6
Italy	11.0	147.3	37.1
Luxemburg	0	0	6.9
Netherlands	31.0	48.0	88.4
Norway	20.7	7.6	5.8
Portugal	4.5	3.0	1.6
Spain	705.5	40.0	68.3
Sweden	3.4	15.9	56.5
Switzerland	13.7	61.1	104.0
Great Britain	1,217.9	2,885.9	94.9
Total	**10,236.3**	**4,674.8**	**1,493.2**

Source: Energy subsidies in Europe, Greenpeace report

Discovery of the new alternative energy sources and increase in the efficiency of known renewable energy conversion systems represents a reliable source for limiting greenhouse gas emissions in the atmosphere, and for gradual healing of Terra.

According to Greenpeace data, historical subsidies have been done: between 1950 and 1990, EU and national governments paid around 32 billion USD for the development of nuclear energy; between 1978 and 1990, 8.6 B$ have been paid for the development of renewable energy.

Greenpeace considers the following immediate measures to be promoted by EU and national governments with regard to environmental policies:

- To stop subsidies for nuclear and fossil fuel energy industries;
- To allocate these funds to programs that would accelerate the development of renewable energy technologies, their marketing and energy conservation;
- Presentation of all direct and indirect energy subsidies.

An important step for environmental improvement was the negotiation of Kyoto Protocol in 1997, regarding the decrease of greenhouse gas emissions and other polluting substances that are "guilty" of global climate change on Terra. Regretfully, USA, the country with the highest greenhouse gas emissions, hasn't signed Kyoto Protocol, the validity of which ends in 2012. Nowadays, the second Kyoto Protocol is being prepared, which will probably have a significant importance

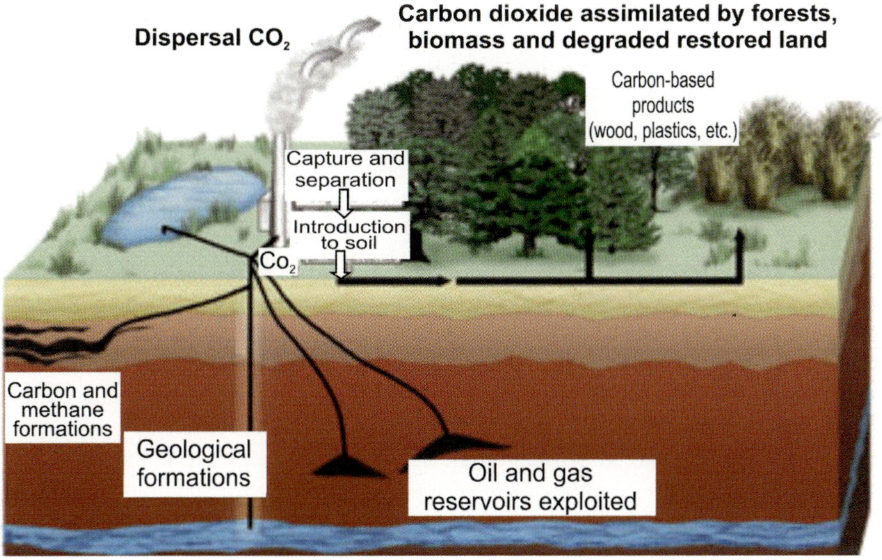

Fig. 2.34 Carbon dioxide dispersion diagram

for global environmental policy. A system of ecological education and instruction is implemented for all educational levels, including university and post-university levels. This book is helpful for this education.

Another measure for environmental protection, which is efficient in the case of impossibility to temporary interrupt fossil fuel burning to obtain electricity, is the development of modern systems for carbon dioxide collection and dispersing (Fig. 2.34).

With account of the existing global energy and ecological conditions, it can be concluded that energy savings obtained via energy efficiency programs can be considered a source of energy, totally non-pollutant and free.

Energy efficiency concept or energy saving concept is not to be mistakenly considered as lack of consumer's comfort. Optimal energy consumption defines a large range of solutions: reduction of energy losses (both at production and transmission, distribution and consumption phases), replacement of units and equipments physically and morally worn, and rather energy consuming, elimination of irrational and unjustified energy consumption, etc. Ecological aspect should not be neglected. It is an important topic of energy saving and conservation programs envisaged by the new world energy policy, as the largest amount of noxious quantities annually emitted to the atmosphere as a result of human activity is due to fossil fuel combustion (over 22.5 billion tones of CO_2 annually at world level). Saving 1% of energy consumption would lead to the reduction of CO_2 quantities by about 2.3 billion tones.

2.3 Renewable Energy and Sustainable Development

2.3.1 Theoretical, Technical and Economical Energy Potential

In contrast to exhaustible fossil and nuclear fuels, which are basically stored energy sources, deposited along many million years, renewable energy sources (RES) are defined as *"energy obtained from existing fluxes in the environment and which have a permanent and reproducing character"* [21]. Unlike renewable energy, fossil fuels energy is embedded (closed) and it can be released as a result of human activity. By releasing the energy stored in fossil or nuclear fuels the environment is polluted by wastes, the greenhouse effect increases, and thermal pollution of the environment occurs too. These two properties are shown in Fig. 2.35. Renewable energy flux has a closed character while the fossil energy has an open character. In the case of RES use the energy flux generated by the environment is transformed with the help of conversion unit into another form of energy necessary for the consumer. Then it reverses (according to energy conservation law the quantity of energy remains unchanged) in the same environment and its thermal balance remains unvaried.

When using a fossil source of energy, the energy stored in the fuel is released to the power unit, used by the consumer and then emitted into the environment, producing thermal pollution. At the same time, carbon dioxide is eliminated as product of carbon burning, stored into fossil fuels during millions of years.

Solar energy – the main renewable energy source: Depending on the origin, RES are divided into groups include solar energy and its derivatives – wind, hydraulic, biomass energy, tide energy, thermal energy of the planet's ocean. Figure 2.36 lists the main forms of solar energy: thermal and photovoltaic (PV), energy obtained as the result of solar radiation direct conversion into heat, and electricity, respectively, and the other forms of energy obtained indirectly from solar energy.

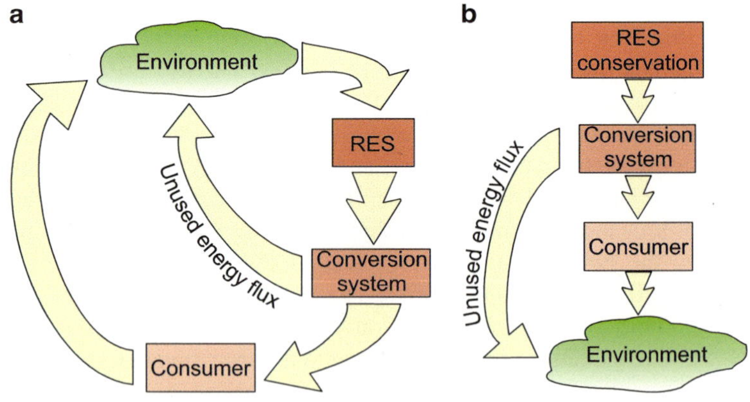

Fig. 2.35 Energy fluxes circulation: (**a**) renewable; (**b**) fossil sources

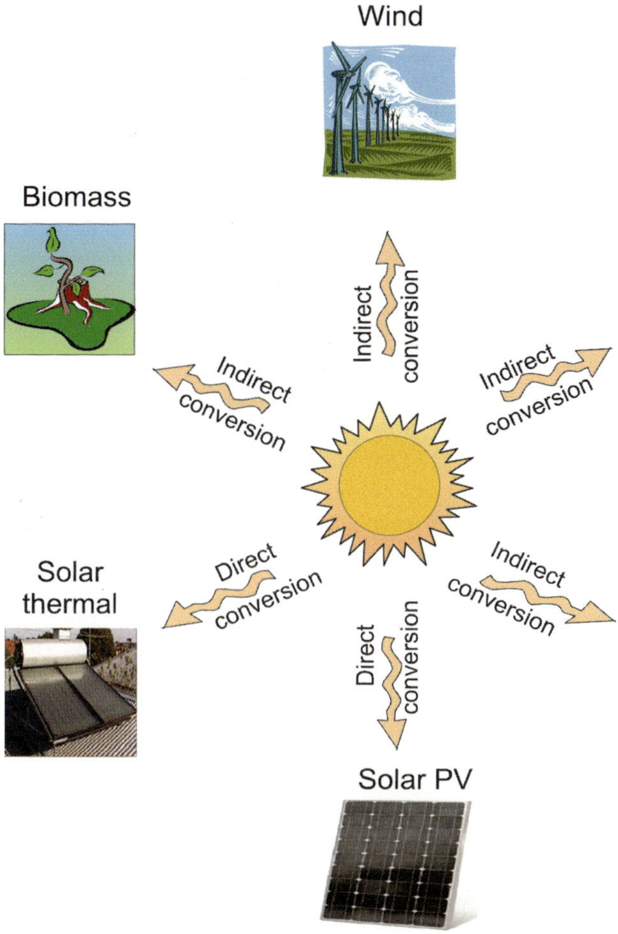

Fig. 2.36 Main forms of solar energy

The thermal energy of the planet's ocean is not described because its technology is in the inception phase of development. The Sun, as an energy source, the characteristics of solar radiation in the outer space and on the Earth's surface, methods for estimating the available solar radiation are described in Chap. 3. The second group of RES is not of solar origin and includes only two forms of energy: geothermal and tide energy. Solar radiation, absorbed directly by solar collectors, can produce heat water, can heat buildings, and can dry medicinal herbs, fruits and vegetables. Buildings can be designed and built in such a way so as more solar energy is captured for heating and lighting. This concept lies at the basis of the so-called technology for solar energy passive use. Concentrated by special reflectors, solar radiation can generate thermal energy with temperatures higher than 300°C, that, in its turn, can be used to produce electricity. Such solar thermal plants are in commercial use in the USA. Conversion technology in which solar radiation is transformed directly into thermal energy is often called thermal-solar energy. Issues relating to this technology are described in Chap. 3.

Solar radiation can be transformed directly into electrical energy with the help of photovoltaic modules. The last 10 years, PV solar technology developed dynamically, with an annual growth rate varying between 25% and 40%. Costs of PV modules are decreasing. New technologies for production of PV cells and modules integrated into buildings' roofs will change in the next years the modern concept of living houses electricity supply. Photovoltaic conversion of solar energy is examined in Chap. 3.

The difference of air mass temperature leads to a difference of pressure, and, as consequence, significant air currents develop, directed to the Polar Regions or, otherwise said, the wind blows and it can be transformed into mechanical energy by means of wind turbines. Wind energy is described in Chap. 4. The last 20 years wind technology developed at a large scale and is considered the most advanced, compared to all so-called conversion technologies of "new" renewable energies.

Hydraulic energy exists under two forms: potential energy (water falling, usually with dams building) and kinetical energy (water flowing without building dams). Hydraulic energy is described in Chap. 5. During the last years, water kinetic energy conversion systems advanced in their development, in particular, concerning the optimization of conversion efficiency parameters of the working element (the rotor).

Friction forces between air currents and the water surface of the seas and oceans generate waves with kinetical energy. The technology of *wave energy* conversion is at the beginning of its development. In some countries (Great Britain, France, etc.) demo projects have been designed and executed already.

Renewable energy sources are the most important alternative energy sources considered as the *"energy of the future"*. When scientists will define the limits of possibilities for renewable energy technologies, the efficiency and cost of conversion systems, it will be possible to quit the use of energy generated by fossil fuels burning. The question is: what are the world reserves of renewable energy? According to estimates, the reserves of renewable energy are enormous (Fig. 2.37) [22]. But, only part of this energy can be converted.

There is the concept of power potential: theoretical, technical and economical. The highest theoretical potential is given to solar energy, reaching an enormous quantity – 89,000 TW. The wind energy – a derivative of solar energy is on the second place with 370 TW. Global energy of sea waves reaches the value of about 200 TW, compared to current global consumption of 15 TW. Global energy needs could be met by only 0.0002% of solar energy; 0.04% of wind energy; and 0.01% of sea energy. The technical potential of the mentioned renewable energy sources equals the maximum quantity of converted energy provided by the technical systems, and is determined by recent degree of system efficiency. The economic power potential equals the quantity of converted energy, justified from the economical point of view.

Renewable energy conversion systems possess multifunctionality, in the case of remote consumers, particularly. Figure 2.38 shows both possibilities – for

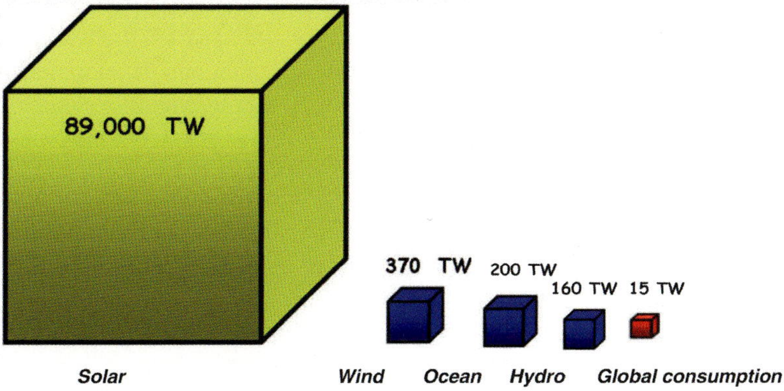

Fig. 2.37 Solar and wind power potential

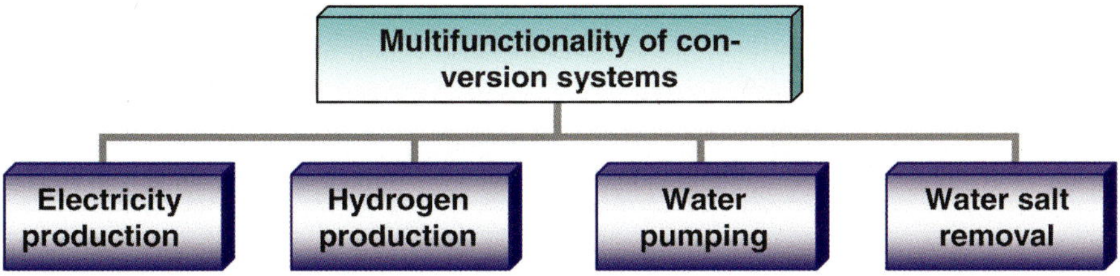

Fig. 2.38 Possible fields of renewable energy use

renewable energy transformation into other forms, and for its storage (for example, water pumping into storage tanks when electricity is not demanded, hydrogen production, etc.).

2.3.2 Global Renewable Energy: Its Present and Future

Renewable energy has been used by man since the oldest times. The burning of biomass for heating and lighting was practiced from prehistoric times, without mentioning the use of organic products as energy for survival. Wind mills and water mills employed natural resources during many decades, as earliest source of energy production for agriculture and small-scale industrial processes.

Modern technologies for renewable sources conversion in Denmark have different stories. The development of wind technologies began at the end of the nineteenth century. The interest in these technologies increased highly during the two world wars because of limited access to fossil fuels. Since the 50s photovoltaic cells (solar) have been developed due to investment as result of their fulminating use in space flying systems, in materials technology and science, followed by the reduction of prices till the level accepted by the consumers. The basic motivation for the expansion of renewable energy was the oil crisis of 1973 and 1979–1980. Due to the support of political leaders in various countries, research and development of new technologies increased. James E. Carter was the first politician to welcome the use of solar energy as response to the energy crisis of 1973. Wind, sea waves and solar technologies were supported by investments with the increase of their application level.

As well, the European Union is in a difficult situation, as by imports of 82% of oil and 57% of gas it is the world leader in this respect. With a balance "*reserves/consumption*" equal to 3.0 (a very low coefficient according to the world standards), the European Union is exposed to an energy vulnerability, which fact has determined it to seek ways of improving its energy security.

Renewable energy is the energy derived from regenerative sources that for all practical applications cannot be exhausted. Nowadays, renewable energy sources have an 18.4% share in the world energy consumption. The primary source of renewable energy is solar radiation, i.e. solar energy.

Solar, wind and hydraulic energy are traditionally used widely both in developed and under-developed countries. However, significant electricity production from renewable energy sources started relatively not so long ago, reflecting major treaties on climate change and pollution, fossil fuel depletion and social, political and environmental risks related to fossil fuels and nuclear energy consumption. Many countries and organizations promote renewable energy by subsidizing it and reducing taxes.

Transition to technologies based on renewable energy is dictated both by the ongoing increase of oil and gas prices (with no chance of their decrease in the future), and by understanding issues of the world climate change. During the last 30 years, solar and wind energy systems have developed rapidly, significantly reducing capital costs and the cost of generated energy, continuing improvement of system performances. In fact, the cost of fossil fuels and renewable energy, as well as growing social and environmental costs favored large-scale rapid evolution of dissemination and development of renewable energy market.

Development and use of renewable energy sources bring a diversity to energy consumption markets and contribute to long-term reliability of sustainable energy supply, as well as to the reduction of emissions into local and global atmosphere, and propose attractive commercial options for specific services promotion in satisfying energy needs, particularly in under-developed countries and in rural areas, helping to develop new opportunities for the labor market.

How renewable energy is divided by forms of renewable sources? Figure 2.39 portrays this distribution. Nowadays, renewable energy is dominated by micro hydropower and biomass, used as fuel for cooking and heating, especially in the under-developed countries from Africa, Asia and Latin America. New renewable energy sources (solar, wind, geothermal and micro hydropower) contribute with just 2%. Undertaken studies and designed scenarios researched the contribution of renewable energy sources to the world energy needs supply, underlining that in the first half of twenty-first century the contribution of RES will grow from 20% to 50%.

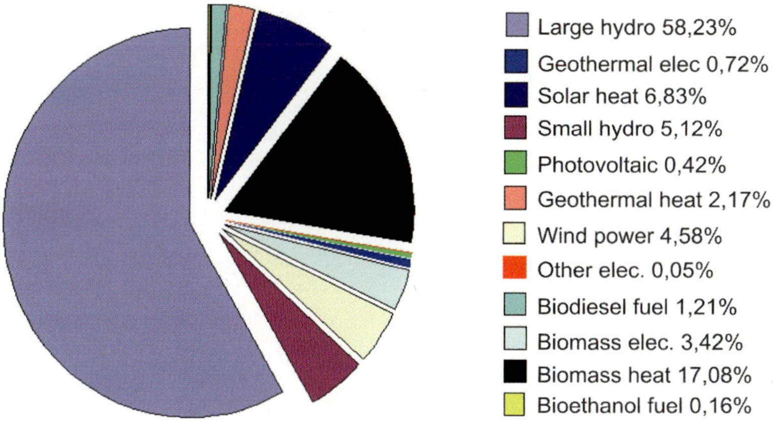

Fig. 2.39 Global renewable energy in 2005

Table 2.6 Renewable electricity production top countries (2000)

No.	Hydro-	Geothermal	Wind	PV solar
1	Canada	USA	Germany	Japan
2	USA	Philippines	USA	Germany
3	Brazil	Italy	Spain	USA
4	China	Mexico	Denmark	Australia
5	Russia	Indonesia	India	Netherlands

The 80% of energy demand of Western industrial societies is focused on building's heating and maintenance, and on vehicles (cars, planes, trains) operation. The majority of renewable sources are used to generate electricity. Iceland is the world leader in renewable energy due to plenty hydro- and geothermal energy sources. About 99% of country's electricity is produced from renewable sources, and most urban home heating is of geothermal origin. Leaders of "*green*" energy production are given in Table 2.6. One can see that USA is on positions 1–3 in all 4 nominations: hydro-, geothermal-, wind- and photovoltaic solar energy. This fact is partially justifying its burden as biggest energy consumer and the most important factor of increasing greenhouse gas concentration.

Denmark is the initial leader in the generation of wind energy and remains the nation with the highest level of wind energy production per capita. Germany has started to develop its wind capacities later – in the middle of the 1990s by injecting generous subsidies and now it has about one third of the world capacity generating wind electrical energy. Spain was mainly on the second position in wind electrical energy generation, but in 2002 was outrun by USA which became the second nation with the highest level of wind energy installed capacity. Table 2.7 lists the top five programs for the use of green energy launched by US Department of Energy. Joint renewable energy source in the top five programs is wind energy.

These encouraging actions of US Department of Energy resulted in the production of 6% of total USA energy from renewable energy sources in 2004. In

Table 2.7 Top five programs of green energy use, USA Department of Energy

No.	Program	Resources used	Production, mln. (kWh/year)
1.	Austin Energy	Wind, natural gas	435.1
2.	Portland General Electric	Geothermal and hydro-wind	339.6
3.	PacifiCorp	Wind, biomass, solar	234.2
4.	Florida Power and Light	Biomass, wind, solar	224.6
5.	Sacramento Municipal Utility District	Wind, natural gas, micro-hydro, solar	195.1

Table 2.8 Geothermal energy installed generation capacities

Country	1990 (MWe)	1995 (MWe)	2000 (MWe)	2005 (MWe)
China	19.2	28.78	29.17	28
San Salvador	95	105	161	151
Iceland	44.6	50	170	202
Indonesia	144.75	309.75	589.5	797
Italy	545	631.7	785	790
Japan	214.6	413.71	546.9	535
Kenya	45	45	45	127
Mexico	700	753	755	953
New Zealand	283.2	286	437	435
Nicaragua	35	70	70	77
Philippines	891	1,227	1,909	1,931
Russia (Kamceatka)	11	11	23	79
USA	2,774.6	2,816.7	2,228	2,544
Total	**5,831.72**	**6,833.38**	**7,974.06**	**8,912**

Israel most living houses are supplied with hot water generated from solar energy and new technologies are being developed for the generation of renewable energy from wastes. Dynamics of geothermal energy increase in 1990–2005 is shown in Table 2.8.

2.3.3 Experience of European Countries in RES Promotion and Use

Certain countries like Denmark, the Netherlands, France, and Germany have long traditions in renewable energy conversion, in particular wind and hydraulic ones. As a whole, the European Union focuses on the use of renewable energy as an alternative to conventional energy. In the 90s of the last century the European Union launched a global strategy for its countries concerning the development, promotion and implementation of RES. Lack of coherent and transparent strategy with a well-defined and ambitious objective is a serious impediment for RES use.

They could not influence the Community energy balance in any way. A first step in the development of the strategy was the presentation of the first version of Strategy, the so-called Green Paper in 1996 [23] "*Énergie pour l'avenir: les sources d'énergie renouvelables*", that was presented to large public debates beginning with 1997. The Green Paper generated multiple reactions from Community institutions, governments and national organizations, companies and agencies, interested in RES development. During this period of consultations the European Commission has organized two conferences to discuss a number of formulated issues and developed proposals. After public debates on the Green Paper, the Strategy was amended and finally published in the White Paper [24]: Livre Blanc "*Énergie pour l'avenir: les sources d'énergie renouvelables. Une stratégie et un plan d'action communautaires*".

EU strategic objectives: The objectives and EU policy on energy and necessary resources to reach these objectives are described in the EU White Paper. Three main objectives for energy policy are envisaged:

- Competitiveness capacity building;
- Security of energy sources supply;
- Environmental protection.

RES promotion is underlined as one of the decisive factors for achievement of the mentioned objectives. Being indigenous, RES will have an important role in decreasing the level of dependence on import and will be quite positive in increasing the security of supply. In 1995, the EU dependence on energy sources import was 50%, by 2020 it will be 70% if no measures taken.

The main objective of this Strategy is to provide 12–15% of the gross national consumption by 2010 from renewable sources. Compared to 1997, the share of RES will double; 23.5% of total electricity will be produced by RES. The highest growth rate will have: photovoltaic energy – by 130 times, wind energy – by 19 times, solar thermal (solar collectors) – by 15.4 times. Three forms of RES (biomass, hydro and wind) have the highest share both in gross electricity consumption and in electricity production (see Tables 2.9, 2.10). Compared to 1995, it has been envisaged the renewable energy will reach 675 TWh by 2010, which means 23.5% of the total energy production, that is a growth of about 70% (Table 2.9).

The achievements of these countries are partially based on their geographical advantages. It should be noted that Germany has no good wind energy resources (for example, in Great Britain these resources are bigger but achievements are more modest). Some other factors played an important role in the achievement of this level of conversion of wind energy and other renewable energy.

Today, renewable energy conversion might be more extensive than fossil fuel burning. The technologies employing fossil fuels are well-known, but renewable technologies are new. Politicians and scientists actively discuss which the optimal mechanism/mechanisms are to achieve the goals of renewable energy policy implementation.

Table 2.9 Current and future production of renewable electricity in EU, European Commission, 1997

	1995		2010	
Form of energy	TWh	% of total	TWh	% of total
Wind energy	2.00	0.20	83.00	2.80
Hydraulic energy	307.00	13.00	355.00	12.40
Solar energy	0.03	–	3.00	0.10
Biomass energy	22.50	0.95	230.00	8.00
Geothermal energy	3.50	0.15	7.00	0.20
Total renewable energy	337.00	14.30	675.00	23.50
Total	**2,366.00**	–	**2,870.00**	–

Table 2.10 Renewable energy gross consumption in the European Union, Mtep

	Consumption 1995				Consumption forecast for 2010			
Forms of energy	Mtep	%	Substitution value	%	Mtep	%	Substitution value	%
Gross total consumption	**1,366**	**100**	**1,409**	**100**	**1,583**	**100**	**1,633**	**100**
Wind	0.36	0.02	0.9	0.06	6.9	0.44	17.6	1.07
Hydro:	26.4	1.9	67.5	4.8	30.55	1.98	78.1	4.78
macro	23.2	–	59.4	–	25.8	–	66	–
micro	3.2	–	8.1	–	4.75	–	12.1	–
Photovoltaic	0.002	–	0.006	–	0.26	0.02	0.7	0.05
Biomass	44.8	3.3	44.8	3.12	135	8.53	135	8.27
Geothermal	2.5	0.2	1.2	0.1	5.2	0.33	2.5	0.15
electricity	2.1	–	0.8	–	4.2	–	1.5	–
heat	0.4	–	0.4	–	1.0	–	1.0	–
Solar thermal	0.26	0.02	0.26	0.02	4	0.25	4	0.24
Total RES	**74.3**	**5.44**	**114.7**	**8.1**	**182**	**11.5**	**238.1**	**14.6**

The number of those who consider that renewable energy is not cost efficient is quite big. This is due to the fact that fossil fuel prices do not include current and future costs related to global ecological changes [25]. As well, the efforts undertaken to extract oil from big depths are increasing but the cost of renewable energy technologies will reduce together with the increase of investments and capacities expansion [26].

In Romania, the explored hydropower potential of the biggest rivers is relatively high (Fig. 2.40). The hydropower potential estimates 40 TWh in Romania, of which 12 TWh are exploited. Three hundred and sixty two hydropower plants with an installed capacity of 6,120 MW amount 27.9% of the total installed capacity of Romanian power system. It is expressed in macro and micro power units (under 10 MW/hydro units):

- Big power plants (34,000 GWh/year);
- Small power plants (6,000 GWh/year).

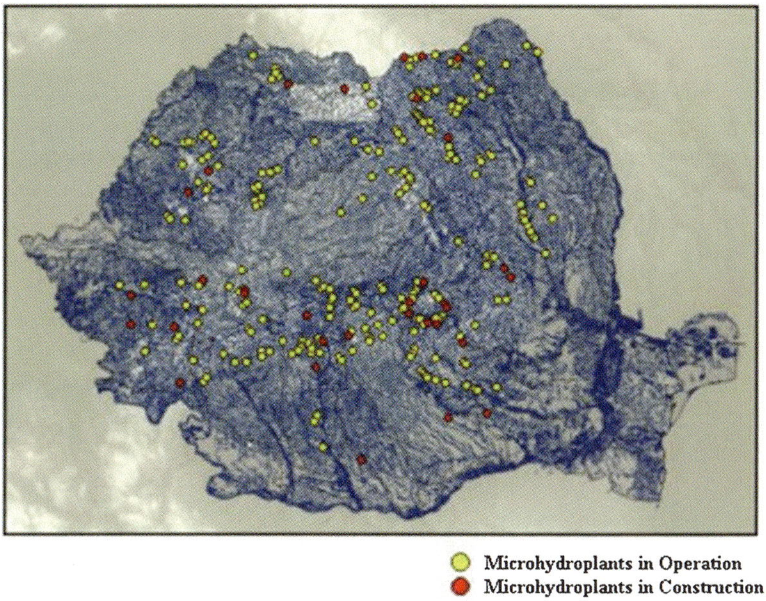

Fig. 2.40 Romania hydropower potential

Small hydropower plants – up to 3.6 MW – divide into:

- Big hydropower plants (HPU) – hydropower units of more than 3,600 kW or equal;

– Hydropower under 3,600 kW, divided into three subcategories:

  - Small hydropower units (SHPU) with installed capacity between 200 and 3,600 kW;
  - Micro hydropower plants (MHPP) with installed capacity between 20 and 200 kW;
  - Artisan hydropower units (AHPU) with installed capacity under 20 kW.

Opportunities for hydropower development in Romania are very high. About 5,000 locations from Romania are favorable for the development of small-scale hydropower.

The majority of renewable energy policies envisage market stimulation that, in its turn, demands technologies with increased efficiency and reduced cost. Many instruments of energy policy have been applied, most employed being the share instrument, contractual and fixed-tariffs instruments and low-credit taxes. The share instrument, known as Renewables Standard Portfolio (RSP), advises governments to oblige companies and other consumers to use a part of consumed electricity from renewable sources. Companies avoiding this obligation should pay fines for each unit of electricity. This leads to an increase in renewable energy capacities and, as a

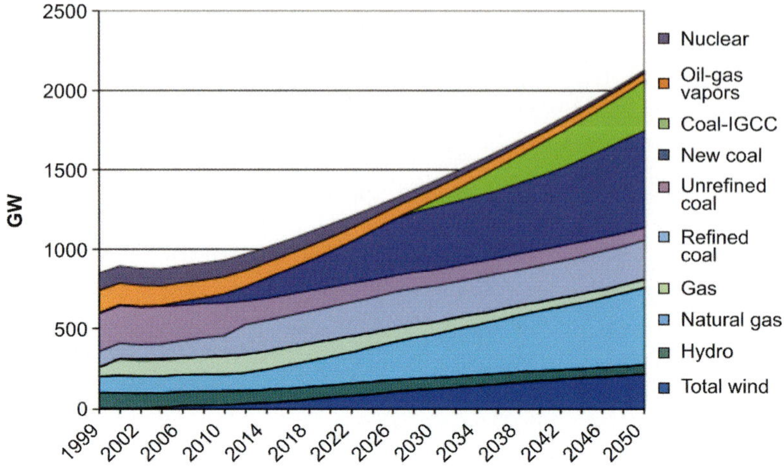

Fig. 2.41 Dynamics of global energy development by sources till 2050

result – to the decrease of prices. This instrument is used by majority of states in the USA, in Great Britain, Italy, Belgium and other European countries. In USA this instrument is applied at federal level as well.

Contractual instruments assist governmental authorities in obliging companies to accept renewable electricity on the basis of a governmental contract. Examples of such instruments in practice are UK Non-Fossil Fuel Obligation, Alternative Energy Demand in Ireland and EOLE in France.

Regarding the tariff instrument the government fixes the price for each unit of produced electricity by means of technologies classified as renewable. The fossil fuels are still cheaper; the price is higher for renewable electricity. Tariffs can be defined for various technologies. The government can subsidize or can impose beneficiary companies to buy the produced electricity without transferring its costs to consumers.

The main advantage of all these instruments is to impel renewable energy technologies, which fact will reduce the impact of human activity on the environment.

According to the forecast given in Fig. 2.41 [27], renewable energy sources will grow considerably starting with 2000. It is expected that total volume of renewable energy will increase about three times in 2050 compared to 2000. The increase of wind and hydraulic energy use is shown in the diagram above.

EU Strategy concerning RES use is supplemented with an action plan that coordinates, defines and mobilizes both Community and member countries' activities. In fact, the action plan is a supporting instrument for RES harmonizing Community actions under energy market liberalization. These measures refer to common rules for internal electricity market and are described in [28]: "*Directive 2003/54/CE du Parlement européen et du Conseil du 26 juin 2003 concernant des règles communes pour le marché intérieur de l'électricité*". This Directive envisages only one instrument for favorable approach to RES electricity, according to article 11, 3rd paragraph: *"A member state can require the owner of distribution grid, in the case this owner chooses production equipment that priority is given to renewable energy sources, to sources based on waste use or to sources of electrical and thermal energy cogeneration"*.

This provision is an exception from the fundamental rule defined in the same article, paragraph 2: "*The selection of production units and the use of interconnections are done on the basis of criteria with account of economic priorities concerning electricity production*".

References

1. Stöcker H (1999) Toute la physique. Dunod, Paris, 1180 p
2. Hinrichs RA, Kleinbach M (2002) Energy: its use and environment, 3rd edn. Thomson Learning, Australia, 590 p
3. Energy Balance of the Republic of Moldova in 2003. Department of Statistics and Sociology, Republic of Moldova. Ch.: 2004, 37 p
4. Boyle G (2004) Renewable energy: power for a sustainable future. Oxford University Press, Oxford, New York, p 452
5. Boyle G, Everett B, Ramage J (2003) Energy systems and sustainability. Oxford University Press, Oxford, New York, p 593
6. Rogner H-H (1997) An assessment of world hydrocarbon resources. Annu Rev Energ Environ 22:217–262
7. Bentley RW (2002) Global oil & gas depletion: an overview. Energ Pol 30:189–205
8. Flavin C, Lenssen N (1996) Energy wave. Guide to the impending energy revolution. Tehnica Publishing House, Bucharest
9. Cavallo AJ (2004) Hubbert's petroleum production model: an evaluation and implications for world oil production forecasts. Nat Resour Res 13(4):211–221
10. Deffeyes KS (2002) Hubbert's peak: the impending world oil shortage. Princeton University Press, Princeton. ISBN 0-691-09086-6
11. Peak oil theory: world running out of oil soon. Could distort policy & energy debate
12. Ramaswany V (2001) Solar forcing of climate. In: Climate change 2001: working group I: the scientific basis. p 351. http://www.grida.no/publications/other/ipcc_tar/?src=/climate/ipcc_tar/wg1/244.htm. Retrieved 10 Mar 2005
13. Solar Variations and Climate Change. *Solar influences on global change*. Washington, D.C: National Academy Press. 1994. p 23. ISBN 0-309-05148-7. http://books.nap.edu/openbook.php?record_id=4778&page=23
14. Crommelynck D, Fichot A, Lee RB, Romero J (1995) First realisation of the space absolute radiometric reference (SARR) during the ATLAS 2 flight period. Adv Space Res 16:17–23
15. Etheridge DM, Steele LP, Langenfelds RL, Francey RJ (1998) Historical CO_2 records from the Law Dome DE08, DE08-2, and DSS ice cores. In: Trends: a compendium of data on global change. Carbon Dioxide Information Analysis Center, Oak Ridge National Laboratory, U.S. Department of Energy, Oak Ridge, Tenn., U.S.A. [http://cdiac.esd.ornl.gov/trends/co2/lawdome.html]. Retrieved 5 Oct 2005
16. The elusive absolute surface air temperature (SAT). Goddard Institute for Space Study. http://data.giss.nasa.gov/gistemp/abs_temp.html. Retrieved 15 Oct 2009
17. Tremblay A, Varfalvy L (2005) Greenhouse gas emissions – fluxes and processes. Springer, New York, Berlin, p 732 p. ISBN 3-540-23455-1
18. www.ipcc.ch/pub/reports.htm. Retrieved 20 May 2005

19. Jones PD, Moberg A (2003) Hemispheric and large-scale surface air temperature variations: an extensive revision and an update to 2001. J Climate 16:206–223
20. Meehl GA, Washington WM, Ammann CA, Arblaster JM, Wigleym TML, Tebaldi C (2004) Combinations of natural and anthropogenic forcings in twentieth-century climate. J Climate 17:3721–3727
21. Twidell J, Wir A (1990) Renewable energy sources (trans from English). Energoatomizdat, Moscow, 392 s
22. Fröhlich C (2006) Solar irradiance variability since 1978: Revision of the PMOD composite during solar cycle 21. Space Sci Res 125:53–65, doi:10.1007/s11214-006-9046-5
23. Livre vert: Énergie pour l'avenir: les sources d'énergie renouvelables. COM(96)576 du 20.11.1996
24. Livre blanc: Énergie pour l'avenir: les sources d'énergie renouvelables. Une stratégie et un plan d'action communautaires. COM(97)599 du 26.11.1997
25. Babe RE (2006) Culture of ecology: reconciling economics and environment. University of Toronto Press, Toronto
26. The economics of solar power for California: a white paper. Akeena Solar, 2005
27. Smil V (2003) Energy at the crossroads: global perspectives and uncertainties. MIT Press, Cambridge. ISBN 0-262-19492-9
28. Directive 2003/54/CE du Parlement Européen et du Conseil du 26 juin 2003 concernant des règles communes pour le marché intérieur de l'électricité

Chapter 3
Solar Energy

3.1 The Sun as Energy Source

3.1.1 General Notions

Designing a system for solar energy conversion into thermal energy or electricity is based on accurate assessment of the solar radiation in the given location and on the knowledge of solar radiation properties. The Sun is the closest star to the Earth, at the average distance of 1.5×10^{11}m.

Schematic structure of the Sun is shown in Fig. 3.1, and the geometric relations of the "*Sun–Earth*" system – in Fig. 3.2. The Sun's energy is the result of several nuclear fusion reactions, the main being the process in which hydrogen (four protons) merges and forms helium. The mass of alpha particle (helium nucleus) is less than the mass of four protons, the difference in mass is converted into energy according to Einstein's formula $E = mc^2$.

Sun's diameter is about 110 times bigger than the Earth's diameter: 1.39×10^9 m, and respectively, 1.27×10^7 m. The solar core, with a radius of about $0.23R$ (R – radius of the solar disk) and a volume up 15% of the total, is a thermonuclear fusion reactor. Its temperature is estimated at $(8–40) \times 10^6$ °K and 90% of energy is released. The core has a density 100 times bigger than water density and its mass is 40% of the total mass. At a distance of $0.7\,R$ from the Sun core, the temperature falls to 130,000°K and the density drops to 70 kg/m^3. The area between 0.7 and 1.0 R is called the convection zone (it is considered that the thermal convection processes are the main ones). The temperature falls to 5,000°K and density is very low – about 10^{-5} kg/m^3.

The Earth revolves in orbit around the Sun, that is approximately circular (circumference deviation does not exceed 1.7%). The average distance between the Sun and the Earth (Fig. 3.2), called the astronomical unit is equal to $1{,}495 \times 10^{11}$ m. The Sun, as viewed from the Earth's surface, represents a disk with an angular diameter 32′. Solar radiation can be modelled with that of an absolutely black body

I. Bostan et al., *Resilient Energy Systems*, Topics in Safety, Risk, Reliability and Quality 19, DOI 10.1007/978-94-007-4189-8_3,

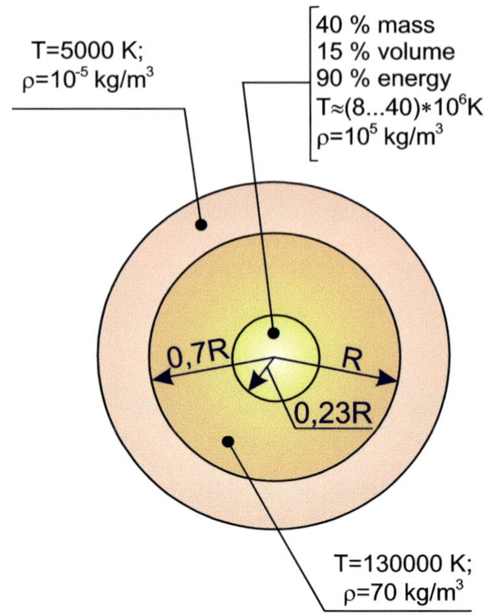

Fig. 3.1 Simplified structure of the Sun

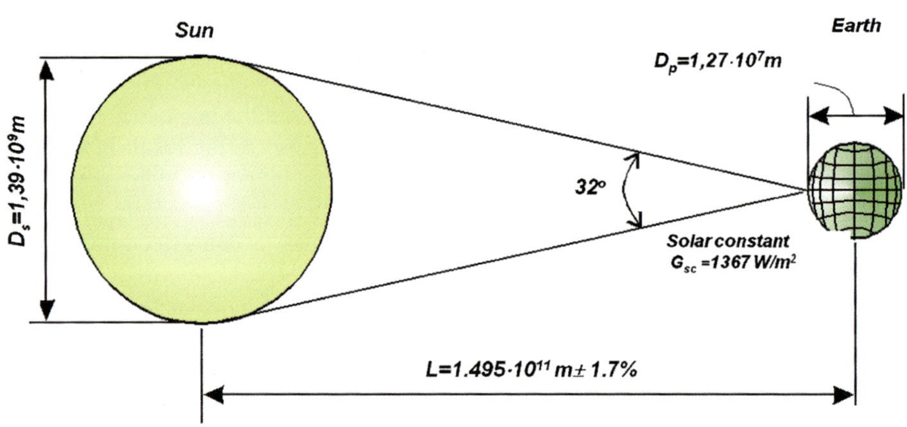

Fig. 3.2 Geometry of the "Sun-Earth" system

with actual (effective) temperature of about 5,777°K (it is a model introduced in physics to characterize the radiation spectrum of a body with a constant temperature T [1]). Calculations of the absolutely black body radiation and measurement results show that 6.4% of energy is carried by ultraviolet band of electromagnetic waves ($\lambda = 0$–0.38 μm), 48.0% – by the visible band ($\lambda = 0.38$–0.78 μm) and 45.6% – by the infrared band ($\lambda > 0.78$ μm). It is important to know the energy spectrum of solar radiation in order to understand the effects of atmosphere on the radiation coming from the Sun and make a correct choice of materials for solar collectors or photovoltaic cells.

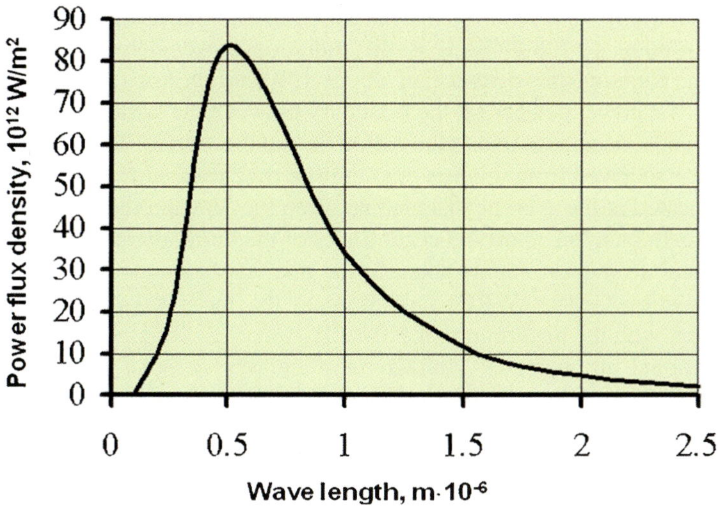

Fig. 3.3 Spectral radiation of the black body

In terms of energy, solar radiation is of interest, which, in fact, is a mix of electromagnetic waves ranging in the wavelength range (0.2–2.5) μm. Wave energy with a length more than 2.5 μm can be neglected. Some definitions, below, will be helpful to properly understand this chapter. *Irradiance* is measured in W/m^2 and is the momentary power flux density of solar radiation. For example, the irradiance equal to 1,000 W/m^2 means that an energy flux equal to 1,000 J falls every second on one square meter of surface. *Irradiation* is measured in MJ/m^2 or kWh/m^2 and is the *energy density* of solar radiation. It is obvious that irradiation is the irradiance integral within a defined period of time – 1 h, day or month. In most cases in the literature, the first term is replaced with *power density of solar radiation,* and the second – by *solar radiation: hourly, daily, monthly or yearly*.

In the calculation of photovoltaic systems the energy density of solar radiation, measured in kWh/m^2, is often expressed as *peak Sun hours*, which means *the time* in hours with a power density 1 kW/m^2 required to produce a daytime solar radiation equivalent to that obtained after the integration of energy density during the day.

Analytical relationship between the power density, temperature and wave length is given by Planck's formula [2]:

$$W_\lambda = \frac{2\pi hc^2}{\lambda^5} \cdot \frac{1}{e^{hc/\lambda kT} - 1}, \ \mathrm{W}/\mathrm{m}^2, \tag{3.1}$$

where

$h = 6.63 \times 10^{-34}$ Ws^2 is the Planck's constant;
$k = 1.38 \times 10^{-23}$ is the Boltzmann's constant;
$c = 299{,}722{,}458$ m/s is the speed of light in vacuum.

Analytical expression (3.1) allows modelling of solar radiation with high accuracy. Figure 3.3 shows the power spectral density on the surface of the Sun, calculated in accordance with Eq. 3.1 for the effective temperature of the absolutely black body equal to 5,800°K. It is the radiation power density on the surface of the Sun. Running the distance of about 150 million km (see Fig. 3.2), total extraterrestrial power density (at the boundary between the Earth's atmosphere and the outer space) decreases to a value called the solar constant. The solar constant S is the energy received from the Sun per a unit of time for a surface perpendicular to the Sun, located at the average distance between the Sun and the Earth, outside the atmosphere. In fact, due to orbital eccentricity of the Earth, extraterrestrial radiation varies. Based on measurements taken in the late 90s of the twentieth century, the World Radiation Centre (WRC) has accepted the mean solar constant equal to 1,367 W/m^2, with the uncertainty of 1.0% [3].

Conventional air mass, *m,* characterizes the way of solar beam through the atmosphere to the sea level. For the extraterrestrial space or if the land would not have the atmosphere, $m=0$. At the Equator, while the Sun is in its zenith, the solar beam runs the shortest distance, $m=1$. For zenithal angles θ_Z (zenithal angle – the angle between the vertical circle and direction to the Sun), between 0° and 70°, the air mass m can be calculated by the expression

$$m = \frac{1}{cos\theta_Z}. \tag{3.2}$$

If θ_Z is equal to 60°, the air mass $m=2$, i.e. the solar ray will run a way through the atmosphere twice bigger than if $\theta_Z=0°$. Obviously, in the second case, the solar beam will be more mitigated and it will carry less energy. This explains the decrease of solar radiation intensity in the northern hemisphere, and in the southern hemisphere, respectively, compared to the equatorial zone.

Beam or direct radiation is the received radiation from the Sun without being scattered by the atmosphere (see Fig. 3.4). The shadow of an object appears only when direct radiation occurs. Further, direct radiation will be marked with B.

Diffuse radiation will be denoted by D. The solar ray passing through the atmosphere is scattered, i.e. diffused in all directions (see Fig. 3.4). Diffuse radiation is always present, even on a clear day this component is about 10%. In this case, solar rays are scattered by the molecules of oxygen, carbon dioxide, dust particles, etc., and the sky becomes blue. If the sky is overcast then direct radiation is zero, and the effect of diffuse radiation is present only. Because of the diffuse radiation, the light penetrates even through a window facing the north.

Global solar radiation: The sum of the two components makes the global radiation ***G*** on some surface. In most cases, it is measured and used as the notion of global radiation on a horizontal surface. From the definition follows:

$$G = B + D \tag{3.3}$$

Albedo or reflected radiation is denoted by R. Usually, one operates with the radiation reflected by the Earth's surface, which falls on the solar collector or photovoltaic panel. In most cases, this component is not taken into consideration,

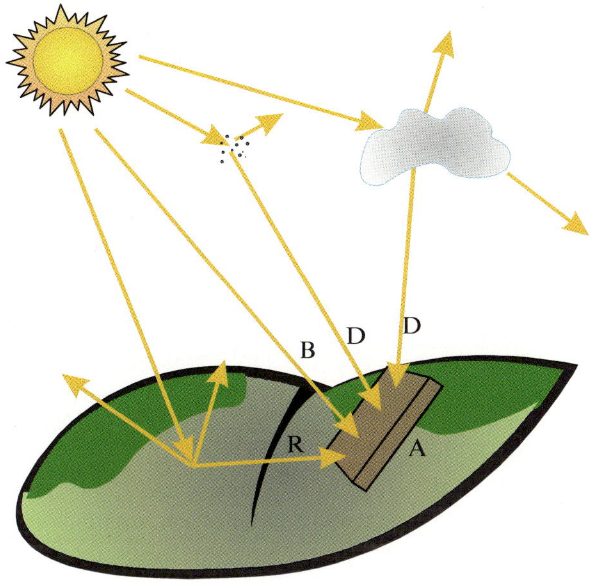

Fig. 3.4 Components of solar radiation at absorbing surface *A*: *B* – direct; *D* – diffuze; *R* – reradiation

except for collectors or bifacial photovoltaic panels (both surfaces are operating – the one oriented to the Sun and another oriented to the Earth's surface). Thus, the total radiation incident on the surface of a body would be equal to the sum of direct radiation that is diffused and reflected (Fig. 3.4):

$$G = B + D + R \tag{3.4}$$

3.1.2 Solar Radiation on the Earth Surface

Above it was mentioned that the radiant power density S in the outer space is constant and equal to 1,367 W/m^2. Each period, the Earth gets the same amount of energy calculated by multiplying the surface S with the surface exposed to irradiation and with the period, we are interested in (a second, a minute, an hour etc.). The surface exposed to irradiation is equal to πR^2, where R is the radius of the Earth, and the total area of the globe – with $4\pi R^2$. Thus, the average radiant power density S_{med} collected by the Earth will be:

$$S_{med} = \frac{S}{4} = \frac{1367}{4} = 342\,\mathrm{W/m^2} \tag{3.5}$$

Further the analysis of solar radiation properties on the Earth surface will be done, on the assumption that the average radiant power density at the boundary between the Earth's atmosphere and the outer space is constant. The solar rays passing through the atmosphere are subject to significant changes. Some are absorbed by the molecules of air, others – scattered (diffuse component) and some enter the atmosphere without being affected (direct component) and are absorbed or reflected (reflected component) by objects on the Earth's surface (see Fig. 3.4). Two major effects influence very much the extraterrestrial radiation that runs the atmospheric layer:

- Diffusion of Sunlight by molecules of air, water and dust.
- Absorption of Sunlight by molecules of ozone O_3, water H_2O and carbon dioxide CO_2.

Diffusion of solar radiation is caused by the interaction of electromagnetic wave with wavelength λ and the molecules of air, water and dust. The intensity of interaction depends on the length of the pathway run by the beam through the atmosphere defined by the air mass m (see expression [3.2]), the number of particles and their size in comparison with the wavelength λ. According to Rayleigh's theory [4], the diffusion coefficient, which is due to the interaction of the Sun ray with the air molecules, is proportional to λ^{-4} and is significant for the wavelengths less than 0.6 μm. Phenomena related to ray diffusion by molecules of water, dust, etc., are quite complicated, their detailed description being given in [5, 6]. The absorption of Sun rays by the atmosphere is due to ozone, oxygen, water and carbon dioxide molecules. Spectral radiant power distribution (Fig. 3.5) is presented for two cases: (1) $m = 0$, extraterrestrial radiation, i.e. the upper boundary of the atmosphere and (2) at the sea level for $m = 1.5$ (azimuthal angle $\theta_z = 48°$, angle of Sun elevation above the horizon, $\alpha = 42°$). There is a strong absorption band of ultraviolet radiation with a wavelength less 0.3 μm by the molecules of ozone (the ozone layer is the protection screen of the biosphere from the killing ultraviolet rays), a high transparency of the atmosphere in the visible band $0{,}4 < \lambda < 0{,}76$ μm, and a strong absorption of spectrum band infrared radiation. In other words, our atmosphere is transparent to visible band radiation and is opaque – in the infrared band. Because of these two effects, solar radiation on the Earth's surface decreases significantly and the irradiation or instantaneous power density peaks about 1,000 W/m^2, which takes place at noon, under clear sky.

3.1.3 The Sun and the Global Energy Balance

The Sun is the main source of energy that defines the climate on the Earth. In accordance with expression (3.5), every square meter of the outer covering of the atmosphere receives 342 W, of which 31% or 106 W is immediately reflected into the space by the clouds, the atmosphere and the land surface (see Fig. 3.5). The rest, i.e. 236 W/m^2, is absorbed by the land, ocean surface water and, partially, by the atmosphere, warming them. The Earth's surface delivers into space the same quantity of energy (236 W/m^2), but already in the infrared wavelength band: partly, the atmospheric cover, the clouds and water vapours, which, condensing, become

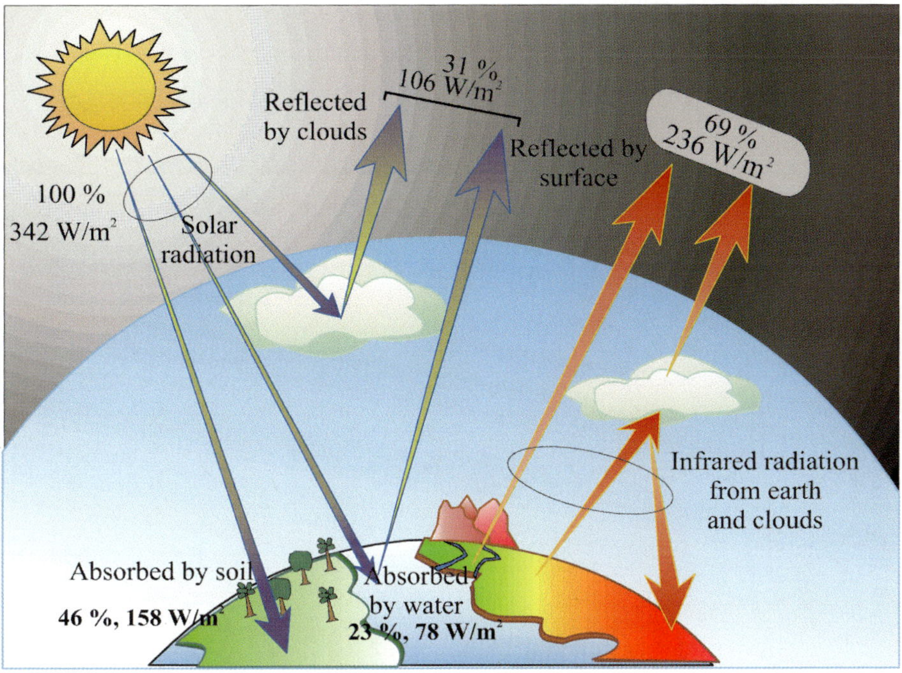

Fig. 3.5 Simplified balance of energy fluxes in the "Earth surface-atmosphere" system

cooler and transmit heat to the atmosphere (see Fig. 3.5). Due to the exchange of energy between the Earth's surface, the atmosphere and the outer space, a constant average global temperature of about 15°C maintains at the sea level, which decreases rapidly with height increasing, reaching –58°C in the upper troposphere. A natural question arises: to which physical phenomena this balance is due to and can it be damaged? The answer is found addressing again the theory of absolute black body, which is modelling either the Sun with actual temperature of 5,777°C or the Earth with effective temperature equal to T_e.

3.1.4 Greenhouse Effect Simulation

The Earth's atmosphere contains several gases called greenhouse gases (GHG): carbon dioxide CO_2, methane CH_4, nitrogen oxide N_2O, and water vapours H_2O. A common property of GHG lies in their absorption by its molecules in the infrared band radiation emitted by the Earth's surface, atmosphere and clouds. The certain molecule begins to vibrate and, in its turn, emits the same spectrum of infrared radiation which is absorbed by other greenhouse gas molecules. Thus, the phenomenon of absorption – emission – absorption leads to energy conservation in the lower layer of the atmosphere. In other words GHG has the same role as

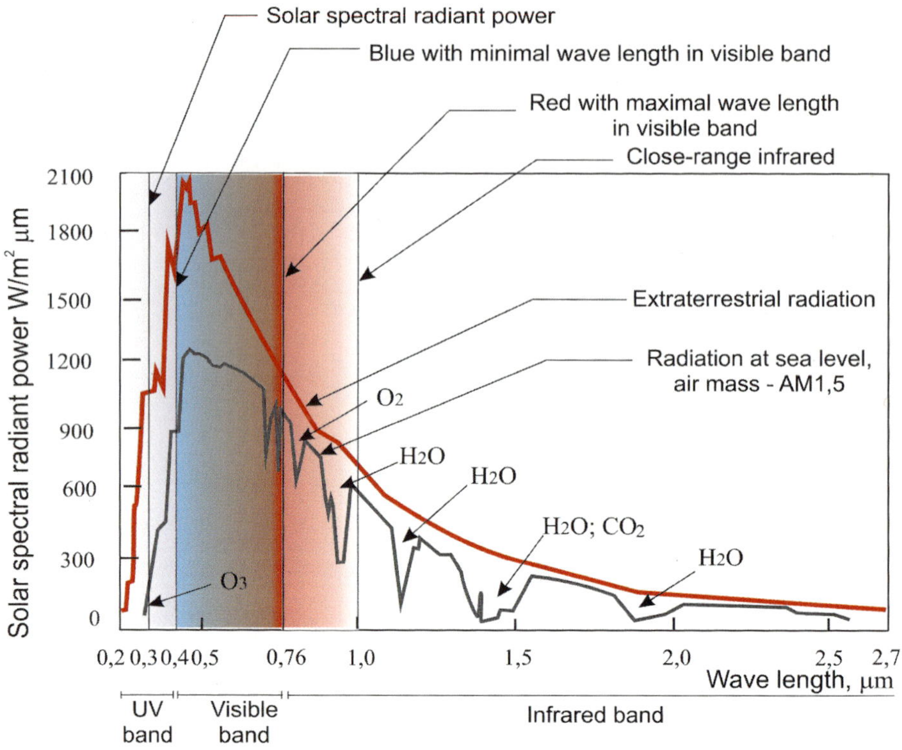

Fig. 3.6 Spectral distribution of power flux (radiant) density

the glass roof of a greenhouse. The term "greenhouse effect" was introduced by meteorologists from the early nineteenth century [7], at that time had no negative implication and was used to describe the natural greenhouse effect due to which the average temperature on the Earth remains constant and is about 15°C. The greenhouse effect has a crucial role in maintaining life on Earth. If the greenhouse effect would not occur then the temperature on the Earth's surface would be determined by the balance between the quantity of energy released by the Earth to space and the received one from solar radiation. In accordance with the Stefan-Boltzmann's law [1] the amount of energy emitted by a body with effective temperature T_e is determined by the relationship

$$We = \sigma \cdot A \cdot T_e^4, \tag{3.6}$$

where $\sigma = 5.67 \times 10^{-8}$ W/m^2 K^4 is Stefan-Boltzmann's constant; A – body area.

Equation 3.6 is obtained as a result of the integration of expression (3.1) of the spectral power density. In the case of Earth $We = 4\pi\ R^2\sigma T^4{}_e$. In addition, the Earth receives from the Sun a quantity of energy equal to 0.69 S/4 (see Eq. 3.5 and Fig. 3.6). In condition of thermal balance, the quantity of the received energy equals the amount of energy emitted from where we get the actual temperature of the Earth's surface T_e:

$$T_e = \sqrt[4]{\frac{0.69S}{4\sigma}}. \tag{3.7}$$

For $S = 1{,}367\ \mathrm{W/m^2}$, from Eq. 3.7 we obtain the effective temperature of the Earth's surface $T_e = 254°\mathrm{K}$ or $-19°\mathrm{C}$. It is clear that life on Earth in such conditions as it exists would not be possible.

Since the second half of the twentieth century and, especially, nowadays, it is often spoken of the so-called *greenhouse effect*, but already with a negative sense, linked primarily to the phenomenon of *global climate change*. From the point of view of environmental professionals that is due to the anthropogenic impact on the natural balance.

Heat balance of the Earth can be damaged by either natural or human activity. Volcanic eruption causes a sudden increase in dust in the upper layers of the atmosphere, burning fossil fuels leads to increased GHG emissions, massive cutting of forests, urbanization, introduction of new agricultural land lead to changing the refraction coefficient (albedo). Because of these effects, the variations of energy flows received and transferred by the Earth's surface occur. These variations can be negative or positive. In both cases, the climate system will respond to establish a new thermal balance. Positive changes will tend to establish a higher temperature of the Earth's surface and negative changes will tend to opposite effects. Recent studies show a predominance of positive change and global warming occurs.

3.2 Solar Energy Potential

3.2.1 Global Solar Energy Potential

The greatest theoretical potential of renewable energy is solar energy (about 89,000 TW, see Fig. 1.40). Among the components of solar energy, direct normal solar radiation is most important in terms of energy conversion.

Direct normal radiation is affected by both climatic conditions and the distance from the Sun through the atmosphere. Figure 3.7 shows a typical chart of diurnal and seasonal distribution of direct normal radiation (DNR) in the Mojave Desert, California, USA [8].

Global solar theoretical energy potential is very big, but various factors, such as areas with limited Sunlight, weather conditions and performance of conversion systems dramatically reduce the energy potential explored. Still the explorable energy potential remains very high, being comparable to global energy production (Table 3.1). The most favourable areas, in terms of solar energy conversion, are those at the equator, and in particular, in the desert, where such a negative effect (for solar energy conversion) as rains rarely occurs. Also, some solar energy

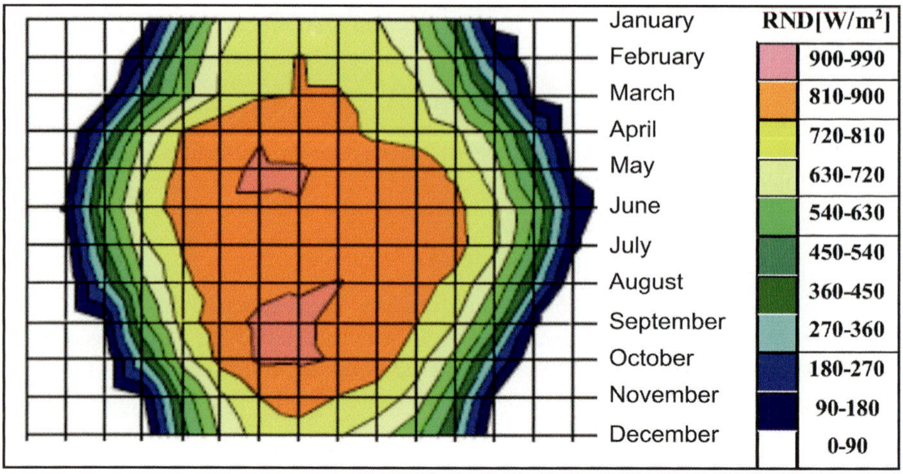

Fig. 3.7 Distribution of daily and seasonal normal direct radiation in Mojave desert, California, USA

Table 3.1 Potential of solar electrical energy generation by thermodynamic conversion

Characteristics	Energy quantity (TWh/an)
Overall solar radiation at the Planet's surface	240×10^6
Desert surfaces occupy 7% of the total Planet's surface	17.0×10^6
Available direct fraction of the radiation (approx. 70%)	11.2×10^6
Annual average efficiency of solar energy conversion into electrical energy (about 15%)	11.2×10^6
Part of semi-arid and arid locations for solar collectors units in living areas and with access to infrastructure (1% of the desert surface)	16.8×10^3
Overall production of electrical energy in 2000	15×10^3

conversion systems require large areas, which, in cases of developed agricultural areas are cut out of the agricultural circuit. From this point of view, the biggest solar energy conversion systems are located in the deserts of the USA, Australia, arid areas of Spain, etc.

3.2.2 *Estimation of Available Solar Radiation*

3.2.2.1 General View

To design a facility for solar energy conversion into heat or electricity it is necessary to know the available solar radiation on the collecting surface either of the solar collector or of the PV module (panel). However, solar radiation is a variable function, which depends on several factors: *latitude and altitude of the location,*

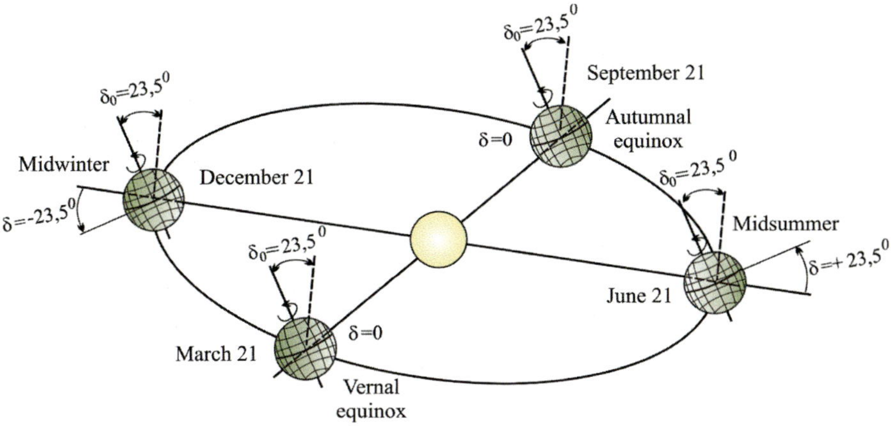

Fig. 3.8 Earth's orbit and the declination angle δ

season, day, hour, degree of cloudiness, dust content, water vapours and aerosols in the atmosphere. It is obvious that there cannot be one or more mathematical expressions, which would allow calculation of solar radiation, taking into account all these factors, in particular, those with random nature – the cloudiness, content of dust, aerosols, etc.

Designing a solar plant and forecasting its future productivity is based on systematic measurements of solar radiation components carried out in the past and available in the publications of Hydrometeo State Department (see Annex A1). In most cases, the measurement results are used for a horizontal surface of the direct and diffuse radiation for a period of an hour, a day or a month. Calculations based on the duration of an hour are more accurate, but require a heavy workload. The methodology is described in detail in [4]. In this manual, daily and monthly data on solar radiation and the isotropic model of the celestial vault will be used. To understand the essence of the calculations, at first, some notions of astronomy, in particular, the analytical relations that describe the apparent motion of the Sun around the sky will be addressed.

The Earth rotates in an elliptical orbit around the Sun during 1 year with the Sun in one of the focal points. Also, the Earth rotates around its axis, making one revolution in 24 h. The Earth's axis of rotation has a fixed direction in space and is inclined at angle $\delta_0 = 23.5°$ towards the normal line in the orbit plane (Fig. 3.8).

The angle between the Sun direction and the equatorial plane is called the declination angle δ and varies throughout the year from +23.5° when the summer solstice (June 21) to −23.5° at the winter solstice (December 21). On March 21st, and respectively, September 21 the declination $\delta = 0$ and the day time and the night time are equal. According to [4] the declination can be calculated using the formula:

$$\delta = 23,45 Sin\left(360\frac{284+n}{365}\right), \tag{3.8}$$

Table 3.2 Number of the average day for every month and its values n [4]

Month	n for i day of the month	For the average day of the month: Date	n, number of days per year	δ (degree)
I	i	17	17	−20.9
II	$31+i$	16	47	−13.0
III	$59+i$	16	75	−2.4
IV	$90+i$	15	105	9.4
V	$120+i$	15	135	18.8
VI	$151+i$	11	162	23.4
VII	$181+i$	17	198	21.2
VIII	$212+i$	16	228	13.5
IX	$243+i$	15	258	2.2
X	$273+i$	15	288	−9.6
XI	$304+i$	14	318	−18.9
XII	$334+i$	10	344	−23.0

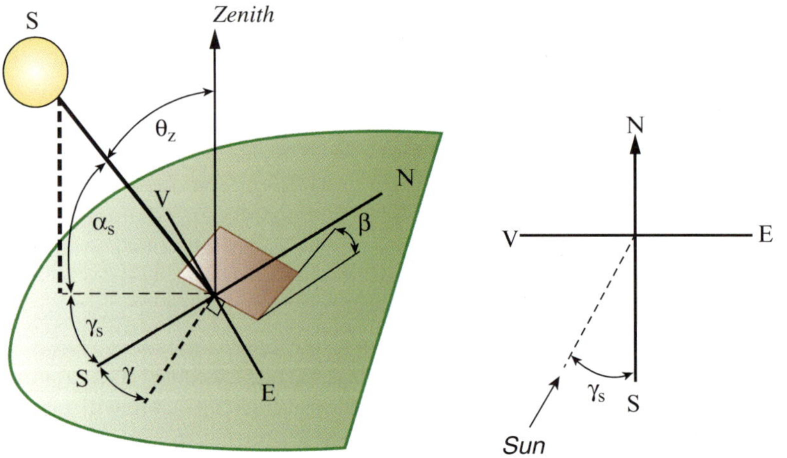

Fig. 3.9 Explications of angles: angle of dip β, azimuthal angle γ; solar azimuth angle γ_s; solar climbing angle α_s; zenithal angle θ_z

where n is the number of days of the year, considering as the first day – January 1st.

If calculations are made based on monthly average variables, it is recommended to use the average number of days for each month and the number n of the day presented in Table 3.2. Geometric relationship between an arbitrarily oriented plane to the Earth and the direct Sunlight that falls on this plane at any moment, the Sun's position against this plane can be described in terms of several angles. The definition of one angle – the angle of declination – has been given above. Definitions of other angles and other necessary analytical expressions are presented below. Geometric interpretations correspond to Figs. 3.9 and 3.10.

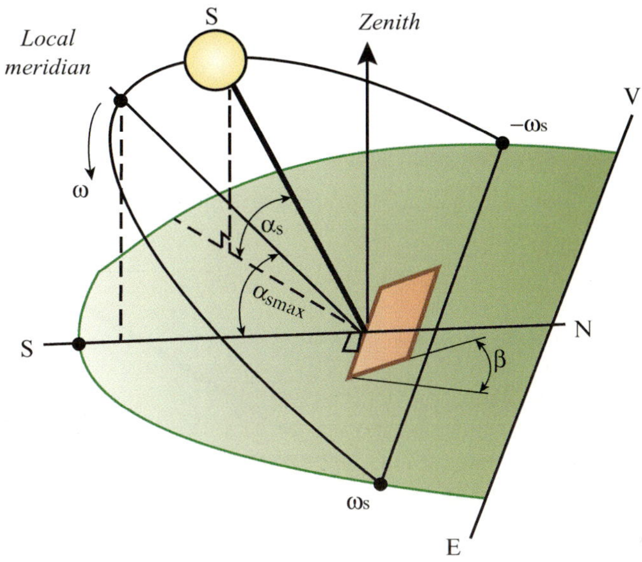

Fig. 3.10 Hour angle ω, sunrise angle ωs and sunset angle −ωs

Latitude Φ is the angle measured from the equator to the point of interest on the Earth's surface. It is considered positive for the northern hemisphere and negative – for the southern hemisphere.

Angle of dip β of the plane is the angle between the plane surface in question and the horizontal surface, $0 \leq \beta \leq 180$ (Fig. 3.9). For normal solar units the maximum value does not exceed 90°.

Azimuthal angle $\boldsymbol{\gamma}$ is the angle between the horizontal plane projection of the normal line on the plane and the local meridian (Fig. 3.9); it is equal to zero for the plane that faces the south, negative – to the east, positive – to the west; $-180 \leq \gamma \leq 180$.

Solar azimuthal angle $\boldsymbol{\gamma_s}$ is the angle between the southern direction and the projection on the horizontal plane of the direct radiation of the solar beam (Fig. 3.9 to the right); the angles measured from the southern to eastern direction are negative, the measured angles to the west are positive.

Solar climbing angle α_s is the angle between the horizon and the line joining the Sun and the point of interest, i.e. the solar beam incident at the point of interest (Fig. 3.9).

Zenithal angle θ_z is the angle between the vertical and the line joining the Sun and the point of interest, i.e., the angle complementary to angle α_s (Fig. 3.9).

Hour angle **ω** determines the position of the Sun in the sky at the given moment. It is equal to zero when the Sun crosses the local meridian, in other words, at noon, positive – to the east and negative – to the west (Fig. 3.10). Accordingly, $+\omega_s$

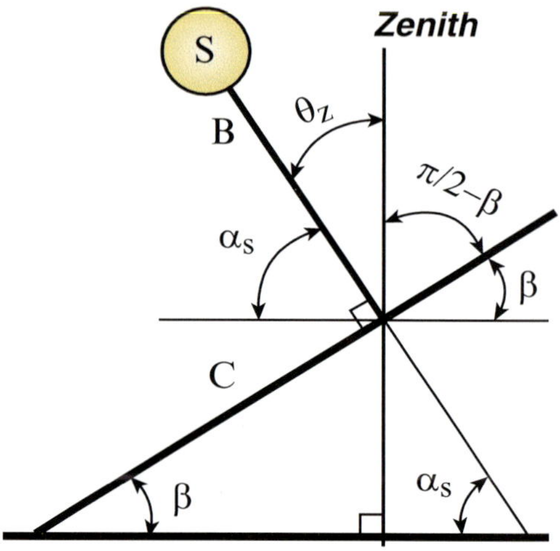

Fig. 3.11 Direct solar radiation on an inclined plane at noon: ω = 0; γ = 0

corresponds to the Sunrise angle, and ω_s – to the Sunset angle. It is clear that within an hour the Sun runs in the sky an angle equal to 15°, and its position at any time T is determined by the expression:

$$\omega = 15(12 - T). \tag{3.9}$$

If δ, Φ and ω are known angles, then the position of Sun in the sky can be determined easily in the point of interest for any time and any day, using the expressions [9]:

$$Sin\alpha_s = Sin\delta Sin\Phi + Cos\delta Cos\Phi Cos\varpi = Cos\theta_z. \tag{3.10}$$

$$Cos\gamma_s = \frac{Sin\alpha_s Sin\Phi - Sin\delta}{Cos\alpha_s Cos\Phi}. \tag{3.11}$$

For any day of the year, the declination δ can be determined from Eq. 3.8, the hour angle ω is determined from Eq. 3.9 for that time zone T and knowing the latitude of the location Φ, the Sunrise angle α_s is determined.

Based on the definitions and the analytical data described, responses to a series of questions can be given, such as: under what angle to the horizon should a solar collector be installed to obtain a maximum amount of energy in the warm or cold periods of the year? Or, generally: what will the solar radiation be in an arbitrarily oriented plane? The example that follows will convince the reader of the usefulness of the information presented above. It is also necessary to familiarize yourself with useful recommendations in the operation of solar systems. Figure 3.11 shows a solar collector C directed towards the south (angle $\gamma = 0$, Fig. 3.11). The surface of plane C is inclined under an angle β to the horizon. What value must have the angle β to obtain a maximum amount of energy if it is operating during the period April to September? Solar radiation on the collector's surface will be the highest at midday

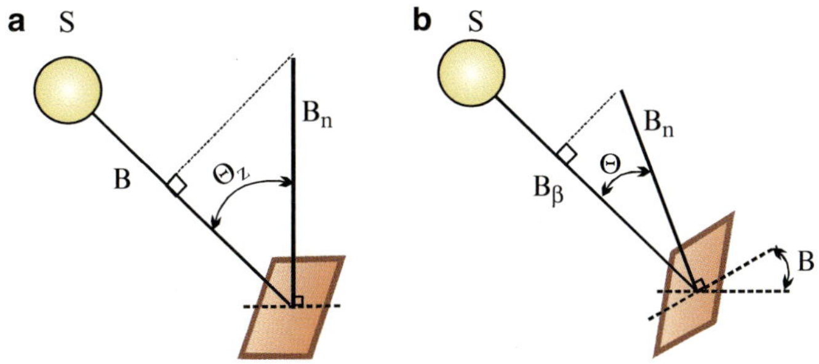

Fig. 3.12 Direct solar radiation on an horisontal (**a**) and inclined (**b**) planes

when the Sunrise angle, α_s, (Fig. 3.10) will be maximum, and the road map of the solar ray will be minimum and the hour angle $\omega = 0$. This situation will occur when direct radiation falls perpendicularly on the surface of collector C. From Fig. 3.11 it results that $\beta = 2.13$, and from the expression (3.10):

$$Cos\theta_z = Cos\Phi Cos\delta + Sin\Phi Sin\delta = Cos(\Phi - \delta), \tag{3.12}$$

From which we obtain:

$$\beta = \Phi - \delta. \tag{3.13}$$

Table 3.2 lists that, for the period researched, the average value of declination is equal to 15° and for the central area of the Republic of Moldova we get $\beta = 47 - 15 = 32°$. For the cold season of the year declination is equal to 15° and the angle of inclination should be $\beta = 47 + 15 = 62°$. It is obvious that the period can be taken arbitrary – an hour, a day, a month or a certain season. Thus, the issue of estimating solar radiation on an inclined plane in the general form, occurs. The solution is given in the next paragraph.

3.2.2.2 Solar Radiation on an Inclined Plane

In most cases, we rely on information about the solar radiation on a horizontal plane. At the same time, it is necessary to know the direct and diffuse components of radiation in terms of a solar collector or photovoltaic panel. Equation 3.13 determines the angle of inclination of the collector or panel for a period of 6 months – the warm or cold periods. The best solution would be to follow the apparent motion of the Sun in the sky, so that direct radiation falls perpendicular to the surface of interest. But the tracking mechanism is complicated and expensive, and usually used in facilities with solar radiation concentration.

Flat solar systems are built with no tracking, are oriented to the south and are mounted at an angle to the horizon optimal for the given location and for the exploitation duration during the year. Figure 3.12 presents the direct solar radiation

B on the horizontal plane (a) and B_β on an inclined plane towards horizon with the angle β (b). Perpendicular radiation on the plane surface is denoted by B_n. It is necessary to determine the ratio between the B_β and B. The ratio of direct radiation on an inclined plane and on the horizontal plane by R_b should be noted. Thus, $R_b = B_\beta/B$. From Fig. 3.12 we have:

$$B = B_n Cos\theta_z;\ B_\beta = B_n Cos\theta. \tag{3.14}$$

the ratio R_b:

$$R_b = \frac{B_n Cos\theta}{B_n Cos\theta_z} = \frac{Cos\theta}{Cos\theta_z}, \tag{3.15}$$

where θ is the solar altitude angle – the angle between the normal line of the plane to the direction of the solar beam. For the horizontal plane (Fig. 3.14) $\theta_z = \theta$.

For an arbitrarily set plane the functions $Cos\theta$ and $Cos\theta_z$ are expressed by combinations of functions of the angles discussed above – δ, Φ, β, γ and ω [4]:

$$Cos\theta = Sin\delta Sin\phi Cos\beta - Sin\delta Cos\phi Sin\beta Cos\gamma + Cos\delta Cos\phi Cos\beta Cos\varpi + \\ + Cos\delta Sin\phi Sin\beta Cos\gamma Cos\varpi + Cos\delta Sin\beta Sin\gamma Sin\varpi. \tag{3.16}$$

From which for $\beta = 0$, we obtain $Cos\theta_z$:

$$Cos\theta_z = Cos\phi Cos\delta Cos\varpi + Sin\phi Sin\delta. \tag{3.17}$$

In most cases, the solar collector or the photovoltaic panel are installed to south with the active side, and $\gamma = 0$ for the northern hemisphere and $\gamma = 180°$ – for the southern hemisphere. Substituting in Eq. 3.15 we get:

$$R_b = \frac{Cos(\phi - \beta)Cos\delta Cos\varpi + Sin(\phi - \beta)Sin\delta}{Cos\phi Cos\delta Cos\varpi + Sin\phi Sin\delta}. \tag{3.18}$$

Equation 3.18 is used to determine the optimal angle β for periods of 1 h or 1 day. When operating with monthly values of solar radiation, Eq. 3.18 is modified:

$$\bar{R}_b = \frac{Cos(\phi - \beta)Cos\delta Cos\varpi_s' + (\pi/180)\varpi_s' Sin(\phi - \beta)Sin\delta}{Cos\phi Cos\delta Cos\varpi_s + (\pi/180)\varpi_s Sin\phi Sin\delta}, \tag{3.19}$$

where ω'_s is the hour angle of the Sunset towards the inclined plane; ω_s – is the hour angle of the Sunset towards the horizontal plane.

Angles ω'_s and ω_s are determined from the relations:

$$\omega_s = \arccos(-tg\phi \cdot tg\delta), \tag{3.20}$$

$$\omega_s' = \min\{\omega_s; \arccos[-tg(\phi - \beta) \cdot tg\delta]\} \tag{3.21}$$

Table 3.3 Typical values of the albedo (reflection coefficient), ρ

Land characteristics	ρ
Ploughing	0.2
Land covered with verdure	0.3
Sandy desert	0.4
Snow	0.7

The calculated values of ratio R_b are presented in [4], in graphical and tabular form, depending on the difference of angles $\phi - \beta$ (over each 5°) and the latitude of the location ϕ (over each 5°). For the ratio value R_b calculated, the direct component of solar radiation on an inclined plane is determined:

$$B_\beta = R_b \cdot B. \tag{3.22}$$

Other components of solar radiation, diffuse and reflected, are determined from the isotropic model of the celestial vault proposed by Liu and Jordan in 1961 and modified by Klein in 1977 [4]. Diffuse radiation on an inclined plane D_β is calculated by formula:

$$D_\beta = \frac{1}{2}(1 + Cos\beta)\, D, \tag{3.23}$$

where D is diffuse radiation on a horizontal plane (Annex A1).

Reflected radiation on an inclined plane R_β is calculated by formula:

$$R_\beta = \frac{1}{2}(1 - Cos\beta)\, \rho G, \tag{3.24}$$

where

ρ is the reflection coefficient of the Earth's surface (see Table 3.3);

G – is global radiation on a horizontal surface

In consequence, global radiation on an inclined plane is equal to the sum of the three components–direct, diffuse and reflected radiation on the same plane:

$$G_\beta = R_b B + \frac{1}{2}(1 + Cos\beta)\, D + \frac{1}{2}(1 - Cos\beta)\, \rho G. \tag{3.25}$$

3.3 Solar Energy Conversion Systems

3.3.1 General View

Technologies for wider use of renewable energies must be simple and reliable, affordable in developing countries with low technological development, which have explorable and technical solar potential, but resources of materials are limited. They must produce more energy during their life, should not deplete natural resources, and should not create products that have a negative impact on the society or the environment.

Solar energy can be converted in two ways:

- By thermal conversion (solar energy is converted into heat energy, used for heating, domestic water or electricity production);
- By using the photovoltaic effect (directly convert solar energy into electricity).

There is a wide variety of technical means to convert solar thermal energy, which can be divided into two groups:

1. Systems for direct and diffuse solar energy conversion into thermal energy at low temperatures. These do not require tracking the Sun and have a simple construction. This group includes:
   - Plane solar collector for water and air heating;
   - System of solar plane collector – flow turbine (solar towers);
   - System of accumulated heat solar energy plane collector – flow turbine (solar towers).

   The last two systems are used to generate electrical energy.
2. Systems for Sun beam concentration, which convert the direct component of solar radiation into thermal energy under average and high temperatures (from several hundred to several thousand degrees). This group includes:
   - Cylindrical-parabolic trough systems;
   - Parabolic systems with different types of engines (steam turbine, Stirling engine, etc.);
   - Heliostat systems.

In most cases concentration systems are used to produce electricity. Cylinder-parabolic systems track the Sun as a single coordinate. A distinctive feature of solar electric systems (SES) consists in multiple energy conversion: solar radiation is converted into heat, then, using Carnot thermodynamic cycle or others, the heat is converted into mechanical energy. Finally, the electromechanical conversion is used to obtain electricity. From this point of view, photovoltaic conversion is superior compared to SES – solar radiation is converted directly into solar energy.

In all primary systems, solar radiation can be absorbed by a flat collector or can be concentrated optically using mirrors and lenses. Optimal choice depends on the technological-energy needs, location and weather conditions. Weather effects and solar angle have a major impact on the efficiency of concentrating collectors, which use only direct rays of the Sun. The combined effect of the Sunrise angles (zenith angle) has a major impact on SES efficiency (Fig. 3.13) [8].

Currently SES with Solar Energy Concentration is the most efficient and effective in terms of cost to generate electricity from solar energy (Table 3.4). However, hundreds of megawatts of SES generating capacities will be put into operation in the next few years and will make an important contribution to meeting the energy needs.

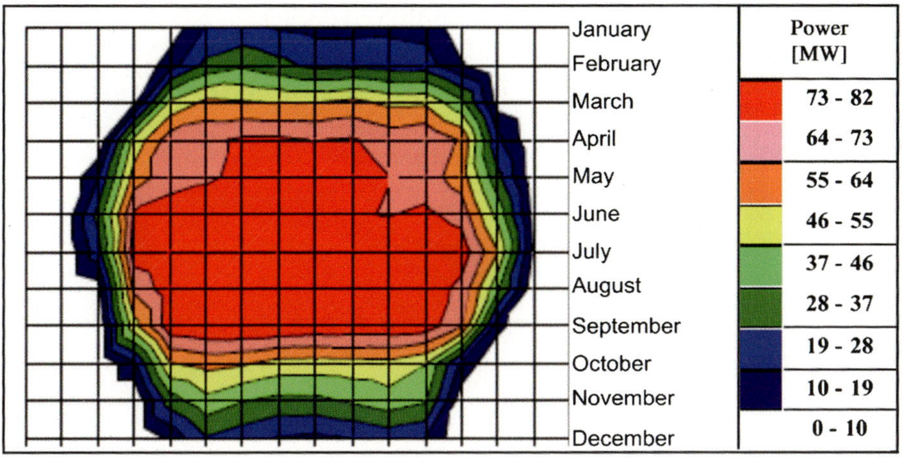

Fig. 3.13 Typical average production of solar electrical energy by SES

Table 3.4 Cost of solar electrical energy

Technology	Capital investments cost ($/kW)	Electrical energy cost (cent/kWh)
PV photovoltaic systems (crystalline silicon)	7,500–8,500	51.0
Technologies for SECS		
Stirling paraboloidal	2,650	16.7
Cylindrical-parabolic systems	2,877	13.4
Solar towers	2,713	9.0

What would be further development of these systems particularly advantageous in economic, environmental and social aspects?

Figure 3.14 shows two possible scenarios for the development of SES till 2020. Under both scenarios, SES cumulative installed capacity will increase substantially: in 2020 rates could reach 2,100 MW (under scenario 1), and respectively, 4,000 MW (under scenario 2).

3.3.2 Solar Systems for Water Heating

3.3.2.1 Brief History

Thermal conversion is the oldest and most widespread form of solar energy utilization. The first attempts to use solar energy for heating were made in Greece in the fifth century before Christ. At that time, the Greeks had already succeeded to exhaust their forestry resources and firewood was imported from Macedonia,

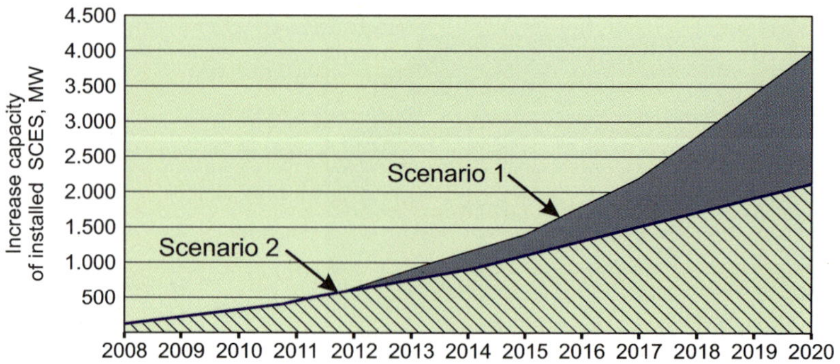

Fig. 3.14 Developing scenarios for SECS till 2020

Cyprus and the Middle East. With the discovery of glass by the Romans, the development of bioclimatic concepts in buildings' construction started, which purpose was the efficient use of natural lighting, aeration (ventilation) of spaces in hot regions or heating the spaces in areas with moderate climates, using solar energy. The discovery of the greenhouse effect at the end of the eighteenth century, followed by the experiments undertaken by Horace de Saussure's (Switzerland, 1767), Augustine and Abel Mouchot Pifre (France, 1870), contributed to the first successful commercial systems for hot water production by using solar power (California, early twentieth century), then in Florida (between 1920 and 1950), water pumping, Frank Shuman (Egypt, 1912) [9–11].

Any black surface exposed to beams, called absorbing surface, transforms solar energy into heat. This absorbing surface is the simplest example of direct converter of solar radiation into thermal energy, called "*flat solar collector*". Solar thermal energy conversion includes several technologies: water heating with flat or vacuum collectors, drying agricultural products and medicinal plants, and in semi-finished wood processing, solar refrigeration, distillation of water, electricity generation, using thermodynamic process, etc. In this paragraph, the emphasis is on three technologies: hot water production, drying of agricultural products and residential space heating.

Technological aspects: Over the centuries, engineers and architects have improved the design of buildings for the most rational use of natural light and heat source – the Sun. From this point of view, the walls of buildings are heat accumulators, and the windows and rooms – are heat collectors, which allow penetration of visible band (short wave) solar radiation in the inside and do not allow infrared radiation (long wave) to leave the building. This technology, and, respectively, the systems used to achieve it were later called passive technologies, systems or methods to use solar energy. Passive systems are distinguished by the following two distinct features:

1. The processes for collecting, storing and using energy are integrated into the building structure. In other words, Sun beams heat the consumer directly;

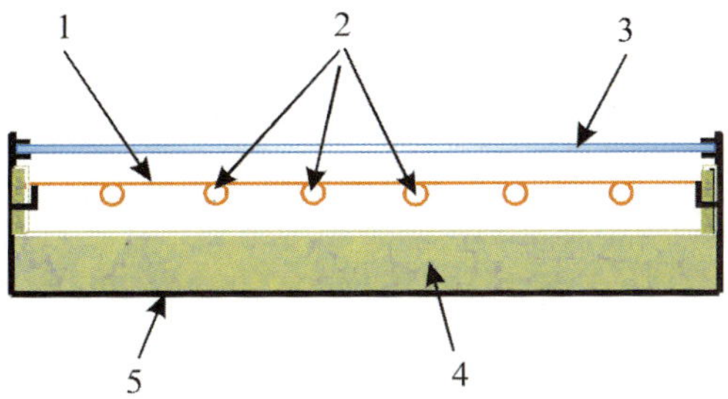

Fig. 3.15 Constructive schema of the solar collector

2. Passive systems do not require mechanical energy to transport heat to the consumer or storage tank. Motion of air flows takes place under temperature differences between different layers.

On the contrary, active technologies or systems for solar energy conversion are equipped with special solar collectors in which solar radiation is converted into heat, and then through a heat carrier (usually water or air) is transported to the consumer or stored in the tank. The most common types of solar collectors are flat solar collectors not concentrating the radiation used to obtain maximum temperatures 150°C.

3.3.2.2 Flat Solar Collector for Water or Air Heating at Small Temperatures

Solar collector is a special heat exchanger that transforms the energy of solar radiation into solar energy. Also, the solar collector differs from most conventional heat exchangers (e.g., liquid-liquid heat exchangers), in which the heat transfer by radiation plays an insignificant role. In the solar collector, on the contrary, energy transfer to the liquid or gas is carried away through solar radiation with the wavelength ranging between 0.3 and 3 μm, and maximal power density of 1,000–1,100 W/m^2.

Flat-shaped solar collector can be designed to supply hot water at low temperatures of about 40–150°C. It uses both components of solar radiation – direct and diffuse, it does not require tracking the Sun in the sky, it generates low operating costs and has a much simpler construction compared to collectors with solar radiation concentration. This type of collector is the most widespread, as it is part of any water heating system, of residential areas, of solar drying chambers and refrigeration systems. It is technically and technologically highly advanced, and has a developed market and economic prospects. Figure 3.15 shows the design scheme of a solar collector for water heating. Its main components are: the so-called "*black box*" (5) with thermal insulation (4) of three walls, coated on the front by transparent surface (TS) (3).

Heat exchanger is metal plate-pipe type, absorbent surface (AS) (1) and pipes (2), respectively. In the case of the air collector, the heat exchanger is metal plate-air channel. The solar collector operates based on two physical phenomena: absorption of solar radiation by a black body (Fig. 3.15, absorbing surface AS) and the greenhouse effect performed by the transparent surface TS. Artificial greenhouse effect is made in the case of the solar collector. TS surface is transparent to Sunbeams and opaque to infrared radiation that is heat, emitted by the absorbing surface AS. AS temperature increases and heat is transmitted to water running through pipes 2. Plate-pipe type heat exchanger is the main element of the collector shown in Fig. 3.15. Over the years, various technical solutions have been proposed for plate (1) – pipes (2) combination. The most viable solutions are shown in Fig. 3.16: coil (a), parallel tubes (b), channels consisting of two metal plates by contact method welding(c) or channels formed inside a plastic plate (d).

For the first two schemes the pipe-plate constructive contact is of particular importance. This must be done at the lowest thermal resistance. Next, the three usual technical solutions are presented:

(a) Contact is performed by traditional welding of plate (1) and of pipe (2). The use of 1.5–2.0 mm thick steel plates is recommended. Often, welding is the main cause of obsolescence. Length of service does not exceed 5 years.
(b) Contact is performed through plate deformation, so as to fold the pipe. It is a simple, highly productive at manufacturing and reliable solution. Over time, due to various linear thermal expansion of plate and pipe, a gap appears between them, increasing the thermal resistance of the plate-pipe contact, and, respectively, decreasing the heat transfer efficiency.
(c) A copper plate (1) with a thickness not exceeding 0.2 mm is cold welded to the copper pipe 2 with inner diameter of 6 mm and outer diameter of 8 mm. Ultrasonic welding is used at wave frequency of 20 kHz and tool amplitude of 150 μm. The solution provides an 11 m/min productivity, a good quality welding (5) and a 20 years service life of the heat exchanger. Modern solar heat exchangers are manufactured using this technology.

In order to reduce heat loss through the space between the AS and TS, the heat exchanger of plate-pipe type is mounted in a glass cylinder (tube) out of which the air is removed. Thus, a significantly lower heat transfer by convection between the AS and TS areas occur and the collector efficiency increases.

The first vacuum tube collector has been proposed by Speyer in 1965. Figure 3.17 shows two design schemes of vacuum collectors. Tightly connected AS 2 and pipe 3 are placed into the glass tube 1. There is an essential difference between these two schemes. In the (a) scheme cold water enters through nozzle 5, is being heated and through nozzle 4 it is transported to the storage tank.

Both nozzles must be tightly connected to the glass tube. Tube (1) and pipe (3) expand differently, thus causing the loss of tightness between nozzle-glass tube connection. In the second scheme there is only one tight connection – the end (6) of tube (3). Heat transfer is done in the heat exchanger (8), to which end pipe (6), serving as condenser, and the pipe (7), through which water circulates, are mounted. Pipe (3) is partially filled with a liquid with a relatively low evaporation temperature. Under the action of heat absorbed by plate (2), the liquid evaporates, pressure

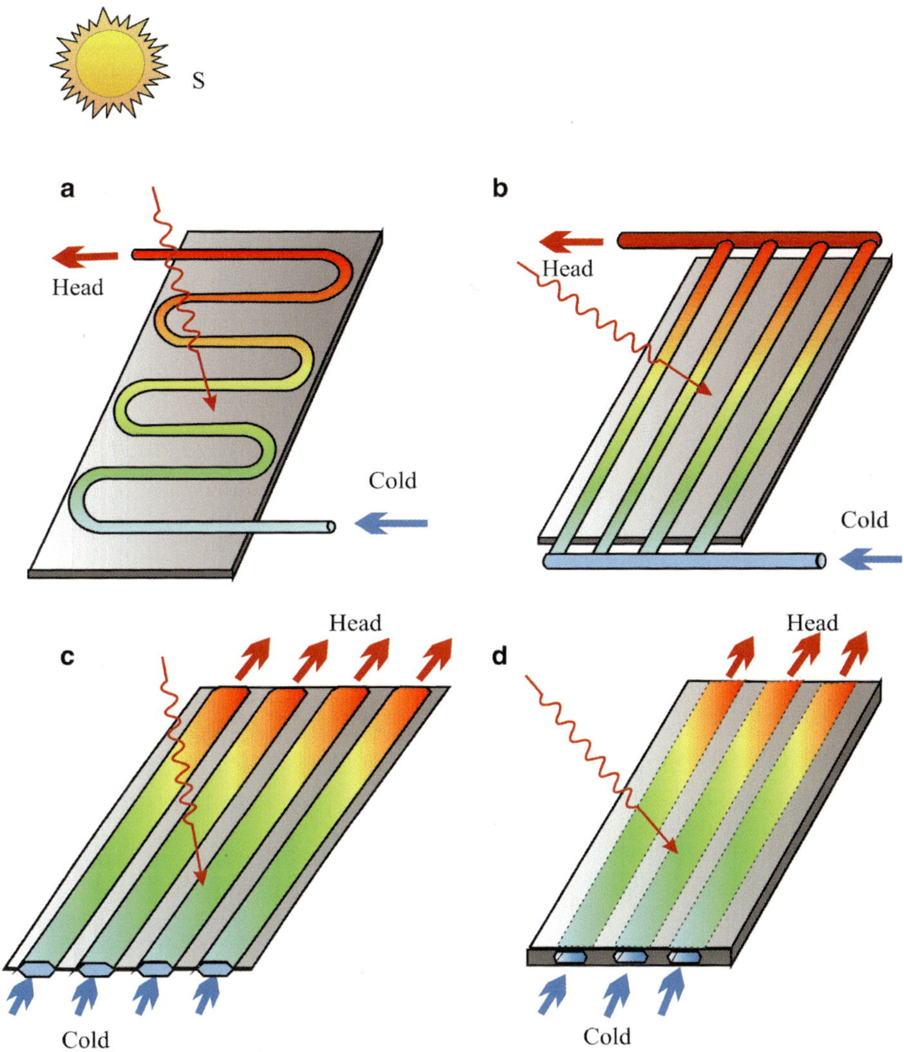

Fig. 3.16 Heat exchangers for solar collectors

increases and vapours move to the capacitor condenser – the end of pipe (6). Here vapours condense giving heat to water flowing through the pipe (7). The liquid from the condenser flows in the opposite direction through pipe (3). Vacuum tube solar collector contains several tubes connected in parallel and mounted into a common frame to form a register.

Disadvantages of vacuum solar collectors: they are about 1.5 times more expensive, have a greater mass; there is a risk of deterioration of tight connections that cannot be repaired under service conditions.

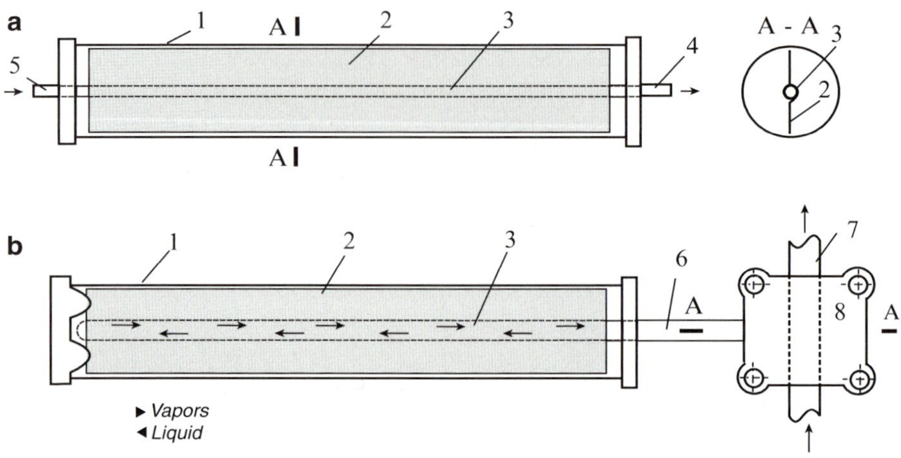

Fig. 3.17 Design of vacuum tubes

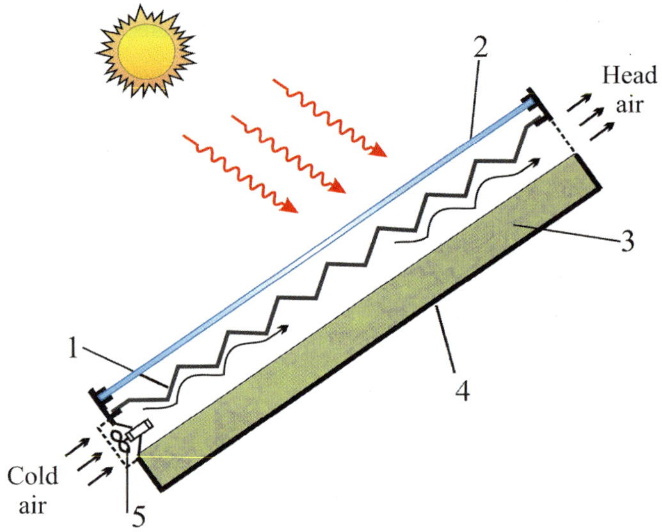

Fig. 3.18 Solar collector for air heating

Scheme design of air heating solar collector is similar to the water collector (see Fig. 3.18). Its main components are: absorbing surface (1), transparent area (2), the thermal insulation (3) and frame (4). Heat transfer takes place between AS and the airflow moving through the channel between TS and AS or between AS and thermal insulation layer, or both.

Air density is about 900 times smaller than that of water and a more intensive circulation of air is required. To this end, the fan (5) is used to transport cold air to AS and further to the consumer. Thermal conductivity of air is about 25 times

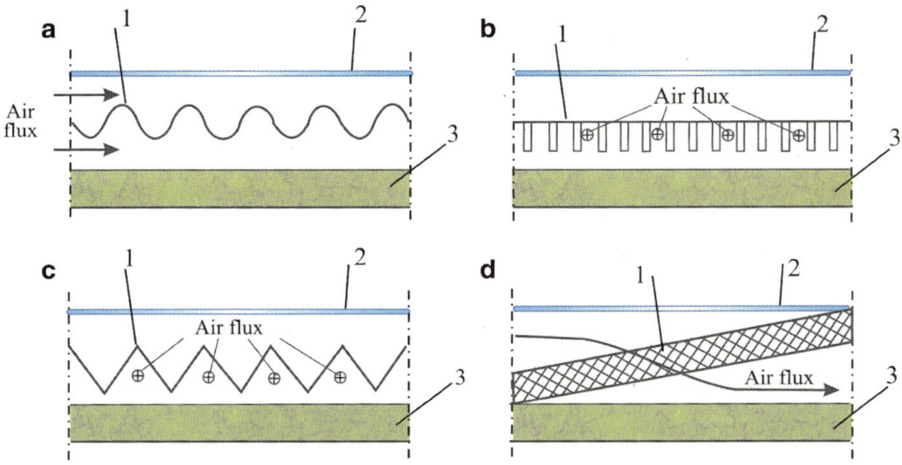

Fig. 3.19 Solar collectors for air heating: design of absorbing surfaces

smaller than that of water and it is necessary to substantially increase the contact area between TS and the airflow to achieve the same heat transfer. Air heating solar collector differs from that discussed above only by the design of absorbing surface 1.

In Figure 3.19 there are four design options of TS, whose major goal is to increase contact area between air and TS, creation of turbulent air movement and, consequently, increase of heat transfer efficiency:

- Wavy surface, air flow moves through both channels;
- The airflow moves through channels formed from metal plates welded to the rear of AS, forming a register (Fig. 3.19, airflow direction is perpendicular to the surface of the page);
- Differs from the previous version of the triangular shape of the channels;
- AS is composed of metal wire mesh or filled with granular material, through which air flows.

3.3.2.3 Thermal and Design Characteristics of the Solar Flat-Plate Collector

Standard flat-plate collector efficiency. Figure 3.20 presents a simplified energy balance of a standard flat-plate collector. Dimensions of the building elements are conventional. From short-wave direct and diffuse global solar radiation G, incident on the transparent area TS, an important part τG determined by the transparency coefficient τ, reaches the absorbing surface AS, where it is converted into heat. Transparent surface TS reflects radiation ρG into space and absorbs αTG, these values being quantitatively determined by reflectance ratio ρ and, respectively, by the material absorption coefficient αT of material TS. Part of the radiation τG incident on the absorbing surface AS is reflected and the biggest part is converted into heat.

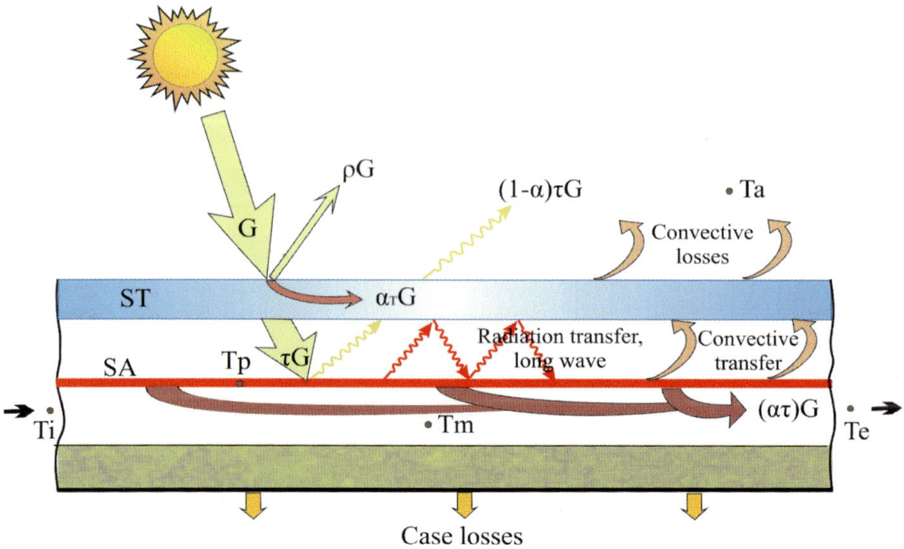

Fig. 3.20 Simplified energy balance of the solar flat-plate collector

It is obvious that for TS the sum of τ, ρ and αT coefficients is:

$$\tau + \rho + \alpha_T = 1. \tag{3.26}$$

Absorbing surface AS absorbs solar radiation, temperature T_p increases to 40–100°C and AS will also radiate energy, but already in the long-wave infrared band, for which TS is opaque. Arriving to TS, this radiation is partially absorbed, heating the TS, and is partially reflected to the surface AS. Thus greenhouse effect is simulated by means of TS which prevents the emission into space of infrared radiation emitted by AS.

Efficiency: Efficiency of converting solar radiation into heat is determined by the absorption coefficient α of the surface AS. Only a fraction of global solar radiation G will turn into heat, due to material properties of transparent surface TS and of the absorbing surface AS:

$$P_{SA} = (\alpha\tau)G. \tag{3.27}$$

P_{SA} power must cover energy losses from the surface TS, which take place by convective transfer, as well as by casing losses. These losses are in the first approximation proportional to the temperature difference between T_p of AS and that of the environment T_a:

$$\Delta P = U_P(T_p - T_a), \tag{3.28}$$

where U_P is the total loss coefficient, W/m^2·°C, ranging from 1 to 30 W/m^2·°C.

Table 3.5 Characteristics of main flat-plate collector types

		Overall loss factor (W/m^{2}°C)	
Collector type	$(\alpha\tau)$	Usual interval	Average
Selective surface, void tubes, SS–V	0.7	2–3	2.5
Selective surface, glass sheet, SS–1	0.85	3–5	4.0
Black surface, two glass sheets, SN–2	0.75	4–6	5.0
Black surface, one glass sheet, SN–1	0.85	6–8	7.0
Without transparent surface, FST	0.95	15–30	20.0

Useful power generated by solar collectors is determined by the expression, known as Hottel-Whillier-Bliss (H-W-B)

$$P_U = (\alpha\tau)G - U_P(T_P - T_a) \tag{3.29}$$

and thermal efficiency

$$\eta = \frac{P_U}{G} = (\alpha\tau) - U_P\frac{T_P - T_a}{G}. \tag{3.30}$$

From the expression (3.30) follows:

- When $U_P = const.$ and $(\alpha\tau) = const$, the efficiency decreases linearly depending on $(T_P–T_a)/G$;
- Maximal efficiency is when $T_P = T_a$ and it depends only on the optical properties of the TS and AS material;
- Global radiation G decrease leads to efficiency decrease.

Factor $(\alpha\tau)$ characterizing optical properties of TS-AS unit (transparent surface – absorbing surface), provides a classification of solar collectors in terms of efficiency and of overall loss coefficient U_P. Table 3.5 lists the main types of solar collectors, usual values of $(\alpha\tau)$ factor, and, overall loss coefficient, respectively. Figure 3.21 presents the evolution of efficiency in terms of temperature difference $T_a–T_P$. The calculation was performed using the expression (3.30) for mean overall loss coefficient and the value of global solar radiation $G = 800$ W/m^2, and the temperature $T_a = 20$°C.

Common collectors' characteristics: Evolution of efficiency from Fig. 3.21 and the characteristics listed in Table 3.5 provide information for comparative analysis of different flat-plate solar collectors' construction:

1. The characteristics of SN-1 collector with a black absorbing surface and a glass layer occupy an intermediate position. It is the most frequently used and in the above conditions it allows water heating up to 60°C temperature with efficiency between 45% and 50%. This type of collector is considered, usually, a reference one;
2. By excluding the transparent surface TS, the FTS collector is obtained, called ***energy roof***, which increases the factor $(\alpha\tau)$, but simultaneously increases the

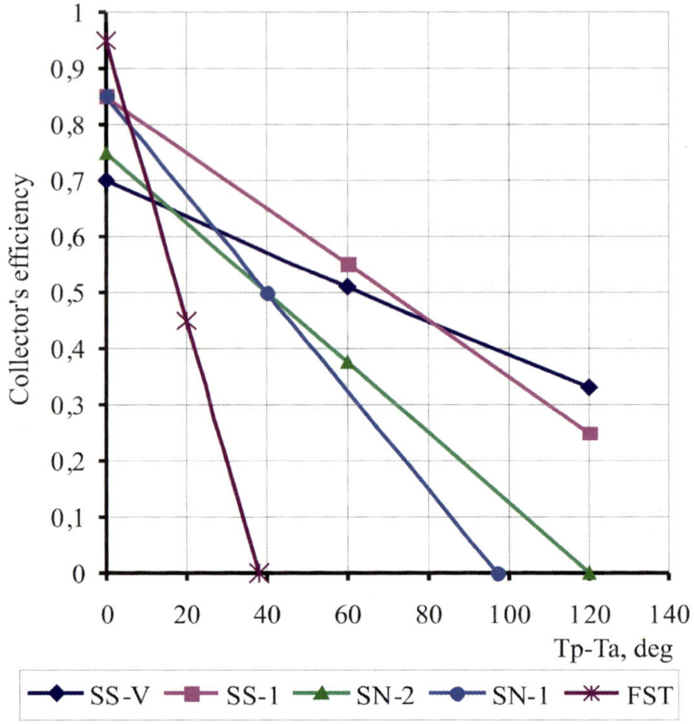

Fig. 3.21 Efficiency evolution of various types of solar collectors

overall loss coefficient U_P due to direct wind exposure and convective heat transfer enhancing. Provides air heating to 40–50°C with efficiency between 30% and 50%. It is commonly used for drying agricultural products (hay, grain, fruit, vegetables, etc.);

3. By installing two glass layers the collector SN-2 is obtained, which reduces the convective and radiative losses, but decreases the factor $(\alpha\tau)$ due to transparency diminution. In the usual $T_P{-}T_a$ temperature differences zone, between 30°C and 60°C, a significant increase of efficiency does not take place, and the collector is harder and more expensive in return.
4. Radiative losses can be significantly reduced in SS-1 collector with selective absorbing surface (next paragraph) and a single glass layer. Here convective losses are dominant;
5. A radical solution to minimize convective losses is to use vacuum tubes. It takes place in the selective surface SS-V collector. This type of collector allows water heating to temperatures above 100°C.

Irrigation factor, Equation H-W-B depending on the fluid temperature: In Eqs. 3.29 and 3.30 useful power delivered and efficiency are expressed in terms of the difference $(T_P{-}T_a)$ between temperature of the absorbing plate (surface),

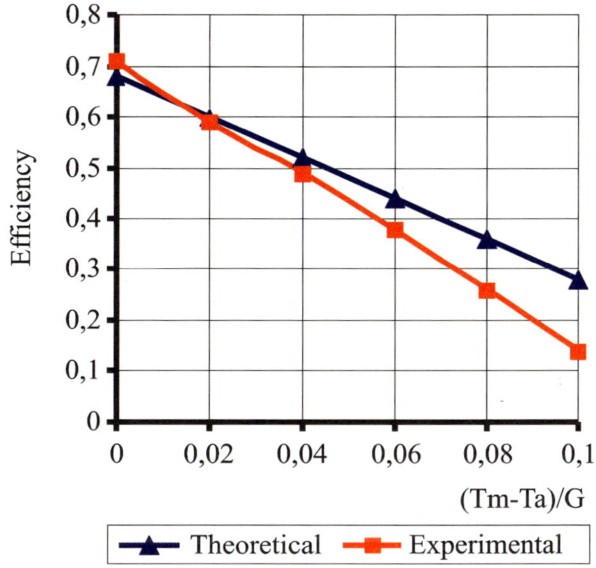

Fig. 3.22 Efficiency of collector, type SS-1

and environmental temperature, respectively. In reality, the net output is lower and proportional to the difference $(T_m - T_a)$ between the average temperature of the heat delivering liquid $T_m = (T_i - T_e)/2$ and environmental temperature, where T_i and T_e (Fig. 3.20) are the fluid input- and output temperatures, respectively. Power or energy transmitted by the plate to heat delivering liquid depends on several factors:

- Its nature: water, air, antifreeze, etc.;
- Specific mass flow per unit area, kg/h m^2;
- Liquid specific heat, J/kg · °C;
- Average temperature of the heat delivering liquid T_m.

To take into account the above stated, into efficiency Eq. 3.30 factor $F' < 1$ is inserted, known in English as *Heat Removal Factor*.

$$\eta = F' \left[(\alpha\tau) - U_P \frac{T_m - T_a}{G} \right] \tag{3.31}$$

or

$$\eta = \eta_0 - U_m \frac{T_m - T_a}{G}, \tag{3.32}$$

where $\eta_0 = F' \cdot (\alpha\tau)$, $U_m = F' U_P$.

Heat Removal Factor F′ is of thermal nature and can be calculated prior, having as input data heat exchanger construction (Figs. 3.16, 3.17). These calculations are described in detail in [4]. For pipe-sheet heat exchanger (Fig. 3.16, b), the F′ factor varies between 0,8 and 0,85, and for the plate-channel type (Fig. 3.16, d) it is equal to 0,98. Figure 3.22 presents SS-1 solar collector efficiency in terms

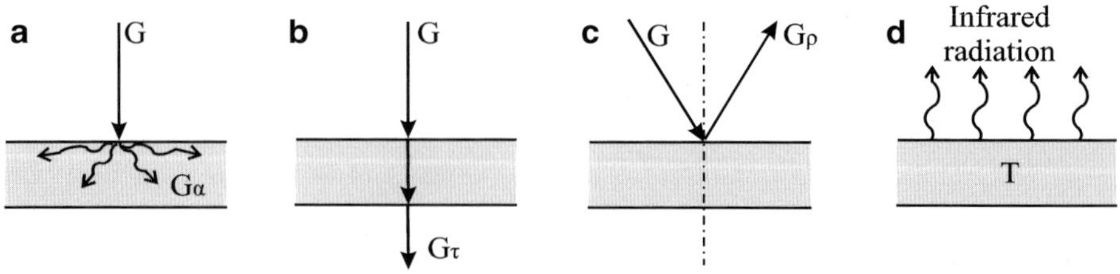

Fig. 3.23 Absorption (**a**), transmission (**b**) and reflection of short wave radiation (**c**). A body (surface) with temperature T is emitting long wave radiation into the space (**d**)

of $(T_m - T_a)/G$ calculated according to (3.31) ($F' = 0.8$) and the average efficiency obtained experimentally according to the American standard ASHRAE 93–73 [4]. There is a good approximation in the water temperatures between 30°C and 60°C and a relatively large deviation at temperatures above 60°C. For a properly designed and constructed collector type SN-1 or SS-1, the following efficiency average values are recommended for technical and economic calculations:

- Summer conditions – from 0.5 to 0.55;
- For the entire warm season (April to October) – from 0.40 to 0.45;
- Throughout the year – from 0.30 to 0.35.

3.3.2.4 Optical Properties of the Materials for Transparent Absorbing Surfaces

As outlined above, maximum efficiency of solar collector or factor ($\alpha\tau$) depends only on the properties of materials used for the absorbing surface or plate AS, respectively, for the transparent area TS. The solar radiation entering any surface can be absorbed, transmitted through material or reflected (Fig. 3.23).

The concept of spectral absorption coefficient α_λ equal to the ratio of the wavelength λ radiation absorbed and radiation incident on the same wavelength is introduced. Coefficient α_λ represents a property of matter and does not depend on the radiation property, for example, on the incident radiation wavelength. It just shows what part will be absorbed by the interaction of respective electromagnetic radiation with the material (for solar installations – absorbing surface). Also, the concept of spectral transmittance coefficient τ_λ, and that of spectral reflection coefficient ρ_λ are introduced.

Energy conservation law requires that the sum of these coefficients is equal to 1:

$$\alpha_\lambda + \tau_\lambda + \rho_\lambda = 1. \tag{3.33}$$

The values of these coefficients are approximately constant in the range of the incidence angle θ variation between 0° and 60° and decreases sharply for angles greater than 70°.

Given that solar radiation has a broad spectrum of electromagnetic waves, the terms below are used for practical purposes, which reflect the interaction of the material and electromagnetic radiation throughout its range:

1. *Absorbance* α, defined as the ratio between the power of the absorbed- and incident solar radiation:

$$\alpha = \frac{G_\alpha}{G}; \tag{3.34}$$

2. *Transmittance* τ, defined as the ratio of radiated power transmitted through the respective material and the incident radiated power:

$$\tau = \frac{G_\tau}{G}; \tag{3.35}$$

3. *Reflectance* ρ, determined as the ratio of reflected radiated power to incident radiated power:

$$\rho = \frac{G_\rho}{G}. \tag{3.36}$$

Equation 3.34 can be written as follows:

$$\alpha = \frac{\int_0^\infty \alpha_\lambda G_{\lambda\alpha} d\lambda}{\int_0^\infty G_\lambda d\lambda}. \tag{3.37}$$

Similarly coefficients τ and ρ can be expressed. The relation (3.32) is also respected

$$\alpha + \tau + \rho = 1.$$

The coefficients α, τ and ρ characterize the behaviour of transparent or absorbing surface at solar radiation action, which represents a spectrum of *short waves* between 0.3 and 3 μm. They are also called *optical* or *solar* coefficients. At the same time, the respective area is heated and generates into space infrared radiation in the range between 3 and 20 μm. This process is characterized by spectral ***emission*** coefficient ε_λ, called *radiative* and defined as the ratio of delivered radiative power flux $W_{\varepsilon\lambda}$ with the wavelength λ and the flow of power delivered by an absolutely black body $W_{\lambda n}$ at the same wavelength λ and temperature T.

According to Kirchhoff's law for any wavelength λ and temperature T the following relationship occurs:

$$\alpha_\lambda = \varepsilon_\lambda, \tag{3.38}$$

in other words, a body with temperature T absorbs and emits the same amount of electromagnetic radiation with wavelength λ.

As defined above, the emission factor or *emittance* for the entire spectrum of radiation is determined similarly to the absorption α (v. 2.36):

$$\varepsilon = \frac{\int_0^\infty \varepsilon_\lambda W_{\varepsilon\lambda} d\lambda}{\int_0^\infty W_{\lambda n} d\lambda} = \frac{W_\varepsilon}{W_n}, \tag{3.39}$$

where W_ε, W_n is the flow of real power delivered by the body with temperature T, and, respectively, the flow of power delivered by an absolutely black body at the same temperature.

In accordance with expression (3.6), an absolutely black body of A area will emit an equal power flux $W_n = \sigma \cdot A \cdot T^4$ and, therefore, the flow of radiative losses will be:

$$W_\varepsilon = \varepsilon \cdot \sigma \cdot A \cdot T^4. \tag{3.40}$$

Absorbing surface materials: The ideal material has the following optical properties $\alpha = 1, \tau = \rho = 0$. It is obvious that such a material does not exist. Absorbing surface, being opaque, has transmittance τ equal to zero. The major part of solar radiation is absorbed and converted into heat and only a small fraction of solar radiation is reflected.

Table 3.6 shows absorption and emission characteristics collected from [4, 11, 12] for various materials and surfaces. The absorbance of material or the absorbing surface of solar radiation in the short wave range (0.3–3.0 μm) was noted with α_{short}, and ε_{long} is emittance of the material or surface in the infrared wave range (bigger than 3.0 μm). We conclude that pure metals are not suitable for making absorbing surfaces. Also, aluminium is not recommendable because it may induce very fast electrical corrosion phenomena into the installation circuit. Steel plate covered with black paint has a satisfactory absorption, is cheap, but is not resistant to corrosion.

A special class is formed by the selective materials and surfaces possessing a high absorption coefficient for short waves, and, respectively, a lower emission coefficient for long waves. For these materials, the report $\alpha_{short}/\varepsilon_{long} \gg 1$.

Figure 3.24a, shows the spectral characteristics for various absorbing surfaces; and Fig. 3.24b, shows the explanation of the selective surface effect. In the spectral range between 0.3 and 3 μm absorbance and emittance of ideal selective surface is equal to 1, and in higher frequencies range over 3 μm – are zero.

Copper oxide (selective material) characteristic differs from the ideal one, but there was noticed a small value of emittance over the spectrum zone bigger than 10 μm. Selective surface absorbs approximately 85–95% of solar radiation (short wave), which turns into heat and is transmitted to the copper plate. However, both copper and copper oxide emit into space only about 10% of radiation that could

Table 3.6 Absorbance and emittance of some materials and absorbing surfaces

Type of material or absorbing surface	Absorbance, α_{short}, short wave	Emittance, ε_{long}, infrared waves	Ratio $\alpha_{short}/\varepsilon_{long}$
Traditional materials			
Pure iron	0.44	0.11	4.0
Pure aluminium	0.10	0.10	1.0
Polished copper	0.35	0.08	8.8
Oxidized steel sheet	0.74	0.82	0.90
Steel sheet plated with black paint	0.95	0.95	1.0
Graphite (black lead)	0.78	0.41	1.90
Carbon black	0.96	0.89	1.08
White paint	0.12–0.18	0.93	0.13–0.19
Selective materials and surfaces			
Black chromium on a nickel surface	0.95	0.09	10.7
Porous ceramic son a steel surface	0.96	0.16	6.0
Black nickel oxide (green oxide) on an aluminium surface	0.85–0.93	0.06–0.10	14.5–15.5
Copper oxide, Cu_2O, on a copper surface	0.90	0.16	6.0

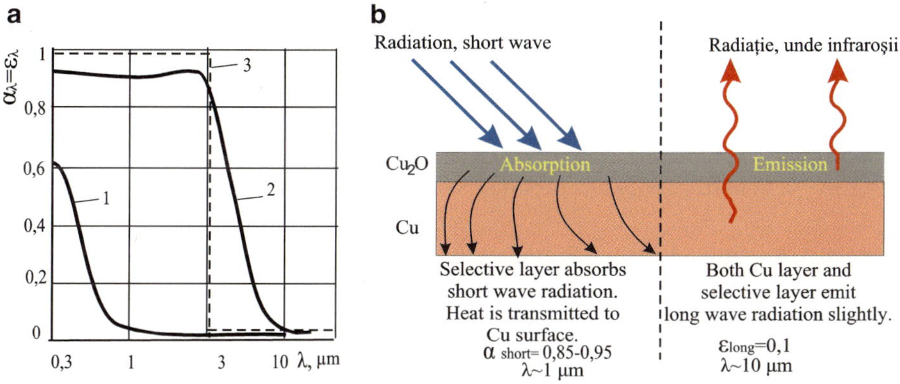

Fig. 3.24 Spectral characteristics of different materials (**a**) and explanation of the selective surface effect (**b**)

be emitted by an absolutely black body at the same temperature. In other words, selective surface creates the same greenhouse effect as a glass layer covering the solar collector surface. This explains the solar collector efficiency increase with selective surface (Fig. 3.21).

Transparent surface materials: Ideal transparent material transmittance is $\tau = 1$, its absorbance is $\alpha = 0$ and reflectance $\rho = 0$. Characteristics of real materials which can be used to manufacture transparent surfaces are shown in Table 3.7. The best quality material for transparent surface is low iron oxide glass or ordinary

Table 3.7 Characteristics of transparent materials

Type of material	Transmittance (τ)	Absorbance (α)	Reflectance (ρ)
Window glass (6 mm)	0.80	0.12	0.08
Float glass (4 mm)	0.87	0.07	0.06
Glass with reduced content of iron rust	0.91	0.01	0.08
Polycarbonate	0.70	–	–
Polyethylene film	0.82	–	–
Plexiglass (acrylic glass) (3 mm)	0.80	–	–
Tedlar	0.88	–	–

glass. Glass transmittance drops abruptly for infrared radiation ($\lambda > 3.0\ \mu m$), and, therefore is opaque for radiation emitted by the absorbing surface, thus the greenhouse effect taking place.

Plastics can also be used. They are lighter and cheaper than glass, but age quickly under the action of ultraviolet rays and hence must be replaced periodically (after every 6–12 months). Another plastics disadvantage is low radiative properties. Thus, the polyethylene has an infrared radiation transparency coefficient of 0.8 and therefore will not provide adequate greenhouse effect.

The characteristics shown in Table 3.7 are presented for Sunbeam incidence angles $\theta < 40°$ – typical angles for most collectors. Reflectance rises sharply and transmittance reduces for incidence angles bigger than 60°.

3.3.2.5 Schematic Diagram of Solar Water Heating Systems

The most widespread solar water heating systems (SWHS) are shown in Fig. 3.25. The basic components of SWHS are: flat solar collector, storage tank (water reservoir), heat exchanger, circulating pump, auxiliary energy source. The difference of the presented diagrams consists in the interconnection and structural arrangement.

SWHS with natural circulation: Figure 3.25a shows a SWHS with natural circulation. To ensure a steady water circulation the storage tank should be mounted higher than the collector. In both components, in the collector and in the water reservoir a difference of temperatures in their upper and lower parts is settled, that leads to a difference of density between the water layers (hot water is easier than cold water) and finally to a pressure difference that ensures water circulation. Pressure difference depends on the temperature difference and thus the water flow in the system depends on the useful power captured by the collector that causes this temperature difference. In these circumstances, SWHS with natural circulation are self-adjustable – the increase in the captured output leads to the increase of the water flow that circulates in the system.

Theoretical and experimental studies carried out in the 70s of the last century have demonstrated [4] that for a wide range of SWHS with natural circulation

Fig. 3.25 Schematic diagram of solar systems for water heating: (**a**) with gravity (natural) circulation; (**b**) with pump circulation; (**c**) with two profiles

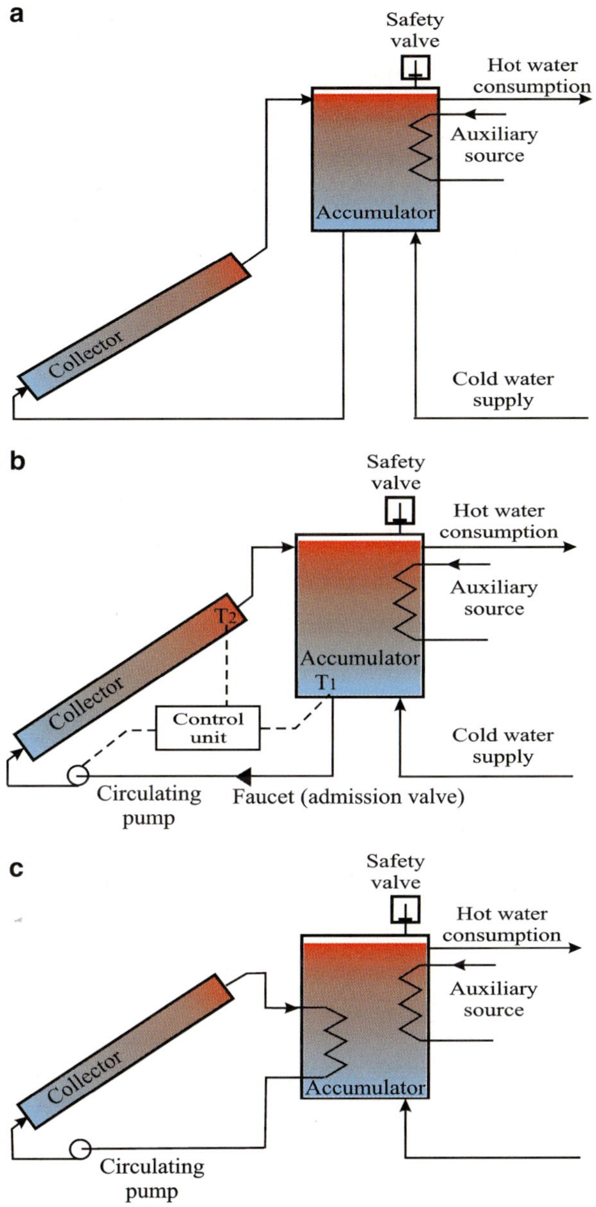

the temperature difference of water input and output from the collector is about 10°C and remains constant during the day (see Fig. 3.26). Specific water discharge is 50–60 l/m^2h. For example, in a SWHS with the collector's surface 4 m^2 water discharge is about 200 l per hour. If storage tank volume is 200–300 l then during a day this volume of water will circulate in the system several times, heating. It is

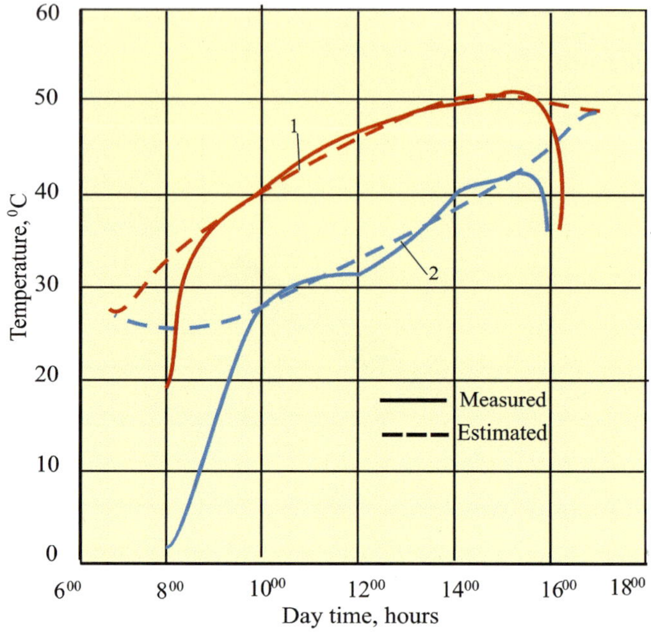

Fig. 3.26 Variation of water temperature during the day: *1* – at outlet; *2* – at inlet

obvious that the temperature difference between the upper and lower layers water in the tank will be minimum and water circulation will stop (see Fig. 3.26). SWHS with natural circulation is widely used in Israel, Australia, California and Florida in the USA where there is no danger of frost. The construction of this system is simpler, does not demand additional cost and is designed to heat water at the temperature of 65°C. In Moldova, SWHS with natural circulation can be operated April to September. To avoid deterioration of collector it is recommended to empty it from water in the cold period of the year. The auxiliary electrical energy source or the source based on natural gas use allows coverage of hot water over consumption. It can be paralleled with the solar collector (as shown in Fig. 3.25a or in series). In the last case, SWHS operates as preheater. It should be mentioned that SWHS with natural circulation will operate with higher efficiency if main hot water consumption will occur in the morning hours. Or, the water reservoir must be filled with cold water before maximum solar radiation occurs.

SWHS with pump circulation: SWHS with pump circulation is shown in Fig. 3.25b. The circulating pump will depend on the temperature difference T_2 in the upper part of the collector and temperature T_1 in the lower part of the reservoir. Temperature is controlled by a balanced relay and the pump will operate only when the temperature difference will overrun the predetermined value. To exclude the reverse circulation of water at night a unidirectional valve is envisaged.

In the regions with danger of frost the diagram of SWHS will be modified and will have two circulation contours. The first contour (Fig. 3.25c) comprises a collector, a circulating pump and a heat exchanger, and it is filled with antifreeze solution. The second contour is the hot water circuit itself. Heat transfer between the first and the second contours is done via a liquid-liquid heat exchanger (see the coil in the storage tank).

This diagram is widely spread in the countries of Central and Northern Europe and in the North America. A standard SWHS comprises 2–3 collectors (their size is 1×2 m), water reservoir volume is 200–400 l, specific water discharge is 50–60 l/m^{2}h. Solar collector is made of single-layer glass, the absorbing surface is selective, and in some cases is a black surface.

An important role in SWHS manufacturing belongs to the degree of water column (water height) layering in the reservoir. The higher the degree of layering, the bigger the solar cut fraction in covering annual hot water supply. In this context, it is rational to decrease the specific discharge. The studies carried out in Sweden in the 90s of the last century showed that the optimal discharge (output) in a SWHS with pump circulation is 7–20 l/m^{2}h.

3.3.2.6 Practical Aspects of Solar Water Heating Systems

Transparent surface (TS) and absorbing plate (AS): The importance of optical characteristics of materials for manufacturing transparent surfaces and absorbing plates was underlined in Sect. 3.3.2.4. Properties $(\tau\alpha)$ and ε have a direct impact on thermal performances of the solar collector. The selected materials should have steady properties in time as their degradation can affect long-term properties of the collector. The surface of the absorbing plate should resist at maximum possible temperatures in the period of exploitation. This can be determined easily from the Eq. 3.30 under the condition that useful power $Q_u = \eta G$ is equal to zero (there is no water circulation). The fluid temperature T_m will equal the plate temperature. Obviously, the selective material or the applied paint should resist to these temperatures. In the case of transparent surfaces there is danger of their destruction by hailstones. On the basis of some real experiments it was concluded that the risk of destruction of the collector covered with tempered 3 mm glass is to be neglected. Thus, in 1979 in Colorado, USA, due to a storm lasting several minutes the size of hailstones was 2–3 cm in diameter, then after a small period of silence, for 1–2 min, the size of falling hailstones was 3–10 cm in diameter. One thousand and ten solar collectors, mounted under angles between 32° and 56°, were exposed to hailstones and the glass was broken in 11 collectors only [4].

The combination of glass with the casing should be done without edges or fins that are pointed out above the transparent surface. This would facilitate water flow and snow sliding from the surface of the collector.

Heat insulation: Transparent surface (glass) and the air case between AS and TS have the role of heat insulation in the frontal part of the collector. The air case should be within 25 and 40 mm. The thickness 28 mm is considered optimum. The other sides of the collector (the underside and the lateral sides) should be isolated with a 5–10 cm layer of fibre glass or other insulating material with similar heat insulation parameters. Fibre glass has the following advantages:

- Is relatively cheap;
- Has excellent insulating properties ($\lambda = 0.05$–0.06 W/m K);
- Is temperature resistant, over 100°C;
- Has very small specific gravity (unit/volume weight) ($\rho = 150$–200 kg/m^3).

A disadvantage of fibre glass is loss of insulating features if it becomes moist. It is recommended to use panels of fibre glass and thus settling will be avoided at inclined placement of the solar collector.

A reasonable solution related to cost-quality can be heat insulation combined with a 4 cm fibre glass panel and with another 3 cm panel made of polystyrene expandant. Fibre glass is more resistant to higher temperatures, and is placed immediately after the absorbing plate, but polystyrene is placed after the fibre glass panel. In its turn, polystyrene is more water resistant. To improve heat insulation of the collector, it is recommended to introduce a thin aluminium sheet between the insulation and the absorbing plate. The sheet will have the role of a screen to infrared radiation that will be reflected to the absorbing plate. An air case between the absorbing plate and the sheet will have the role of additional heat insulator.

Collector pressurization: To prevent water and dust intrusion into collector, it is necessary to pack it. In conditions of cloudy sky water vapours condense on the internal surface of glass that will lead to decreasing its transparency and efficiency. At next starting of the collector, condensation will disappear only when the glass will become sufficiently warm for its damping. Thus, the collector, which internal air is damp, will start operation later and stop it earlier. If the collector is not packed the dust will intrude into it and will deposit on the interior side of glass and on the transparent plate. If economically it is difficult to perform perfect packing then it is recommended to have internal ventilation of the collector. Aiming at this, several holes (2–3 mm diameter) are made between TS and AS that are sheltered from rain.

Casing: It supports the unit and ensures collector's pressurization. The mostly recommended material is the anodized aluminium angle and zinc steel sheets. The structure of casing does not need high mechanical strength. Usually the collectors are placed on house roofs (Fig. 3.27) or are retained by a support made of steel angle profile. This support will take over wind stresses too.

The casing is simple in construction and presents a box with flat or undulated ground (for a higher rigidity) and four sides. The height of sides should correspond to the mounting requirements for heat insulation, transparent and absorbing surfaces and an air case, a total height of 10–15 cm. It is important that the casing and collector's weight is small in order to ease its mounting.

Collectors' mounting will be in series or parallel (Fig. 3.28). At series connection water flows consecutively through collectors and heats more and more. The collectors will operate under various temperatures, with a decreasing efficiency from input to output. Hydraulic resistance is increasing. Due to this, it is recommended for pump circulation systems.

Fig. 3.27 An example of solar water heating system: solar collectors are integrated in the roof of the house directed to the south

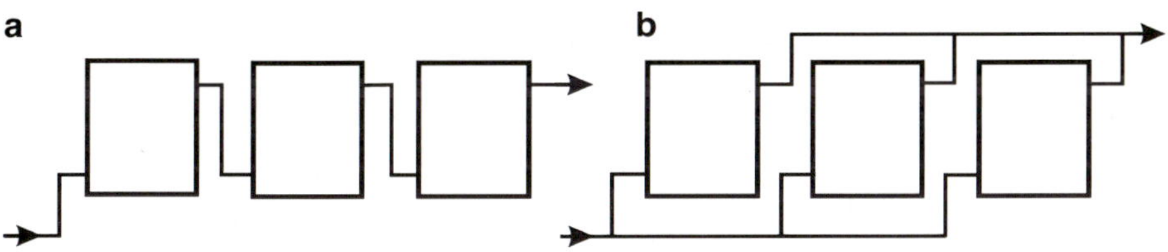

Fig. 3.28 Connection of solar collectors, in series (**a**) and parallel (**b**)

When mounted in parallel, each solar collector will ensure only a part of the discharge and will have about the same temperature. Hydraulic resistance decreases and the system can operate under natural circulation (thermo siphon) Also hybrid systems with series/parallel or parallel/series connection might be applied.

An autonomous solar water heating system is shown in Fig. 3.29. Its solar collectors are mounted on metallic supports on the roof. Solar collectors are fixed and oriented to Sun at maximum position of solar radiation.

3.3.2.7 Dimensioning of a Solar Water Heating System

Previously it was stated that SWHS efficiency decreases together with the increase of temperature between the absorbing plate and the environment. It is important to

Fig. 3.29 Solar system for water heating

estimate correctly the dimension of solar collector's surface and the volume of water reservoir depending on the:

- Available overall radiation on the collector's surface;
- Daily hot water consumption and temperature;
- Distribution of water consumption over the day.

Further on, a simple method for dimensioning is described. This method allows preliminary SWHS estimation, in particular, at the first stage of designing solar energy systems, inclusive feasibility studies.

SWHS diagram is shown in Fig. 3.25a. It has a single contour with natural circulation. The dimensioning estimates refer to the following sizes:

- Optimum angle of dip of solar collectors for the given period: March to October, central regions, e.g. Moldova;
- Average global radiation for the given period incident on collector's surface and monthly average global radiation;
- Daily consumption of hot water and its distribution for 24 h;
- Solar collector surface;
- Volume of hot water storage reservoir;
- Minimum water temperature per month with smaller solar radiation.

The optimum angle of dip of solar collectors is chosen so as in March and October more solar energy is radiated to the collector's surface, and care is taken that in summer this quantity of energy does not decrease under the required level.

Overall average radiation incident on the collector's plane: Average daily values of direct and diffuse solar radiation on a horizontal surface for the period from March to October are taken from local characteristics. Diffuse radiation is determined as

Table 3.8 Overall radiation at the collector surface for the studied period

Month		III	IV	V	VI	VII	VIII	IX	X
R_b	–	1.57	1.15	0.93	0.85	0.88	1.06	1.38	1.94
B	$MJ/m^2 \cdot zi$	4.22	7.34	10.26	13.03	12.55	11.24	8.51	4.78
D	$MJ/m^2 \cdot zi$	5.57	7.98	9.32	10.05	9.55	8.06	6.14	4.3
$B_\beta = R_b B$	$MJ/m^2 \cdot zi$	6.63	8.44	9.54	11.07	11.04	11.91	11.74	9.27
$D_\beta = 1/2\ (1 + Cos\beta)$	$MJ/m^2 \cdot zi$	4.86	6.96	8.13	8.76	8.33	7.03	5.35	3.75
$G_\beta = B_\beta + D_\beta$	$MJ/m^2 \cdot zi$	11.49	15.4	17.67	19.83	19.37	18.94	17.09	13.02
	$kWh/m^2 \cdot zi$	3.20	4.31	4.95	5.55	5.42	5.30	4.79	3.64

Table 3.9 Twenty-four hours consumption of home hot water

Hour	Consumption (%)	Hour	Consumption (%)	Hour	Consumption (%)
0–1	0.0	8–9	4.6	16–17	4.2
1–2	0.0	9–10	4.4	17–18	4.5
2–3	0.0	10–11	3.9	18–19	5.7
3–4	0.0	11–12	4.0	19–20	7.8
4–5	0.0	12–13	4.6	20–21	9.1
5–6	3.0	13–14	4.6	21–22	9.1
6–7	5.2	14–15	4.6	22–23	6.5
7–8	5.4	15–16	4.0	23–24	4.7

Table 3.10 Daily distribution of hot water consumption, *C(t)*

Hour		0	4	6	8	10	12	14	16	18	20	22	24
C(t)	%	0.0	0.0	3.0	13.6	22.6	30.5	39.7	48.3	57.0	70.5	88.7	99.9
	l	0.0	0.0	6.0	27.2	45.2	61.0	79.4	96.6	114.0	141.0	177.4	200.0

the difference between the overall and direct radiation. Global radiation on the collector's surface is determined from the expression (3.25), neglecting the reflected component:

$$G_\beta = R_b \cdot B + \frac{1}{2}(1 + Cos\beta)\,D. \qquad (3.41)$$

The obtained results are included in Table 3.8.

Daily consumption of hot water and its distribution for 24 h: Daily demand for hot water and its distribution for 24 h depend, to a great extent, on the consumer's specific characteristic measure. It is different for houses and industrial buildings. Assuming as consumer a family of four persons, the specific consumption q is 50 l/day per capita at the temperature 55°C. The distribution of water consumption for 24 h is shown in Table 3.9 [13].

The data in Table 3.9 and the daily consumption of 450 = 200 l/day defines the distribution of hot water consumption *C*(*t*) during the day by summing the values for each hour. The results are introduced in Table 3.10, graphic presentation is given in Fig. 3.30. The same graph shows cumulative production of hot water from 7^{30} to 18^{00}.

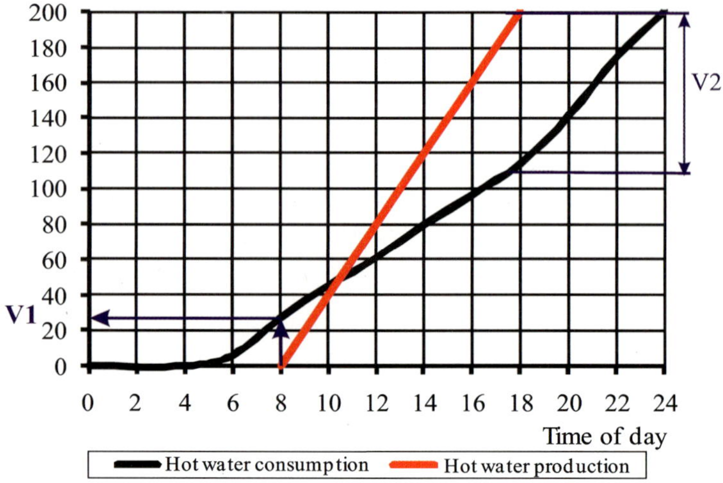

Fig. 3.30 Explanation of hot-water tank volume determination

Solar collector surface: The energy demanded to heat the daily water volume $C_z = 200$ l from 15°C to 55°C is estimated:

$$E_{nec.} = C_z \cdot c_{apei} \left(t_{cald\breve{a}} - t_{rece} \right) = 200 \cdot 4.173 \cdot 10^3 \cdot 40 = 33.39 \text{ MJ /day}$$

where $c_{water} = 4.173 \cdot 10^3$ J/°C kg shows the water specific heat.

Capture surface of the solar collector:

$$S_{col} = \frac{E_{nec}}{\eta_{col} \cdot G_{\beta med}} = \frac{33.39}{0.45 \cdot 16.6} = 4.5\text{m}^2,$$

where

η_{col} is collector average efficiency for the period of operation;

$G_{\beta med}$ – daily global solar radiation determined as the average for the period from March to October (see data in Table 3.8)

Volume of hot water storage tanks: Figure 3.30 shows that the storage tank should supply the hot water production deficit from 5 to 8 o'clock in the morning, and from 18 to 24 o'clock in the evening. Thus, the storage tank volume is:

$$V_a = V_1 + V_2 = 30 + 90 = 120 \text{ l}.$$

The temperature of hot water in March when solar radiation is the smallest:

$$t_{cald\breve{a}} = t_{rece} + \frac{C_\beta^{III} \cdot S_{col} \cdot \eta_{col}}{C_z \cdot c_{apei}} = 10 + \frac{11.69 \cdot 10^6 \cdot 4.5 \cdot 0.45}{200 \cdot 4.173 \cdot 10^3} = 38.4°\text{C}.$$

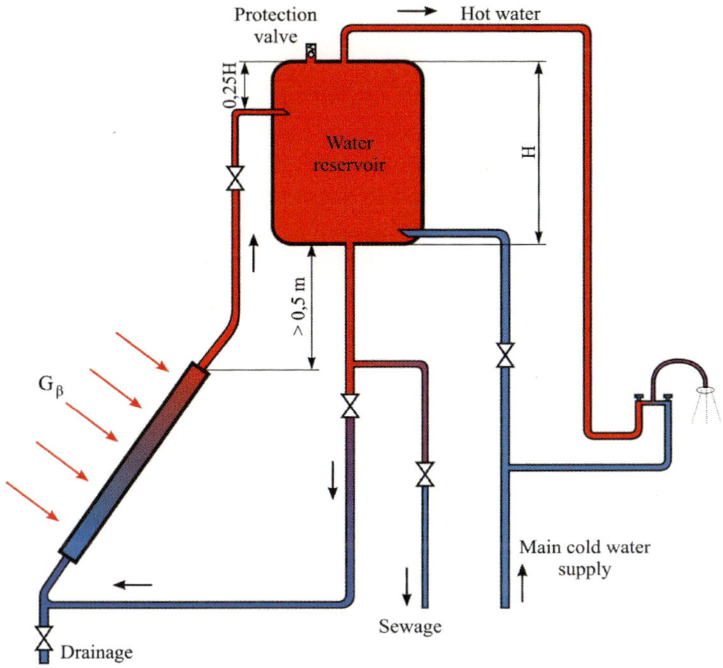

Fig. 3.31 Schematic diagram of a SWHS with natural circulation supplied with cold water from the main apeduct

So, one can see that in March a comfortable temperature of hot water for residential purposes is also supplied. Figure 3.31 shows the diagram of SHWS supplied with cold water from the supply main. In this case the water tank and the solar collector should resist to the main water pressure. If admissible pressure is overrun, the water tank is endowed with a protection valve. To ensure a good layering of water in the tank, hot water is injected to the upper part, and cold water to the lower part of the tank. The system described in Fig. 3.31 can be produced on the basis of the solar collectors manufactured at Incomaş–S.A. enterprise (Chişinău). The collector surface is 2.2 m^2. Two collectors connected in parallel are necessary.

Quantity of thermal energy produced from March to October is determined from the expression:

$$E_{III-X} = N_z \cdot G_{\beta med} \cdot S_{col} \cdot \eta_{col} = 245 \cdot 16,6 \cdot 4,4 \cdot 0,45$$
$$= 8053 \text{ Mj} \qquad \text{Or about 2 Gcal.}$$

3.3.3 Systems for Solar Thermal Energy Conversion into Electrical Energy by Combining the Greenhouse Effect and Air Pressure Difference

3.3.3.1 Brief History

In 1903, for the first time, Isidoro Cabanyes, a Spanish colonel, proposed a solar energy tower in the journal *La energia electrica* [14]. The solar tower with air currents has three main components: the solar collector, the tower and the wind turbine. One of the first descriptions of the solar energy tower was published in 1931 by Hanns Gunther, a German author [15]. Since 1975, Robert E. Lucier obtained several inventions for the electric generator with solar tower. From 1978 to 1981 these patents were valued in Australia, Canada [16], Israel and the USA. For the first time, solar energy came into the attention of society after the energy crisis of the 70s in the last century, on large scale.

In 1972 an operational model of small dimension thermal solar tower was developed under the supervision of engineer Jorg Schlaich, of German origin, in Manzanares, Ciudad Real, and located 150 km to the south from Madrid, Spain. The project was financed by the German Government [17–21]. The tower was 195 m high, and its diameter was 10 m. It had a collecting surface (greenhouse) of 46,000 m^2 (diameter – 244 m). This pilot-plant operated successfully about 8 years. Some experiments have been carried out [22, 23] and then it was closed up in 1989. Solar towers were advertised by an Australian TV programme *After* 2000 in 1984, under the aspect of their adaptability for electrical energy generation in the under populated regions of Australia with no impact on animal breeding.

Haaf presented the testing results and described theoretically the solar tower prototype from Manzanares, Spain [17, 18]. Transferability of the obtained results in Manzanares was discussed by Schlaich, etc. (1990). In 1997 Kreetz has introduced the concept of water tank use that is mounted under the roof for heat collecting. Gannon and Backström have presented the analysis of solar tower thermodynamic cycle and the analysis of turbine characteristics, as well, in 2000 [19, 24]. Ruprecht [25] and others have described the results of dynamic design of the fluid and turbine for a 200 MW solar tower in 2003. *Serag-Eldin MA* (2004) [26] Weinrebe, G. (2000) [14, 26] have performed ample computer simulation of the air currents in the solar tower. *El-Haroun AA* (2002) [25] *researched air velocity at tower output. Thermal and technical analysis was calculated computer assisted* (*described by* Dos Santos Bernardes et al. in 2003) [4, 27]. A similar solar tower is developed recently in Australia, where conditions are favourable for such solar energy units, like: high level of solar radiation, vast available areas, demanded for electricity generation increase and existence of a governmental body – Mandatory Renewable Energy Target (MRET), that has demanded 9,500 GWh of renewable energy in 2010–2020 [17].

Economic estimations based on the experience and knowledge have demonstrated that large solar towers ($\geq$100 MW) are capable to generate electrical energy

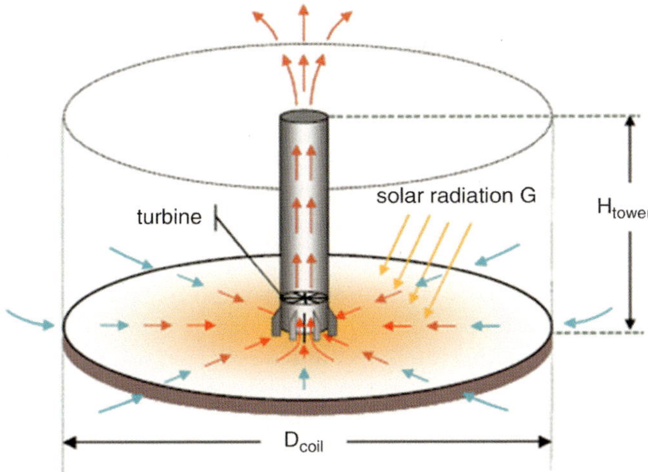

Fig. 3.32 Operation principle of solar tower

at the same cost as energy produced by conventional energy units [28–30]. In future, solar towers will contribute to low cost and environment friendly electrical energy generation in Sunny regions.

3.3.3.2 Solar Tower with Collector and Air Turbine: Theoretical Issues

Solar towers are solar thermal plants that use the combination of solar collector greenhouse effect and the pressure difference in the tower for generating convective air currents that drive the turbine to produce electricity. The main components of a solar tower are: air collector, tower (furnace) and the wind turbine.

These systems are destined for power scale applications ≥100 MW. The operating principle is shown in Fig. 3.32 [20]. The air heated by direct and diffuse solar radiation through a transparent circular shelter, open at periphery: the shelter and the natural soil form a solar collector. A tower with large entrance at its basis is placed in the middle of the shelter. As warm air is easier than the cold air, pressure difference occurs in the tower.

Thus, solar radiation is the reason of a constant pressure drop in the tower. Continuous operation for 24 h can be ensured by placing the tubes or tanks filled with water under the roof. Heat accumulated during the day is delivered during the night. The energy contained in the air flow created by the pressure difference is converted into mechanical energy through one or more turbines installed in the base of tower.

Next, the fundamental relations and influence of the basic parameters on the quantity of energy are presented in a simplified form. Generally speaking, the output power $\boldsymbol{P}$ of a solar tower can be calculated as input power Q multiplied by the efficiencies of the collector, the tower and the turbine:

$$P = Q_{soil} \cdot \eta_{col.} \cdot \eta_{tower} \cdot \eta_{turbine} = Q_{soil} \cdot \eta_{plant.} \tag{3.42}$$

Solar energy input into the system can be represented as the product of horizontal global radiation $\boldsymbol{G_o}$ and the collector's surface $\boldsymbol{A_{col}}$:

$$Q_{soil} = G_o \cdot A_{col} \tag{3.43}$$

The tower transforms the warm currents produced by the collector into kinetic energy (convection currents) and the pressure action on the turbine. This difference in air density caused by temperature rise in the collector, works as a driving force. The column of warm air from the tower is connected with the lower atmosphere at the base (entry into the collector) and the atmosphere at the top of the tower, thereby creating a lifting force. Pressure difference ΔP_{sum} is produced between the base of the tower (collector's inlet) and the environment (tower's outlet):

$$\Delta P_{sum} = g \int_{0}^{H_{tower}} (\rho_0 - \rho_{tower}) dH_{tower}. \tag{3.44}$$

This difference of pressure ΔP_{sum} increases together with the tower height. The difference of pressure ΔP_{sum} can be divided as the static component and the dynamic component, neglecting the friction losses:

$$\Delta p_{sum} = \Delta p_s + \Delta p_d, \tag{3.45}$$

where:

$$\Delta p_s = P_a - P_{tower},$$

where:

P_a is the atmospheric pressure (at collector's inlet)

P_{sum} – pressure at tower outlet.

The difference of static pressure reacts on the turbine, and the dynamic component describes the kinetic energy of air currents. For a total difference of pressure and the volume of air currents when $\Delta P_s = 0$, power P_{sum} contained in the currents is:

$$P_{sum} = \Delta P_{sum} \cdot v_{tower.max} \cdot A_{col}. \tag{3.46}$$

Efficiency of the solar tower can be defined from the relation:

$$\eta = \frac{P}{Q_{sol}}. \tag{3.47}$$

Splitting the difference of pressures in static and dynamic components depends on the energy taken from the turbine. In the case of the tower without the turbine, maximum speed of air currents is reached, and all pressure difference is used to accelerate the air and is converted into kinetic energy:

$$P_{sum} = \frac{mv^2_{tower\max.}}{2}. \tag{3.48}$$

Using Boussinesq approximation, the speed achieved by convection currents can be expressed as:

$$v_{tower.\max.} = \sqrt{2gH_{tower}\frac{\Delta T}{T_o}}, \tag{3.49}$$

where ΔT is the temperature increase between the environment and the collector's outlet.

Efficiency of the solar tower will be defined from the relation:

$$\eta = \frac{g \cdot H_{turn}}{c_p T_o}. \tag{3.50}$$

where c_p is specific heat at constant pressure, J kg^{-1} K^{-1}.

This simplified representation explains one of the basic characteristics of the solar tower – solar tower efficiency depends on its height. For the 1,000 m height, the deviation from the exact solution due to the Boussinesq approximation is to be neglected. Using Eqs. 3.42, 3.43 and 3.50 it can be defined that the generated power is proportional to the collector's area and the tower height. Since the generated electricity is proportional to the volume of the tower's height and collector's area, the same power can be generated by a tower with a bigger diameter and small collector or vice versa. This refers to the operation of the solar tower during the day. But what happens at night? For the tower to operate 24 h out of 24 the water tanks with the black surface are installed on the ground collector. How these tanks operate? The tanks are sealed so that evaporation does not occur (Fig. 3.33) [20]. The volume of water in the reservoirs is calculated so as to correspond to a layer of water with the depth equivalent to 5–20 cm depending on the expected power characteristics (Fig. 3.34) [20]. At night, when the air in the collector is cooler, the water tanks deliver the accumulated heat. Heat capacity of water is approximately five times higher than that of the soil.

Solar tower does not convert the total amount of solar energy into electric energy. Relatively low conversion rate of solar tower is balanced by investment costs reduced to 1 m^2 of the collector [31]. According to the calculation, a solar thermal power plant (solar tower) with 200 MW capacity requires a collector with a 7 km diameter (total area about 38 km^2) and a tower at 1,000 m height [32].

Solar tower performance can be degraded by such factors as atmospheric winds [32] or by the brake induced by the arms used as tower carriers [32]. Another factor is the reflection of Sunlight by the material of the roof. The diagram (Fig. 3.35) shows that the solar beam PO breaks in point O the interface between

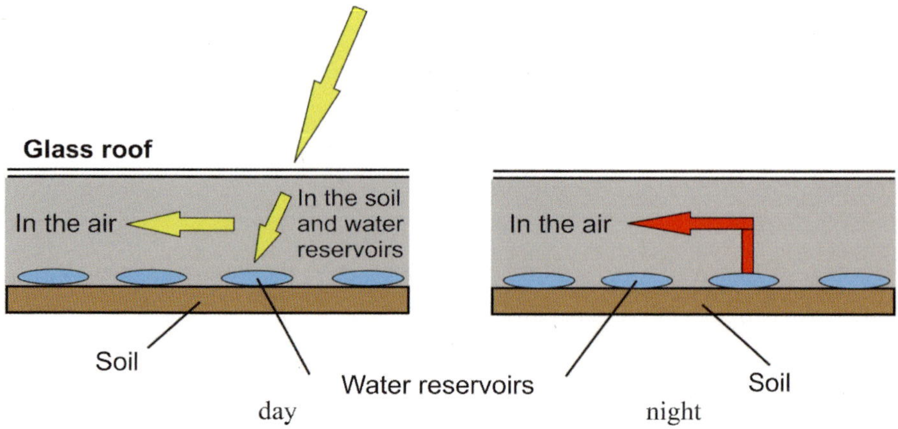

Fig. 3.33 Thermal energy storage in water-filled pipes

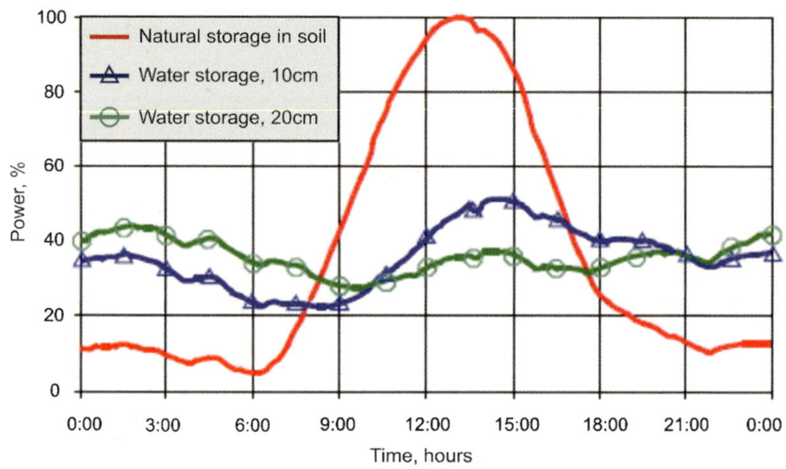

Fig. 3.34 Effect of heating storage in the collector, using waterfilled black surface storage tanks (simulated by Kreetz, 1997)

two environments with refractive indices n_1 and n_2. Some of the rays (OQ) are reflected, and some (OS) are refracted. Angles between the rays (incident, reflected and refracted) towards the interface normal are respectively θ_i, θ_r *and* θ_t.

Relations between these angles are set by the law of reflection and Snell's law. Fraction of intensity of the incident beam, reflected by the interface (in our case by the transparent roof), is presented as the coefficient of reflectance ρ_a. The refracted fraction is represented by the transmittance coefficient τ_a. Fresnel equations, based on the assumption that both materials are not magnetic, can be used to calculate the coefficients:

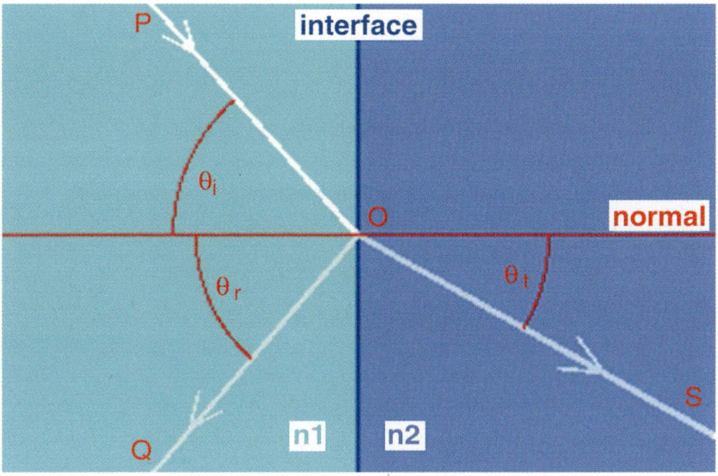

Fig. 3.35 Diagram of incident solar ray reflection and refraction

$$\rho_a = \left[\frac{\tan(\theta_t - \theta_i)}{\tan(\theta_t + \theta_i)}\right]^2 = \left[\frac{n_1\cos(\theta_t) - n_2\cos(\theta_i)}{n_1\cos(\theta_t) + n_2\cos(\theta_i)}\right]^2 \tag{3.51}$$

and

$$\tau_a = 1 - \rho_a. \tag{3.52}$$

According to Fresnel Eqs. 3.51 and 3.52 solar rays reflection can reach about 7.7% when the roof is made of glass. Also, an influence up to (15–20)% is exercised by the plant location. Solar towers located at a high latitude (Canada for instance) cannot produce more than 85% compared with a similar station located at the equator [33].

The following is an example of determining the basic parameters for a case study of a solar tower, which was to be built in a village in northern China, where solar radiation is more intense than in other regions. As a result of an analysis the site was chosen for the power plant construction, taking into account the location and solar radiation regime. A Solar tower with a diameter of 10 m, 200 m height and 500 m diameter of the collector is able to produce (110–190) kW of electricity per month the year round. Several parameters such as tower height, diameter of the solar collector, temperature of the environment, solar radiation and wind turbine efficiency, influencing energy generation performance, were also analyzed. Next, a list of all parameters considered for the solar tower design is presented:

A_c *plan area of solar tower,* m^2;
A_{col} *solar collector area,* m^2;
C_p *air specific heat,* kJ/kg.°C;

g	*free fall acceleration,* m/s^2;
G	*solar radiation,* W/m^2;
H_{sc}	*solar tower height,* m;
$\dot{m}$	*air current mass per unit of time,* kg/s;
P_{tot}	*air flux total power,* kW;
$P_{wt,max}$	*maximum mechanical power generated by the turbine,* kW;
P_e	*electrical power generated by the solar tower,* kW;
$\dot{Q}$	*air received energy in the collector,* kW;
T_0	*environmental temperature,* °C;
V_c	*air velocity in the solar tower,* m/s;
$(\tau\alpha)$–	*derivative of transmittance and absorbance coefficients for the Earth's surface;*
β	*coefficient of losses,* W/m^2.K;
η_{col}	*solar collector efficiency;*
η_{sc}	*solar tower efficiency;*
η_{wt}	*turbine efficiency;*
ρ	*air density,* kg/m^3;
ΔP_{tot}	*pressure difference between the lower and upper parts of the tower,* Pa;
ΔT	*temperature increase at output from collector, compared to input,* °C

3.3.3.3 Towers with Solar Collectors and Air Turbine: Practical Issues

Solar Energy Tower is a special power facility designed for renewable energy conversion. Air is heated in a large circular collector, and the resulting convection causes increased air speed when it passes through the tower. Moving, the air drives the turbine (or turbines), which produces electricity.

In the last 30–40 years a wide range of solar air collectors power plants have been invented. The operating principle of a solar air collector tower is described in the invention [34] presented in Fig. 3.36. It operates based on the principle of converting solar heat into electricity by means of accumulating unheated air in the air collector (1), which covers a large area of land. Heated air due to pressure difference circulates through a very tall tower (2) of large diameter, located in the centre of the roof. In the centre of the tower there is a support (3), on which an electric generator (4) driven by a turbine (or several turbines) (5) excited by the cold air passing through the tower, is mounted.

In order to increase the air velocity a pair of truncated cones is used, one of which (6), being mounted in a holder and passed by air currents, works as a convergent nozzle. The second truncated cone (7), installed at the outlet, works as a diffuser.

Generation capacity of a solar tower depends largely on two factors: the size of the collector and the tower height. The larger the area of the collector the greater amount of air is heated (collector diameter up to 7 km). At a big tower height (up to 1,000 m) the pressure difference leads to an increase of the furnace effect (Fig. 3.38).

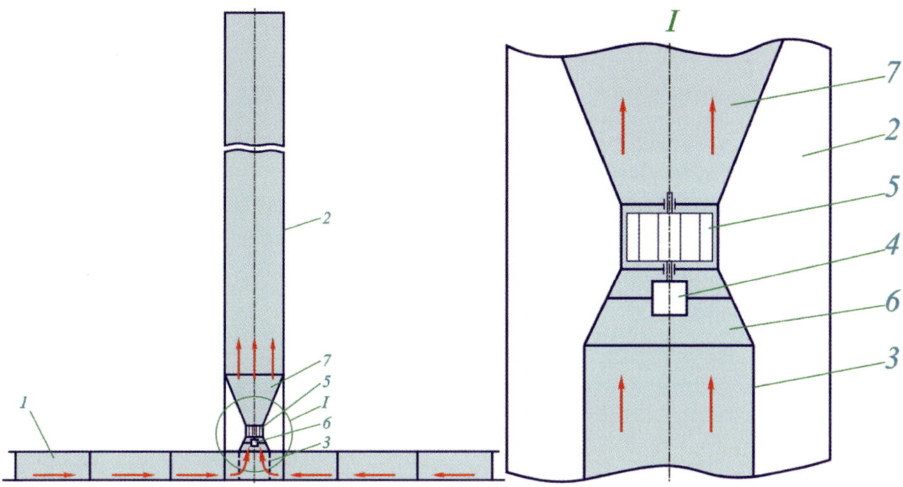

Fig. 3.36 The principle of converting solar heat into electricity by means of accumulating unheated air in the air collector

The combination of the effects of these two important parameters determines the energy facility productivity. Heat accumulated during the day may be partially stored in water pipes (reservoirs) placed on the collector's land area. The air velocity is approximately directly proportional to the increase of temperature difference (ΔT) in the collector and at tower's outlet. In multi-megawatt solar towers, the collector causes a 20°C temperature increase. This produces an increasing of air velocity up to 15 m/s. Solar towers do not produce carbon dioxide, but harmful materials are used in the construction, particularly cement. Payback period is 2–3 years [35, 36].

Turbines can be installed in a row at the basis of the horizontal axis tower (as in the Australian project) or a single vertical axis turbine can be installed inside the tower as in the prototype from Spain. Turbines in solar towers do not work due to air velocity only, but similarly to hydropower stations, static pressure is converted into rotary motion, using the turbine of the case study. Solar tower turbine, unlike wind turbine, operates in a closed air flow. Therefore the ratio of power to the rotor area is about 10 times higher than the same ratio of the wind turbine. Air velocity before the turbine and after it is about the same. The power generated is proportional to the product of volume currents per time unit and pressure differential on the turbine. To increase conversion efficiency, the propeller blades is adjusted during operation in order to regulate energy production by means of varying the speed of air and air currents. If the blades are parallel to air currents and allow air to flow freely through them, pressure on the blade is not created and electricity is not generated. Between these two extremes the optimal position of the blades is situated: generated power is the highest if the pressure drop on the turbine is about 80% from available pressure differential. Optimal fraction depends on the plant characteristics as pressure loss due to friction (Fig. 3.37).

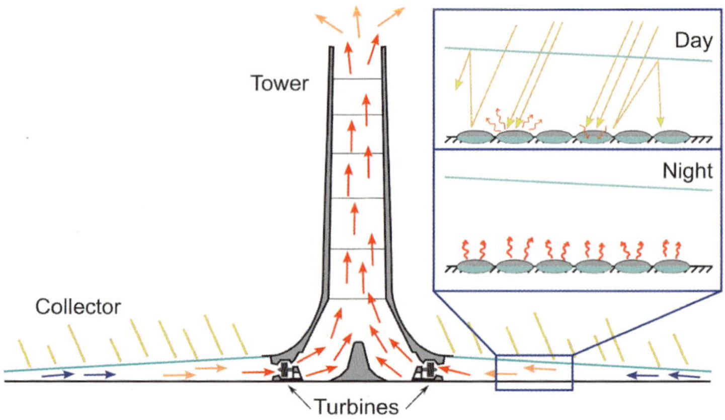

Fig. 3.37 Operation principle of the solar tower with circular collector

Fig. 3.38 Solar tower prototype's Manzanares, Spain, 1981

Detailed theoretical investigations and conducted extensive aerodynamic tunnel experiments [20] allowed in 1981–1982 the construction of an experimental 50 kW peak production facility in Manzanares district, Union Electrica Fenosa utility (Fig. 3.38) with funds from the German Ministry of Research and Technology (BMFT). Table 3.11 gives basic dimensions and technical parameters of the Manzanares prototype.

Table 3.11 Basic dimensions and technical parameters of Manzanares prototype

Tower height	194.6 m
Tower radius	5.08 m
Average collector radius	122.0 m
Average height of roof	1.85 m
Number of turbine blades	4
Turbine blades profile	FXW-151-A
Velocity ratio of blade at velocity of air conveying	1:10
Operation mode	Individual or grid connected
Increase of collector's air temperature	$\Delta T = 20$ K
Rated output	50 kW
Collector coated with plastic membrane	40,000 m^2
Collector coated with glass	6,000 m^2

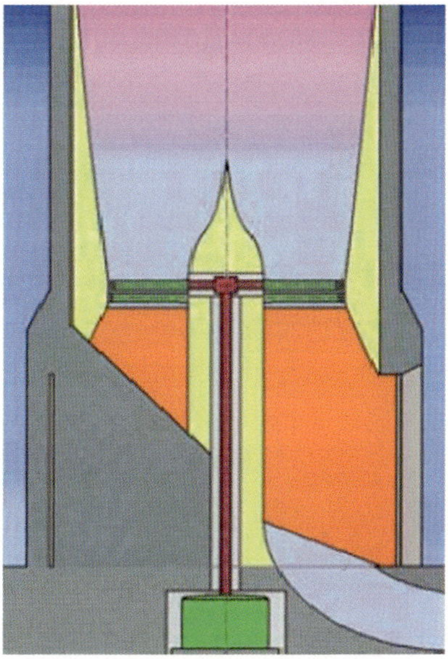

Fig. 3.39 Operation principle of the air turbine

The operating principle of wind turbine is presented in Fig. 3.39. The turbine (Fig. 3.40) [20] is installed separately in the solar tower in a steel frame at a 9 m height above ground level. Initial 2.5 m/s vertical wind speed may reach up to 12 m/s during turbine operation.

The roof of solar tower collector is usually manufactured from a transparent material: glass (Fig. 3.41) or plastic membranes. The roof must not only be transparent, but also sufficiently durable and have a reasonable price. Plastic membrane costs are lower than those of glass. Figure 3.42 [20] presents measurements made on June 8, 1987 in Manzanares: air speed in the tower and a typical day power production. Figure 3.43 [20] presents a comparison between the measured and calculated energy,

Fig. 3.40 Air turbine of the power plant prototype

Fig. 3.41 Glass roof of the power unit prototype in Manzanares

produced at Manzanares facility. A disadvantage of solar air collector tower is large collector area and relatively low efficiency. Decrease of this disadvantage can be achieved by increasing the height of the tower.

An ambitious project is the prototype solar power plant to be built in Australia by Germany. Solar tower will have 1 km height (Fig. 3.44).

Another interesting project, which is under development stage, is Monohans Solar Tower project, U.S. (Fig. 3.45) [37]. Monohans Region is a national leader in oil production, and officials and residents from this region want to start this renewable energy project. A greenhouse made of transparent material with a diameter of

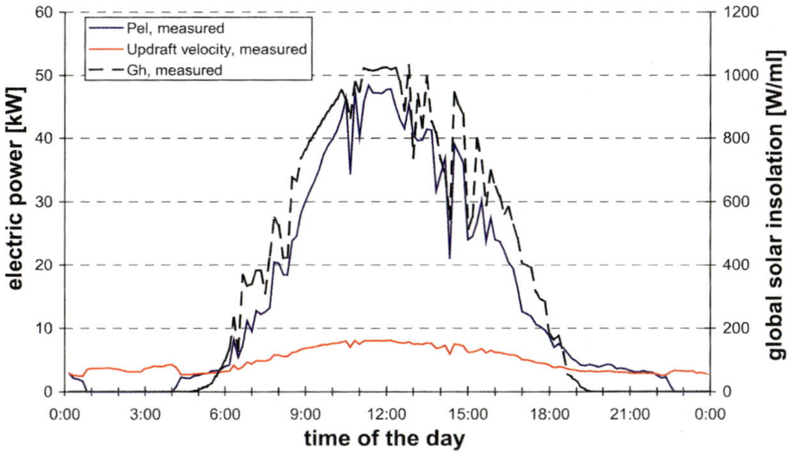

Fig. 3.42 Measurements made on June 8, 1987 in Manzanares: air speed in the tower and a typical day power production

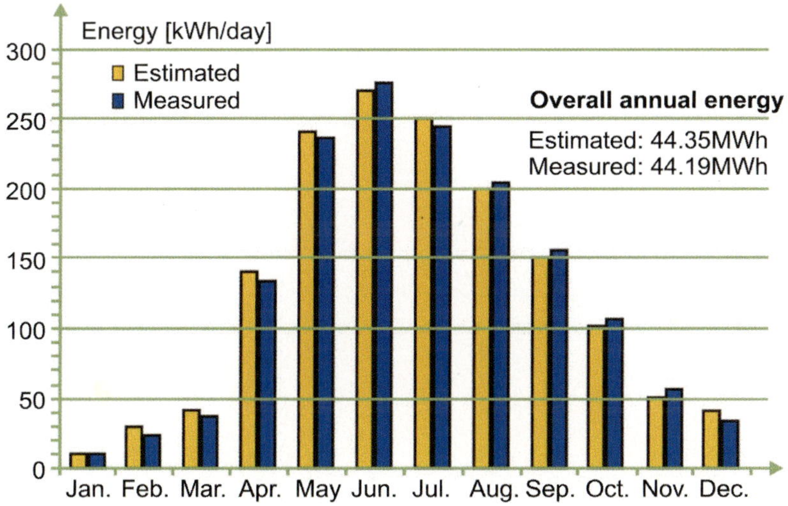

Fig. 3.43 Comparison of measured and estimated energy, produced by Manzanares power unit

2.5–3 miles shall collect warm air that will flow through 32 turbines located around the tower base. The difference in temperature between the turbine interior and the environment will ensure a steady air flow that will produce electricity. Solar tower will have a height of about 1,100 m and will produce 200 MW of green electricity, enough to supply approximately 200,000 houses. Project cost is estimated at approx. 350 million dollars.

To assess the energy potential of solar air collector tower, Table 3.12 [20] presents four typo dimensions of solar towers. Investment costs were calculated based on the data from Table 3.12. From Table 3.13 [20] it follows that the cost of

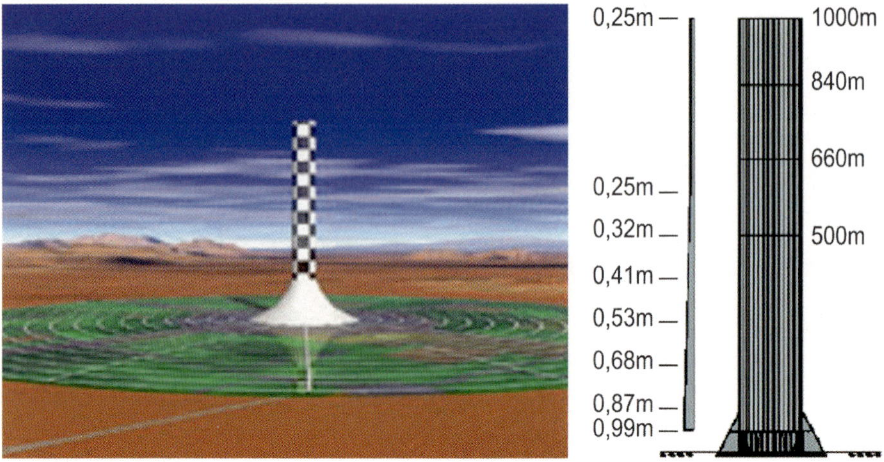

Fig. 3.44 Australian design of the solar electrical tower of 1 km height

Fig. 3.45 Design of the solar tower in Monohans, USA

Table 3.12 Specific size and power capacity

Capacity	MW	5	30	100	200
Tower height	m	550	750	1,000	1,000
Tower diameter	m	45	75	110	120
Collector diameter	m	1,250	2,900	4,300	700
Electricity generation (production)[a]	GWh/a	14	99	320	680

[a]For locations with annual overall solar radiation 2,300 kWh/(m^2year)

Table 3.13 Investment costs

Capacity	MW	5	30	100	200
Cost of tower[a]	Mln euro	19	49	156	170
Cost of collector	Mln euro	10	48	107	261
Cost of turbine[b]	Mln euro	8	32	75	133
Engineering, testing	Mln euro	5	16	40	42
Total	**Mln euro**	**42**	**145**	**378**	**606**
Annual instalment investments	Mln euro	2.7	10.2	27.1	43.7
Annual operation and maintenance costs	Mln euro	0.2	0.6	1.7	2.8

[a]Labour cost – 5 euro/h
[b]Benefit rate – 6% and duration – 30 years

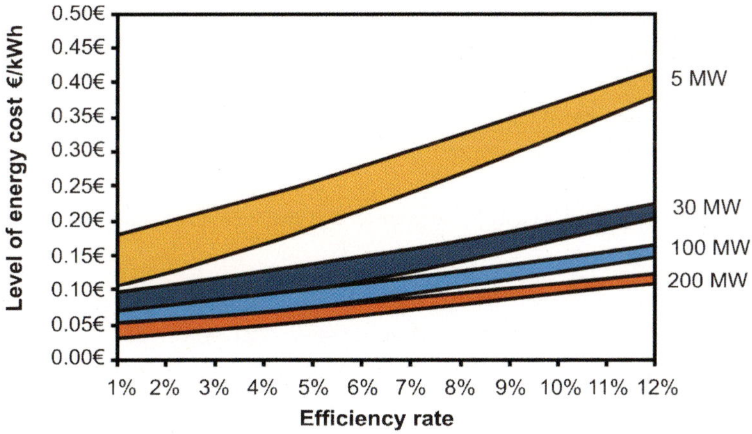

Fig. 3.46 Leveling cost of electricity depending on the selected efficiency ratio and on the solar tower

electricity for small solar towers is relatively high, compared to those of PV systems. Increasing the plant size is associated with significant decrease of generated energy cost reaching minimal cost values of 0.07 €/kWh for 200 MW plant.

Changes in the electricity costs depending on selected efficiency level, lifetime and installed capacity of solar tower, are presented in Fig. 3.46 [20]. The top line of graphs is calculated for 20 years lifetime, and the lower – for a lifetime of 40 years. As shown, the cost depends on the desired efficiency ratio and the lifetime (life time or depreciation period). For example, for a 12% efficiency rate and a 20-year depreciation, the cost for 200 MW solar tower is 0.12 €/kWh. For a 6% efficiency rate and a 40-year depreciation period, the cost decreases to 0.06 €/kWh, that is, in half.

Figure 3.49 considers a comparative analysis for the costs of electricity produced by coal-fired power plants and by a solar tower. In the first years of operation the cost of electricity generated by solar towers, is greater than that produced by coal-fired power plants. Costs are the same after 20 years of operation. Further, the solar

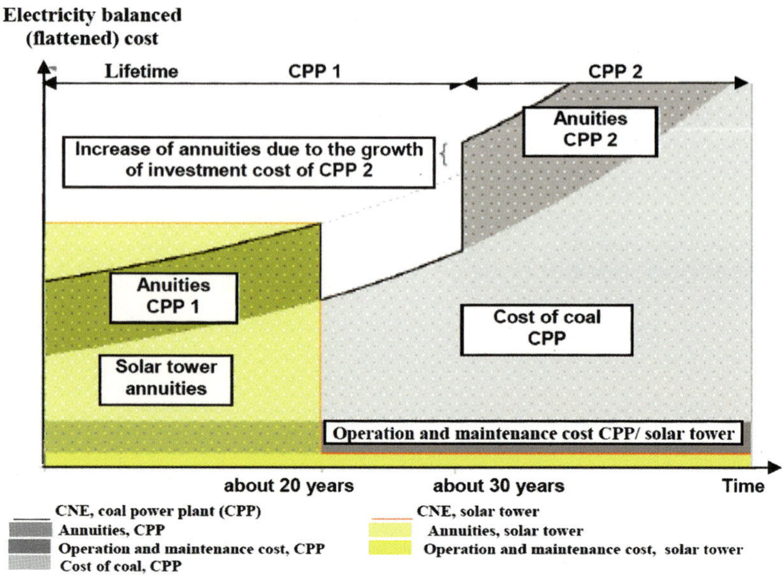

Fig. 3.47 Electricity generation cost for solar tower and coal thermal power plant

tower will generate electricity at low cost (including only operating and maintenance costs). In contrast, electricity generation costs of coal power plants are relatively high and imposed by the permanently growing cost of coal. This fact does not take into account the costs for environmental protection (meeting an approved CO_2 emissions level) for coal power plants.

In Moldova, where the Sun shines 2,200 h a year (in the south – 2,300 h/year) and annual amounts of radiation are about 1,350 kW/m^2 [38], there are favourable conditions for the development of such solar energy capture systems, which can be combined with plant cultivation technologies in green houses (Fig. 3.47) [20].

3.3.4 Solar Systems with Solar Rays' Concentration

3.3.4.1 Brief History

The power density of solar radiation on the Earth's surface rarely exceeds 1,000 W/m^2. Using plane collectors analysed above, it becomes impossible to obtain temperatures greatly exceeding 100°C. Temperatures of hundreds or even thousands degrees can be obtained only through the concentration of solar radiation direct component.

Legend says that in 212 BC Archimedes used for the first time the effect of Sunlight concentration. With polished-to-shine soldier shields (thus becoming genuine bronze mirrors) Sunlight was reflected and concentrated in such a manner that allowed burning the Roman ships besieging the Syracuse fortress. In order to verify the legend, in 1973 the Greek Naval Authority showed how 60 people, each with a mirror sizing 1×1.5 m ($1.5\ m^2$), burned a wooden ship situated at 50 m distance [4]. If during the experiment the radiation density was about 800 W/m^2, than in the centre it was about $0.8 \times 1.5 \times 60 = 72\ kW/m^2$, thus exceeding 1.5 times the power density on the surface of an electric heating plate.

This way of capturing solar energy always is in the attention of researchers. L. Stoddard and others have conducted extensive studies on the economic, environmental and energy benefits of Sunlight concentrating systems [39, 40]. The largest group, which focuses on research and use of concentrated Sunlight solar systems, is U.S. Solar Energy Generating Systems (SEGS), which developed and built in the Mohave Desert from southern California paraboloidal systems with a total capacity of 354 MW. These systems functioned very well during 15–20 years. Today SEGS has a number of commercial projects under development or under construction, including a 64 MW paraboloidal system in Nevada and several 50 MW systems in Spain. Integrated systems are at various planning stages in Southern California, India, Egypt, Morocco, Mexico and Algeria. A paraboloidal system has recently been built for Arizona Public Service [41].

In 2005, Southern California Edison Company publicly announced the completion of contract negotiations with Stirling Engine Systems (SES) Company for the purchase of 1,182–2,010 GWh per year for a period of 20 years. As a result of large-scale development of these paraboloidal systems with Stirling engines significant reduction of capital and operating, as well as maintenance costs are expected [41]. Detailed information on Sunlight concentration solar systems can be found in [41, 42].

In general, three basic types of solar power plants with Sunlight concentration are known: cylindrical-parabolic systems (with gutters); with parabolic mirrors and Stirling engine; solar towers with heliostat concentrator.

3.3.4.2 Thermal Conversion of Solar Energy at High Temperatures: Theoretical Issues

Operation of solar concentrators is based on two phenomena studied by physics: reflection and refraction of light. If a surface is not rough or the linear sizes of a surface are much smaller than the wavelength of an incident light beam, this surface reflects. Surface reflection capacity is characterized by the reflectance factor. Reflectance values for common materials are:

- Galvanised silver 0.96;
- Pure aluminium 0.91;
- Silver deposited on the inner surface of the glass (mirror) 0.88;
- Silver deposited on the outer surface of the glass 0.93;
- Posterior surface aluminized acrylic film 0.85.

Fig. 3.48 Parabolic mirror

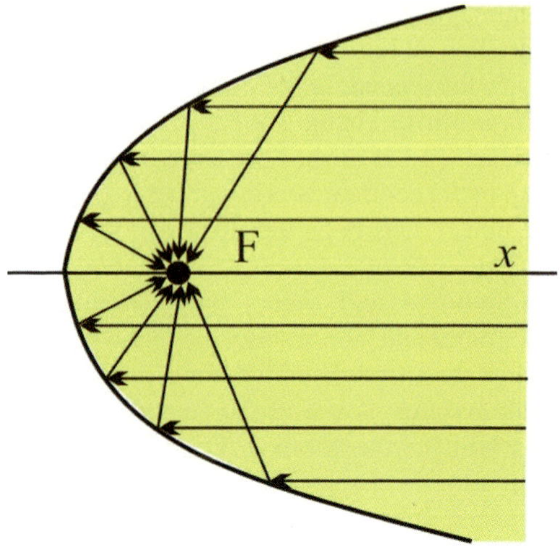

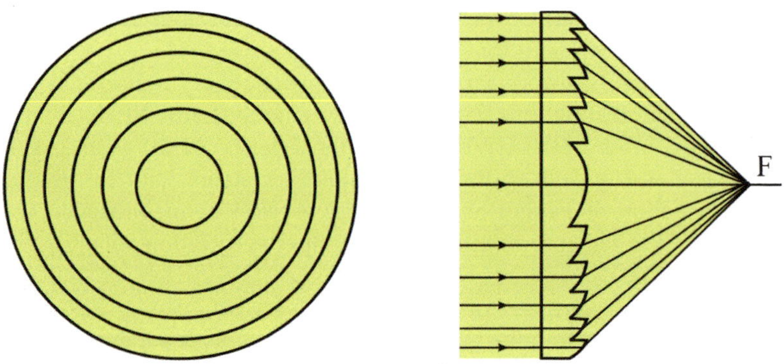

Fig. 3.49 Fresnel lens

Parabolic mirror. This is a concave mirror (Fig. 3.48), which surface is described by the parabola $y^2 = 2cx$ around the optical axis x. Figure obtained is called paraboloidal. A cylindrical-parabolic mirror is obtained if parabola is moved in the direction of z axis (axis perpendicular to the surface of the page). Rays of light parallel to the optical axis x, incident on reflective surface, concentrate in point F, called focus point

Fresnel lens: This is a unit composed of several concentric lenses (Fig. 3.49), which functions as an integral lens. Material and mass costs of a Fresnel lens are much smaller than those of a conventional one. Parallel light flux concentrates in focus point F.

The three types of solar energy concentrators used in thermal conversion are composed of three main components: the reflector that receives Sunlight and directs in to the focus point; the receiver, located in the focus point and that turns Sunlight

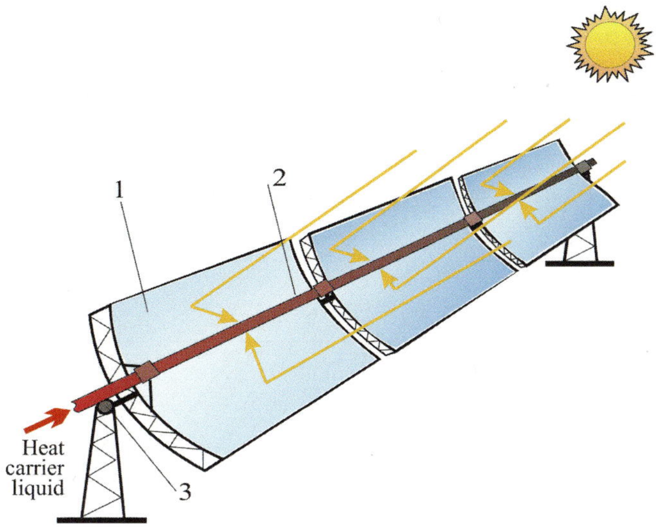

Fig. 3.50 Operating principle of cylindric-parabolic concentrator

into heat; Sun position tracking system. Fresnel-type concentrators are used in the photovoltaic technology of solar energy conversion only.

The basic parameters that characterize a solar concentrator are concentration coefficients. Geometric concentration coefficient is the ratio between surface area open to Sunlight, or aperture A_a, and receiver's surface area, A_r:

$$C_g = \frac{A_a}{A_r}. \tag{3.53}$$

Optical concentration coefficient is the ratio between the power density of direct radiation on the receiver's surface, B_r and power density of direct radiation on the aperture B_a:

$$C_r = \frac{B_r}{B_a}. \tag{3.54}$$

For an ideal concentrator $C_g = C_r$, in fact $C_r < G_g$. Taking into consideration the distance between the Earth and the Sun disk diameter, the geometric concentration factor cannot be bigger than:

$$G_g < \left(\frac{2L}{D_s}\right)^2 = 45000. \tag{3.55}$$

Figure 3.50 shows a design diagram of the concentrator with three cylindrical-parabolic reflective mirrors (1) (in the form of a gutter). Solar radiation receiver (2)

Fig. 3.51 Operating principle of solar concentrator with paraboloidal mirror

represents a pipe through which the thermal agent fluid, usually water, circulates. The receiver (2) is mounted in the focus point of the parabolic cylinder, representing a line crossing the optical focus point F of the parabola from Fig. 3.48.

An important feature of the cylindrical-parabolic concentrator is the tracking of only one coordinate from Sun trajectory – elevation angle α_s (Fig. 3.49). In the construction from Fig. 3.50, tracking is done by rotating mirror 1 around the axis 3. Receiver temperature reaches 400–500°C.

The second concentrator type (Fig. 3.51) has a paraboloidal mirror (1) (in the form of a plate), serving as a reflector. The receiver (2) is mounted in the focus point of the paraboloidal. The orienting mechanism (not shown) must ensure tracking of two angles – elevation α_s and azimuth γ (Fig. 3.9). Receiver temperature can reach over a thousand degrees.

The heat collected is usually used directly by thermal engine mounted on the receiver, which moves with the plate unit. Concentrators of this type are used for melting and obtaining of most pure metals in ground conditions or for electricity generation. Stirling or Brayton heat engines are usually used as heat engines. Such modular systems with total capacity of up to 5 MWe have been manufactured. Maximal power registered for a module is about 50 kWe, achieving an efficiency of up to 30%.

The main advantages are:

- Rotation mechanism around two axes allows to collect the maximum Sunlight quantity;
- Easily transported and assembled modular construction;

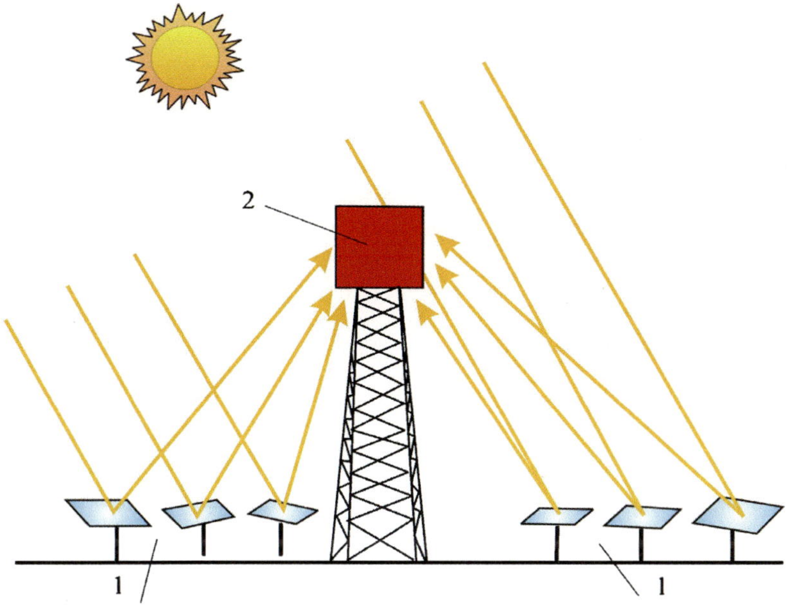

Fig. 3.52 Operating principle of the concentrator with heliostate

- Highest efficiency compared to any other generating solar technologies;
- Requires limited areas;
- Long operation period.

Heliostat concentrator is the equivalent of a very large parabolic concentrator (Fig. 3.52). Heliostats 1, small flat-shaped reflectors, are located around the entire circumference of the central tower forming the so-called heliostat field. The reflector 2 is mounted on the tower in the focus point of the paraboloidal. Each heliostat is remote-controlled and follows the two coordinates of the Sun position. The first heliostat concentrator was built in 1949 under the supervision of the French professor, Felix Trombe, in Mont-Luis area in the Pyrenees [43]. The heliostat field was situated on 130 m^2, receiver power reached 50 kW, and the temperature reached 3,000–4,000°C. The facility aimed the study and creation of new materials, in particular of composite materials.

To estimate the temperature of the receiver we use Fig. 3.53, which depicts a simplified diagram of parabolic mirror concentrator: 1 is a cylindrical-parabolic or paraboloidal mirror, 2 – receiver placed in the focus point of the parabola or of the paraboloidal, 3 – receiver screen. Cylinder-parabolic mirror has length l and opening D, and in the case of paraboloidal mirror – aperture diameter is equal to D. The receiver of cylinder-parabolic concentrator represents a pipe with diameter d and length l, or has a spherical shape with diameter d for paraboloidal concentrator. Solar disk angular size is equal to $2\theta_s$, where $\theta_s = r_s/L$ (Fig. 3.2).

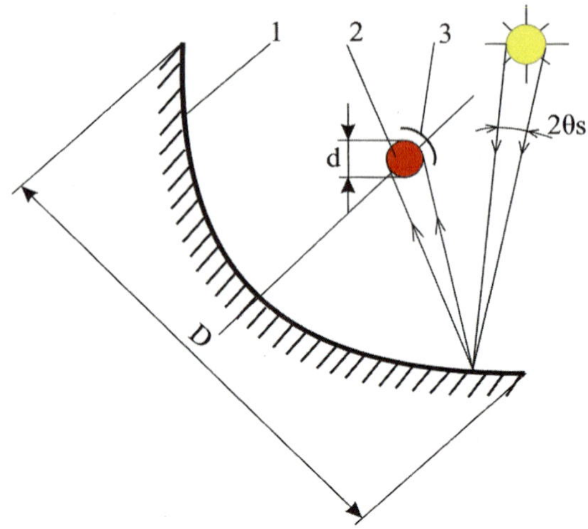

Fig. 3.53 Simplified diagram of the parabolic concentrator

Temperature of the cylindrical-parabolic concentrator receiver: The power of solar radiation absorbed by the receiver:

$$P_{abs} = \rho_c \alpha l D B, \tag{3.56}$$

where ρ_c is the reflectance of mirror surface, α – absorption of receiver; B – direct solar radiation.

The receiver, having the temperature T_R, emits radiant power in space, which, according to Stefan-Boltzmann law, is determined by the expression (3.6):

$$P_{rad} = \varepsilon \sigma T_R^4 \pi d l (1 - \zeta/\pi), \tag{3.57}$$

where ε is the receiver emissivity; σ – Stefan-Boltzmann's coefficient, ξ – shielding factor of receiver, usually $\xi = 2/3\pi$, d – diameter of the receiver, determined by the opening D: $d = D\theta_s$.

At steady heat condition absorbed power is equal to the radiated power (other power losses of the receiver are neglected), from where the maximum temperature of the receiver is determined:

$$T_R = \sqrt[4]{\frac{\rho_c \alpha B}{\varepsilon \sigma \theta_s}}. \tag{3.58}$$

For $B = 600\ \mathrm{W/m^2}$, $\alpha/\varepsilon = 1$, $\rho_c = 0.8$, $\sigma = 5.67\ \mathrm{W \cdot m^2 \cdot K^{-4}}$, $\theta_s = 0.00465$ Rad, maximum temperature of 1,162°K 889°C is obtained. In usual systems with cylindrical-parabolic concentrators temperatures up to 700°C are obtained. Temperature decrease is due to mirror surface deviation from the ideal parabolic shape and to the fact that liquid thermal agent circulating through the receiver, leads to P_{rad} decrease.

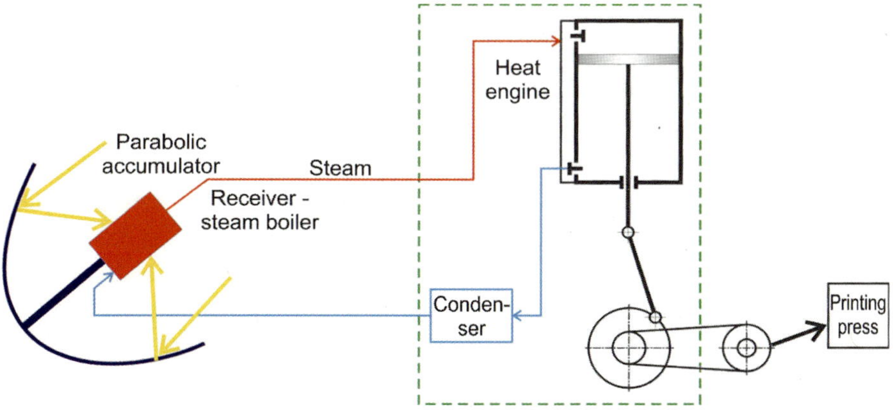

Fig. 3.54 Schema of solar thermal engine elaborated by Augustin Mouchot and Abel Pifre in 1872

Temperature of paraboloidal concentrator receiver. A sphere-shaped receiver with a diameter d is considered, other sizes and assumptions remain the same as in the previous case. From the thermal balance condition we obtain:

$$T_R = \sqrt[4]{\frac{3\rho_c \alpha B}{8\varepsilon\sigma\theta_s^2}}. \quad (3.59)$$

For the same amount of direct solar radiation $T_R = 3{,}480°K$ or 3,208°C are obtained. In real plants temperatures up to 3,000°C can be obtained.

3.3.4.3 Heat Engines Used at Solar Thermal Plants

Solar thermal system with steam engine: working principle: Conversion of solar thermal energy into mechanical energy or electricity was the goal of experiments for over a century. At 1872 Paris exhibition professor of mathematics Augustin Mouchot and engineer Abel Pifre presented the first system converting solar energy into mechanical energy – a printing press driven by a steam engine powered by a solar parabolic concentrator (Fig. 3.54).

Later, in 1913, American entrepreneur Frank Shuman applies the same principle in Egypt, manufacturing the first solar unit for irrigation [43, 44]. The steam for the thermal engine was generated by five cylinder-parabolic collectors with an 80 m length and 4 m aperture each. The receiver – a cast iron pipe situated in the focus point, provided transportation of the steam to the engine. The unit developed a mechanical power of about 45 kW, which was used to pump water from river Nile for irrigation. Due to the high price of coal in that area at that time, system recovery period did not exceed 4 years.

It is necessary to note one very important disadvantage of solar heat engines – low efficiency. It follows from the low power density of solar radiation and the fundamental principles of thermodynamics. A more understandable interpretation of the second law of thermodynamics states: thermal or heat energy cannot be fully converted into other forms of energy, such as mechanical or electrical. The efficiency of a piston- or turbine heat engine depends on the warm source temperature, i.e. input temperature T_{in} and cold source temperature (condenser) T_{ies}, between which heat exchange occurs. Maximum efficiency or maximum Carnot efficiency can not exceed the value obtained from the expression:

$$\eta_C = 1 - \frac{T_{ieş}}{T_{in}} \tag{3.60}$$

For example, a turbine powered by a steam parabolic concentrator at a 350°C temperature and a 30°C condenser temperature shall have a theoretical efficiency equal to $1-(30+273)/(350+273)=0.51$ or 51%. Real efficiency, due to energy losses shall be much lower than about 25%.

To obtain water vapours a temperature of at least 100°C is required. If temperature obtained from the solar collector is lower, than a heat engine employing Rankine cycle can be used. Here, as fluid thermal agent, organic substances are used that boil at below 100°C and used in refrigerators and heat pumps. Such a heat engine shall have an even lower efficiency. For example, a heat engine efficiency operating on 85°C vapours, obtained from a plate solar collector, and at 30°C condensing temperature shall not exceed 15%.

Solar thermal system with steam turbine: World War I caused the widespread use of internal combustion engines operating on petrol or diesel. Cheap oil era began and solar thermal systems have been forgotten for a period for over 50 years. They returned after the start of the 1973 oil crisis. A group of U.S. engineers led by G. Alexander performed in 1979 a new solar irrigation project in the Gila Bend area [4].

Figure 3.55 shows the operation scheme of this system. Primary energy is obtained from a 537 m^2 cylindrical-parabolic solar collector, which heats water to a temperature of 150°C at about 7 atm pressure. The hot water circulates through the heat exchanger, working as a boiler too, pre-heater and receiver. In the secondary circuit (not shown in the diagram) of the heat exchanger, the working fluid Refrigerant 113 with a low boiling temperature is used. The vapours with 138°C temperature and about 9 atm pressure are used to power a turbine which drives the pump. Part of vapours is used to preheat the water in the primary circuit. After the turbine vapours are used in the regenerator to heat the water from the preheater circuit. The regenerator vapours condense in the condenser, where part of the water pumped is used for cooling and at a temperature of about 32°C is returned to the tank. In the first year of operation, the system worked for 323 h at a capacity between 240 and 570 l/s or 364 and 2,052 m^3/h. In the second year it worked for 188 h, delivering 1.24×10^5 m^3 [4] or at a capacity of 660 m^3/h.

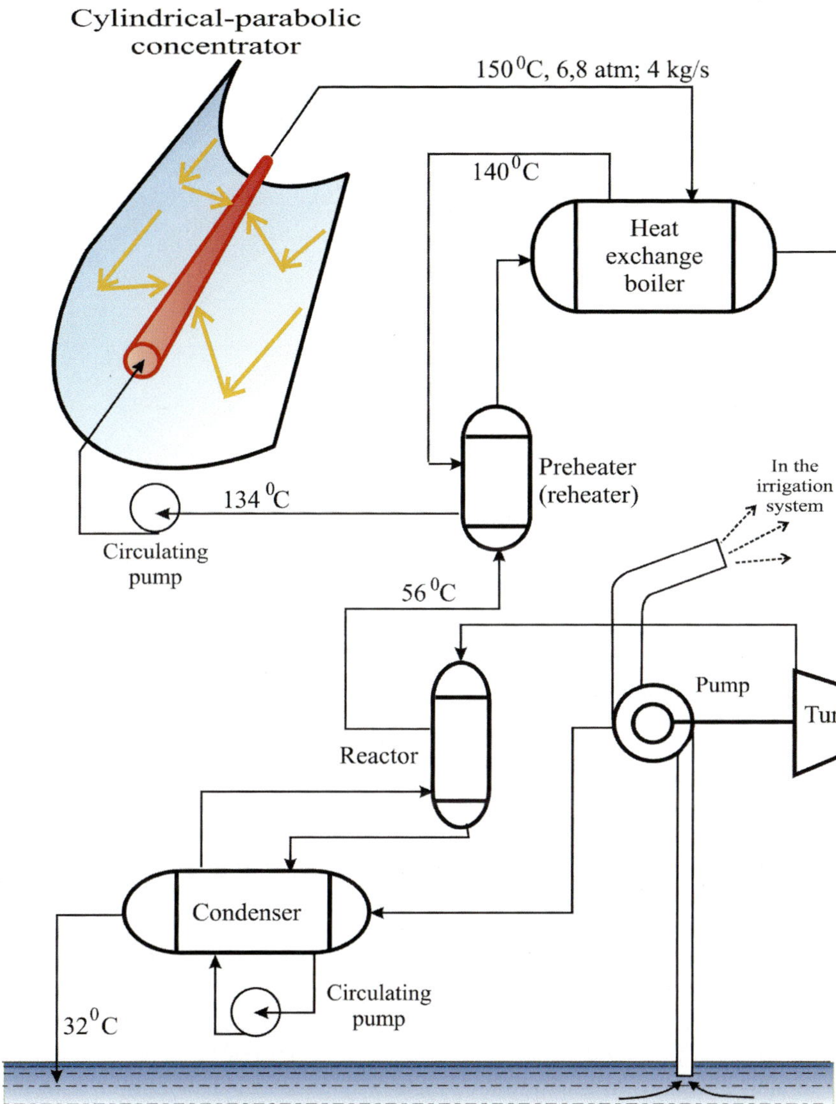

Fig. 3.55 Solar pumping system, manufactured by G. Alexander in 1979

Stirling engine solar thermal systems. For more than 20 years Stirling engine parabolic solar technology is in the focus of researchers. Figure 3.56 shows structural drawing of a paraboloidal concentrator solar thermal system (plate) and Stirling engine. The system includes a parabolic concentrator (plate) (1), heat receiver (2) and a heat engine/heat generator (3), which are installed on a support 4 with a concentrator orientation mechanism towards the Sun 5 (Fig. 3.56). Stirling engine is used as heat motor/generator.

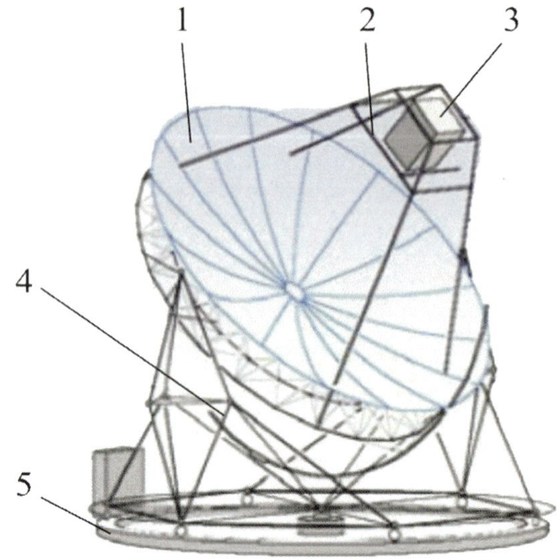

Fig. 3.56 Operating principle of the solar unit with Stirling engine

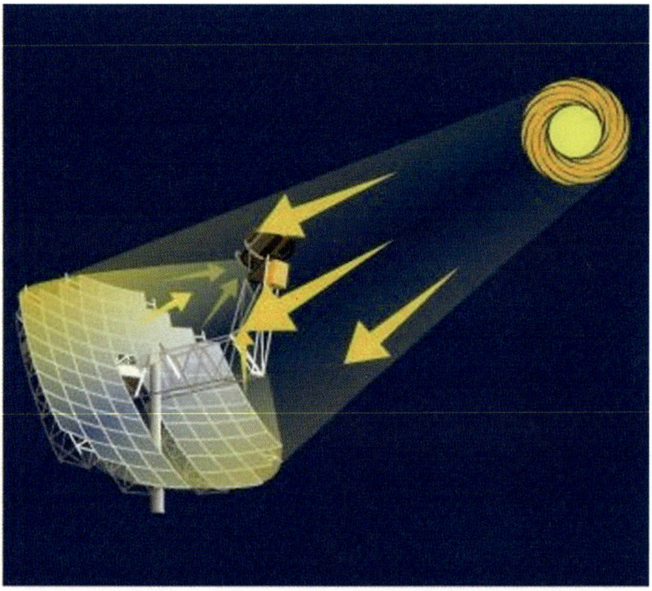

Fig. 3.57 Solar rays concentration principle in the Stirling engine unit

The operation principle of Stirling engine parabolic solar system and of Stirling engine is shown in Fig. 3.57. Parabolic concentrator reflects solar radiation received by the cavity of the receiver, located in the focus point of the concentrator. Generated temperature heats Stirling engine working agent and makes it operational. To better

Fig. 3.58 The first Stirling engine

understand the issue, a brief history, construction and operating principle of the Stirling engine is presented.

On 27 September 1816 Robert Stirling patented his Economizer in Edinburgh, United Kingdom (Fig. 3.58). At that time Robert Stirling was a minister in the Church of Scotland, where he continued to work until the age of 80. During this time, he built his heat engine at home. Lord Kelvin used one of his models for demonstrations at the university. Simple and elegant dynamics of the engine has been explained for the first time in 1850 by Professor A. McQuorne Rankine. Approximately 100 years later the term *Stirling engine* was used by Rolf Meijer to describe all types of closed-cycle regenerative engine.

This engine is a unique heat engine because its theoretical efficiency is almost equal to the maximal theoretical efficiency, known as the Carnot efficiency. Stirling engine (Fig. 3.59) is driven by the expansion of the heated gas, followed by its compression when cooled. Stirling engine contains a fixed quantity of gas which is transferred back through the hot and cold areas. The distribution piston moves the gas between both ends of the cylinder and the power piston changes the internal volume of gas through compression and expansion. The gas used in the Stirling engine does not leave the engine during operation. Unlike petrol and diesel engines, Stirling engine contains no valves, no explosion occurs. Stirling cycle uses an external source of heat, which can be anything – from gasoline to solar energy. No combustion takes place in the Stirling engine. Solar radiation is absorbed by a heat exchanger (receiver) and it heats the working gas (helium or hydrogen) of

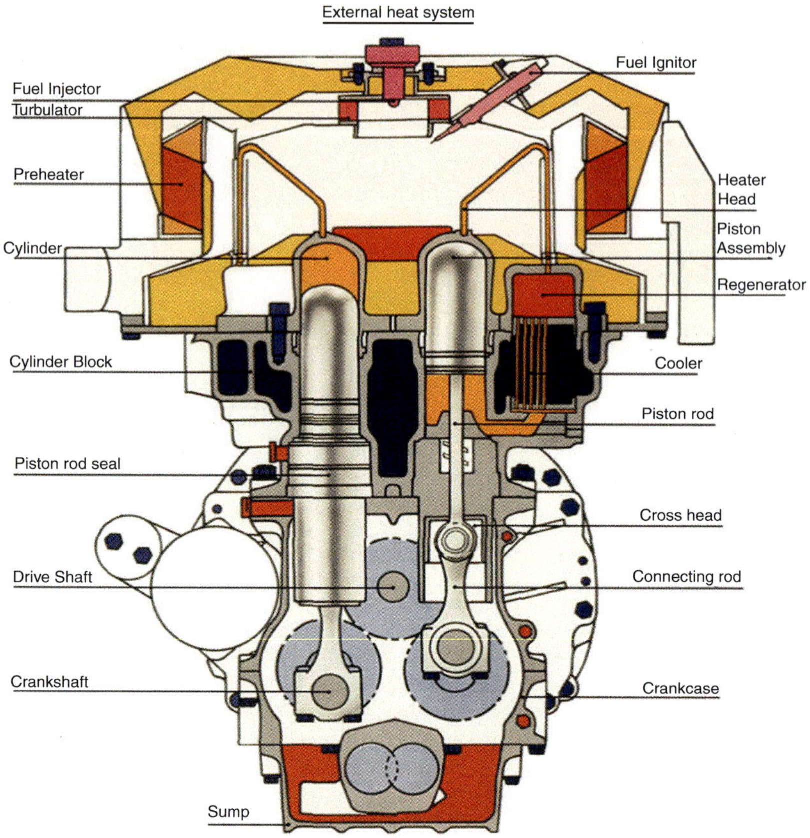

Fig. 3.59 Operating principle of Stirling motor

the Stirling engine up to temperature of about 650°C (Fig. 3.60). Stirling engine converts this heat into mechanical energy.

Stirling engine shaft turns mechanical energy into electricity. Theoretically possible efficiency of the Stirling heat engine is about 60%, compared with 45% for the diesel engine and no more than 30% for standard Otto engine. Parabolic modules with Stirling engine can produce between 5 and 40 kW. A network of modules can be used for producing electricity in large quantities.

3.3.4.4 Solar Thermal Power Plants (STPP) with Cylindrical-Parabolic Concentrators

The conversion of solar thermal energy into electricity is similar to the traditional conversion, based on fossil fuels where energy stored in the fuel is released through combustion, and is converted into potential energy of compressed and heated to

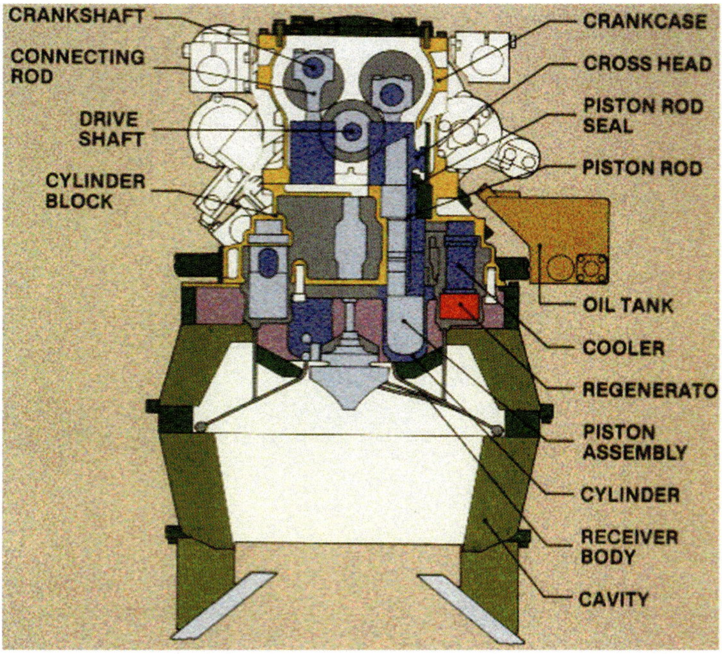

Fig. 3.60 Heat receiver embeded in Stirling motor

500–600°C steam. In the turbine, the steam expands, the potential energy turns into kinetic energy which, in its turn, transforms into electricity. In solar systems, fossil fuel is replaced by solar radiation, boiler combustion chamber – with a concentration solar collector: cylindrical-parabolic, with parabolic mirror or with heliostat and central tower. Solar thermal collectors with focused Sunlight are mainly used to generate electricity. There are various systems that differ in the way how they collect and concentrate solar rays, but the final step of generating electricity is identical to that of conventional power plants. Finally, the heat engine is used to convert heat into electricity. The solar system is designed to focus solar rays and transform them into heat, which is used in the heat engine. In a simple way, solar thermal plant is a conventional power unit that uses solar energy as primary source of heat.

In the last two decades of the previous century in the U.S. several STPP pilot projects have been successfully completed, which proved their technical and technological viability and paved the way towards new projects with hundreds of MW capacity. Next, a brief description of certain commercial projects with cylindrical-parabolic concentrators is given [4, 44]. The largest solar thermal plant in the world has a maximum power of 354 MW and is located in the Kramer Junction area, California, USA [45]. It was built by Luz International Company in 1985–1991 and consists of nine units with an electric capacity between 14 and 30 MW. Later this type of plant was called LUZ solar system. Figure 3.61 shows a part of

Fig. 3.61 Fragment of field with cylindric-parabolic solar collectors of the thermal power plant Kramer Junction (**a**) and a separate line (**b**), California [46]

cylindrical-parabolic concentrators' field. By 2001 the facility supplied in the public network of California 9 TWh of electricity. The plant is equipped with an auxiliary circuit, operating on natural gas and allowing expansion of energy production by 25% either on cloudy weather or during peak hours. The plant was designed to provide electricity during peak hours, when the cost of delivery is the highest. In California this period lasts from June to September, 12:00–18:00 h. Main plant components are shown in Fig. 3.62. Solar collector field is composed of cylindrical-parabolic concentrators with a total area of about 1.75×10^6 m^2. The active part of the concentrator – the reflector is composed of low-iron glass and coated with silver at the back. Glass is mounted on a metal sectioned frame, which allows the formation of different length collectors. The collectors are mounted in parallel rows in the south–north direction. The Sun is tracked using hydraulic drives. The solar radiation receiver has a 70 mm steel tube in diameter covered with a layer of selected material (see Sect. 3.3.2.4). To reduce energy losses, the receiver is placed in a vacuum glass tube. At a temperature of 350°C, the surface receptor has an absorbance equal to 0.96 and an emittance of 0.19 only.

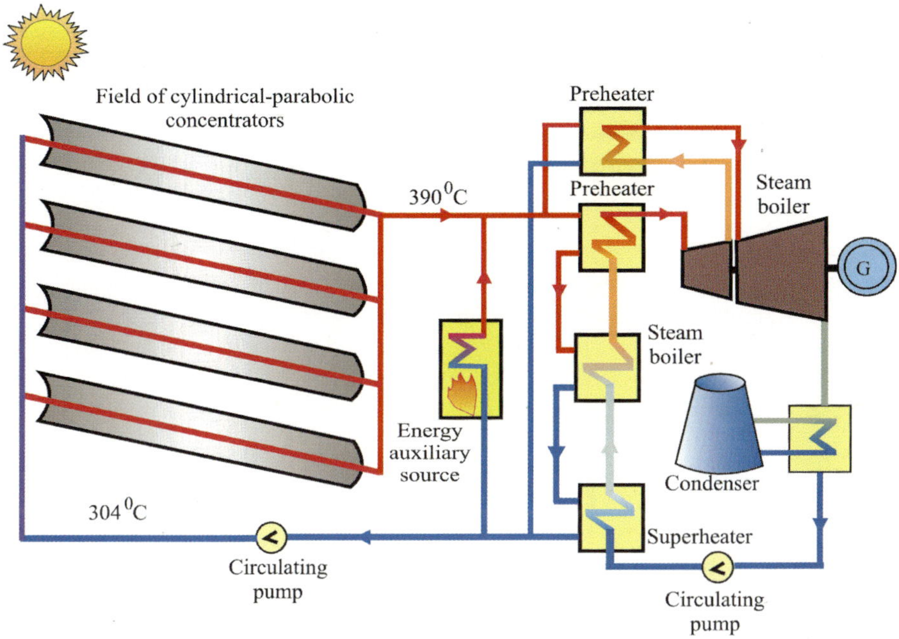

Fig. 3.62 Operating diagram of solar thermal plant from Kramer Junction, California

Reflectance of clean mirrors is equal to 0.94. In the first contour, which includes solar collectors, a synthetic fluid (usually oil) is used as thermal agent. The fluid leaving the collector has a temperature of 390°C, and at the entrance of the collector – about 304°C. In the second circuit water turns to steam into the boiler supplied with thermal energy from solar collectors. The generating unit represents a traditional group 'steam turbine-synchronous generator'. Steam turbine has two pressure stages – the first is supplied with superheated steam and the second – from the preheater.

Solar Energy Generating Systems (SEGS) owns the largest collection of parabolic systems in the world. Nine facilities are located in the Mohave Desert in Southern California. SEGS plants were built during 1985–1991. A large network of 80 MW SEGS plants is located in the Harper Lake area. All SEGS plants are hybrid, using fossil fuel to supplement solar power during periods of low solar radiation. Each plant is allowed to generate 25% of annual energy using fossil fuels. The hybrid SEGS plants capacity factor exceeds 100% in peak-hours for more than a decade, with more than 85% of electricity obtained from solar energy. In general, SEGS plants operate well. New models of heat collector elements increase the plant performance and reduce costs.

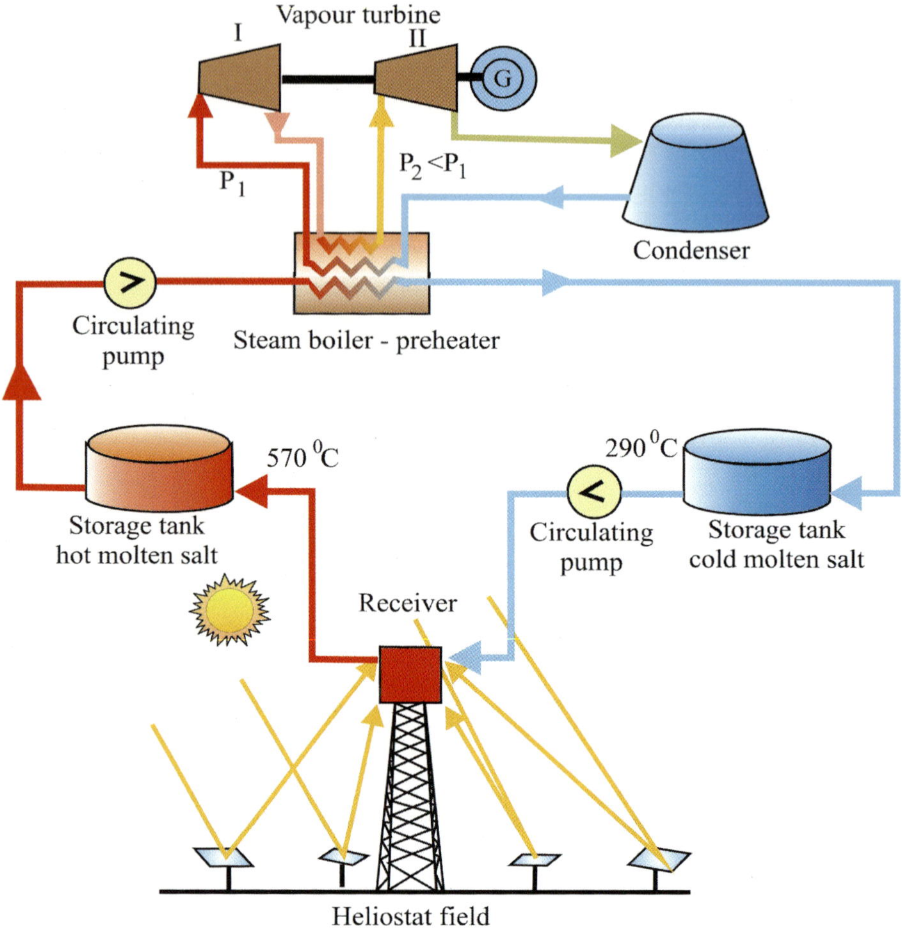

Fig. 3.63 Operation principle of solar thermal plant Solar Two

3.3.4.5 Solar Thermal Power Plants with Heliostats and Central Tower

Solar thermal power plants with heliostats and central tower (or, in short, solar towers) use thousands of mirrors called heliostats, which redirect concentrated solar light to the receiver that is installed at the top of the tower. With most receivers, salt solution heated in the receiver is used to generate steam directly into the receiver, steam that is used by a conventional turbine generator to produce electricity. Molten nitrate salt possesses heat transfer properties and has superior energy storage capacity. Commercial power towers can be sized to produce from 50 to 200 MW electricity. In international practice, systems with air as working fluid in the receiver are used. Figure 3.63 shows the technological scheme of high temperature solar receiver solar tower. In the 90s of the last century several pilot

Fig. 3.64 Solar thermal plant Solar Two [48]

plants with heliostats and central tower were built: in the former USSR – employing 5 MW power; in Italy, Spain and France – employing 1 MW power.

The world's largest heliostat plant was built in 1982 in the U.S., Barstow district, California. The plant was called Solar One and has a 10 MW capacity [4, 45, 47, 48]. The operating principle of the heliostat solar concentrator is described in Sect. 3.3.3.2. A general view of the plant is shown in Fig. 3.64. After 6 years of operation, in 1988 it was decided to rebuild the plant. It was equipped with a thermal energy storage system with molten salt; heliostats control system has also been modernized. Heliostat field includes 1818 reflectors with a total area of 71,100 m^2.

Each heliostat is oriented so as to reflect Sunlight on to the receiver installed in the centre of the field at 90 m above the ground. A heliostat includes 12 concave panels with a total area of 39.12 m^2. Glass coated with silver was used as reflecting material. The receiver is a 13.7 m height- and 7 m diameter cylinder. The receiver cylinder consists of 24 panels, each having a 0.9 m width and 13.7 m height.

An interesting European project, similar to the above, is 3,000 SOLAIR system installed in Almeira region, Spain. Concentrated solar rays are reflected on an open volumetric absorber part of the receiver. Environmental air heated to 680–750°C supplies the steam generator to produce electricity. This system has a 70% rate of converting solar energy to heat at a temperature of up to 700°C. A high efficiency is obtained at a temperature of 590–630°C and at incident solar light of

Fig. 3.65 Stirling paraboloidal system (SES)

310–370 kW/m^2. Being the most efficient system to date, an inherent disadvantage of it is the complexity of the scheme (high maintenance costs) and heat loss in multiple heat exchange processes.

3.3.4.6 Paraboloidal Solar Thermal Power Plants with Stirling Engine

Stirling engine parabolic solar technologies have been demonstrated in a variety of complete systems over the past two decades and can move further towards commercial application over the next 5 years. Several U.S. and European projects are aiming at the use of solar energy this way. Stirling engine paraboloidal units have several common characteristics compared to the wind turbines. Like wind turbines, Stirling solar parabolic units are an intermittent energy source, include a base that can be built within a few days, are small power energy facilities and are modular units. Regarding Stirling parabolic units marketing, some wind turbines market penetration tactics may be used.

Figure 3.65 presents a Stirling paraboloidal system including an air heating system and a mechanism for rotating the parabolic mirror concentrator around two

Fig. 3.66 Stirling paraboloidal system with working agent – hydrogen

axes. This allows the collector to be continuously oriented to the Sun. The system was designed and built by SES (Stirling Energy Systems), which together with the STM Company today are most aggressive in implementing Stirling parabolic systems on the market. They are engaged in developing and building new Stirling parabolic facilities. In SES opinion, engine and collecting system changes will increase peak efficiency of the coming plants by 10%. This is a new efficiency world record for any solar electricity generation technology and it will increase the annual electricity production by 6.3%. Stirling parabolic system (Fig. 3.66) uses an efficient Stirling engine. Solar energy is converted into heat in a closed hydrogen spiral. Heated hydrogen drives Stirling engine-generator.

Due to air cooling it does not require water cooling as in the solar cylinder-parabolic systems or solar towers. Thermal energy storage is not considered a viable option for Stirling parabolic systems. Relatively flat land is preferable to build and maintain these systems. At the same time, land requirements are less important than for cylindrical-parabolic solar systems or solar towers. An individual Stirling parabolic plant produces 10–25 kW. Having high efficiency and modular construction, the cost of these systems is expected to be competitive on the market. According to the estimates of SES, the main developer of Stirling parabolic systems, costs of parabolic concentrators shall dramatically decrease hundreds of times at larger power stations (MW level) or at network connection. Stirling parabolic facilities are not yet marketed, being in the experimental stage. Recently at Sandia National Laboratories (SNL) in Albuquerque (USA) an installation with six parabolic mirrors has been tested. In August 2005, Southern California Edison

Fig. 3.67 *Stirling engine – electrical generator* unit of paraboloid power plant

Company publicly announced the completion of negotiations with SES for the right to use Stirling parabolic facilities of 500–850 MW capacity (annual production between 1,182 and 2,010 GWh/year) over a 20-year period. In September 2005, SES publicly announced a contract with San Diego Gas & Electric to develop solar plants with capacities between 300 and 900 MW.

Purchase price for these rights remains confident information. Development of large scale Stirling parabolic systems will cause drastic reduction of capital and operation and maintenance costs, and will result in increased system reliability. Figure 3.67 shows the "Stirling engine-electric generator" node. Figure 3.68 shows a field with Stirling engine parabolic solar systems located in California, USA.

Further, Figs. 3.69a, b and 3.70a, b present a Stirling parabolic system, developed and patented by the authors [49, 50]. The examined solutions propose to simplify the transformation mechanism of the alternative movement of Stirling engine pistons into rotary motion. Crank mechanism and the cams (see the original construction of the Stirling engine above) of a Stirling engine is replaced with a precessional planetary transmission 2, which reduces the speed of output shaft 3 of Stirling engine. Precessional transmission satellite block 4 is connected kinematically with piston rods 5 of the Stirling engine, which allows direct conversion (without using additional elements) of alternative translational motion of the pistons into reduced rotation of electric generator.

In order to spin paraboloidal concentrator together with Stirling engine in an azimuth and zenith plane precessional drive mechanisms 6 and 7 (Fig. 3.69a, b) are used. These, due to very high transmission ratio (up to 3,600 at single stage) ensure

Fig. 3.68 Field of Stirling (SES) paraboloidal solar systems, California

very slow rotational motions (basically, a rotation around the axis of the tower for 24 h) required for permanent orientation of Stirling paraboloidal system after Sun position. Precessional drive mechanisms developed are compact and reliable, containing a small number of elements compared to other transmission systems, and this provides low cost.

3.3.5 Photovoltaic Systems

3.3.5.1 Brief History of Photovoltaic Technologies

The term "*photovoltaic*" is derived from the combination of the Greek word *"photos"* – meaning *light*, and the name of the electromotive force unit – *volt*. Thus, photovoltaic technology (PV) describes the generation of electricity using light. In 1839, during the industrial revolution, Alexander Edmond Becquerel, the father of Nobel laureate Henri Becquerel, discovered the photovoltaic effect, which explains how electricity can be generated from sunlight. He concluded that "*illumination of an electrode immersed in a conductive solution would create an electric current*" [51].

Despite extensive research, after this discovery, photovoltaic conversion continues to be ineffective. Photovoltaic cells were used mainly for measuring of light intensity proposals. The first report on photovoltaic or photoelectric effect, as it was

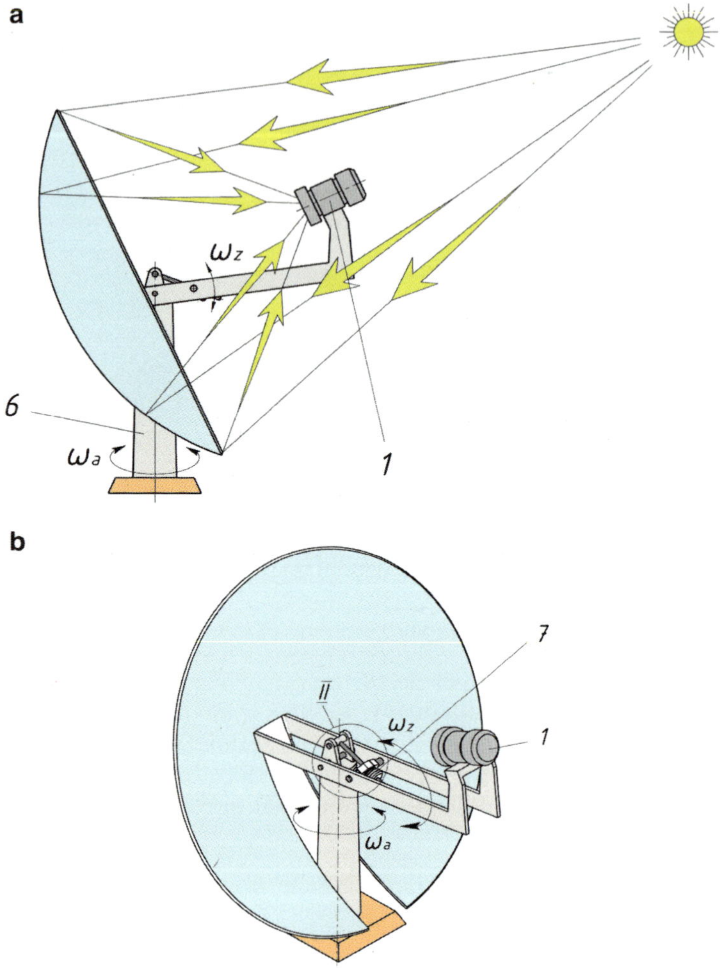

Fig. 3.69 Stirling paraboloidal solar system, author's elaboration [49]

called at that time, was done by Cambridge scientists W. Adams and R. Day in 1877. The report described changes that occurred in light-exposed selenium plate. In his experience, Heinrich Hertz noticed in 1887 that a zinc plate is loaded with positive charge when exposed to ultraviolet radiation. The phenomenon is due to the same photoelectric effect: the action of ultraviolet rays separates electrons from metal, as a result the metal is positively charged.

The first PV cell was built by an American electrician, Charles Fritts, in 1883 using selenium. Construction of the cell was patented in 1884. It should be noted that the construction of the cell was very similar to today's cells. However, cell efficiency was less than 1% and the industrial use has not been realized.

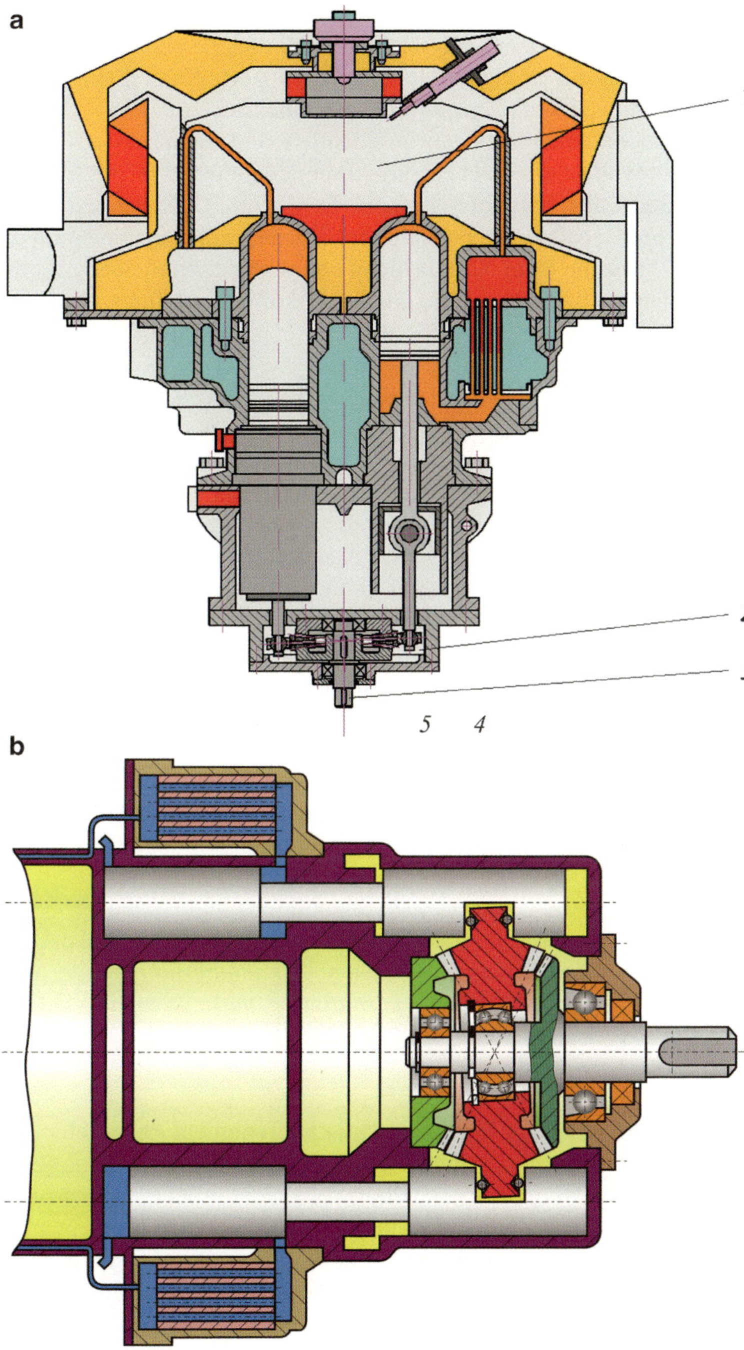

Fig. 3.70 Stirling engine with piston rods linked kinematically to presessional transmission satellite, author's elaboration [50]

After about a century since the first discovery of the effect, Albert Einstein in 1921 received the Nobel Prize in physics for explaining the photoelectric effect that allowed the practical use of photovoltaic cells. In 1946, Russell Ohl invented the solar cell [52], followed by the invention of the transistor in 1947.

In the middle of twentieth century scientists and engineers returned to the study of the photovoltaic effect, occurring in semiconductors. In 1953, a Telephone Laboratories (Bell Labs) team of engineers, D. Chapin, C. Fuller and G. Pearson, creates the silicon PV cell with a much higher efficiency than the selenium cell. The following year, the same team built a silicon cell with a 6% efficiency. At the same time, first consumers of photovoltaic energy – artificial satellites, have appeared. In 1957, PV cells were installed on the first artificial satellite of the earth "*Sputnik 3*", and in 1958 PV cells were installed aboard the U.S. satellite Vanguard 1 and served to power a radio transmitter. Until these days PV cells are the most suitable energy sources for space machinery. Competition between the USA and the former USSR in the 60s of last century in the field of sources for electricity supply of the satellites led to a spectacular development of PV technology and caused a breach in the rigid dependence of decentralized energetics from traditional sources: electrogen units, storage batteries or dry cell batteries. A new competition started – to bring PV generator back to earth. Governments of industrialized countries and many private companies have invested billions of dollars in the development of PV technology.

By using the photovoltaic effect direct conversion of sunlight into electricity occurs. Direct conversion technology excludes intermediate transformations: solar radiation to heat, heat to mechanical energy, and mechanical energy into AC power. Direct conversion is performed using semiconductor materials and the photovoltaic effect. Photovoltaic generator, the so-called photovoltaic cell, unlike the electromechanical generator produces DC electricity. Elimination of intermediate processing from the technological chain, lack of motion, noise and vibration, modular construction, a service life of over 25 years, are arguments in asserting that the future of decentralized energy shall belong to photovoltaic technology. Not accidentally, on 31 August 1991 The Economist magazine mentioned the following about the photovoltaic solar energy conversion: "*Of all alternative energy sources – wind, sea wave, tidal, geothermal – perhaps the most promising solar energy conversion into electricity is photovoltaic one*".

3.3.5.2 Photovoltaic Cell: Characteristics and Technical Parameters

Construction and operating principle: Photovoltaic cell is an optoelectronic device which operation is due to the generation of light by free charge carriers and their separation by the internal electric field of p-n, MOS or Schottky junctions. Crystalline or polycrystalline silicon is used as initial material for manufacturing in which, usually, by various technological methods different conductive layers are formed to obtain *p-n* junction. Basic semiconductor material used to produce PV cells is silicon. Table 3.14 shows the PV conversion efficiency based on three types of photovoltaic materials: crystalline silicon, polycrystalline silicon and amorphous silicon [53, 54].

Table 3.14 Materials comparison for PV cells

Solar cell material	Crystalline silicon	Polycrystalline silicon	Amorphous silicon
Power conversion efficiency (%)	15–22	14–15	7–10

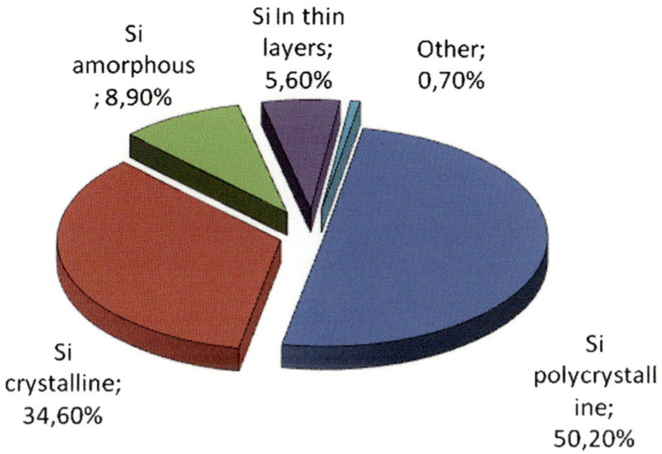

Fig. 3.71 World market distribution depending on the materials for PV cell production

Global market share of different technologies for producing PV cells is shown in Fig. 3.71. Over 84% of world production of PV cells is based on polycrystalline and crystalline silicon. Today, polycrystalline and crystalline silicon technology is the most advanced, ensures the production of PV modules on an industrial scale with a yield of 14–17% and a 30 years lifetime of modules.

But this technology has a crucial disadvantage – limited potential for future decrease of PV cell production costs. Experts in the field believe that the cost of a watt will not drop below 2 euros [55–57]. From this point of view, technology of amorphous silicon and silicon in thin layers is a promising perspective. The costs of 1 W produced with these technologies will decrease to 1 € – the limit cost for PV electrical energy is cheaper than electrical energy produced from fossil sources. It is assumed that for these reasons in recent years there has been a redistribution of the global market in favour of amorphous silicon and in thin layers technology.

Figure 3.72 shows a schematic diagram of the simplified design of PV cell, based on *p*-type semiconductor material. Consider the phenomena that occur when PV cell is exposed to incident radiation (Fig. 3.72). This radiation can be equalized with a flux of photons, which have energy $E = h\nu$, where $\boldsymbol{h}$ is Planck's constant and $\boldsymbol{\nu}$ is photon frequency. If the photon energy is bigger than the energy of the semiconductor power forbidden band, then, from the photon interaction with an atom, the electron from the valence-bond band will pass into the conduction band, becoming free, and generating also in a gap in the valence-bond band. Thus,

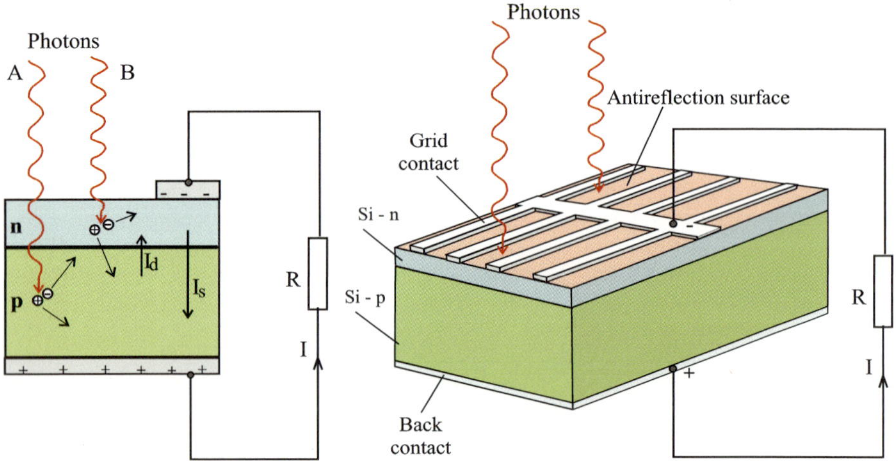

Fig. 3.72 Design diagram of photovoltaic cell

under the action of photons, generation of hole–electron pairs occurs. This is called internal photovoltaic effect.

In Fig. 3.72 on the left, the photon *A* has a lower frequency and therefore a lower energy, and photon *B* has a higher frequency and, correspondingly, a higher energy (low-frequency electromagnetic wave penetrates to large depths of the material and vice versa). Free charge carriers are separated from the electric field of the *p-n* junction, characterized by the barrier potential U_0 and which, depending on the type of semiconductor used, is about 0.2–0.7 V.

Here, the electric field will serve as free load break switch – hole–electron pairs. The electrons will be directed to the *n* zones, the holes – to the cell *p* zone. This is why, under the influence of light, *p* zone is positively charged and zone *n* – negatively charged, which leads to an electric current through the external circuit, caused by photovoltaic conversion of solar radiation. This current (Fig. 3.72 on the left), leads to a voltage drop *U* in the external load *R* connected to the rear contacts and to the front-grid contact (Fig. 3.72 on the right). Voltage *U* compared to *p-n* junction acts against the direct sense and, in its turn, will determine via the junction the diode I_d current against the inverse direction of photovoltaic current I_s which is determined from the known expression:

$$I_d = I_0 \left[\exp\left(\frac{eU}{kT} \right) - 1 \right] \tag{3.61}$$

where: I_0 is the saturation current strength; *k* – is Boltzmann's constant; *T* – absolute temperature; *e* – is the electron charge.

Photovoltaic cell characteristics: The main characteristics of PV cells are as follows: ampere-volt *I(U)* or volt–ampere *U(I)* characteristic and power characteristic

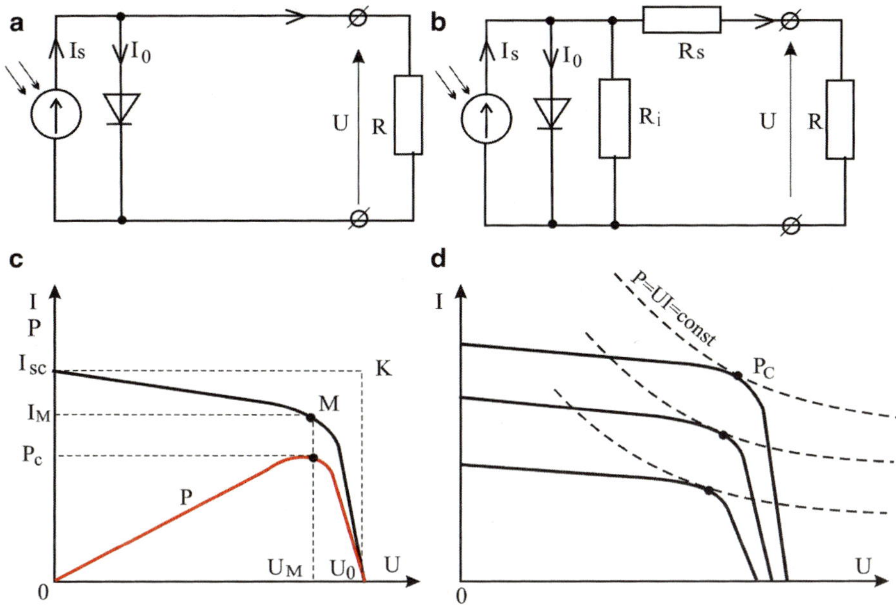

Fig. 3.73 Equivalent diagrams of PV cell: (**a**) simplified; (**b**) complete; (**c**, **d**) cell characteristics

P(U). The current in the external circuit I is determined by the difference between photovoltaic current I_s and the diode current I_d [2, 57–59]:

$$I = I_s - I_d = I_s - I_o \left[\exp\left(\frac{eU}{kT}\right) - 1 \right] \tag{3.62}$$

The simplified equivalent circuit of PV cell, shown in Fig. 3.73a, corresponds to Eq. 3.62. If we take into account the leak resistance R_i of the PV cell *p-n* junction and the cell base-spreading resistance R_s, a complete equivalent scheme of PV cell can be produced (Fig. 3.73b). Modern technologies help to obtain cells with $R_i = \infty$ and $R_s = 0$, so that the simplified equivalent circuit is satisfactory.

Electrical power transferred to load R of a PV cell is:

$$P = UI = U \left\{ I_s - I_o \left[\exp\left(\frac{eU}{kT}\right) \right] - 1 \right\}. \tag{3.63}$$

Maximum value of power is obtained at point M of the current–voltage characteristic, which coordinates are derived from the condition $dP/dU = 0$:

$$U_M = U_0 - U_T \ln\left(1 + \frac{U_M}{U_T}\right),$$

$$I_M = I_s \left(1 + \frac{I_0}{I_s}\right) \frac{U_M}{U_M + U_T}, \tag{3.64}$$

where $U_T = kT/e$.

For a passive load, the optimum value of load resistance is:

$$R_M = \frac{U_M}{I_M}. \quad (3.65)$$

Parameters PV cells and modules: Manufacturers of PV cells and modules show in the technical book of the product the technical parameters for standard conditions:

- Global solar radiation on the cell surface, $G = 1{,}000\ \mathrm{W/m^2}$;
- Cell temperature, $T_C = 25°\mathrm{C}$;
- Conventional air mass, $AM = 1.5$.

It is compulsory that the book lists the following information: short circuit current, I_{sc}, no-load voltage, U_0, maximum or critical power, P_c, voltage and current in the critical point, U_M and I_M. Besides these parameters, additional indicators may be: the Fill Factor, *FF*, efficiency of PV cell or module, normal operating temperature of the NOCT cell, coefficients of variation of no-load voltage and short circuit current with the temperature.

Short circuit current: It occurs at shorting of load *R* terminals as shown in Fig. 3.73. On the *I-U* characteristic this is the point with coordinates $U = 0$, $I = I_{sc.}$ From the expression (3.62) for $U = 0$, we obtain $I_{sc} = I_s$. The engine power is zero.

No-load voltage: Corresponds to the point on *I-U* characteristic with coordinates $I = 0$, $U = U_0$. The rated power in this point is equal to zero. No-load voltage can be determined from Eq. 3.62 for $I = 0$:

$$U_0 = \frac{kT}{e} \ln \frac{I_s + I_0}{I_0} \approx \frac{kT}{e} \ln \frac{I_s}{I_0}. \quad (3.66)$$

For silicon cell the ratio I_s/I_0 is about 10^{10}, factor kT/e, called thermal voltage, is equal to 26 mV. Thus, $U_0 = 0.6$ V.

Critical or maximum power: It is the product of current and voltage in point M of *I-V* characteristic. This parameter is called the *peak power* and is noted by P_C.

$$P_C = U_M \cdot I_M. \quad (3.67)$$

Geometrically, critical power P_C meets the tangent points of hyperbolas $P = UI = const.$ for the ampere – volt *I-U* characteristics (see Fig. 3.73).

Fill Factor is determined as the report between the surfaces of rectangles OU_MMI_M and OU_0KI_{sc} (Fig. 3.73c) or

$$FF = \frac{U_M I_M}{U_0 I_{sc}}, \quad (3.68)$$

Where

$$P_C = FF \cdot U_0 \cdot I_{sc}. \quad (3.69)$$

Fill factor is the measure of the PV cell quality. The lower internal resistance R_s of PV cell the bigger *FF*. Usually $FF > 0.7$.

Efficiency of PV cell or module: is determined with the ratio between the generated power of the PV cell or module in the optimum operating point *M* for a specified temperature and the solar radiation power

$$\eta = \frac{P_C}{A \cdot G}, \tag{3.70}$$

where P_C is the delivered power in W; *A* is the surface of cell or module in m; *G* – global radiation incident on the cell or module surface in W/m^2.

A PV module with an efficiency of 12% and with the surface area of 1 m^2, exposed to solar radiation equal to 1,000 W/m^2 will produce approximately 120 W.

In laboratory conditions crystalline silicon cells have been obtained with an efficiency of 13–25% depending on the cell surface, and in factory conditions – 12% to 14%. Polycrystalline silicon cell efficiency is up to 20% in laboratory conditions. Theoretical limit of the crystalline silicon cell efficiency is 37% and that of amorphous silicon – 28% [60].

Normal operating temperature of the cell: corresponds to the temperature of PV cell in no-load operation conditions at the environment temperature of 20°C, global radiation of 800 W/m^2 and wind speed smaller that 1 m/s. For usual cells NOCT parameter is between 42°C and 46°C. If NOCT parameter is known, it is possible to determine the cell temperature T_C for other operating conditions characterized by the environment temperature T_A and global radiation *G* [59]:

$$T_C = T_A + \left(\frac{NOCT - 20}{0,8} \right) \cdot G. \tag{3.71}$$

The influence of solar radiation and temperature on the characteristics of PV cells and modules: PV cell characteristic for different solar radiation values are presented in Fig. 3.74a.

It is noted that the short circuit current is directly proportional to solar radiation and no-load voltage is varying a little, since according to Eq. 3.66, voltage U_0 depends logarithmically on solar radiation (I_s is proportional to the radiation) and often in practical calculations this variation is neglected. Short-circuit current, for different values of solar radiation *G*, can be determined with a satisfactory approximation of the formula:

$$I_{sc} = \frac{G}{G_{st}} \cdot I_{scst}, \tag{3.72}$$

where I_{scst} is the cell short circuit current corresponding to the standard radiation $G_{st} = 1{,}000$ W/m^2. PV cell temperature significantly influences the no-load voltage and less – short-circuit current (see Fig. 3.74b). With increasing temperature, no-load voltage decreases. For silicon cells the voltage variation coefficient of

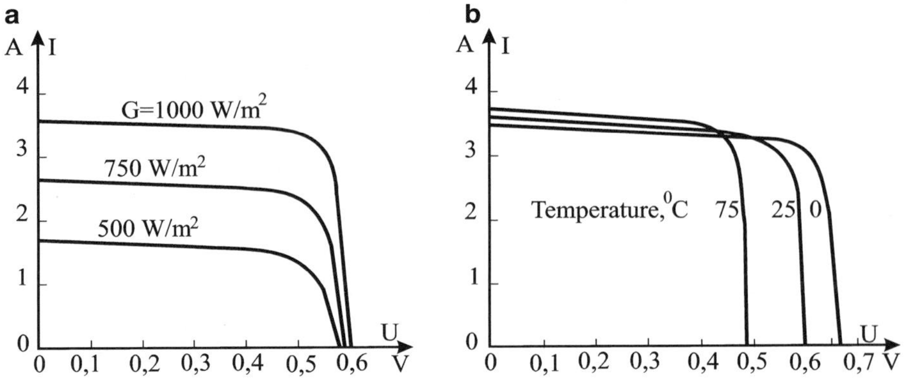

Fig. 3.74 PV cell characteristics at variation of solar radiation (**a**) and temperature (**b**)

temperature K_T is equal to 2.3 mV/°C. Thus, the parameter U_0 for the temperature different from the standard is calculated by following equation:

$$U_0 = U_{025} - 0.0023(t - 25) \tag{3.73}$$

where U_{025} is the PV cell short circuit current corresponding to the standard temperature; t – current temperature of the cell, °C. In the design calculations variation of short circuit current and of the filling factor FF depending on the temperature is neglected.

3.3.5.3 Photovoltaic Modules

Modern photovoltaic cells produce electrical energy which does not exceed 1.5–2 W power at voltages from 0.5 to 0.6 V. To obtain voltage and power necessary for the consumer, PV cells are connected in series and/or parallel. The smallest photoelectric installation, consisting of PV cells connected in series and/or parallel and encapsulated to obtain greater strength and protect cells from environmental action, is called photovoltaic module. A number of PV modules assembled mechanically as one larger unit and electrically connected is called panel or field of modules. In accordance with the standards of the International Electrotechnical Commission (IEC) the term "array" is used, which means system or network. The terms "photovoltaic module", "photovoltaic panel" or "field of modules" very often have the same meaning. When designing PV modules take into account the frequent use of PV modules for charging electric batteries, whose voltage is 12–12.5 V. Thus, in standard radiation conditions, voltage U_M must be 16–18 V, and no-load voltage – 20–22.5 V. A single cell generates in open circuit approximately 0.6 V and it is necessary to connect 33–36 cells in series to obtain the necessary voltage. The module power will vary between 50 and 100 W. PV module construction (Fig. 3.75a) is usually rectangular.

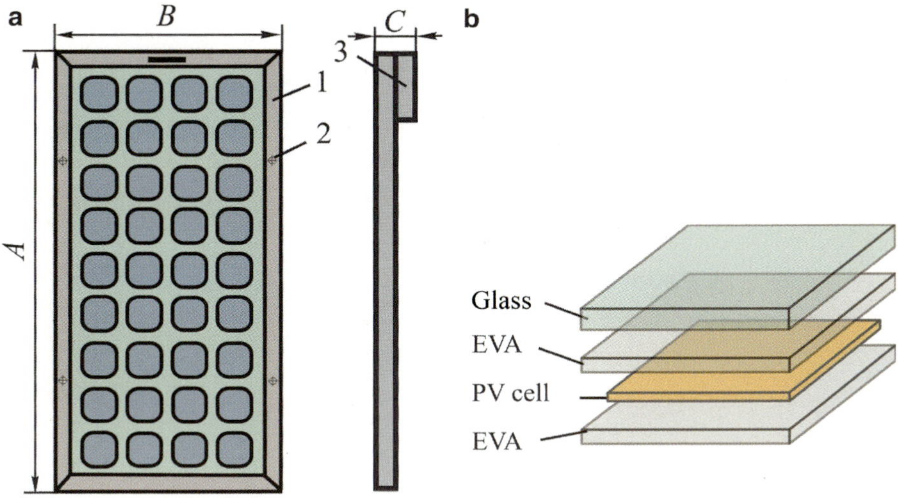

Fig. 3.75 Design of PV module (**a**) and PV cell package (**b**): *1* – support; *2* – panel assembly holes; *3* – terminal box

The support is made of anodized aluminium rod and separated from the cell laminated structure by lining that prevents moisture intrusion. PV cells are protected from unfavourable conditions impact that may occur during operation: rain, hail, snow, dust, etc.., by a system which consists of a layer of glass and at least two layers (front and rear) of ethylene vinyl acetate EVA or PVB polyvinyl butyral (Fig. 3.75b). Figure 3.76 shows PV modules of various powers, manufactured by Kyocera Company, and in Table 3.15 – their basic characteristics. To obtain the necessary voltage and power for the consumer of electrical energy, PV modules can be connected in series, parallel or in series–parallel (see Fig. 3.77a, b, c). At series connection of two identical PV modules the current delivered to the consumer remains the same, and voltage increases twice. In Fig. 3.79, PV1 and PV2 modules connected in series charge GB storage battery. Operating point of the system "PV modules-GB" is the point of intersection *M* of these characteristics: two modules connected in series and a storage battery. Diodes VD1 and VD2, called bypass diodes or bypass connects in parallel with each module or group of modules connected in parallel (see Fig. 3.77a). Bypass diode limits the reverse voltage, if a circuit module in a row is less efficient or is shaded and avoids thermal overstress. In normal operating mode diodes VD1 and VD2 do not consume energy.

Diode VD, called anti-return, is connected in series with the load. This diode prevents the situation when the PV module can become energy consumer, if the generated voltage will be less than the battery voltage. It is obvious that it introduces a voltage drop of about 0.5 V and, accordingly, loss of energy. Figure 3.77b shows a parallel connection of two identical modules. The voltage output remains the same and the current increases twice.

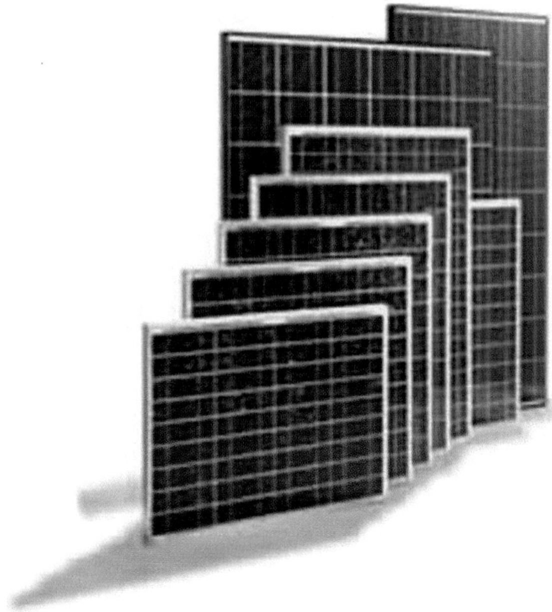

Fig. 3.76 PV module manufactured by Kyocera, Japan [61]

Table 3.15 Technical data of PV modules, manufactured by Kyocera Company [61]

Technical data	Type of module KC35	KC40	KC45	KC50	KC60	KC70	KC80	KC120
Maximum output (W_C)	35.0	40.0	45.0	50.0	60.0	70.0	80.0	120.0
Maximum voltage (V)	15.0	16.9	15.0	16.7	16.9	16.9	16.9	16.9
Maximum current (A)	2.33	2.34	3.00	3.00	3.55	4.14	4.73	7.10
Idle running voltage (V)	18.8	21.5	19.2	21.5	21.5	21.5	21.5	21.5
Short-circuit current (A)	2.50	2.48	3.10	3.10	3.73	4.35	4.97	7.45
Length (mm)	471	526	573	639	751	865	976	1,425
Width (mm)	652	652	652	652	652	652	652	652
Thickness (mm)	52	52	54	54	52	56	52	52
Mass (kg)	4.0	4.5	4.5	5.0	6.0	7.0	8.0	11.9

Operating point of the system "PV modules – resistance R" is the point of intersection M of the volt-ampere characteristics of the module and consumer – $I = (1/R) \cdot U$. Anti-return diodes VD11 and VD12 do not allow a module or a group of modules connected in parallel to pass under the consumer regime, when they are not identical or when they are shaded.

In Fig. 3.77c modules PV1-PV2, PV3-PV4 and PV5-PV6 are joined in series, but between them – in parallel. Thus, we obtain a double increase of the voltage and a triple increase of the current. Obviously, the unit power increases six times.

VD1-VD6 are bypass diodes and VD12, VD34, VD56 – are anti-return diodes. The parameters of a PV module are determined by the cell parameters.

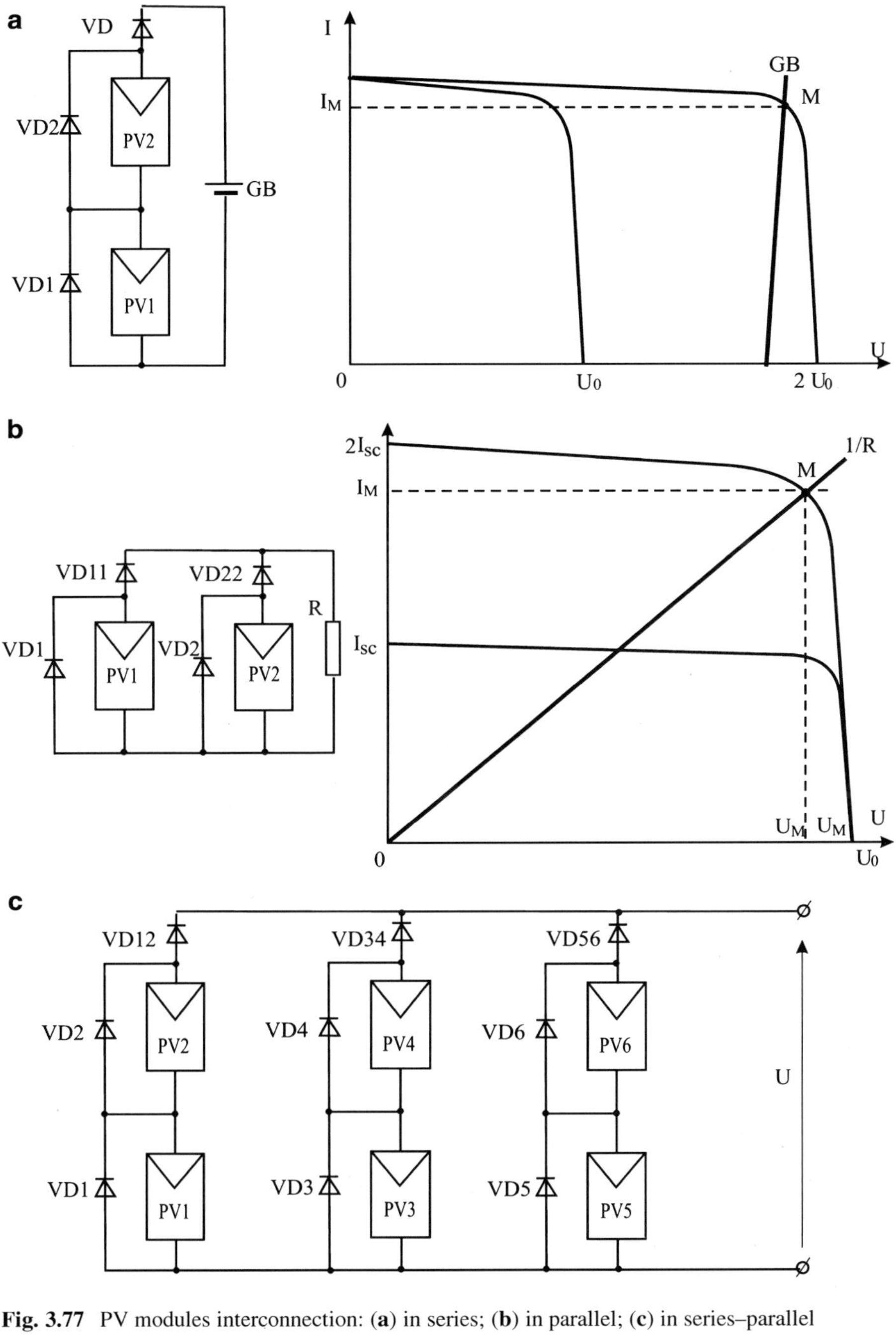

Fig. 3.77 PV modules interconnection: (**a**) in series; (**b**) in parallel; (**c**) in series–parallel

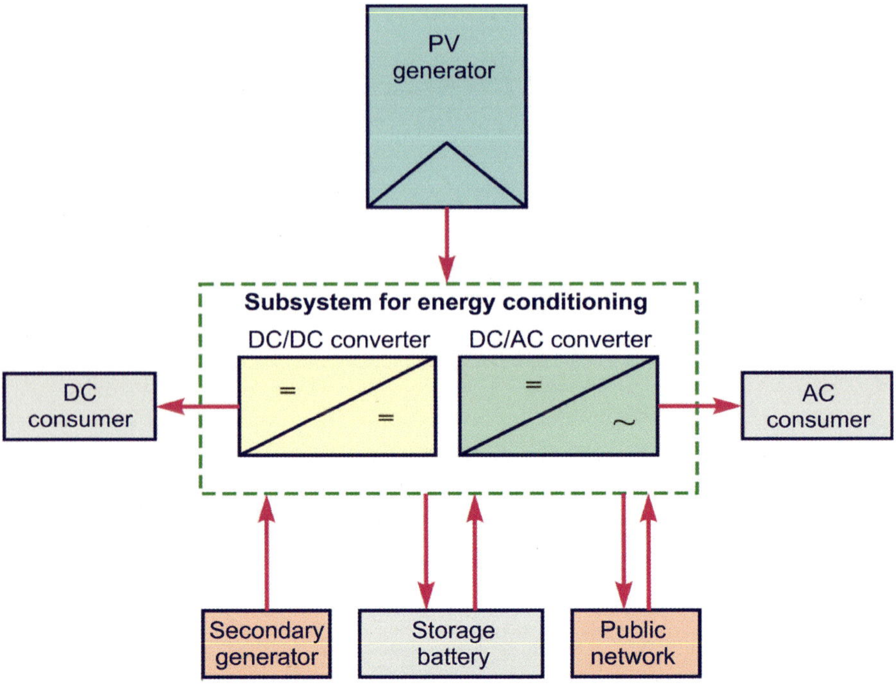

Fig. 3.78 Photovoltaic system structure

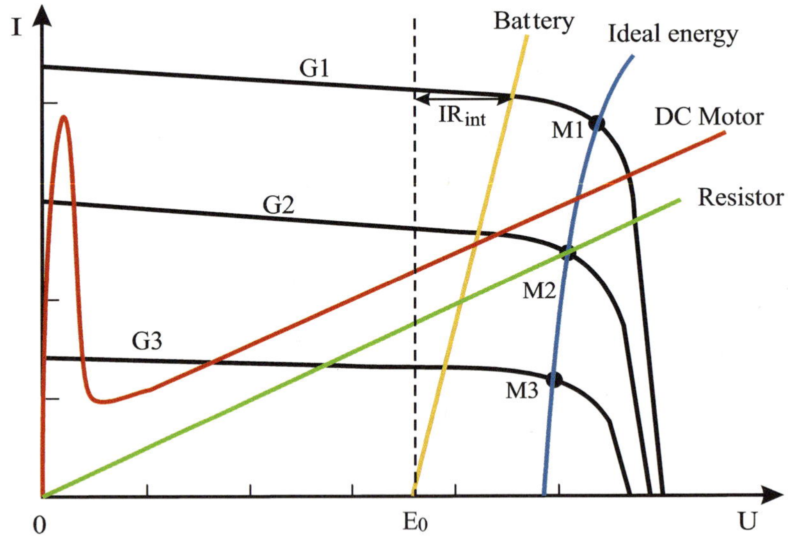

Fig. 3.79 I–V characteristics of three of the most widespread consumers

Further analysis will determine a numerical example for PV module parameters, operating in specified weather conditions

Numerical example: Determine the parameters of PV module consisting of 36 cells. The module operates under the following conditions: global radiation $G = 800$ W/m^2, environment temperature $T_a = 30$°C. Manufacturer of PV module guarantees the following parameters in the standard operating conditions:

- *Short-circuit current,* $I_{scst} = 3$ A;
- *No-load run voltage,* $U_{0st} = 20.5$ V;
- *Critical (maximal) power,* $P_{Cst} = 50$ W;
- *Cell normal operating temperature, NOCT* = 45°C.

Solution:

1. *Short-circuit current:* According to Eq. 3.72

$$I_{sc}(G) = (G/G_{st}) \cdot I_{scst} = (800/1000) \cdot 3 = 2.4 \text{ A}.$$

2. *Cell temperature*. According to Eq. 3.71:

$$T_C = T_A + \left(\frac{NOCT - 20}{0,8} \right) \cdot G = 30 + 25 = 55°\text{C}.$$

3. *No-load run voltage*. The following expression is applied Eq. 3.73:

$$\begin{aligned} U_0(55°C) &= U_{0st} - 0,0023 \cdot n_c \cdot (T_C - 25) \\ &= 20.5 - 0,0023 \cdot 36 \cdot 30 = 18.0 \text{ V}. \end{aligned}$$

4. *Fill factor.* According to Eq. 3.68:

$$FF = \frac{P_C}{U_{0st} I_{scst}} = \frac{50}{3 \cdot 20.5} = 0.81.$$

5. *Maximal power.* Determined if the FF factor does not depend on solar radiation and PV cell temperature:

$$P_C = FF \cdot U_0(55°C) \cdot I_{sc}(G) = 0,81 \cdot 18 \cdot 2,4 = 35 \text{ W}.$$

3.3.5.4 Photovoltaic Systems

The structure of a photovoltaic system: PV cells or modules are not the only components of a PV system. To provide continuously electricity to the consumer, PV systems include more power batteries. PV module has a DC generator (DC), but often the consumer is AC power. PV electricity has a variable character. The alternating day/night process and clear/overcast sky causes a wide variation in the energy flux and stress generated by the PV module. Thus, there is need for conditioning the power flux using electronic converters: DC/DC, which also has monitoring function of the charge/discharge process of the battery, DC/AC – to transform DC into AC. To avoid over sizing of the photovoltaic system, an auxiliary power source often is used: a generator or a wind generator or even public electricity network.

All these components must be interconnected, designed and specified to operate in a single system, the so-called photovoltaic system. Figure 3.78 shows the structure of a PV system. Its main components are:

- Module, panel, module field or otherwise said photovoltaic generator. PV generator characteristics and parameters have been analysed in Sect. 3.3.5.2;
- Battery charging system;
- Subsystem for electrical energy conditioning including measuring, monitoring, protection, etc. elements;
- Auxiliary energy source, for example, an electrogen group (*back-up generator*), that works on gasoline or diesel. In this case, the PV system is called hybrid photovoltaic systems.

PV systems are divided into two main categories: connected to the grid (*grid-connected*) or operating in parallel with the public electricity grid and autonomous PV systems (*stand – alone PV system*). The simplest system is the PV system for pumping water, the DC pump motor being used. This system does not contain electric batteries (water tank serves as a battery) or AC/DC converters.

Grid-connected PV systems can be divided into: PV systems, in which the public electricity network serves as a secondary source of energy (*grid back – up*); PV systems, in which the excess PV energy is supplied to the network (*grid interactive PV system*) and PV power stations (*multi-MW PV system*), which provides all the energy produced in the network.

PV module load operation: In Sect. 3.3.5.2 it was noted that PV cell, the PV module respectively, has the best performance at point M (see Fig. 3.73), and where the load delivered power is maximum. However, variation of overall radiation and temperature changes cause changes of PV module I-V characteristics.

Also, different consumers have different I-V characteristics. As a result, the operating point of PV module subsystem – load (the intersection of I-V characteristics of the module and load) will not coincide with the point M. In Fig. 3.79, I-V characteristics of three of the most widespread consumers are presented: resistor, DC motor with permanent magnet and a battery. The properties of an ideal consumer are presented for which the point of operation coincides with the optimal point M. I-V characteristics are described with the following analytical expressions:

- Resistor
$$I = \frac{1}{R}U; \tag{3.74}$$

- DC motor
$$I = \frac{U - E}{R_i} = \frac{U - k\Omega\Phi}{R_i}; \tag{3.75}$$

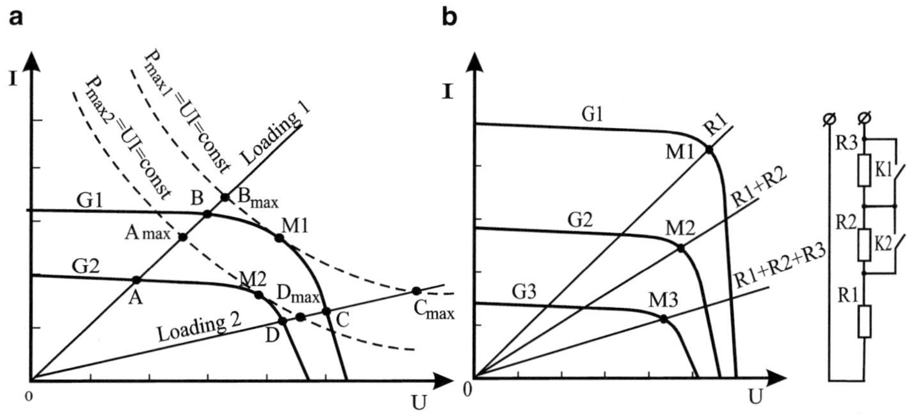

Fig. 3.80 Maximum output diagram: (**a**) employing MPPT technology; (**b**) by modifying load characteristics

- Battery

$$I = \frac{U - E_0}{R_{\text{int}}}, \tag{3.76}$$

- where U is PV module stress; k – motor constant; Φ – flow of excitation; Ω – rotational speed; R_i – armature resistance; E_0 –battery no-load voltage; R_{int} – internal resistance of battery.

When starting the DC engine, the input current drawn from the module is up and is close to the short circuit. Although the induced voltage is minimal its starting occurs due to the product created $k\Phi I_{sc}$. If $U = E_0$, the battery is charged and will not consume power, otherwise charging current will increase with increasing global radiation, and with the voltage respectively. Voltage drop IR_{int} increases with increasing load current.

Figure 3.79 shows that resistor type load or DC motor will not operate optimally in the optimum point of changing radiation. You will need to amend *I-V* characteristics of PV module or of the load to track the optimal operating point. To this end, they use DC/DC electronic converters called MPPT (Maximum Power Point Tracker).

MPPT connects between the PV module and the load to change the output voltage so as to ensure optimum operating point tracking. Figure 3.80 demonstrates two cases of the maximum point tracking – using MPPT technology (Fig. 3.80a) and by changing the load (Fig. 3.80b). In the first case, we have two loads with different I-V characteristics, which, for simplicity, linear allowed. For both tasks we find an essential deviation of operating points *A, B* and *D, C* from the optimal points *M1* and *M2*. Hyperbolas $I = P_{max.1}/U$ and $I = P_{max.2}/U$ are drawn in the same coordinates. At any point of mentioned hyperbolas, power $P_{max.1}$ or $P_{max.2}$ are constant sizes and respectively equal to the maximum rated power in point $M1$ or $M2$.

Table 3.16 Performance parameters of main types of invertors (rectifiers) [59]

Parameters	Rectangular voltage	Quasi-sinusoidal or unit step	Time impulse modulation
Rated output (kW)	Up to 1,000	Up to 2.5	Up to 20.0
Overload factor	Up to 20	Up to 4	Up to 2.5
Efficiency (%)	70–98	>90	>90
Harmonic (wave form) distortion (%)	Up to 40	>5	<5

It is considered that the subsystem PV module – work load 1 operates in point B in conditions of global radiation equal to $G1$. To get the maximum power from the module load I-V characteristics should be modified, so as to cross in point $M1$. The same result can be achieved if the voltage decreases and current increases in comparison with $M1$ point, moving on the hyperbola in B_{max} point, similarly if radiation reduces from $G1$ to $G2$. For task 2, to track the maximum point it is necessary to do the opposite – to increase the voltage and to decrease the current (C_{max} will be compared to C or D_{max} with D). Electronic converter MPPT should vary the voltage and current so that their product at outlet to be constant and equal to the maximum power generated by PV module exposed to global radiation G. In some specific cases, the maximum power point tracking can be achieved by changing the load *I-V* characteristics, as shown in Fig. 3.80b. For maximum solar radiation, equal to *G1*, the subsystem PV module-load $R1$ will operate at the point $M1$; in which case contacts $K1$ and $K2$ are closed. $K2$ contact opens at the average value of solar radiation equal to $G2$, the load *I-V* characteristics changes and the subsystem will operate at point $M2$. If solar radiation continues to decrease, contact $K1$ opens and the subsystem will operate at point $M3$. Subsystem *"PV module – battery"* does not require the use of MPPT technology, because if *I-V characteristic* moves to the right (Fig. 3.79) it will be close to that ideal. On the contrary, the battery requires automatic monitoring of the degree of loading and unloading to avoid its damage.

Designer's decision to use or not use MPPT technology will be taken as a result of economic estimation. The following should be taken into account: MPPT converter cost, energy losses in MPPT (modern DC/DC converters efficiency is 90–95%) and power gain under optimal MPPT operation subsystem. According to available data [60], the maximum power point tracking in pumping PV systems increases the rated output by at least 20%.

Inverter: Inverter is part of the power conditioning subsystem of PV system (see Fig. 3.78) and is the main component of the converter DC/AC. Inverter converts DC energy generated by PV modules or stored in the batteries into AC energy of predetermined frequency. Converters have been already designed which provide power quality parameters at the same level as public networks: steady frequency and voltage, sinusoidal form of the voltage wave and current. Depending on load requirements on the voltage waveform, the overload factor, efficiency, different types of inverters are available, which parameters are presented in Table 3.16.

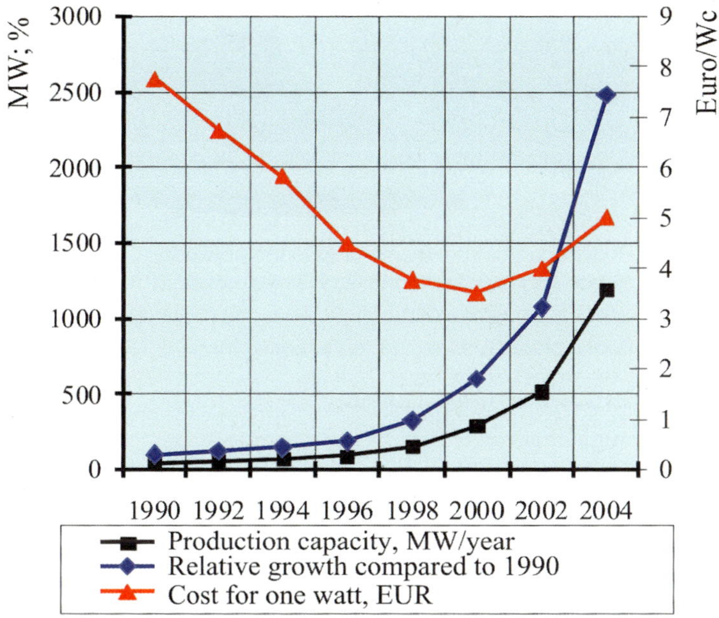

Fig. 3.81 Evolution of PV module world production

Indicated efficiency corresponds to inverter's operation at a load of 75–100% of rated output. When choosing the inverter it is important to know the characteristic of efficiency as a function of load. Motors require a starting current much higher than the nominal one. It is important that the inverter's overload factor meets this requirement.

Rectangular wave inverter has the simplest scheme, a relatively good efficiency, is the cheapest, but it causes the highest harmonic distortion, which causes engine overheating. This type of inverter is recommended for use in low power PV systems for lighting, heating at other than DC voltages, also in the composition of DC/DC converters, electromagnetic actuators. Quasi-sinusoidal wave inverter is more complicated, but relatively efficient. The pulse duration modulation technology is newer. Inverter control scheme is much more complicated, the cost of the inverter is higher, but also ensures high efficiency and minimum harmonic distortion.

3.3.5.5 Development Potential of Photovoltaic Systems

Figure 3.81 shows the evolution of global production of PV cells and modules and the cost of a watt in the period 1990–2004. After 1996 there is an extraordinary increase in worldwide production. In the years 2002–2004, growth compared with the previous year was equal to 28.2; 79 and, respectively, 60.5%. Over a period of 15 years, global production capacity of PV modules has increased about 25 times.

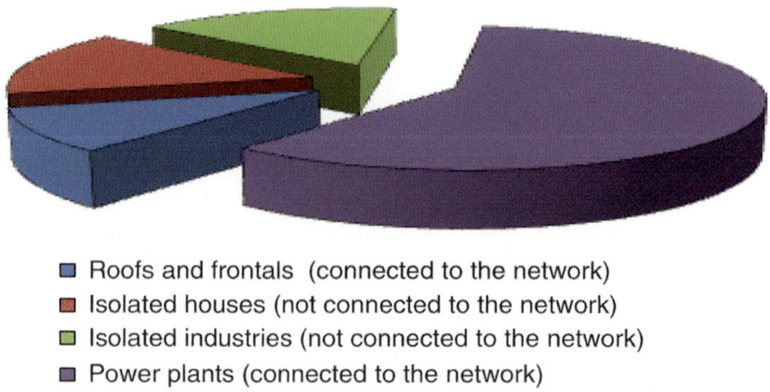

Fig. 3.82 Use of PV systems in developed countries

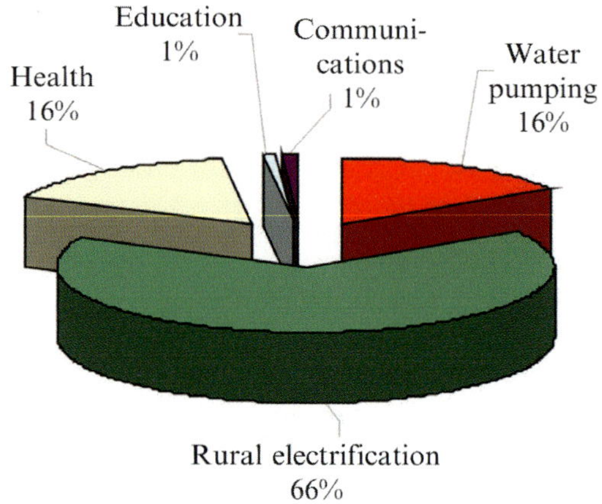

Fig. 3.83 Use of PV systems in under-developed countries

This trend will increase in the coming years as more oil companies (Shell, British Petroleum) will be involved in the new national programs. Along with the growth of production, photovoltaic cells become lower in cost. Over a period of 10 years the cost of a watt has decreased by 2.35 times. After 2000 there is an increase in the cost of a watt, which is explained by the formation of a gap between global production capacity of cells and assembly of PV modules and the production of raw material – pure silicon. Currently, worldwide there has been a shortage of raw material.

Areas of use in developed and developing countries are different (Figs. 3.82, 3.83). For example, in EU the PV systems are connected to the grid by 68%, in and developing countries there are three main areas of PV electrical energy use: rural electrification, health and water pumping. All these systems operate in autonomous mode, in other words, are decentralized and dispersed territorially.

Installed power worldwide was about 1,194 MWe in 2004, the first three places were occupied by Japan with 51.8%, followed by the EU with 25.8% and the U.S. with 11.5% (Fig. 3.84).

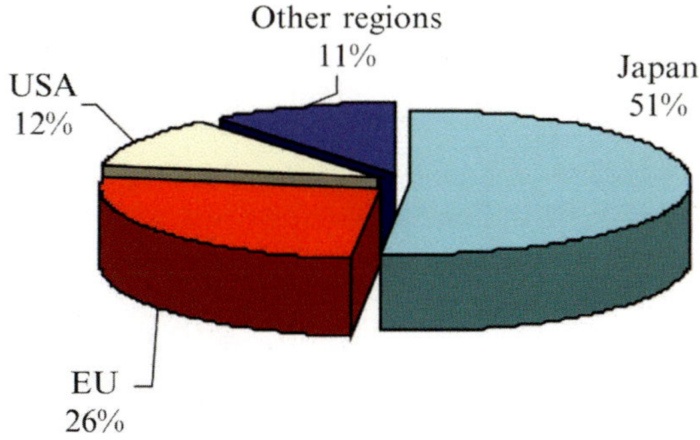

Fig. 3.84 PV installed capacity distibution in 2004 at world level

Table 3.17 Main national programs for promotion of PV energy and allocated subsidies

Country	National program	Acquisition price (Euro/kWh)	Subsidies (Euro/W_c)
Japan	70,000 solar roofs	0.3	1.13
USA	1,000,000 solar roofs	N/A	Differs by state
Germany	100,000 solar roofs	0.5	0.75–1.0
Great Britain	70,000 solar roofs	N/A	N/A 32 mln. Euro envisaged
Austria	–	0.72	Differs by region
Spain	–	0.47	5.5 autonomous systems 2.6 grid connected
Italy	10,000 solar roofs	0.35–0.40	70% of cost
France	–	0.15–0.3	4.6
Switzerland	–	0.3–0.6	1.3–2.7

N/A information not available

PV electrical energy is more expensive and in most cases do not compete with energy from fossil sources. Exceptions are pumping PV systems and systems for small and territorially dispersed power consumers, a case that can be compared with electrogen groups or with the extension of public power grids. Therefore, in various countries of the world, in national programs promoting PV energy, various methods have been developed to support and subsidize PV electrical energy (see Table 3.17).

The cheapest PV electrical energy can be obtained from the systems connected to public electricity networks, in other words, operating in parallel with the network. The system has the simplest configuration (modules field plus frequency converter), and there is no need for electric batteries. In 2005, photovoltaic electricity generation costs reached 50 cents/kWh in the Central Europe, reducing by 25 Eurocents/kWh in areas with high solar radiation. This cost can be compared to the cost of electricity produced from fossil sources, which in 2005 varied in the world between 0.04 and 0.05 $/kWh. Note that the cost also depends on the duration of solar cells life. C-Si solar cells have 25 years warranty, but can be used 35 or even more years. Electricity costs spent on PV cells production can be recovered in the first 2 years (Fig. 3.85) [63].

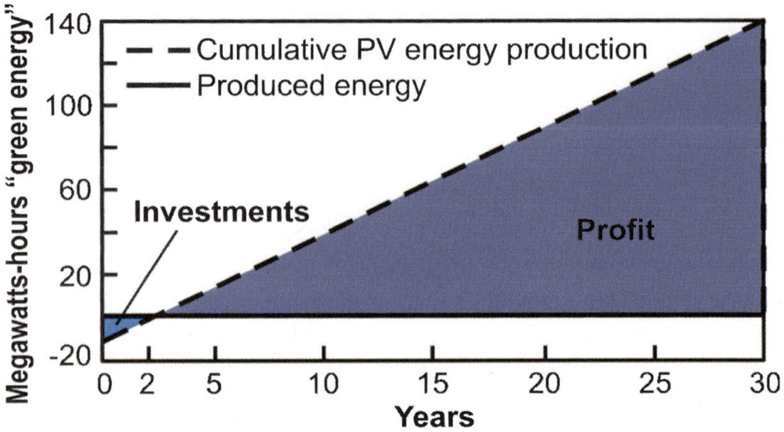

Fig. 3.85 Cumulation of PV energy production

Photovoltaic systems are evolving in terms of reducing manufacturing costs. Research is carried out on different semiconductor materials, and on PV cells manufacturing technologies. From this point, the evolution of the most efficient solar cells over the past 30 years (period 1975–2005) for different materials and technologies is of interest, according to NREL (National Renewable Energy Laboratory, USA) research. Areas where ice-photovoltaic cells find use on a more wide-scale are isolated consumers of electrical energy. Solar energy technologies use sunlight to produce heat, lighting, hot water and even air conditioning for residential and industrial areas. Solar panels are one of the most popular alternative energy sources used for private and industrial power systems.

According to the report of the International Energy Association (IEA) in 2004, it was an important year in terms of increase in the installed capacity of photovoltaic systems – from 770 MW, at the beginning of the year, to 2.6 GW by the end of 2004. The absolute majority of those capacities (94%) were installed in Germany, Japan and the U.S. (Table 3.18). Figure 3.86 presents the increase in photovoltaic electricity installed capacity since 1992 and the separation of this capacity between the two primary forms of using them.

Cumulative market in the countries with PV energy systems has achieved a growth rate of 42% between 2003 and 2004 and over 37% last year. A very large market for photovoltaic systems has been recorded in Germany (137%). Analysis of historical perspectives of the installation of photovoltaic systems in the countries with PV systems (Table 3.19) shows a spectacular development of PV systems in Germany (an increase of more than 140 times in 2004 compared to 1992) and Japan (over 60 times.). This tremendous growth continues to be developed for market support mechanisms, which initially was aimed at domestic applications in urban or suburban environments.

Table 3.18 PV installed capacity in AIE countries at the end of 2004

Country	Cumulative PV power (kW) Domestic	Cumulative PV power (kW) Non-domestic	Total PV installed capacity (kW)	Total installed capacity per capita (W/capita)	PV installed capacity in 2004 (kW)
Australia	15,900	29,640	**52,300**	2.60	6,670
Austria	2,687		**19,180**	2.37	2,347
Canada	5,291	8,081	**13,884**	0.44	2,054
Switzerland	2,810	290	**23,100**	3.12	2,100
Germany	26,000		**794,000**	9.62	363,000
Denmark	65	190	**2290**	0.43	400
Spain	14,000		**37,000**	0.87	10,000
France	12,500	5,800	**26,300**	0.44	5,228
Great Britain	193	585	**8,164**	0.14	2,261
Israel	653	210	**886**	0.13	353
Italy	5,300	6,700	**30,700**	0.55	4,700
Japan	1,136	83,109	**1,131,991**	8.87	272,368
Korea	461	4,898	**9,892**	0.21	3,454
Mexico	14,169	4,003	**18,182**	0.17	1,041
The Netherlands	4,769		**49,079**	3.01	3,162
Norway	6,438	375	**6,888**	1.50	273
Portugal	1,657	569	**2,643**	0.25	574
Sweden	3,070	602	**3,866**	0.43	285
USA	77,900	111,700	**365,200**	1.24	90,000
Total estimated	**170,730**	**281,021**	**2,595,545**		**770,270**

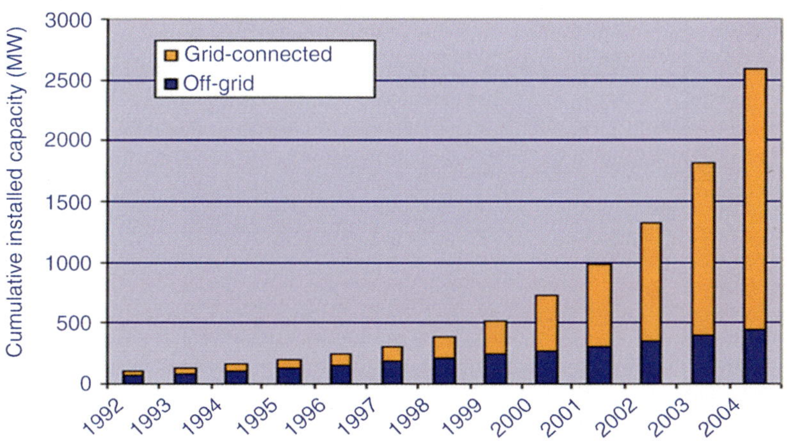

Fig. 3.86 PV power installed capacity, grid connected and separated in the countries from Table 3.18

Table 3.19 PV cumulative installed capacity in AIE countries

Country	1992	1993	1994	1995	1996	1997	1998	1999	2000	2001	2002	2003	2004
Argentina	7.3	8.9	10.7	12.7	15.7	18.7	22.5	25.3	29.2	33.6	39.1	45.6	52.3
Australia	0.6	0.8	1.1	1.4	1.7	2.2	2.9	3.7	4.9	6.1	10.3	16.8	19.2
Austria	1.0	1.2	1.5	1.9	2.6	3.4	4.5	5.8	7.2	8.8	10.0	11.8	13.9
Canada	4.7	5.8	6.7	7.5	8.4	9.7	11.5	13.4	15.3	17.6	19.5	21.0	23.1
Switzerland	0	0.1	0.1	0.1	0.2	0.4	0.5	1.1	1.5	1.5	1.6	1.9	2.3
Germany	5.6	8.9	12.4	17.7	27.8	41.8	53.8	69.4	113.7	194.6	278.0	431.0	794.0
Denmark	4.0	4.6	5.7	6.5	6.9	7.1	8.0	9.1	12.1	15.7	20.5	27.0	37.0
Spain	0.9	1.0	1.2	1.3	1.5	2.0	2.2	2.3	2.6	2.7	3.1	3.4	
France	1.8	2.1	2.4	2.9	4.4	6.1	7.6	9.1	11.3	13.9	17.2	21.1	26.3
Great Britain	0.2	0.3	0.3	0.4	0.4	0.6	0.7	1.1	1.9	2.7	4.1	5.9	8.2
Israel	0.1	0.1	0.2	0.2	0.2	0.3	0.3	0.4	0.4	0.5	0.5	0.5	0.8
Italy	8.5	12.1	14.1	15.8	16.0	16.7	17.7	18.5	19.0	20.0	22.0	26.0	30.7
Japan	19.0	24.3	31.2	43.4	59.6	91.3	133.4	208.6	330.2	452.8	636.8	859.6	1,132
Korea	1.5	1.6	1.7	1.8	2.1	2.5	3.0	3.5	4.0	4.8	5.4	6.4	9.9
Mexico	5.4	7.1	8.8	9.2	10.0	11.0	12.0	12.9	13.9	15.0	16.2	17.1	18.2
The Netherlands	1.3	1.6	2.0	2.4	3.3	4.0	6.5	9.2	12.8	20.5	26.3	45.9	49.1
Norway	3.8	4.1	4.4	4.7	4.9	5.2	5.4	5.7	6.0	6.2	6.4	6.6	6.9
Portugal	0.2	0.2	0.3	0.3	0.4	0.5	0.6	0.9	1.1	1.2	1.7	2.1	2.6
Sweden	0.8	1.0	1.3	1.6	1.8	2.1	2.4	2.6	2.8	3.0	3.3	3.6	3.9
USA	43.5	50.3	57.8	66.8	76.5	88.2	100.1	117.3	138.8	167.8	212.2	275.2	365.2
Total	**110**	**136**	**164**	**199**	**245**	**314**	**396**	**520**	**729**	**989**	**1,334**	**1,829**	**2,596**

3.3.5.6 Photovoltaic System Dimensioning

The general principle underlying the PV system dimensioning is to always respect the balance between the energy produced by PV generator and the consumed energy. This balance is achieved for a defined period, usually a day or a month. Presence of the storage (secondary or storage cells) battery allows compensation of the shortfall between the produced and consumed energy, which may be due to cloudy weather or overloading on behalf of the consumer.

Sizing a PV system requires the following basic steps:

1. Calculation of available solar radiation on the surface of PV module;
2. Daily electricity consumption calculation – E_c;
3. Calculating the amount of electrical energy needed to be produced by PV module – E_p;
4. Calculation of critical power of PV module – P_c and its choice;
5. Calculation of battery capacity – C and choice of batteries;
6. Checking electrical energy consumption and generation balance.

Figure 3.87 shows the procedure for sizing a PV system with storage battery.

Calculation of available solar radiation on the surface of PV module is made in accordance with the method described in Sects. 3.2.3.2 and 3.3.2.7. PV module inclination angle to the horizon β is determined from the condition ensuring the equilibrium *"consumption – electrical energy production"* in the months with the lowest solar radiation.

Daily electricity consumption calculation: To this end, for every consumer of DC and AC nominal power and hours of daily consumption are determined. Electrical energy consumption E_C is determined as the output of nominal power to the number of hours:

$$E_C = \sum_{i=1}^{k} \frac{P_{ni}^{cc} \cdot t_i}{\eta_R \cdot \eta_{Ac}} + \sum_{j=1}^{m} \frac{P_{nj}^{ca} \cdot t_j}{\eta_{CF}} \tag{3.77}$$

where k is the number of DC consumers; m is the number of AC consumers; P_{ni}, P_{nj} is the nominal power of Dc and AC consumers; t_i, t_j – operating period of the given consumers; η_R, η_{CF}, η_{Ac} – efficiency of battery charge – discharge controller and the frequency converter. For preliminary calculations $\eta_R = 0.95$–0.98, $\eta_{Ac} = 0.85$–0.90, $\eta_{CF} = 0.85$–0.95. Electrical equipment nominal powers are specified in the technical book. However, they can be made available to the designer by the company producing these machines. The values of operating times per day of the equipment are obtained from the beneficiary's stated needs or are determined from statistical data.

Some technical data of household consumers: [59, 64, 65]. Further on, some technical data on the most usual estimates of household electricity consumers follows: rated output, efficiency, lifetime, etc. (Tables 3.20, 3.21).

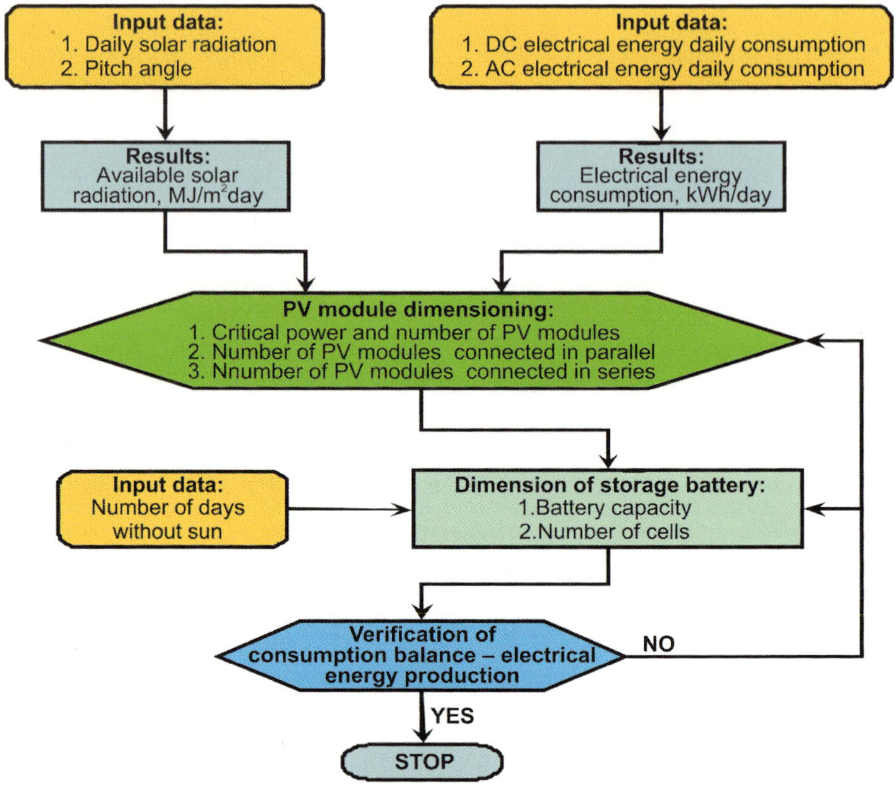

Fig. 3.87 Photovoltaic system dimensioning

Table 3.20 Lighting source

Lighting source	Power (W)	Efficiency (lm/W)	Lifetime (h)
Electric bulb	25	9.0	2,500
Electric bulb	40	9.0	1,000
Electric bulb	75	13.0	1,000
Electric bulb	100	16.0	1,000
Electric bulb (crystal, quartz)	50	19.0	2,000
Compact fluorescent lamp (CFL)	4 8 13 18	45.0	6,000–10,000
Fluorescent (luminous) lamp T-8	n/d	75–100	12,000–24,000
Halide lamp	n/d	80–115	10,000–20,000
Low pressure sodium lamp	35	128.0	5,000
LED surface	3.6	130.0	>100,000
High pressure sodium lamp	n/d	90–140	10,000–24,000

Table 3.21 Estimative values of functioning period of electrical energy consumers for an isolated house

		h/day		
		Months		
Consumer	Rated output (capacity) (W)	XII,I,II	III,IV,V,IX,X,XI	VI,VII,VIII
Lighting kitchen	2 × 13 LFC	4.0	3.5	2.0
Lighting bedroom	3 × 9 LFC	1.0	1.0	1.0
Lighting living room	2 × 20 LFC	1.0	1.0	1.0
Lighting bathroom	1 × 18 LFC	1.0	1.0	1.0
Vacuum cleaner	1,200	0.5	0.5	0.5
refrigerator	100	7.0	7.0	7.5
Colour TV, 54 cm	60	4.0	4.0	4.0
Stereo	60	2.0	2.0	2.0
Microwave oven	600	0.5	0.5	0.5
Water pump	200	1.0	1.0	1.0

Calculating the amount of electrical energy needed to be produced by PV module: Electrical energy to be produced by PV module:

$$E_P = \frac{E_C}{K} \tag{3.78}$$

Where K factor takes into account the uncertainty of weather data, the losses in cables, the deviation of functioning point of the subsystem "*PV module – load*" from the optimum one, etc. According to [64], the value of K factor for PV systems with storage batteries is between 0.75 and 0.85.

Critical power of PV module is determined from the relation:

$$P_C = \frac{E_P}{G_\beta} = \frac{E_C}{K \cdot G_\beta} \tag{3.79}$$

where G_β shows the average global solar radiation during the period of interest in this locality for the angle of inclination β of the PV module. In the formula (3.79) G_β is equal numerically to the number of hours per day of standard solar radiation, that is equal to 1,000 W/m^2 and is noted by *HRS*.

Depending on power P_C, the power of PV module and the number of modules connected in series are selected:

$$N_S = \frac{U_{cc}}{U_m} \tag{3.80}$$

where U_{cc} is the nominal stress of AC consumers; U_m – nominal stress of a PV module that equals 12 V.

The number of PV modules connected in parallel is defined in the following way: the average current of load per day is estimated:

$$I_{med} = \frac{E_P}{24U_{cc}} \tag{3.81}$$

At the same time, respecting the condition of energy balance it is possible to write:

$$24 \cdot I_{med} \cdot U_{cc} = HRS \cdot I_{PV} \cdot U_{cc} \text{ or } I_{PV} = \frac{24 I_{med}}{HRS} \quad (3.82)$$

where I_{PV} is PV panel current.

The number of PV modules connected in parallel will be:

$$N_P = \frac{I_{PV}}{I_{sc}} \quad (3.83)$$

where I_{sc} is the short circuit current of a PV module that is considered almost equal to the current in point M (see Fig. 3.86).

Calculation of battery capacity: It is determined from the relation:

$$C = \frac{n \cdot E_C}{K_D \cdot U_{cc}} \quad (3.84)$$

where n is the number of days without sun; K_D – battery discharge rate (0.5–0.6 for Pb–acid and 1.0 for Ni–Cd).

The number of batteries connected in series:

$$N_{As} = \frac{U_{cc}}{U_A} \quad (3.85)$$

Where U_A is the battery nominal stress, usually equal to 12 V.

Checking the balance of electrical energy consumption and generation: Verification is done by comparing the amount of electricity, E_i, that will be produced by the PV panel in a day for each month of the period of interest to the amount of electricity needed calculated as 2.77. Calculations made from the relationship:

$$E_i = HRS_i \cdot P_C \quad (3.86)$$

where HRS_i is the number of hours per day of standard solar radiation equal to 1,000 W/m^2 for the given month.

3.3.5.7 Fields of Use of Photovoltaic Electrical Energy

Solar energy is the most accessible of renewable energies. Anyone can use this free energy. Photovoltaic systems are often used in combination with other energy sources: wind, hydro or energy produced from burning fossil fuels.

A very important area for the use of photovoltaic systems is cosmic flight devices. Solar photovoltaic systems are practically the only source of electrical energy for

Fig. 3.88 PV systems installed on the first artificial satellite *Sputnik 3*

these devices. Soviet satellite *Sputnik 3*, launched on 15 May 1957, was the first satellite, which used photovoltaic systems (Fig. 3.88).

The first U.S. satellite, which used solar panels, was the U.S. Army Forces satellite "*Explorer* 1", launched on 1 January 1958. Dr. Hans Ziegler, who worked for 30 years (1947–1976) in the field, may be called the father of solar energy systems for spacecrafts [66]. Since the device should have small dimensions, this fact limited the power that could be produced [67]. The requirements to solar panels were very rigid. Solar panels must include a driving mechanism for orientation to the sun (at the stage of electrical energy production) and removal from the sun when electrical energy needs were lower than the production capacity. Gallium arsenide was used as solar cells. Efficiency of gallium and silicon arsenide multi-junction cells reaches almost 29% in ideal conditions [68]. Such solar cells were also used in solar panels installed on "*Stardust*" spacecraft (Fig. 3.89), launched on 7 February 1999 to investigate the comet Wild 2 and returned on 15 January 2006. *Stardust* was the first cosmic mission returned to Earth after collecting a sample of cosmic dust.

European Space Agency (ESA) has developed a perspective project with real chances of its fulfilment, which investigates the possibility of launching satellites – solar power plants that will generate electrical energy in the space, transported afterwards to the earth by laser or microwaves. These outstanding achievements have become a cosmic challenge for the governments in the industrialized countries (Japan, Germany, U.S. and others) to bring solar cells on Earth. Terrestrial photovoltaic

Fig. 3.89 Solar panels installed on the cosmic spacecraft Stardust (NASA image)

systems have been developed in two directions: using simple photovoltaic cells (without concentrating sunlight) and systems with focus sunlight on PV cells (CPV).

Today, there are two companies in the world that deal with developing concentrating photovoltaic systems (CPV): Amonix, in Torrance, California, USA and Solar Systems Pty Ltd, in Australia. Figure 3.90 shows a photovoltaic system, developed by Amonix Company, rotating around two axes (in zenithal and azimuthal planes), producing 25–35 kW on a mobile system (depending on the number of PV modules installed on the system). The Amonix system uses thousands acrylic Fresnel lenses that concentrate sunlight on solar PV cells. Annual Efficiency of Amonix module is 15.5%. Amonix systems were used in the public services with a total capacity of 547 kW in Arizona. 10–20 MW systems are scheduled for use in the near future in Spain. *Solar Systems Pty Ltd* uses paraboloidal concentrators to concentrate sunlight on the PV receiver. Figure 3.91 presents such a system, which achieves an efficiency of 15–16% and has a capacity of 220 kW, with the possibility of increasing the load up to 720 kW. More MW capacity contracts are coming in the near future. A new generation of approx. 50 MW of CPV systems will be produced in California, USA.

Solar panels with concentration of sunlight are widespread, especially in cases of isolated energy consumers. Figure 3.92 shows the use of a photovoltaic system on a yacht for charging 12 V and 9 A batteries. The solar panels have found wide use in

Fig. 3.90 Amonix system with concentrator with acrylic lenses and PV silicon cells (Source: NREL)

Fig. 3.91 PV paraboloidal concentrator: Solar Systems Pty LTD (Source: NREL)

Fig. 3.92 Photovoltaic system installed on a yacht to charge 12 V and 9 A cells

meeting domestic energy needs. Figure 3.93 shows a variant of photovoltaic systems connected to the electrical grid of a house, and Fig. 3.94 portrays an example of installing solar panels on the roof of a house.

At present decentralized rural electrification is accelerating fast by using solar energy, in particular, in the developing countries (LDCs). Although PV technology is considered expensive, the growth rate of world production of PV cells in 2000 constituted 60.5% (see Sect. 3.5.5). Why developing countries do not follow the same way passed by the European countries in electrification? The reasons limiting the application of traditional solution – thermal conversion of fossil fuel into electricity and its distribution via networks – are as follows:

1. Majority of rural consumers of electrical energy are located far away from the existing electrical networks and their density is low;
2. Electricity demand is low and falls within the range of 10–20 kWh/day;
3. Usually electrical energy is used for lighting, telecommunications and electric drive of small motorized equipment: electric pumps, small power workshop tools etc.

In these circumstances, traditional electrification is an inefficient solution in terms of both economic and environmental points: increase of investments and operating costs, increase of specific GHG emissions per capita calculated, as the efficiency of transport and distribution of 1 kWh of electrical energy decreases.

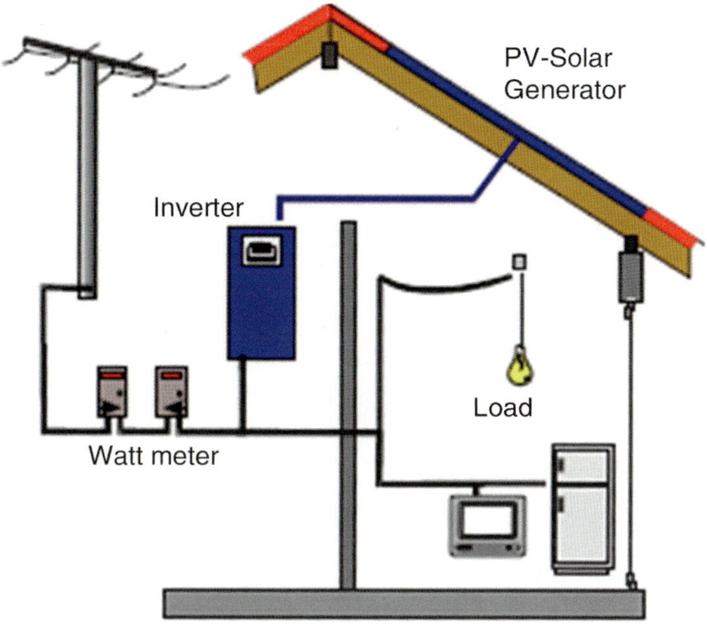

Fig. 3.93 PV grid connection in a living house

Fig. 3.94 Solar panels on the roof of a residential house

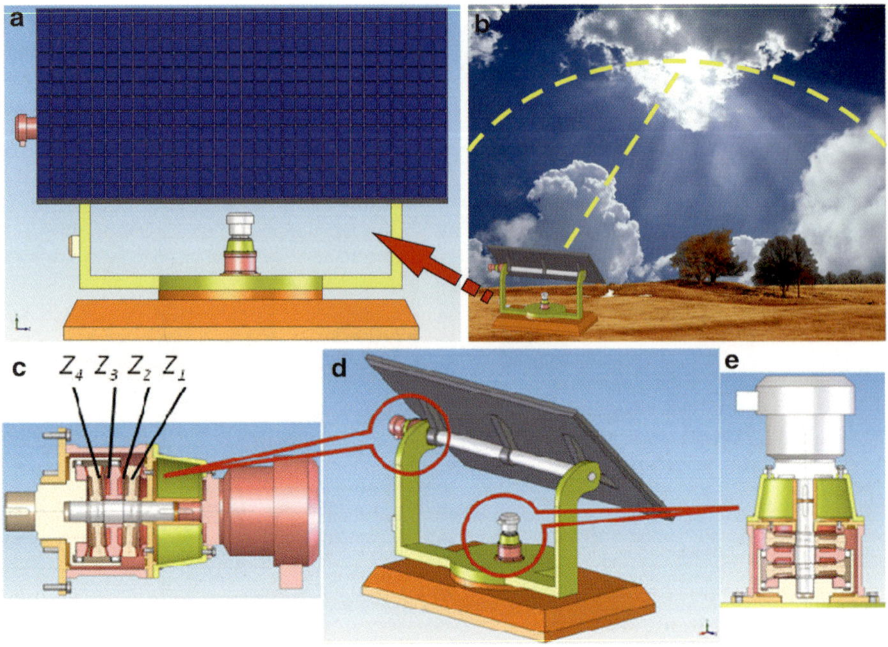

Fig. 3.95 Photovoltaic solar system orientation unit: author's elaboration

Photovoltaic panels orientation systems: It is known that the efficiency of a photovoltaic system depends largely on the orientation of the solar panels toward the sun standing both in azimuth and zenith. For this purpose they use different driving mechanisms. Because the motions of rotation around the two axes must be very slow (in the azimuthal plane the device must revolve from 10 to 16½ h), the gear ratio of these mechanisms (if done on the basis of gears) must be very high. Planetary gears (most common for such cases) should include 3–4 steps, which considerably increases the number of elements and hence their cost. In order to simplify construction and reduce costs of these orientation mechanisms, the research team of the Department of Theory of Mechanisms and Machine Parts, at the Technical University of Moldova, designed and patented the self-orientation solar unit conceptual diagram [69]. The photovoltaic panel is oriented to the sunlight, being rotated in azimuthal and zenithal planes by driving mechanisms with precessional transmission (Fig. 3.95). Very broad cinematic options of precessional transmissions ensure very slow rotation around the two axes (one revolution in 24 h). In the developed system, slow motions of rotation, necessary for the revolution of the solar panel in azimuthal and zenithal planes, are performed by precessional planetary transmissions in two steps: for the rotation of the solar system in the zenithal plane (Fig. 3.95c) and in the azimuthal one (Fig. 3.95e), subject to the motion of the sun in the sky (actually the earth rotating around the sun). Precessional transmission gear ratio in one step is determined from the relation:

$$i = \frac{Z_2 Z_4}{Z_1 Z_3 - Z_2 Z_4}, \tag{3.87}$$

where Z_1, Z_2, Z_3, Z_4 is the number of crown wheel serrated teeth (see Fig. 3.95c). A simple calculation shows that, to ensure rotation of the solar system in azimuthal plane within 24 h with a 1,500 min^{-1} servomotor, reducer gear ratio must be equal to $i = 2{,}160{,}000$. To achieve this transmission ratio, it is recommended to choose the following gear ratios of the precessional gear reducer: for the first step less required $i = -2{,}115$ ($Z_1 = 46$; $Z_2 = 47$; $Z_3 = 46$; $Z_4 = 45$); for the second step $i = -1{,}023$ ($Z_1 = 32$; $Z_2 = 33$; $Z_3 = 32$; $Z_4 = 31$). To achieve this transmission ratio, a 5-speed planetary gear would be required, which includes at least 25 gear wheels; while the precessional gear reducer in two steps includes only four gear wheels and two satellite wheels with two toothed crowns. Relatively simple construction of the driving mechanism (to obtain a high transmission ratio, dictated by the necessity of very slow rotation, using a small number of elements), and high reliability ensures relatively low cost of the solar photovoltaic installation and long period of operation.

Based on the research, the authors have developed, designed and manufactured four types of photovoltaic systems for use in various fields. Figure 3.96*a, c,* shows a photovoltaic system endowed with four modules of solar cells with summary power of 700 W, installed in a common housing mounted on the support by the sun orientation mechanisms 4 and 5 (Fig. 3.96d). Figure 3.96*b* shows another option of assembling a photovoltaic system with two modules of solar cells. Orientation of photovoltaic panels to the Sun is done by two axes – the horizontal one with the angular extension of 95° through a solar motor Sun Tracer Pump 4 operated by the Time derived Astronomical Positioning System (TdAPS). Photovoltaic panels are oriented on the azimuth with angular extension of 75° through the drive mechanism 5 with 2 K-H type precessional gear with gear ratio $i = 144$, driven by a solar motor Sun Tracer + electric motor step by step. In this case the rotational motion of the drive shaft reduced by the gear ratio of planetary precessional transmission is forwarded to body on which the housing with solar panels is assembled (Fig. 3.96).

In order to simplify construction and reduce costs of these orientation mechanisms to the sun, the authors propose a tracking mechanism based on the use of shape memory material properties (Fig. 3.97 [62]). The orientation of paraboloidal solar unit includes a dead abutment 1, on which the rotating part 2 is installed with the paraboloidal concentrator 3. Receiver 4, the reflector 5 (mirror) and the electric generator (power unit) 6 are also linked to the rotating part 2. Disk 7 is fixed rigidly on the rotating part 2; on disk 7 whose area is equal to 180° curvilinear elements 8 are fixed, made of shape memory material (for example, nitinol alloy – NiTi), which were printed in the form of rectilinear memory. The free ends of the curvilinear elements 8 contact consecutively with asymmetrical teeth 9, designed on the front part of wheel 10, rigidly fixed to a support 1. Inside the rotating part 2 is located rod 11 with the bolt (pin, finger) 12, one end of which is located in the groove 13, designed on the area of 180° of the cylindrical inner surface of the rotating part 2, so that half of it is made with ascending angle and the second

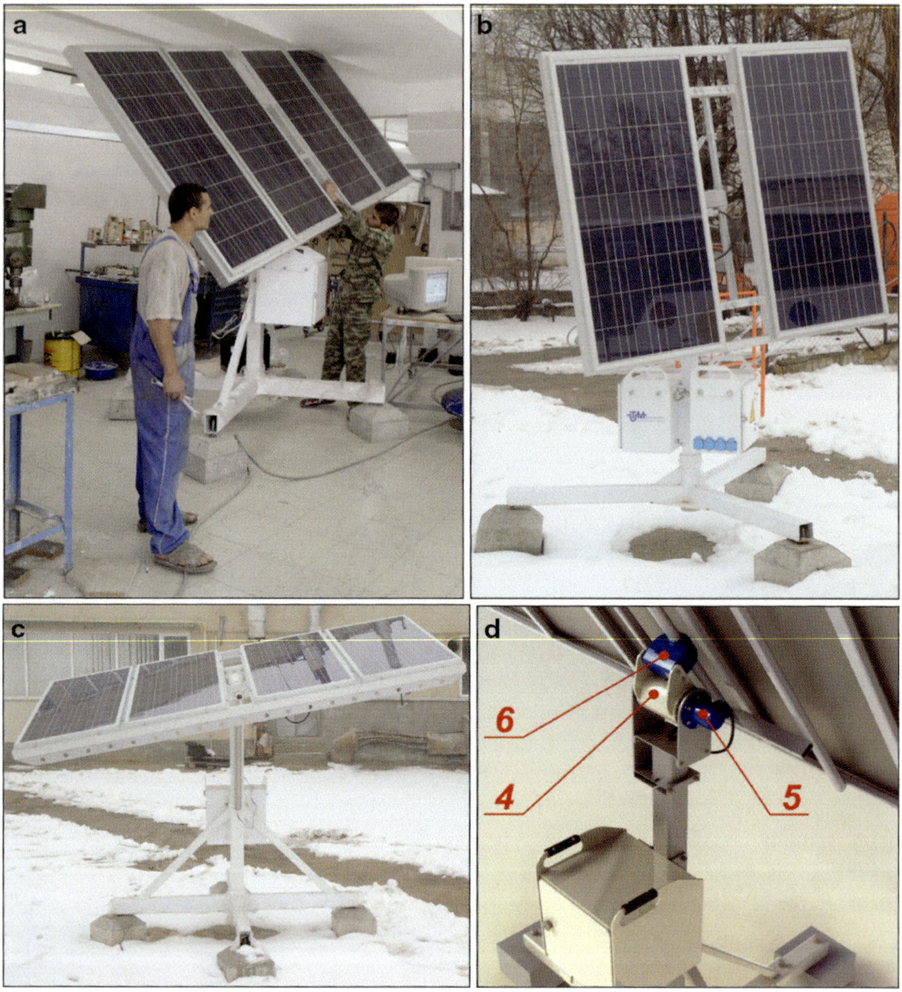

Fig. 3.96 Photovoltaic system with solar astronomical orientation, assembled at the Centre for Renewable Energy Development, TUM

half – with the descending angle. The rod 11 is connected via levers 14 with the paraboloidal concentrator 3 housing Inside support 1 an electric motor 15 and a reducer 16 are placed, which shaft is connected to the rotating part 2 and the photocell 17. During the day solar unit occupies three extreme positions: in the morning, at sunrise, the parabolic concentrator is in the position when the angle of rotation (azimuthal) $\varphi a = 0°$and the zenithal angle $Qz = 0°$; at noon, when the azimuthal angle $\varphi\text{a} = 90°$, and the zenithal angle is maximum Qzmax; at sunset, when the position of the azimuthal angle $\varphi a \approx 180°$ (do not take into account deviations of about $\pm 20°$ related to the season: in summer $\varphi a = 180° + 20°$, in

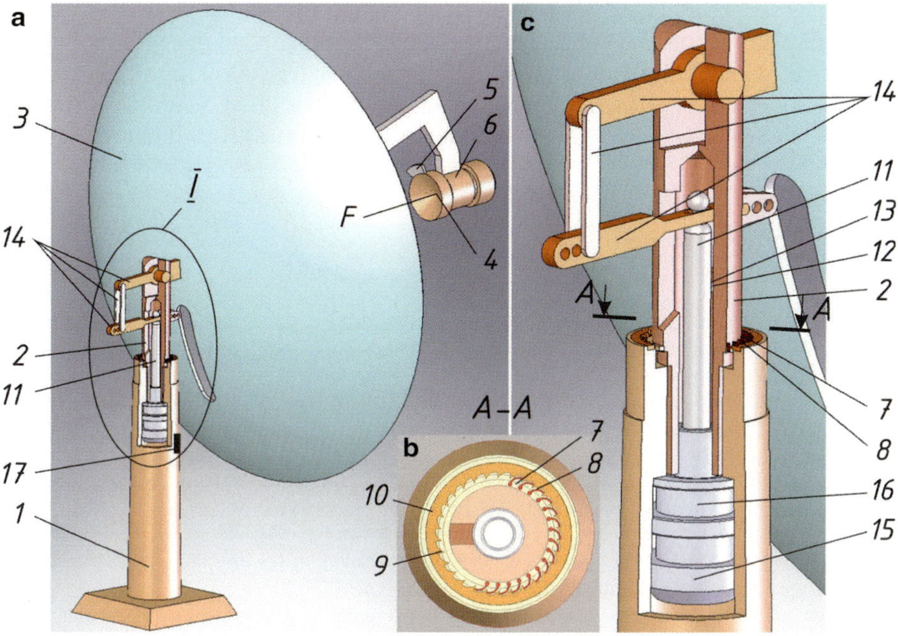

Fig. 3.97 Parabolic solar system orientation unit: author's elaboration

winter $\varphi a = 180°–20°$) and the zenithal angle is equal to $Qz = 0°$, as well. To ensure the rotation of the paraboloidal concentrator in azimuth and zenith planes, the orientation system includes curvilinear elements with the shape memory 9. Number of curvilinear elements with shape memory 9 is chosen depending on the degree of discretion allowed for rotating motion in the azimuthal plane $\varphi ai = 180o/i$, where i is the number of curvilinear elements with shape memory 9. Angle φai depends on the distance crossed by the focal point within the receiver's 4 length (when the rotating part is stationary). When the focal point goes outside the limits placed on the length of the receiver and mounts on reflector 5, the reflected concentrated rays are oriented towards curvilinear elements 8, warming them up to temperature of intercrystalline changes (for nitinol is equal 149°C) – a return to the memorized shape (rectilinear). The new rectilinear form, resumed by elements 8, leads to the rotation of the rotating part 2 and, also, of paraboloidal concentrator with angle φai. To ensure placement of focal point F in the working space of reflector 5 and the zenithal plane, to the rotation of the rotating part 2, the bolt (finger) 12 moves in groove 13, ensuring lifting (or lowering) of the rod 11 that, through levers 14, rotate the paraboloidal concentrator in the zenithal plane. Bringing the solar paraboloidal unit in the original position (the position before sunrise) is performed by the electric motor 15 and reducer 16, at the signal of the photocell 17. At the same time, the shape memory elements 8 are deformed (to) the initial curvilinear shape. To take into account the factor of season (winter, spring – autumn, summer), when the sun zenithal angles are different, system of levers is made adjustable. The orientation system of the designed paraboloidal solar plant provides orientation to the sun with a minimum of electricity (solar plant is self-controlling, without consuming electricity), increasing the efficiency of solar plant. Simple construction and low cost of the designed orientation mechanism ensures cost-value efficiency.

References

1. Stöcker H (1999) Toute la physique. Dunod, Paris,1180 p
2. Messenger R, Ventre J (2004) Photovoltaic systems engineering, 2nd edn. CRC Press LLC, Boca Raton, 455 p
3. http://en.wikipedia.org/wiki/Solar_radiation. Accessed 26 June 2005
4. Duffie JA, Beckman WA (1991) Solar engineering of thermal processes, 2nd edn. Wiley Interscience, New York
5. Thekaekara MP (1974) Data on incident solar energy. Supplement to the Proceedings of the 20th annual meeting of the Institute for Environmental Science. Pages 21–49, Washington, D.C., April 30
6. Iqbal M (1983) An introduction to solar radiation. Academic, Toronto
7. www.ipcc.ch/pub/reports.htm. Accessed 20 May 2005
8. Experience, prospects and recommendations to overcome market barriers of parabolic trough collector power plant technology status. Report on solar trough power plants. Sponsored by the German Federal Minister for Education, Science, Research and Technology under Contract No. 0329660. ISBN 3-9804901-0-6. Copyright 1996. Pilkington Solar International GmbH
9. Bougard J (1995) Conversion d'energie. Machines solaires. Faculte Politechnique de Mons, AGADIR
10. Hinrichs RA, Kleinbach M (2002) Energy: its use and environment, 3rd edn. Thomson Learning, Brooks/Cole, 590 p
11. Boyle G (2004) Renewable energy: power for a sustainable future. Oxford University Press, Oxford, 452 p
12. Twejdell JU (1990) Renewable energy sources (Trans from English). Energoatomizdat, Moscow, 392 c
13. Dumitraşcu Gh, Macri V, Stadoleanu O (1998) Solar energy use. Timpul, Iaşi
14. Lorenzo E (2002) De Los Archivos Históricos De La Energía Solar. Las chimeneas solares: De una propuesta española en 1903 a la Central de Manzanares (pdf)
15. Günther H (1931) In hundert Jahren – Die künftige Energieversorgung der Welt. Kosmos, Gesellschaft der Naturfreunde, Franckh'sche Verlagshandlung, Stuttgart
16. Lucier RE (1978) Utilization of solar energy. Patent no. 1,023,564 CA. Int.Cl. F03G6/04; F03G6/00
17. Haaf W, Friedrich K, Mayr G, Schlaich J (1983) Solar chimneys. Part 1: principle and construction of the pilot plant in Manzanares. Int J Sol Energ 2(1):3–20
18. Haaf W (1984) Solar chimneys. Part II: preliminary test results from the Manzanares pilot plant. Int J Sol Energ 2(2):141–161
19. Schlaich J, Schiel W (2001) Solar chimneys. Encyclopedia of physical science and technology, 3rd edn. Academic, London. ISBN 0-12-227410-5
20. Schlaich J, Bergermann R, Schiel W, Weinrebe G. Design of commercial solar updraft tower systems. Utilization of solar induced convective flows for power generation. Schlaich Bergermann und Partner (sbp gmbh). Stutgart, Germany. http://www.1000friendsofflorida.org/solar/thesolarupdraft.pdf. Retrieved 23 Mar 2011
21. Torre solar de 750 metros de altura en Ciudad Real (España). Green energy plan to use smaller solar tower, ABC, 2006-06-13

22. Schlaich J (1995) The solar chimney. Edition Axel Menges, Stuttgart
23. Schlaich J, Schiel W, Friedrich K, Schwarz G, Wehowsky P, Meinecke W, Kiera M (1990) Abschlußbericht Aufwindkraftwerk, Übertragbarkeit der Ergebnisse von Manzanares auf größere Anlagen. MFTFörderkennzeichen 0324249D, Stuttgart
24. Gannon AJ, Backström TW (2000) Solar chimney cycle analysis with system loss and solar collector. J Sol Energ Eng 122(3):133–137
25. El-Haroun AA (2002) The effect of wind speed at the top of the tower on the performance and energy generated from thermosyphon solar turbine. Int J Sol Energ 22(1):9–18. doi:10.1080/01425910210000003336
26. Weinrebe G (2000) Solar chimney simulation. In: Proceedings of the IEA SolarPACES task III simulation of solar thermal power systems workshop. Cologne, 28, 29 Sept 2000
27. Dos Santos Bernardes MA, Voß A, Weinrebe G (2003) Thermal and technical analyses of solar chimneys. Sol Energ 75:511–524
28. Pretorius JP, Kröger DG (2006) Critical evaluation of solar chimney power plant performance. Sol Energ 80(5):535–544. doi:10.1016/j.solener, 2005.04.001 DOI:dx.doi.org
29. Fickling D (2002) Real power from nothing but hot air. The Guardian, 19 Aug 2002
30. China invests in solar towers. Asia Times, 7 Oct 2004. Retrieved 9 July 2006
31. Padki MM, Sherif SA (1999) On a simple analytical model for solar chimneys. Int J Energ Res 23:289–294.
32. Schlaich J, Bergermann R, Schiel W, Weinrebe G (2005) Design of commercial solar updraft tower systems–utilization of solar induced convective flows for power generation (PDF). J Sol Energ Eng 127(1):117–124
33. Bilgen E, Rheault J (2005) Solar chimney power plants for high latitudes. Sol Energ 79(5):449–458. doi:10.1016/j.solener, 2005.01.003 DOI:dx.doi.org
34. Lucier R (1981) System for solar heat conversion into electrical energy. Brevet No. 4275309 US. F03D1/04
35. Dai YJ, Huang HB, Wang RZ (2003) Case study of solar chimney power plants in Northwestern regions of China. Renew Energ 28(8):1295–1304. doi:10.1016/S0960-1481(02)00227-6
36. Onyangoa FN, Ochieng RM (2006) The potential of solar chimney for application in rural areas of developing countries. Fuel 85(17–18):2561–2566. doi:10.1016/j.fuel.2006.04.029
37. Monohans solar tower project heart of oil industry could become site of renewable energy project. Scan Syst 19 July 2003. http://www.democraticunderground.com/discuss/duboard.php?az=view_all&address=115x21816
38. Gidrometeoizdat L (1990) Research and applied handbook on climate in the USSR. Series 3: Multiannual data. Leningrad: Gidrometeoizdat. Part 1–6, Vyp. 11, MSSR
39. Executive summary: assessment of parabolic trough and power tower solar technology cost and performance forecasts. Sargent & Lundy LLC Consulting Group Chicago, Illinois NREL Technical Monitor: H. Price. Prepared under Subcontract No. LAA-2-32458-01
40. Solar energy systems. Status report on solar trough power plants, 1996
41. Stoddard L, Abiecunas J, O'Connell R (2005) Economic, energy, and environmental benefits of concentrating solar power in California. May 2005–April 2006. Black & Veatch Overland Park, Kansas NREL Technical Monitor: M. Mehos. Prepared under Subcontract No. AEK-5-55036-01
42. www.stirlingenergy.com
43. Bougard J, Benallou A (1998) Le solaire thermique au service du developpment durable. Sous la direction scientifique de: IEPF, Quebec
44. Frank Shuman's Solar Arabian Dream. http://renewablebook.com/chapter-excerpts/350-2/. Accessed 17 Aug 2005
45. Solar energy system design. www.powerfromthesun.net. Accessed 25 Mar 2012
46. Ruprecht A et al (2003) Strömungstechnische Gestaltung eines Aufwindkraftwerks (Fluid dynamic design of a solar updraft power plant). In: Proceedings of the internationales symposium über Anwendungen der Informatik und Mathematik in Architektur und Bauwesen. Bauhaus–University Weimar, Weimar, 10–12 June 2003
47. Concentrating solar power: energy from mirrors. DOE/GO-102001-1147, FS 128, Mar 2001

48. Solar two central receiver. Consultant report. California Energy Commission, 1999
49. Bostan I, Dulgheru V, Dicusară I. Solar unit with Stirling motor. Patent No. 3348MD. BOPI no. 10/2007
50. Bostan I, Dulgheru V, Nicu T, Ciupercă R. External combustion Engine. Patent No. 2679 MD. BOPI no. 1/2005
51. http://www.enviromission.com.au/
52. Russell O (1946) Light sensitive device. Patent No. 2,402,662 US
53. Dones R, Frischknect R (1998) Life-cycle assessment of photovoltaic systems: results of Swiss studies on energy chains. Prog Photovolt Res Appl 6:117–125
54. Alsema E (2000) Energy pay-back time and CO2 emissions of PV systems. Prog Photovolt Res Appl 8:17–25
55. Systemes solaires. Le Journal des Énergies Renouvelabeles. Novembre–Decembre, No. 134, 1999; Mars–Avril, No. 136, 2000
56. Office for official publications of the European communities (1996) Photovoltaics in 2010. Vol. 1: current status and a strategy for European industrial and market development. Office for official publications of the European communities, Luxemburg
57. Rauschenbach HS (1980) The principles and technology of photovoltaic energy conversion. Litton Educational Publishing, New York
58. Markvart T (ed) (2000) Solar electricity, 2nd edn. UNESCO energy engineering series. John Wiley & Sons. Baffins Lane, Chichester, England, 280 p. ISBN 0-471-98853-7
59. Rauschenbach HS (1980) The principles and technology of photovoltaic energy conversion. Litton Educational Publishing, New York
60. Messenger R, Ventre J (2004) Photovoltaic systems engineering. 2nd ed., CRC Pres LLC
61. www.kyocerasolar.de/products. Accessed 11 Jan 2006
62. Bostan I, Vişa I, Dulgheru V, Dicusară I. Self-orientation paraboloidal solar unit (variants). Patent no. 3975 MD. BI no 3/2010
63. Palz W, Zibetta H (1991) Energy payback time of photovoltaic modules. Int J Sol Energ 10(3–4):211–216
64. Chancelier L, Laurent E (1996) L'électricité photovoltaïque. Collection "Le point sur". Gret, Ministère de la Coopération, 255 p
65. Rodot M, Benallou A (1998) Guide de l'énergie solaire: Electricité solaire thermique au service du développement rural. RIES, 172 p
66. Perlin J (2007) Late 1950s – saved by the space race (HTML). SOLAR EVOLUTION – the history of solar energy. The Rahus Institute. Retrieved 25 Feb 2007
67. NASA JPL publication: basics of space flight, chapter 11. Typical onboard systems, electrical power supply and distribution subsystems. http://www2.jpl.nasa.gov/basics/bsf11-3.php(3of5) Accessed 25 Mar 2012
68. Gaddy EM (1996) Cost performance of multi-junction, gallium arsenide, and silicon solar cells on spacecraft. Photovoltaic specialists conference. Conference record of the twenty fifth IEEE. Washington, DC, 13–17 May 1996, pp 293–296
69. Bostan I, Dulgheru V, Dicusară I. Patent no. 2965 MD. Self-orientation solar unit. B.I. no. 2/2006

Chapter 4
Kinetical Energy of River Running Water

4.1 Energy Potential

4.1.1 Classification of Hydropower Systems

The inevitable increase of global energy consumption and the risk of a major environmental impact and climate change as a result of burning fossil fuels opens wide prospects for the exploitation of renewable energies. Hydropower, as a renewable energy source, will have an important role in the future. International research confirms that the emission of greenhouse gases (i.e. life cycle evaluation) is substantially lower in the case of hydropower compared to that generated by burning fossil fuels. From the economical point of view, the utilisation of half of the feasible potential can reduce the emission of greenhouse gases by about 13%; also it can substantially reduce emissions of sulphur dioxide (main cause of acid rains) and nitrogen oxides.

Hydraulic energy is the oldest form of renewable energy used by man, and has become one of the most currently used renewable energy sources, being also one of the best, cheap and clean energy sources. Hydraulic energy as a renewable energy source can be captured in two extra power forms:

- Potential energy (of the natural water fall);
- Kinetic energy (of the water stream running).

Both extra power forms can be captured at different dimensional scales. Table 4.1 presents a simple classification of hydraulic plants according to the electrical energy output.

I. Bostan et al., *Resilient Energy Systems*, Topics in Safety, Risk, Reliability and Quality 19, DOI 10.1007/978-94-007-4189-8_4,

Table 4.1 Classification of hydroelectric power plants according to electrical energy output

Large scale hydro power	More than 100 MW, usually connected to a large electricity network
Medium scale hydro power	15–100 MW, usually connected to a large electricity network
Small scale hydro power	1–15 MW, usually connected to a large electricity network
Mini hydropower	About 100 kW, often is isolated, but sometimes can be connected to a large electricity network
Micro hydropower	From 5 kW to 100 kW, usually for a small community or rural industry
Pico-hydropower	From several hundreds of watts to 5 kW, usually for remote (isolated) consumers

4.1.2 Global Energy Potential

Hydropower, in general, has become the most important source of clean renewable energy, economically feasible. Hydroelectric power plants, integrated in multifunctional schemes, have performed various works such as irrigation, water pumping, etc. It is clear that hydropower will play an important role in the future, both in terms of ensuring energy supply and water resources development.

Under these options, it is necessary to develop these resources in conformity with the social, economic, technical and environmental standards. It is easy to forecast that global energy needs, especially electricity, will grow significantly during the twenty-first century, not only under demographic pressure, but also because of rising living standards in the underdeveloped countries (78% of total population), which is estimated to be 7 billion people by 2050. Primary energy consumption will grow by mid-century, and growth will be higher for electricity [1]. From the point of view of this situation more alternative energy sources will be required. However, for environmental considerations, an important priority must be given to developing, technically, the full feasible potential of environmentally friendly renewable sources, in particular, hydropower. Of all renewable energy sources, hydro (or energy of the running water) has been mostly exploited, although lately the implementation of hydropower schemes in developing countries was temporarily halted for financial, social or environmental reasons. Currently, only a small part of hydropower potential is used in the developing countries: 5% – in Africa, 8% – in Latin America, and 9% – in Asia. Nowadays, China operates approximately 10% of its enormous exploitable potential (about 378 GW), which is the largest in the world. Taking note of the rise of macro hydropower in the twentieth century, it has had a large development in the countries with considerable hydropower potential. Today, hydropower provides about one fifth of the global electricity needs. If the remaining hydro power potential would have been used, the overall needs of mankind in electricity could be satisfied.

Hydro power potential: Today, hydropower provides about 19% (2,650 TWh/year) of the global electrical energy. Information received from members of the WEC

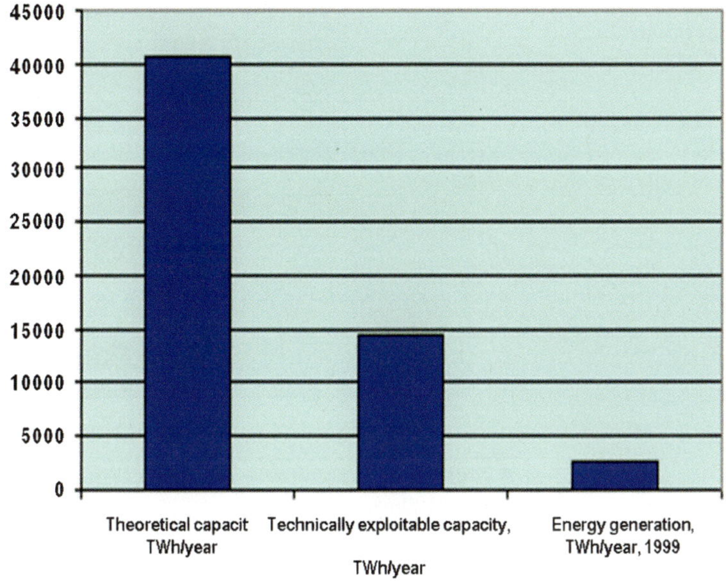

Fig. 4.1 Global hydro power capacities: theoretical, technical explorable, and overall energy production in 1999

(Water Economic Committee), supplemented by data published in *The International Journal on Hydropower & Dams*, show that technically feasible hydropower potential is about 14,400 TWh/year (Fig. 4.1) [2], of which nearly 8,000 TWh/year are considered now economically feasible for development. Hydroelectric generating capacities of about 692 GW have been already installed, with about 110 GW under construction. The remained potential, economically exploitable, is about 5,400 TWh/year: the exploitation of this potential would require the construction of about 1,400 GW of hydropower capacity (double the present installed capacity). Investment of at least US$1,500 billion will be needed to implement such a program.

Considering that a hydroelectric power plant capacity is between 50 MW and 100 MW it will be necessary to build about 20,000 hydroelectric power plants – very large constructions like Three Gorges (China) and Itaipu (Brazil) will not be the norm. A simple analysis shows that, currently, only about one sixth of the technical exploitable hydropower potential is explored. Installed hydropower capacities until 1999, by regions, are presented in Fig. 4.2 [3].

The top regions with the highest installed capacity of hydropower are Europe and North America. Global hydropower capacities (theoretical, technically exploitable and generated energy) for 1999 are listed in Fig. 4.3.

Technically exploitable capacities (output) of hydropower and hydroelectric energy production for 1999 by global regions are shown in Fig. 4.3. The existing hydropower potential is best exploited in North America: Canada and the USA. produce about a quarter of hydroelectric power energy. Canada, with its abundant water resources, has had good opportunities to produce low-cost, clean electricity.

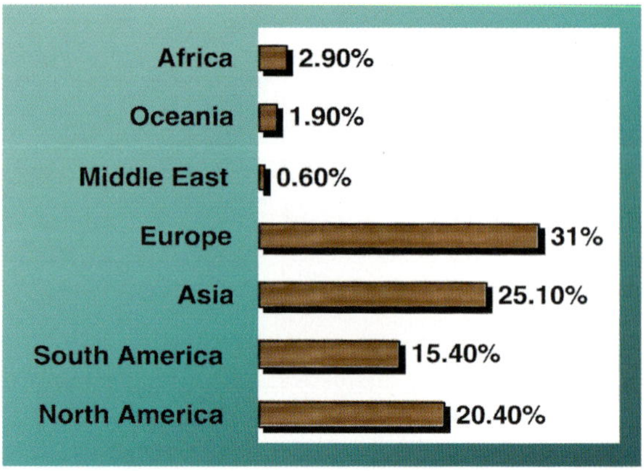

Fig. 4.2 Installed hydropower capacity till 1999, by regions

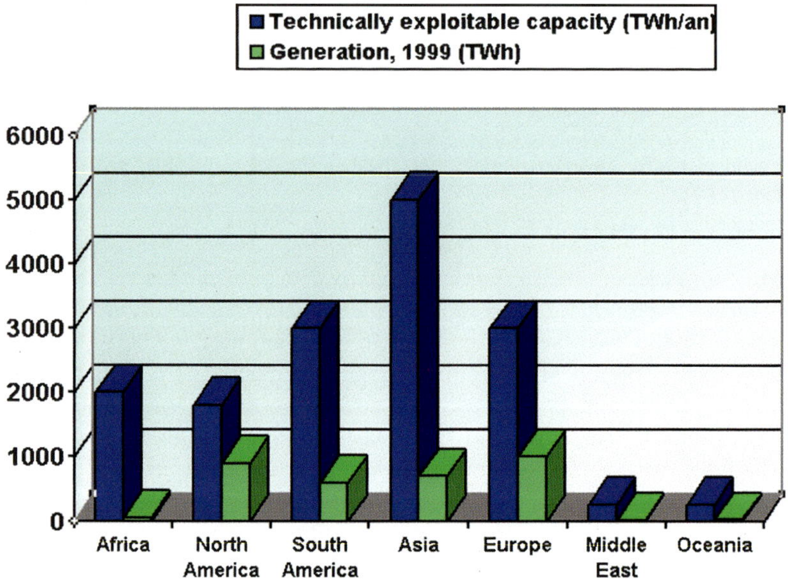

Fig. 4.3 Installed hydropower capacity till 1999, by regions

This fact had an important role in the economic and social development of Canada over the past two centuries. According to the diagram in Fig. 4.4, hydropower is the main source of electricity in Canada, representing about two thirds of the total energy produced. Most (about 59%) comes from large hydroelectric plants [4], such as the giant complex James Bay that borders the province of Quebec; its capacity is over 15,000 MW.

As shown in Table 4.2, in 2002 Canada was the world leader in producing hydraulic power. But Norway is the country where 99% of the electricity produced is of hydraulic origin. In New Zealand this indicator is 75%. Asia, with the highest

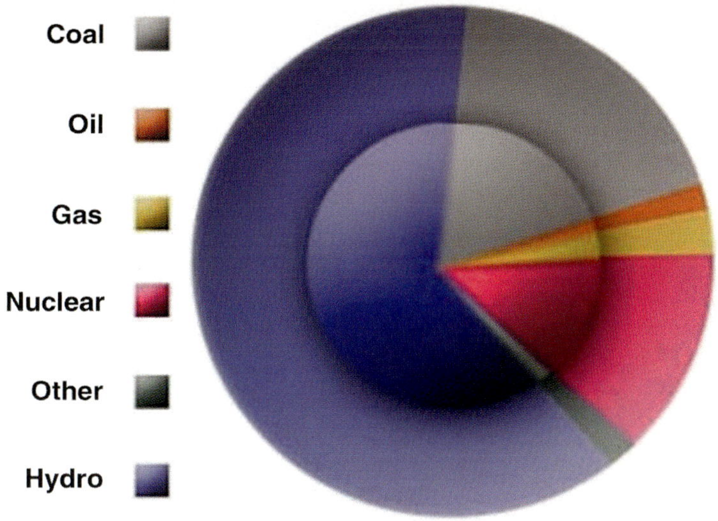

Fig. 4.4 Sources of electrical energy production in Canada

Table 4.2 Comparison of international hydropower, 2002

Country	Production (GWh)	Capacity (GWh)
Canada	353,000	67,100
USA	300,000	76,000
Brazil	300,000	64,000
China	258,000	82,700
Russia	174,000	44,700
Norway	121,000	27,600
World Total	2,740,000	729,000

energy potential in the world, stays behind by the degree of its utilisation. But the recent rapid economic development of China has placed it first in the world in 2006. In this connection, there is interest in information on major countries, which contribute essentially to the global production of hydroelectric power [5, 6]:

- China – 416,700 GWh (128,570 MW installed) (2006);
- Canada – 396,700 GWh (68,974 MW installed);
- Brazil – 285,603 GWh (57,517 MW installed) (1999);
- USA – 260,400 GWh (79,511 MW installed);
- Russia – 169,700 GWh (46,100 MW installed) (1999);
- India – 125,126 GWh (33,600 MW installed) (2006);
- Norway – 180,800 GWh (27,528 MW installed);
- Japan – 88,500 GWh (27,229 MW installed);
- France – 56,100 GWh (25,335 MW installed).

Table 4.3 The largest world hydropower plants

Power plant name	Country	Put into operation time	Total capacity (MW)	Annual maximum production of electricity
Itaipu	Brazil/Paraguay	1984/1991/2003	14,000	93.4 TW/h
Guri	Venezuela	1986	10,200	46 TW/h
Three Gorges Dam (2)	China	2004	9,800(2006) 18,200 (final)	84.7 TW/h
Grand Coulee	USA	1942/1980	6,809	22.6 TW/h
Sayano Shushenskaya	Russia	1983	6,721	23.6 TW/h
Krasnoyarskaya	Russia	1972	6,000	20.4 TW/h
Robert-Bourassa	Canada	1981	5,616	
Churchill Falls	Canada	1971	5,429	35 TW/h
Bratskaya	Russia	1967	4,500	22.6 TW/h
Ust' Ilimskaya	Russia	1980	4,320	21.7 TW/h
Yaciretá	Argentina/Paraguay	1998	4,050	19.1 TW/h
Ertan Dam	China	1999	3,300	17.0 TW/h
Gezhouba Dam	China	1988	3,115	17.0 TW/h
Nurek Dam	Tadjikistan	1979/1988	3,000	
Volzhskaya	Russia	1961	2,541	12.3 TW/h
Ataturk Dam	Turkey	1990	2,400	
Zhiguliovskaya	Russia	1957	2,300	10.5 TW/h
Porţile de Fier	Romania/Serbia	1970	2,280	11.3 TW/h
Asuan Dam	Egypt	1970	2,100	
Tarbela Dam	Pakistan	1976	2,100	
Hoover Dam	USA	1936/1961	2,080	
Cahora Bassa	Mozambique	1975	2,075	
Karun III Dam	Iran	2007	2,000	4.1 TW/h

The largest hydroelectric power plants in the world (see Table 4.3) have a total capacity of 2–10 GW. The largest hydroelectric power plant with a capacity of 14 GW is built on the Amazon River in Brazil.

Hydropower characteristics: The most important characteristics of hydro power can be summed, as follows:

1. Hydropower resources are widely spread around the globe. There is hydropower potential in about 150 countries, and about two thirds of the economically feasible potential have to be developed, especially in the developing countries where these capabilities are urgently required.
2. Advanced technologies are used, based on a secular experience. Modern power stations provide a highly efficient degree of conversion.
3. Hydropower is a clean energy source. It has a major role in reducing the emission of greenhouse gases; prevent annual burning of 22 billion gallons of oil and 120 million tons of coal. Hydropower is a relatively small source of atmospheric emissions compared to fossil fuels.

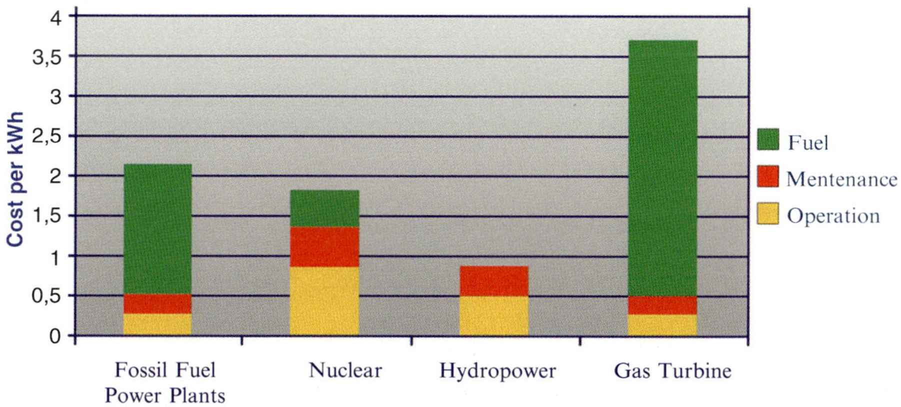

Fig. 4.5 Estimate production expenditures for 1 kW electrical energy

4. Hydropower is the most effective way of generating electricity. Modern hydro turbines can convert up to 90% of the water potential energy into electricity. The best fossil fuel plants have an efficiency of about 50%. In the USA, the cost of produced electricity is approximately 0.85 cents/kWh. This constitutes about 50% of the cost of nuclear electricity, 40% of the cost of electricity produced by burning fossil fuels (except gas), 25% of the cost of energy produced from burning gas. Hydroelectric power has reduced operational costs and long lifetime, compared with other options for large-scale electricity generation. The diagram in Fig. 4.5 shows approximate comparative costs for 1 kWh of electricity produced from different sources. The lowest cost (of the four sources) is for hydropower. The most important is that a basic component is excluded from the complex of expenditures, such as the costs for buying fuel. Once the initial investments have been allocated to the construction of the dam, the lifetime of the hydropower plant can be extended economically by relatively cheap maintenance and periodic replacement of the electromechanical equipment.
5. If the hydropower plant is integrated into multifunctional development, the project can help meet other basic human needs (e.g. irrigation of agricultural lands, providing domestic and industrial water, etc.). Water basins can be used for other purposes such as fisheries, adjustment of water level for navigation.
6. *The fuel* (water) is renewable and does not depend on the market fluctuation costs. Hydropower can provide energy independence to many countries.

4.1.3 Macro-, Mini- or Micro Hydropower?

Global hydro power energy today is about 715,000 MWe, or about 19% of global electricity (16% in 2003). However, macro hydro power is not a major option for future energy production in the developed countries in terms of industrial purposes for various reasons, such as the environmental one.

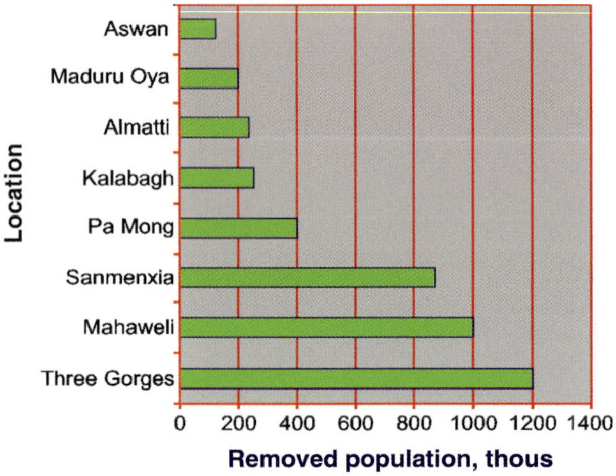

Fig. 4.6 Removal of inhabitants as result of dam construction

Construction of dams on rivers created major environmental and social problems. Development of huge artificial water reservoirs by damming the Earth's major arteries has led to climate and wildlife change in the region, to misbalancing of migration processes of some species of fish, to creating some generating sources of greenhouse gases (the formation and elimination of methane in the atmosphere). More recent studies of large water reservoirs created by hydroelectric dams have shown that the processes of decay of aquatic vegetation can lead to the emission of greenhouse gases quantities in the atmosphere that are equivalent to the emissions from other electrical energy sources [7]. For example, in tropical regions, macro hydro power can lead to greenhouse gas emissions, comparable with emissions of an electric power plant based on fossil fuels. According to Philip Fearnside (researcher at the National Research Institute of Brazil), during the first 10 years of operation, hydroelectric power plants could produce four times more gas emissions than a thermal power plant. These data cast doubts on the plans to build dams in the underdeveloped countries, including the 5 billion US dollars project proposed for the Congo River. On the contrary, small hydroelectric power plants without dams and reservoirs are not sources of greenhouse gases.

The biggest impact of hydroelectric dams is the flooding of vast agricultural lands and forests. Grande Dam project in the James Bay region of Quebec has flooded about 10,000 km^2 of land, which will increase as expansion plans, reaching an area larger than Switzerland. Flooding of large areas bordering the rivers has created major social problems, dozens of villages were flooded and hundreds of thousands of inhabitants from these villages have been removed (Fig. 4.6).

Large dams and reservoirs can lead to water quality deterioration by accumulation of large quantities of mud and various bacteria. Bacteria present in decaying vegetation can also change mercury, present in rocks underlying a reservoir, into a

Fig. 4.7 Three Gorges Dam hydroelectric power plant, Yichang, China

form, which is soluble in water. The mercury accumulates in the bodies of fish and poses a health hazard to those who depend on these fish for food. For example, the Three Gorges Dam hydroelectric power plant with a combined generating capacity of 18.2 million kW (Fig. 4.7), which construction was launched on the Yangtze River, China, in 1993 and lasted 13 years, has a length of 2,309 m and 185 m height. The population that was removed to other places reached 1.2 million people [8, 9].

A more efficient use of hydraulic energy, in terms of environmental and social impacts, is the conversion of kinetic energy of running river water without dams' construction. What are the main advantages of this type of energy? First, the relative simplicity of these energy conversion systems. Also, the density of water is considerably higher than, for example, air density, and, thus, contains a greater amount of energy in itself. The kinetic energy of water is available 24 hours a day. It does not create noise pollution of the environment and doesn't affect aquatic creatures. The new Laws of the environment affected by the danger of global warming consider hydraulic energy obtained from small stations much more relevant. The use of hydropower potential at very small-scale is substantiated and in terms of its cost. The analysis of economic viability of the most widely used types of energy with a capacity of 10 kW, made by the US Office of Technology Support, is presented in Table 4.4. In the case of micro hydroelectric power the negative environmental impacts associated with large hydroelectric power stations are also eliminated [10].

These mini-hydroelectric power plants can meet energy needs of consumers, particularly in rural areas. Local industry should be encouraged to use this power for its sustainable development. This is a technology with enormous potential, which should exploit water resources to meet, in the first place, the needs of consumers

Table 4.4 Analysis of economic viability of various forms of energy of 10 kW capacity

Form of energy	Cost
Micro hydro	0.21$/kWh
Wind	0.48$/kWh
Diesel	0.8$/kWh
Network extension	1.02$/kWh

in rural areas with little access to conventional sources of energy. Table 4.5 shows the state of small scale hydroelectric power development (<10 MW) at the end of 1999. An important success in this respect belongs to the countries of Latin America, which at the end of construction and according to the prognosis will double their annual energy production. According to experts, the cost of 1 kW of installed capacity of micro hydro electric power plants is $400–500 with an efficacy of 40–50% and the redemption period of 1–2 years.

Micro hydroelectric power plants have been used extensively in the past for various practical applications or to supply electricity to towns. Later, due to low cost of fossil fuels, to the economic level of macro-hydroelectric power and massive quantities of energy required, micro hydroelectric power plants have been partially abandoned. In remote areas the installation of diesel units or the network mapping was preferred instead of micro hydroelectric power plants. The basic argument was high investment costs. Today, when the price of fossil fuels and the expenses for environmental protection grow continuously, micro hydroelectric power plants win in the competition related to energy supply, primarily, of remote localities and isolated objects. Also, they offer additional advantages because they are environmentally friendly, does not require big additional civil engineering works, such as access roads, temporary houses for workers, etc.

Not everybody is lucky enough to have a river near his home, but for those who fall into this category, micro hydro turbines are the most reliable and cheap source of alternative energy. A small turbine can produce energy as long as there is water non-stop, regardless of weather conditions.

In this context, the following advantages of micro hydroelectric power plants are highlighted:

- Suitable for small power requirements, decentralized (light industry, farms and private enterprises, rural communities) and for operations external to the main network;
- Require low-voltage distribution networks and, eventually, micro sub-regional networks;
- Can be used as private property, in co-ownership or joint ownership with a semi-qualified labour force required and with a joint or separate administration;
- Short period of construction using local materials and skills of people in the area can have a significant impact on the quality of rural life;
- Their flexibility, particularly in regard to adapting to variable charges depending on the inflow **rate**, making them a prime component in any integrated power systems;

Table 4.5 Hydroelectric power: level of development at the end of 1999 at small scale (<10 MW)

Region	Economically exploitable capacity (GWh/an)	In operation: Capacity (MW)	In operation: Effective production in 1999 (GWh)	In construction or planned: Capacity (MW)	In construction or planned: Annual probable production (GWh)
Africa	351	15	53		
North America	41,157	3,698	12,934	152	570
Canada	41,157	865	4,410	100	570
USA		2,537	7,456	52	
South America		1,566	7,563	1,639	7,897
Asia	2,587	3,722	886	689	3,239
Europe	26,752	8,979	23,730	819	3,615
Middle East	87	107	431	7	24
Oceania		78	370		

Note: Data on small scale schemes are reported only at WEC countries in 2000/2001

- Plants can operate a long period. Some have been built 70 years ago and are still running. Plants that are ready to become operational in the near future can register even longer life and serve consumers over several generations without polluting the atmosphere;
- Small investments in hydroelectric power plants have proved to be safe and reliable for several decades.

Production of electricity, using water as primary source, is an energy conversion process in which water is an effective means of transmission and transformation of the flow gravitational potential energy into mechanical and electrical energy.

4.1.4 Calculation of Water Kinetic Energy Potential

How is the kinetic energy of water captured? In principle, mechanical systems to capture this energy are relatively simple. A pontoon system, consisting of floating bodies, linked through a metal structure, on which a multiblade rotor is mounted, which blades, most often, have a hydrodynamic profile in section. Blade shape and design can be calculated for water kinematic parameters. Summary torque oscillations can be reduced by using the flywheel – or by using blade constructive and orientation solutions. Use of technical solutions for the orientation of blades in optimal positions against the water currents, in terms of flowing water kinetic energy conversion, can lead to increased torque and growth of generated power. It is known that power increases cubically compared to the water flow velocity. Use of the running water high velocity or the generation of such speeds by some systems, can also lead to the generated power increase. A surplus of energy in the theoretical limits of the Betz coefficient (<0.59) can be achieved by using new design solutions for the construction of rotor blades.

Kinetic energy is a superpower phenomenon because it occurs due to motion. It is defined as the work necessary to accelerate a body. Having this energy during its acceleration, the body maintains this kinetic energy till it changes the velocity. The word *kinetic* is of Greek origin and means *motion*. The terms *kinetic energy* and *work,* and their presence in the scientific environment starts in the mid nineteenth century. The earliest understanding of this idea can be attributed to Gaspard-Gustave Coriolis, who published *Du Calcul de l'Effet des Machines* in 1829, defining in this work the mathematical instrument of the kinetic energy. Later, William Tomson (Lord Kelvin) confirmed the use of the term "kinetic energy" in 1849.

The kinetic energy can be better understood by examples that demonstrate how it transforms into other forms of energy. Under non-relativistic aspect, the action represents particle acceleration/deceleration in a period of infinite estimated time that is presented by:

$$F \cdot dx = F \cdot vdt = \frac{dp}{dt} \cdot vdt = v \cdot dp = v \cdot dmv$$
$$= \frac{m}{2} d(v \cdot v) = \frac{m}{2} dv^2 = d\left(\frac{mv^2}{2}\right) \quad (4.1)$$

By integrating the relation (4.1), kinetic energy results

$$E_k = \int F \cdot dx = \int v \cdot dp = \frac{mv^2}{2} \quad (4.2)$$

Under non-relativistic aspect, it is necessary to modify the expression (4.2) for the linear momentum. Integrating by portions, we get:

$$E_k = \int v \cdot dp = \int v \cdot d(m\gamma v) = m\gamma v \cdot v - \int m\gamma v \cdot dv$$
$$= m\gamma v^2 - \frac{m}{2} \int \gamma \cdot d(v^2) \quad (4.3)$$

Substituting $\gamma = \left(1 - v^2/c^2\right)^{-1/2}$, we get:

$$E_k = m\gamma v^2 - \frac{-mc^2}{2} \int \gamma \cdot d\left(1 - v^2/c^2\right) = m\gamma v^2 + mc^2\left(1 - v^2/c^2\right)^{1/2} + C \quad (4.4)$$

or

$$E_k = m\gamma\left(v^2 + c^2\left(1 - v^2/c^2\right)\right) + C$$
$$= m\gamma\left(v^2 + c^2 - v^2\right) + C = m\gamma c^2 + C \quad (4.5)$$

Constants of integration are obtained from the observation that $\gamma = 1$, if $v = 0$, obtaining a common formula:

$$E_k = m\gamma c^2 - mc^2 = \frac{mc^2}{\sqrt{1 - v^2/c^2}} - mc^2 \quad (4.6)$$

The fluid in action contains potential energy, or kinetic energy. Potential energy is present in hydroelectric dams and can be captured by water passing a controlled way. In this case, power is characterized by the torrent of water mass:

$$\frac{E}{t} = \frac{m}{t} gh. \quad (4.7)$$

Substituting *E*/*t* by *P* and expression *m*/*t* in terms of the liquid volume in motion in a time unit (fluid flow rate φ) and water density, a common expression is obtained (4.7):

$$P = \rho\varphi gh. \tag{4.8}$$

For *P* in *W*, ρ is measured in kg/m^3, φ is measured in m^3/s, g is measured in m/s^2, and *h* is measured in meters.

A number of hydropower systems, such as water wheels can generate energy from water flow velocity without the need to change its height. In this case, available energy for conversion is the kinetic energy of the water flow:

$$P = \frac{1}{2}\rho\varphi g v^2 \tag{4.9}$$

where *v* is water flow velocity.

Substituting, $\varphi = A \cdot v$ where *A* is the plan area for the passage of water, we get:

$$P = \frac{1}{2}\rho A v^3 \tag{4.10}$$

Some types of water wheels can capture both types of energy. To collect this energy some physical principles are applied. The impulse of turbine changes the direction of the high speed fluid. As a result, the fluid enters and gets out the turbine with a reduced kinetic energy. Applied to hydraulic turbines, depending on their design implementation, the fluid kinetic energy is converted into mechanical energy, using the turbine impulse or the turbine reaction force (Fig. 4.8).

Reaction turbines develop the torque by the force of reaction to fluid pressure. Fluid pressure drops when it passes through the turbine bladed rotor giving some energy to the blades. This concept is applied in Francis turbines and in most steam turbines. Turbine designers use both concepts based on the nitial conditions. Another form of reaction is the hydrodynamic effect of blade profile used by designers to raise the power efficiency of the turbines.

The classical method of the turbine design was developed in the mid-nineteenth century. At first, graphical methods of calculation were used. Use of calculation relations for sizing the turbine parts allow design of efficient turbines for any fluid flow conditions. Some calculations are based on empirical formulas, and others – on classical mechanics. Modern design methods are inconceivable without computer. Modelling and simulation capabilities of computers equipped with advanced software allows 3D design of turbines, turbine dynamic modelling, numerical modelling and simulation of fluid interaction with turbine blades for various geometrical parameters of the turbine parts, and for kinematical and viscosity parameters of the fluid.

A characteristic parameter of the hydraulic turbine is the load factor, which is the ratio of annual generated energy and annual installed capacity:

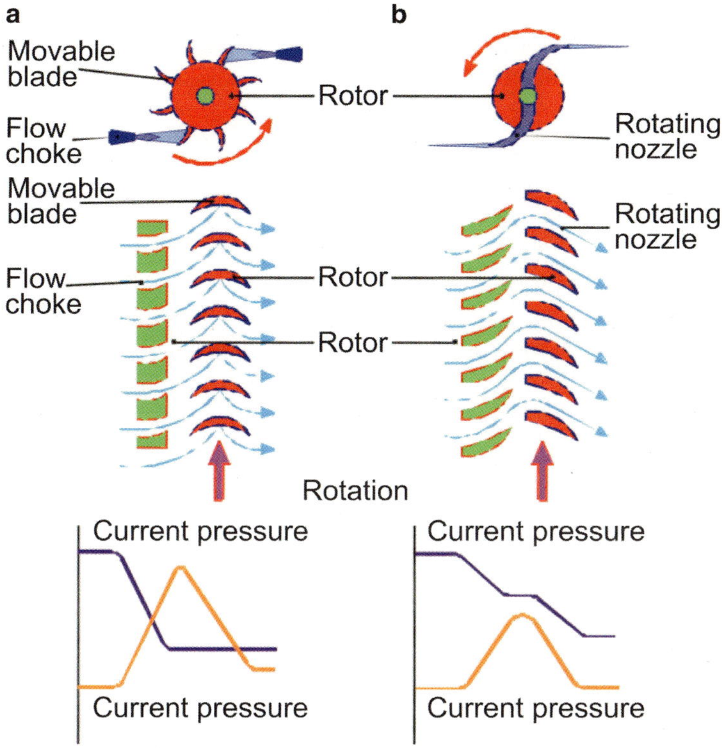

Fig. 4.8 Collection of water kinetical energy by utilising the impulse or the reactive force

Table 4.6 Dependence on the load factor and the designed flow

Designed flow Q_o	Load factor (LF)
Q_{nom}	40%
0.75 Q_{nom}	50%
0.5 Q_{nom}	60%
0.33 Q_{nom}	70%

$$Load\ factor = \frac{Annual\ generated\ energy\ (\mathrm{kWh/year})}{Installed\ capacity\ (\mathrm{kW}) \times 8760\mathrm{h/year}}$$

A first estimate on how the load factor varies depending on the designed flow is shown in Table 4.6.

Annual produced energy is estimated, utilising the load factor, from the relation:

$$Energy\ (\mathrm{kWh/year}) = P(\mathrm{kW}) \times FS \times 8760$$

At the development stage, the designer is forced to make the best choice: to choose a larger turbine that will produce more energy, but will work with a lower load factor, or a smaller turbine that will produce less energy, but with a higher load factor. Load factor for most micro hydroelectric power plants must be within 50–70%, in order to obtain a satisfactory recovery of investments.

4.1.5 Flow Rate Measurement

4.1.5.1 Method 1

Determination of the energy potential of river is an important objective in the design and installation of micro hydropower plants. Measurements are carried out according to [11, 12]. An important parameter is water flow rate. The term of flow rate used in the case of micro hydroelectric power plants, is the volume of water, but not its speed. There are various methods for determining the flow rate. The most simple way of measuring the flow rate consists in the following four steps:

- Measurement of water velocity (m/s);
- Determination of cross-sectional area of water source (in m^2);
- Calculation of flow rate (m^3/s);
- Conversion of flow rate from m^3/s into litres per minute.

Water velocity. Determination of water velocity is performed with advanced measuring instruments. Measurement is performed at different distances from the river banks, at different depths from the sea surface, at different times of the year. The measurements are performed several times and the arithmetic mean is determined. Due to friction losses between the water currents, river banks and bottom, the average measured velocity is multiplied to 0.8.

Flow area. The segment of river where water velocity measurements were made, water depth and width are determined for segments equal to 2 m (Fig. 4.9). Based on the measurements, polygonal section is projected, close to the actual one, and flow area of the river is calculated in m^2.

Calculation of flow rate. Based on measurements and calculations, water source flow rate is calculated:

$$D = V \cdot S \left(\mathrm{m}^3/\mathrm{s}\right) \tag{4.11}$$

To convert (*m³/s*) unit of measurement in *l/min* the obtained value (*m³/s*) is multiplied to 166.

4.1.5.2 Method 2

Another method of measuring the water sources flow rate is the use of hydrological data provided by the National Centres for Ecology and Hydrology. For example, in Great Britain there are 1,300 measuring stations (Fig. 4.10). Data for about 200

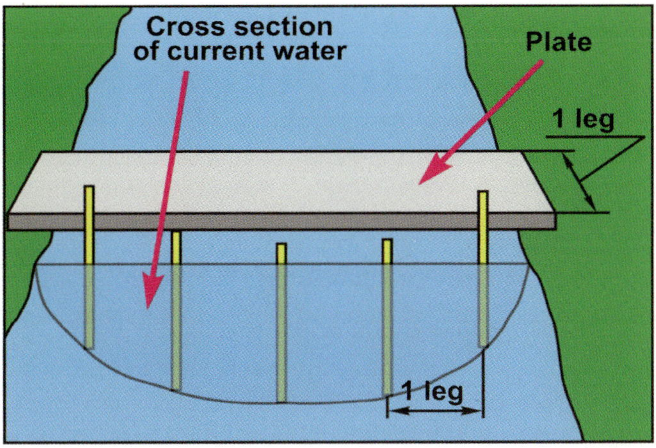

Fig. 4.9 Calculation of flow area of water source

Fig. 4.10 Station of hydrological measurements măsurări hidrologice

water sources are available on the Internet [13]. These data can be used to determine the flow rate in the proposed sites. If necessary data are not available, it is possible to use hydrological methods that are based on long-term measurements, and on the evaporation and elevation data for similar places. This information allows overall assessment of the hydraulic potential without taking any momentum observations.

More sophisticated methods for hydrology assessment and measurement of water sources flow rate are described in [14, 15]. The most precise and affordable method of measuring the water flow rate is to install some means of measurement, which are summed up in the following.

Fig. 4.11 Measuring weirs

Measuring weirs. Flow rate measuring weirs have a weir notch by which all water flows. It is usually applied for flow rates of the order 50–1,000 l/s. Flow rate can be assessed by determining the difference of heights between the bottom of the water currents and the crest of the weir (Fig. 4.11). Calculation relationship of the flow rate for weirs with rectangular weir notch is:

$$Q = \frac{2}{3} C_d \sqrt{2g}(L - 0.2h)h^{1,5} \tag{4.12}$$

where: Q is the flow rate [m^3/s]; $\boldsymbol{C_d}$ – discharge coefficient; L – width of the crest [m]; h – weir deepening to lower water level [m]; g – gravity acceleration [9.81 m/s^2].

Since the current flow varies both day by day and from season to season, measurements must be carried out over a long period of time, preferably several years. If $\boldsymbol{C_d}$ is usually taken equal to 0.6, then the following equation is obtained:

$$Q = 1.8\,(L - 0.2h)\,h^{1.5} \tag{4.13}$$

Flow duration curves (FDC). There are two ways of expressing the variation of the annual river flow rate: annual hydrography and flow duration curve (FDC), which is shown in Fig. 4.12 carrying out annual hydrography is easy to understand, because it shows a daily variation in the flow rate during a calendar year. Furthermore, FDC is more useful to estimate the available energy for the chosen micro hydro power plant scheme. FDC curve shows how the flow rate is distributed over a period (usually, 1 year). The y-coordinate indicates the flow rate, and the x-coordinate shows annual percentage by which the flow rate exceeds the amount shown on the y-coordinate. For example, FDC can indicate the flow rate immediately, which will be available for at least 50% of the year (known as Q_{50}). Flow rate exceeding 95% of the year (Q_{95}) is usually taken as a characteristic value for the minimum river flow rate.

Flow rate compensation. The flow rate portion, called the flow rate compensation will be required for aesthetic and environmental aspects. Abstracting the schemes, when water deviates from the main water course of the river, this flow rate

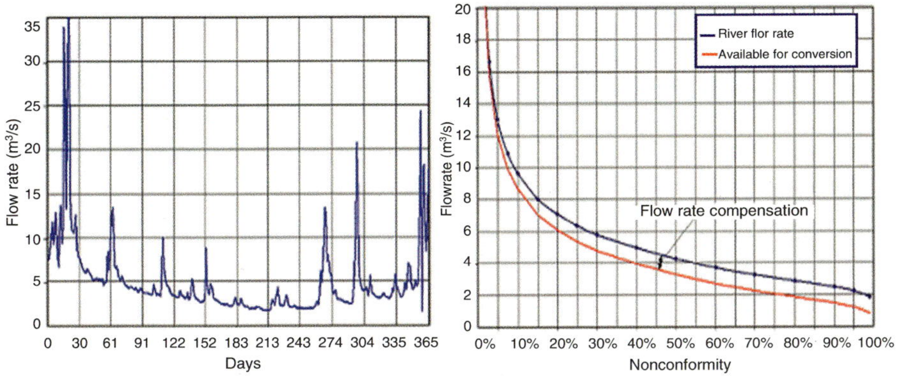

Fig. 4.12 Anual hydrography and flow duration curve

compensation is required to maintain the ecological and aesthetic appearance of the river in the exhausted area. The value of flow rate compensation depends on the nature of the river, and in first approximation, will be within Q_{90} and Q_{99}.

4.1.6 How to Choose a Hydro Turbine?

Operation efficiency of a micro hydroelectric power plant depends, to a great extent, on the right choice of turbine in terms of the characteristics of the water source. If the water source provides a relatively good height drop, then impulse turbines previously reviewed are chosen. If the water source does not provide an acceptable drop height, but has an acceptable rate or flow velocity economically explorable, then reaction turbines, considered above, are chosen. In all cases, it is necessary to undertake more accurate measurements of the water source parameters. Let analyze this problem for both cases examined. *Accuracy is critical* when measuring the level difference, because it will affect not only the power produced, but will define the turbine type too, for example, Francis or Pelton turbines, and, obviously, the blade shape, as well. Using GPS (Global Positioning System) is useful only if we have a highly grade instrument, if not, it is necessary to use a simple and accurate method. Using a simple height gauge in millimetres and a horizontal cam groove it is possible to measure the water fall height with sufficient precision. Mere summing of shares A, B, C and D in the example (Fig. 4.13) will lead to the exact result.

What type of turbine is chosen? The use of the best turbines depends on the difference in level and flow rate. For example, the Pelton turbine is more effective in locations with high-level difference, in contrast to a cross-flow turbine, which is used only in locations with little difference in levels, but high flow rate.

How much does the influent conduit matter? The influent conduit is not only "the guiding" of water on its way to the water turbine, but it also creates the

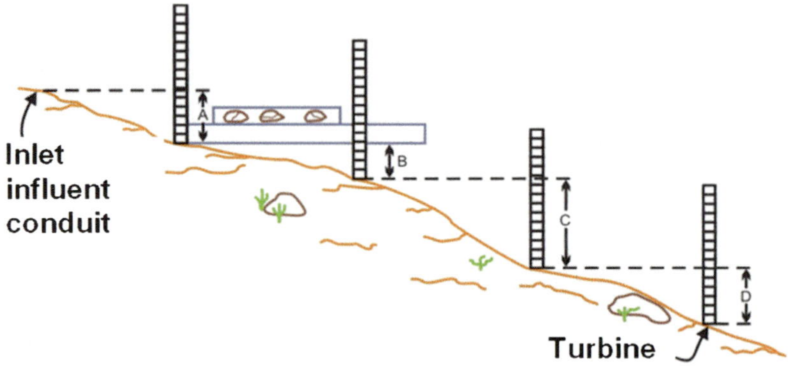

Fig. 4.13 Level difference measuring diagram

Fig. 4.14 Harris microturbine (900 W) systems

necessary pressure. As a result, the influent conduit focuses all pressure created by the difference in level at the base of the pipe, at entering the turbine. Conduit diameter, length, material of which it is made, and the *road map*, influences the turbine efficiency. A small diameter conduit can significantly reduce the energy obtained, even if all available water is used.

What is an alternative energy system composed of? Despite the fact that wind turbines, solar panels or micro hydro turbines are used for alternative power generation for low capacity applications, the system design is similar. As basic components, we need batteries, inverters and charge controllers.

How much current can be produced? Obviously you can not consume more energy than it is accumulated. So, if a 900 W Harris generator is taken (Fig. 4.14) ideally it will produce energy equivalent to 21.6 kWh per day. If a house consumes on average 180–200 KWh per month, it results that this micro hydro turbine can feed three separate houses, monthly production being equivalent to 540 KWh (let consider an operation month of 25 days). If 1 kWh = 0.08 Euro, it results that a month' production is equivalent to 43.2 Euros.

4.2 Hydraulic Energy Conversion Systems

4.2.1 Brief History of Hydraulic Energy Conversion Systems Development

The existence of water on the Earth has conditioned the emergence and development of life. From the times immemorial, man has chosen a place to live near rivers and lakes to meet their natural needs in water, but also for carrying out basic irrigation works. Floating or rowing led human thought by observation, to use water force and energy. Thus, the mechanical power of running water can be considered one of the oldest tools. The means of water use and exploitation have evolved from a historical epoch to another, from one nation to another, in relation to the natural conditions, depending on the level of production relationships and forces. Thus, water energy uses has marked stages of development of the social systems from the primitive to modern society.

Historical research, ancient engravings and writings show that in ancient times in India (by about 4,000 years before Christ) [16] and China (about 5,000 years ago) [17], dams and canals were built thousands of kilometres long, serving for irrigation and navigation. The Chinese hydraulic wheels, invented and used during the Han Dynasty to grind grains, served to use energy supplied by the water velocity from canals and rivers (ancient power wheels – Fig. 4.15). Those hydraulic wheels turned the linear velocity of water V into rotational motion with angular velocity ω of a

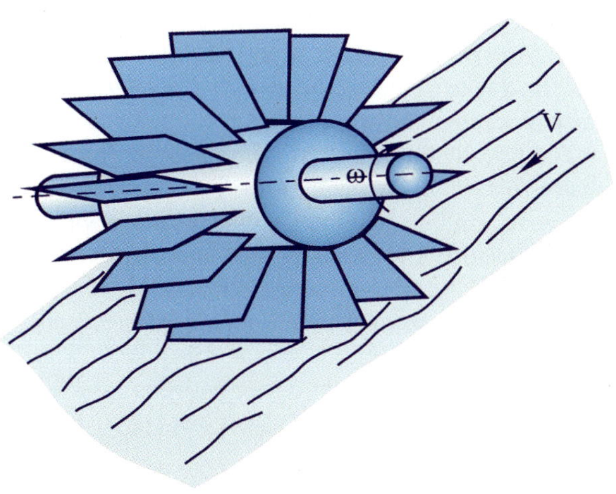

Fig. 4.15 Power wheel, ancient age

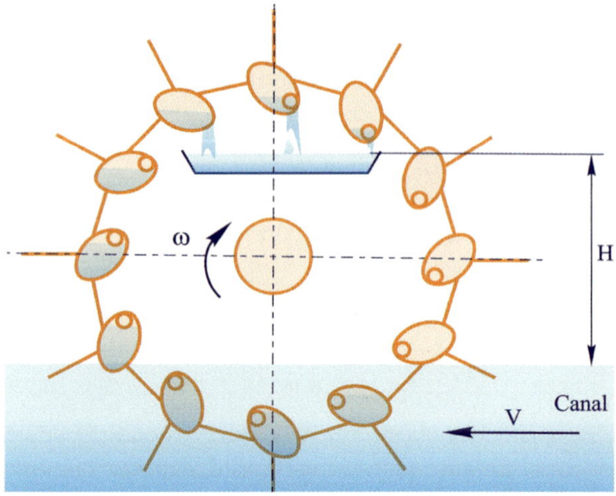

Fig. 4.16 Chinese water wheel ancient age

Fig. 4.17 Ancient tympan

shaft on which paddles were thrusted, at first primitive, then, over time, improved in the form of blades. Last several thousands of years people living in the valleys of the Himalayas have used water mills, or *chakki* for different usages. Water mills are much older than the wind mills.

The first hydraulic transformer was known about 4,000 years ago; it served to raise water at height H with the help of primitive buckets, attached to a paddle wheel (Fig. 4.16). In low position the buckets filled with water and after rotation, reaching the top positions, emptied into a water trough located at H meters above the water trough.

As an improvement the "tympan" was created in ancient Greece (Fig. 4.17), which peripheral paddles rotate the power wheel by ω. The power wheel is composed of two parallel discs, between which curved or polygonal blades rise water (as shown in successive positions 1-2-3). From position 3, the water is emptied

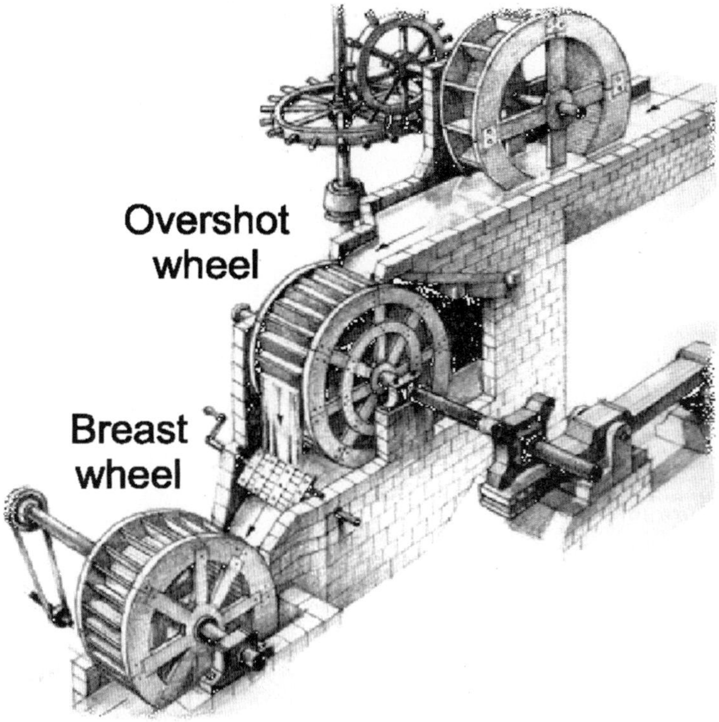

Fig. 4.18 Roman water mill

into the trough that is concentric with the shaft (a). At tympan the water is raised to a height H smaller than that of Fig. 4.16. The pre-Roman experience, accumulated over several thousand years, was described by Heron of Alexandria, renowned inventor of the ancient times, in his known book *Hydraulics*, which served as a guidebook for generations of builders of water wheels from ancient to modern times. During the Roman period the most famous inventor of hydraulic wheels was the famous Vitruvius. During the Roman Empire fixed and floating water mills have been used and widely spread in other countries of the Empire. So-called Greek Mills had vertical axis. They were older and simpler, and operated at high water velocities only with smaller diameter of water wheel. Roman mills had horizontal axis and were more complicated in terms of their construction (Fig. 4.18) [18]. They needed gear wheels for transmitting power from the main shaft to a shaft installed vertically.

The Cistercian Order built in the Middle Ages a complex of water mills in Western Europe. Figure 4.19 shows a water mill with horizontal shaft built in a small town in Belgium in the twelfth-century. Monasteries of the Cistercian Order widely used water wheels to drive water mills for various destinations. Another early example of the use of water wheels in the thirteenth century was the Cistercian

Fig. 4.19 Water mill in Braine le Chateau, Belgium, twelfth century

monastery *Real Monasterio de Nuestra Senor Rueda* in the Aragon region of Spain. Water wheel remains competitive to the steam engine of the Industrial Revolution period.

The use of water wheels on the British Islands dates back to the year 900. But water wheels have developed extensively in England in the eighteenth century by its famous persons, John Smeaton and James Brindley, following the theoretical calculations and practical experiments in France and elsewhere. They have justified the possible areas of use of open and submersible wheels. Open wheels allowed large dimensions of wheels (more than 2 m) associated with small water reservoirs. Half-submersible and submersible wheels can be used only on very deep rivers and reservoirs. Figure 4.20 shows an open water wheel with a diameter of approx. 10 m, operating an old mill in Rome, Georgia, USA.

The Laxey Wheel (also known as Lady Isabella) is a large water wheel built in the village of Laxey in the Isle of Man, United Kingdom (Fig. 4.21). Designed by Robert Casement it has a 22 m diameter and revolves at approximately three revolutions per minute 3 min^{-1}.

Water wheels have accompanied us over thousands of years. They were cheap and easy to build. Thousands and thousands of water wheels were built in America by farmers, millers, mine operators, etc. The water wheel was used practically in every household located on the waterfront. The wheels were of all dimensions and different designs: from very large diameters, to slightly narrow and small. In one of the oldest mills in the mountains of North Carolina [13], restored subsequently, the

Fig. 4.20 Open water mill, Rome, Georgia, USA

water wheel has a 7 m diameter and its width is about 1 m (Fig. 4.22); it is equipped with an observation cabin located at the top. Through a number of gear wheels and a belt transmission it drives a working machine.

Figure 4.23 [18] shows another variant of the narrow water wheel with a large diameter. The book “Water Wheel Factory” written by Robert Vitale gives advice on how to build a real water wheel, using the technical means and materials at hand. The operating principle of the hydraulic system is shown in Fig. 4.24. Until the widespread use of electricity, hydropower was used for irrigation, grain milling, textile manufacturing, etc.

In the 1700s, the mechanical power of running water captured by water wheels was used extensively for water mills and water pumping. In the period of the Industrial Revolution, water wheels have become a new profession – they drove various technological facilities. Their number increased continuously, in the nineteenth century their number reached about 20,000. In 1830, in the era of canal construction, hydropower was used to transport barges. Nowadays, hydropower is widely used to produce electricity. But some rural water mills, located in the northern USA, have operated commercially until the 1960s.

The use of water wheels on the territory between the Prut and Nistru rivers dates back to the times of Dacians before the arrival of the Romans.

Fig. 4.21 The Laxey Water Wheel, Isle or Man, UK

Fig. 4.22 Water wheel linked via mechanical transmissions to a working machine

Fig. 4.23 Narrow water wheel

4.2.2 Modern Water Wheels

The evolution of the modern hydropower turbine began in the mid-1700s when a French hydraulic and military engineer Bernard Forest Belidor wrote *Architecture Hydraulique* in four volumes. During the 1700s and 1800s, water turbine development continued. At the end of 1800, the invention of the Pelton wheel (named after the inventor) encouraged many owners of water mills to replace conventional water wheels with the Pelton turbines.

In 1826, the Frenchman, Jean Victor Poncelet, proposed a machine that included a fully submersible water wheel, where water had to run into the wheel otherwise than along it [19]. Jean Victor Poncelet worked on increasing the efficiency of the submersible water wheel, using the physics of modern hydraulics. Following this concept, the American Samuel Howd patented the first turbine in 1838. James Francis improved its design later by assigning the blades a radius of curvature. Known as the Francis turbine, this is the first water turbine.

In 1879 T. A. Edison demonstrated the incandescent electric bulb. The electric dynamo machine was first used for lighting Michigan in 1880; it was driven by a

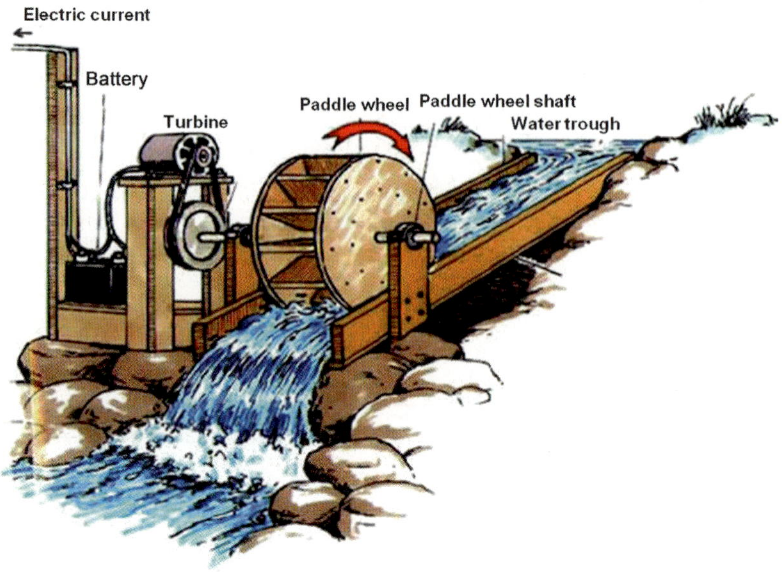

Fig. 4.24 Principle diagram of a water wheel system

water turbine. And in 1881 an electric dynamo connected to a water mill produced street lighting in Niagara Falls, New York. These two projects have established the basis of direct technology of the electric current use. In 1882, the first hydroelectric plant in the world was built in Appleton, state of Wisconsin, USA. Ottawa, in 1885, became the first city in the North America, which signed a contract for lighting the streets.

The hydraulic turbines have replaced quickly the water wheels in driving mills and textile plants. The end of the nineteenth century was the golden era of the hydraulic power. Thousands of low-power hydraulic plants have scattered the riverbeds with thousands of turbines manufactured in the country. In the late nineteenth century a new use of hydraulic turbines was found: the generation of electricity. The hydroelectric power represented about 40% of the electricity produced in the USA in 1920, and in 1940 it was about 40% of the worldwide generated electricity. With the industrial development of other forms of power generation and rural electrification programs the share of hydropower has declined steadily, now constituting about 10% of the electricity produced in the USA. Slowly, the manufacturers of micro hydro power turbines went out of business, the windmills were destroyed and the hydraulic turbines abandoned. Today, we observe a revival of hydro power as a source of clean and renewable energy. Modern hydraulic plants have increased in size from micro hydro power turbines to giant systems with dams, like Hoover Dam, which generate electricity for millions of people daily.

In contrast, to early water wheels and turbines, modern turbines are compact, highly efficient and capable of operating at very high speeds. Hydropower is the best source of renewable non-pollutant energy that can be easily integrated into the irrigation and water supply projects.

China has more than 85,000 small dimension hydraulic plants, producing electricity. Over the past decades, micro hydro power plants have played an important role in the developing countries, in the economic development of rural areas, especially mountainous areas. Micro hydro power plants can provide energy for industrial, agricultural and domestic use via direct use of mechanical energy or by turbine coupling to electrical generators.

4.2.2.1 Classification of Micro Hydro Power Plants

Micro hydroelectric power plants are the most efficient and cheapest generators of electrical energy. If you have a small creek or a river near the chalet or house that can provide at least 5 l/s. water drop from a difference in level of at least 3 m, or 0.5 l/s. from a difference of level from at least 10 m, do not hesitate to use a hydroelectric generator. You'll have clean, free and unlimited energy.

Hydropower is the most important energy source, which has no carbon dioxide, sulphur dioxide, nitrogen protoxide or any other type of pollutant emissions in its composition, and does not produce any solid or liquid waste. Micro hydroelectric power plants can be divided into two basic types: without dams, based on the conversion of kinetic energy (fixed or floating) of river water; with small dams, which use natural or artificial water drop of a river. They encompass the main advantages compared to other energy sources, saving coal, fuel or firewood, being independent at the same time.

A micro hydroelectric power plant without dams practically does not require civil engineering constructions, except the ground foundation, which connects the micro hydroelectric power plant with the shore. The main components of a hydro power plant without dams are as follows (Fig. 4.25) [20]:

- *Shore foundation:* **that** includes a complex connection system (rigid and flexible);
- *Hydraulic turbine:* it is a part of the plant where water power is converted into mechanical energy;
- *Generator:* the mechanical energy transmitted to the turbine maintains the generator rotor velocity, producing electricity in accordance with the laws of electromagnetics;
- *Transformer station and transmission line:* electrical energy is run and converted in order to be connected to the grid for electricity supply to consumers.

The quantity of the produced energy depends on the running water velocity: Main components of a small power hydro plant with dam are as follows (Fig. 4.26):

- *Storage:* is a form of available potential energy storage;
- *Transfer system:* which includes the capture device (flooding valve equipped with grid) and the transfer circuit (trough, sluice gate valve, weir, culvert, off take or discharge) where a part of the available energy is converted into kinetic energy;

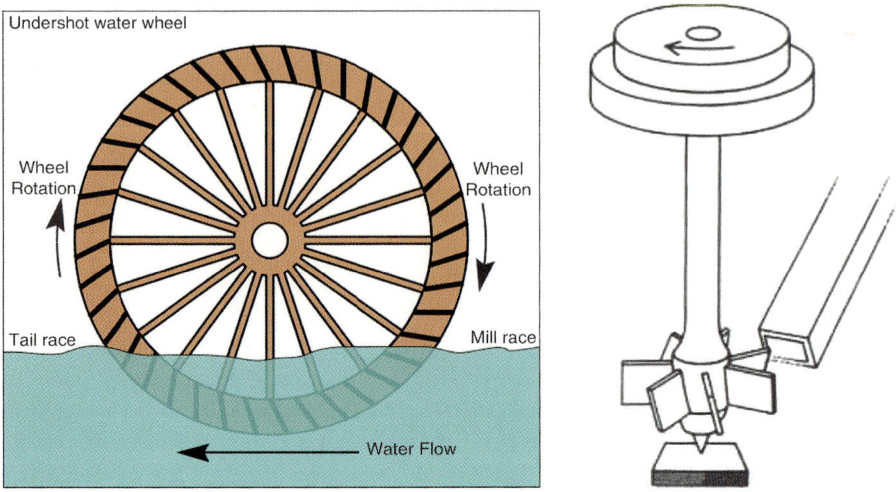

Fig. 4.25 Operation principle of the water wheel engaged by the flow

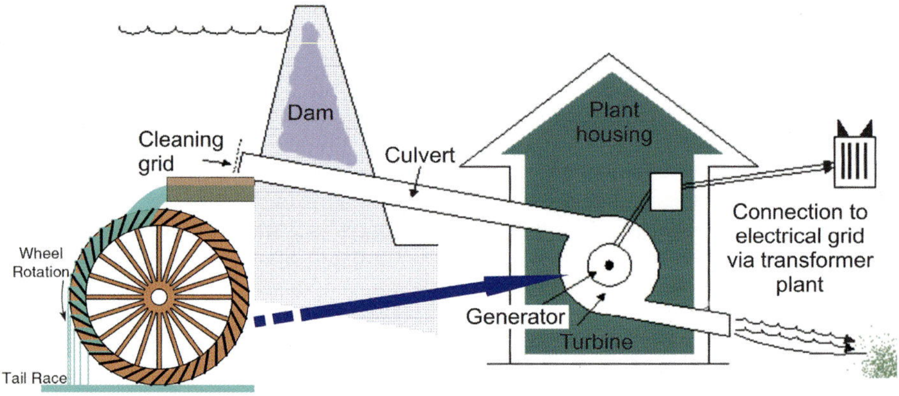

Fig. 4.26 Main components of a micro hydroelectric power plant with dam

- *Hydraulic turbine:* is a part of the plant where water energy is converted into mechanical energy;
- *Generator:* mechanical energy transmitted to turbine maintains the generator rotor velocity producing electrical energy;
- *Transformer station and transmission line:* electrical energy is transmitted and transformed in order to be connected to the grid with further delivery to the consumers.

The quantity of generated energy depends on two factors:

– *Vertical water fall height* (VFH): the higher water fall, the greater the power generated;

Table 4.7 Classification of hydraulic turbines

	Classification by heights of water falling		
Turbine type	High (>50 m)	Medium (10–50 m)	Low (<10 m)
With impulse	Pelton	Banki	**Cross flow**
	Turgo	Turgo	
	Pelton multi-jet	Pelton Multi-jet	
With reaction		Francis (spiral case)	Francis (open) Kaplan with propeller

– *Water flow rate moving through the turbine*: the power produced is proportional to the volume of water passing through the turbine in a unit time (seconds, minutes).

4.2.2.2 Classification of Water Turbines

The potential energy of water can be converted into mechanical energy in the turbine through two fundamental, but absolutely different, mechanisms:

- Water pressure may apply a force on the blade surface, which decreases when passing through the turbine. The turbines operating under this mechanism are called reaction turbines. They are fully submerged in the water flow. Francis and Kaplan turbines fall into this category.
- Water pressure is converted into kinetic energy before entering the turbine. The kinetic energy is in the form of a high speed jet acting on some blades, mounted at the periphery of the turbine. The turbines, operating under this mechanism are called impulse turbines. The most usual turbine of this type is the Pelton turbine.

Micro hydroelectric power systems for the conversion of water potential energy are divided into two major categories of turbines (Table 4.7):

- Turbines for big water heights and small flow rates, impulse turbines.
- Turbines for small water heights and big flow rates, reaction turbines.

There is a wide range of structural and functional varieties of water turbines. The choice of turbine depends primarily on the height available (for the conversion of potential energy) and the parameters of the river (to convert the kinetic energy of running water). According to Table 4.7, in a simple way, the hydraulic turbines can be classified in turbines of high, medium and low height according to the water fall height (VFH).

Turbines for big water heights and small flow rates, impulse turbines (Fig. 4.27). The power produced by an impulse turbine is integrally the result of the moment of water impact on the turbine blades. This water creates a direct or impulse push of blades, hence the name.

Turbines for small water heights and big flow rates, reaction turbines (Fig. 4.28). The reaction turbines are rotated by the water force of reaction hitting the rotor blades. They can operate at very small water heights of up to 0.6 m, but need a much larger quantity of water compared with *the impulse turbines*.

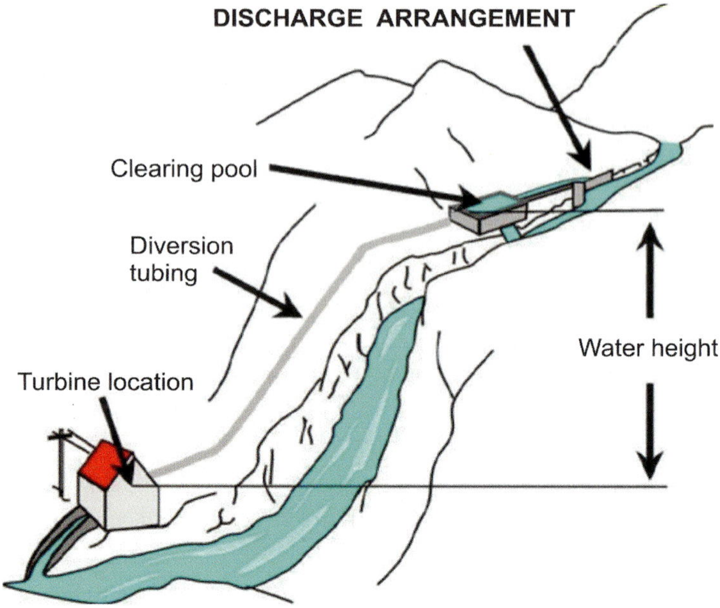

Fig. 4.27 Diagram of micro hydroelectric power plant with big height and small flow rates

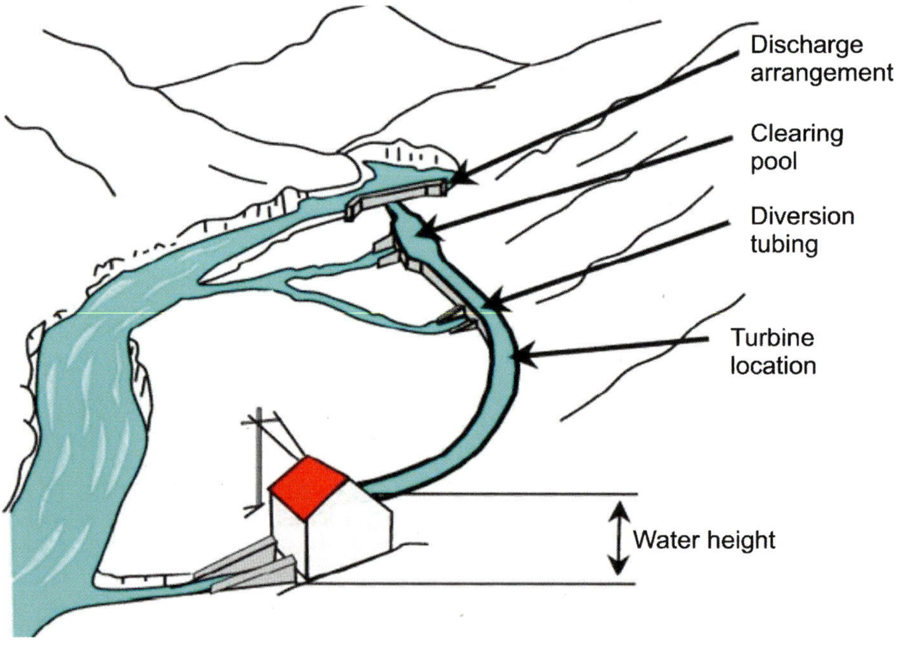

Fig. 4.28 Diagram of micro hydroelectric power plant with small height and big flow rates

The turbines used for small or medium water falls are most frequently reaction turbines and include the Francis and Kaplan turbines with fixed or variable blades. The turbines used for large installations are impulse turbines, including the Pelton, Turgo and Cross-flow (cross flow) turbines. The turbine with cross flow is sometimes called the Cross-flow turbine. It is used for a wide range of falls, covering the areas of Kaplan, Francis and Pelton turbines. It is suitable for water running with high flow rates and small falls. The turbine selection, geometry and dimensions depend mainly on the fall, flow rate and rotor speed. Water *force* is actually a combination of the water fall height and flow rate (or the stream). The water fall height is the pressure created by the vertical distance between the place where water enters the influent conduit and the location of the turbine, and is measured in meters or as pressure.

The flow rate is the *amount* of water (in volume per time) that flows through the influent conduit in a given period of time and is measured in cubic meters per second or litres per minute. Water is collected in a micro basin and then channelled through a pipe directly into the turbine inlet. The vertical water fall (WFH) creates the pressure needed at the lower end of the influent conduit to make the turbine move. The higher the flow or vertical water fall height will be, the more electricity we get. As shown, the values of these two criteria are important for determining the amount of electricity (potential) for a location in order to implement a micro hydro system based on micro hydro power turbines. The difference between the impulse and reaction turbines can be explained in a simple way: impulse turbines can operate in a wide range of vertical heights, and reaction turbines operate in a wide range of flows, but at low heights.

The working principle of the reaction turbines is similar to the principle of the first steam turbine invented at the beginning of our era by the ancient inventor Heron of Alexandria (Fig. 4.29).

The device includes a chamber (spherical or cylindrical) with two diametrically opposite nozzles located in areas perpendicular to the radius of rotation. When the steam (or water) passes through the nozzles under pressure (according to the Newton's third law) the chamber rotates. The reaction turbine rotor is completely immersed in the water currents, and the hydrodynamic profile of blades generate a lift force. Reaction turbines exploit water currents for the generation of hydrodynamic lift forces that rotate the blades. All reaction turbines have a diffuser known as the exhaust pipe through which water is discharged. The exhaust pipe reduces the static pressure of water on the diffuser and thus increases the effectiveness of the turbine. Approximate scales of height, flow and power applied to different types of turbines are shown in the diagram (Fig. 4.30) (up to 500 kW).

These data are approximate and depend on the design and manufacturing accuracy. Figure 4.31 shows the areas of application of impulse (action) and reaction turbines [21]. Mostly used impulse turbines are the Pelton, Turgo and Cross-flow turbines; the reaction turbines are the Francis and Kaplan turbines with propellers. An important factor in the comparison of different types of turbine is their efficiency. To better understand the operation of water turbines, the following is an explanation of their basic parameters [22].

Fig. 4.29 Reaction engine of Heron of Alexandria

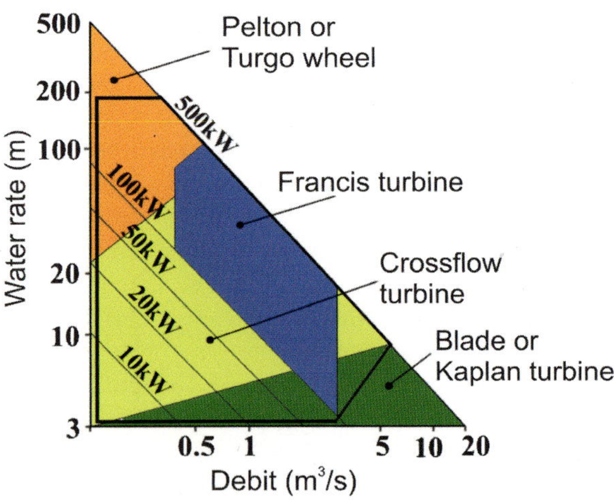

Fig. 4.30 Vertical water fall height-flow rate scale of the micro hydro power turbines

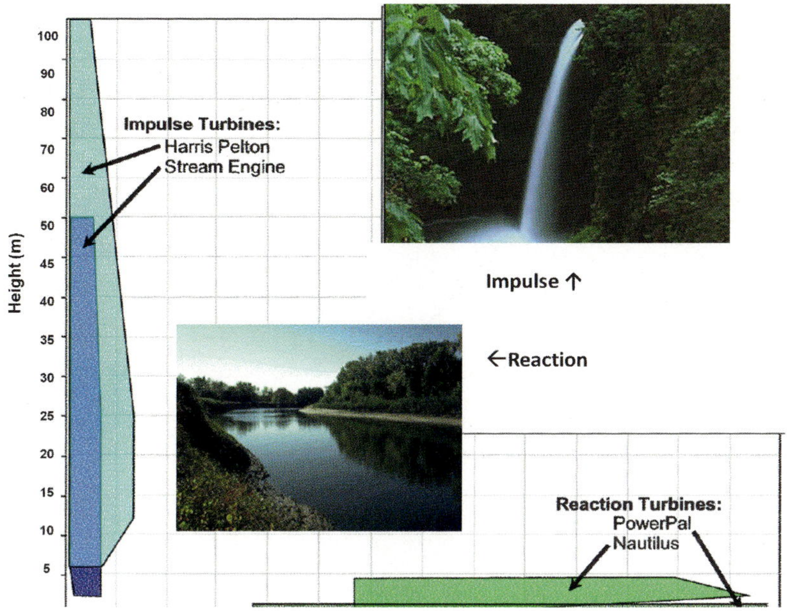

Fig. 4.31 Application areas of water impulse and reaction turbines

Hydraulic power of the turbine is expressed by the relation:

$$Ph = \rho Q \cdot gH\,[\mathrm{W}] \tag{4.14}$$

where:

ρQ is the flow mass rate [kg/s];
ρ – is water specific density [kg/m^3];
Q – is the flow rate [m^3/s];
gH – is the specific hydraulic energy of machine [J/kg];
g – is gravity acceleration [m/s^2];
H – useful height [m].

Mechanical power output of the turbine is determined from the relationship:

$$P_{mec} = Ph\eta\,[\mathrm{W}] \tag{4.15}$$

where: η is turbine efficiency.

Machine specific hydraulic energy is defined by the following:

$$E = gH = \frac{1}{\rho}(p_1 - p_2) + \frac{1}{2}\left(c_1^2 - c_2^2\right) + g(z_1 - z_2)\ [\mathrm{m}] \tag{4.16}$$

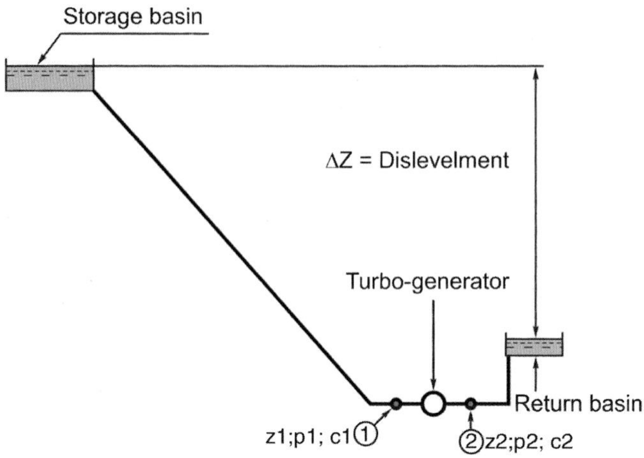

Fig. 4.32 General view of hydroelectric plant diagram and measument sections

where: gH is the specific hydraulic energy of machine [J/kg];

p_x – pressure in section x [Pa];
c_x – water velocity in section x [m/s];
z_x – height of section x [m].

Indices 1 and 2 define the input and, respectively, output measurements of the turbine (Fig. 4.32).

The height is defined as:

$$H_n = \frac{E}{g} \text{ [m]} \tag{4.17}$$

Specific velocity and similarity. Most hydraulic structures are designed and constructed based on the preliminary study of models. Using the similarity theory, the scientific criteria for cataloguing the turbines are determined. The similar geometric model must be the reduced copy of the industrial turbine, following a fixed ratio for all sizes. Physical quantities used in the geometric similarity are the length, area and volume. If the ratio of length is k, then the area ratio is k^2 and the volume will be – k^3.

It is important to note that the testing of models and laboratory developments are the only way to ensure the efficiency of industrial turbines and their hydraulic requirements. All similarity rules are strictly defined by the international standards IEC 60193 and 60041. In accordance with these standards, the specific speed of turbine is defined as:

$$n_{QE} = \frac{n\sqrt{Q}}{E^{1/4}} \tag{4.18}$$

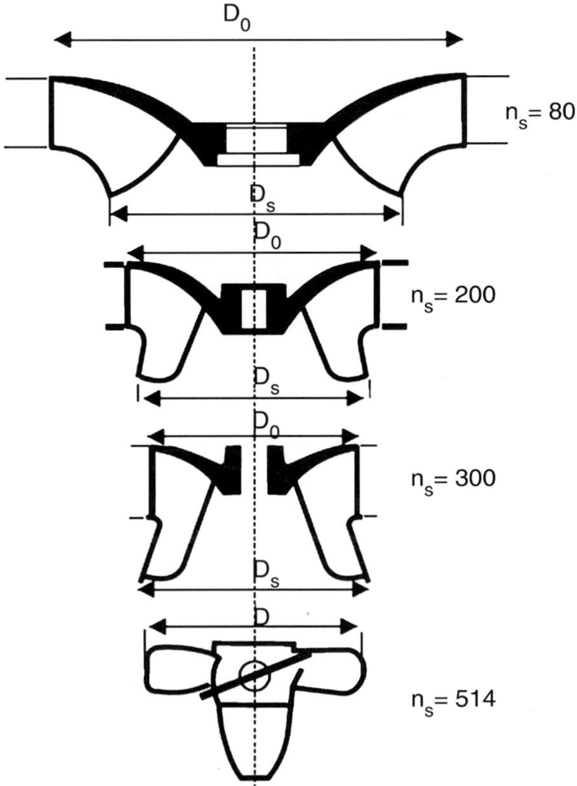

Fig. 4.33 Design of hydraulic wheel depending on specific velocity n_s

where:

Q is the flow rate [m^3/s];
E – is specific hydraulic energy of machine [j/kg];
n – is the turbine speed of rotation [min^{-1}].

Figure 4.33 [21] shows four designs of the hydraulic wheel and their specific velocities, optimised from the point of view of their efficiency. For engineering applications, a great number of statistic studies carried out on a large range of diagrams have allowed reaching the correlation between the specific velocity and the useful height for each type of turbine:

The Pelton turbine (with one nozzle) [23–27]:

$$n_{QE} = \frac{0.0859}{H_n^{0.243}} \tag{4.19}$$

The Francis turbine:

$$n_{QE} = \frac{1.924}{H_n^{0.512}}. \tag{4.20}$$

Table 4.8 Specific velocities for the main turbine types

Pelton with 1 nozzle	$0.025 \leq n_{QE} \leq 0.005$
Pelton with n nozzles	$0.005n^{0.5} \leq n_{QE} \leq 0.025\, n^{0.5}$
Francis	$0.05 \leq n_{QE} \leq 0.33$
Kaplan, propeller	$0.19 \leq n_{QE} \leq 1.55$

Table 4.9 Range of heights

Turbine type	Range of heights (m)
Kaplan and propeller	$2 < H_n < 40$
Francis	$25 < H_n < 350$
Pelton	$50 < H_n < 1{,}300$
Cross-flow	$5 < H_n < 200$
Turgo	$50 < H_n < 250$

The Kaplan turbine:

$$n_{QE} = \frac{2.294}{H_n^{0.486}} \tag{4.21}$$

Propeller (with blades):

$$n_{QE} = \frac{2.716}{H_n^{0.5}} \tag{4.22}$$

The basic dimensions of the turbine can be easily estimated if the specific velocity is known. Table 4.8 lists specific velocities for the main turbine types, and Table 4.9 shows the range of heights.

Further on, some basic laws of similarity are presented:

$$\frac{Q_t}{Q_m} = \frac{\sqrt{H_t}}{\sqrt{H_m}} \cdot \frac{D_t^2}{D_m^2} \tag{4.23}$$

$$\frac{n_t}{n_m} = \frac{\sqrt{H_t}}{\sqrt{H_m}} \cdot \frac{D_t}{D_m} \tag{4.24}$$

Preliminary design. Some statistical formulas are described in order to determine the basic dimensions of hydraulic wheels for Pelton, Francis and Caplan turbines.

For all turbines the first step is the determination of rotational speed:

The Pelton turbine. If we know the speed of the hydraulic wheel, its diameter can be determined from the following equations:

$$D_1 = 0.68 \frac{\sqrt{H_n}}{n} \tag{4.25}$$

$$B_2 = 1.68 \sqrt{\frac{Q}{n_{jet}} \frac{1}{\sqrt{H_n}}} \tag{4.26}$$

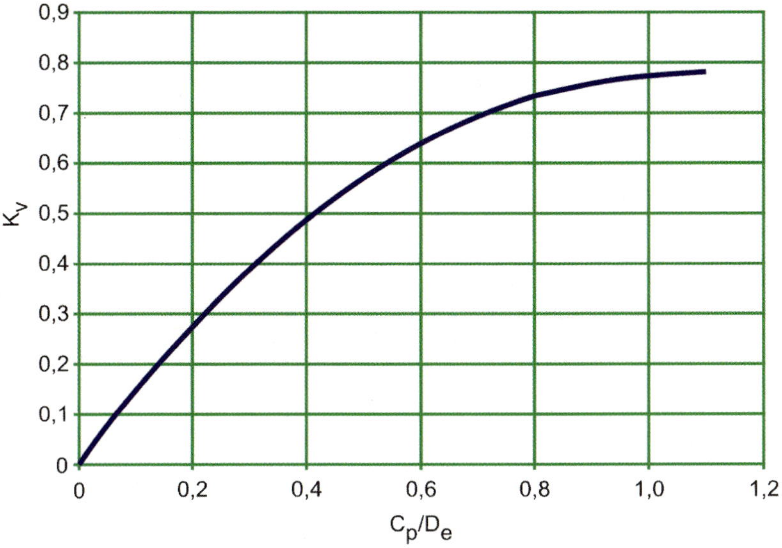

Fig. 4.34 Nozzle characteristics

$$D_e = 1.178\sqrt{\frac{Q}{n_{jet}}\frac{1}{\sqrt{gH}}} \qquad (4.27)$$

where n is rotational speed in *t/s* and n_{jet} – number of jets. D_l is defined as the diameter of circle that describes the blades' centre line. B_2 is the blade width that depends on the number of nozzles. D_e is the nozzle diameter. The general rule is that the ratio D_1/B_2 should always be bigger than 2.7. If this condition is not respected, a new calculation is made for a smaller speed or more jets.

The discharge function of the nozzle opening C_p in a turbine with one nozzle can be defined from the below formula:

$$Q_{jet} = K_v \cdot \pi \cdot \frac{D_e^2}{4} \cdot \sqrt{2gH} \qquad (4.28)$$

where K_v is shown in Fig. 4.34 depending on the relative opening $a = C_p/D_e$ [24].

The basic geometrical parameters of other types of turbines are determined similarly.

Turbine efficiency. Turbine efficiency depends on many factors (cavitation, vibration etc.), that can considerably reduce annual energy production. Therefore, each operator of hydraulic plants must ask from manufacturer an efficiency warranty, based on the laboratory development. It is the only way to be sure that the turbine will be effective. Figure 4.35 shows an example of a real installation developed without efficiency warranty and laboratory testing.

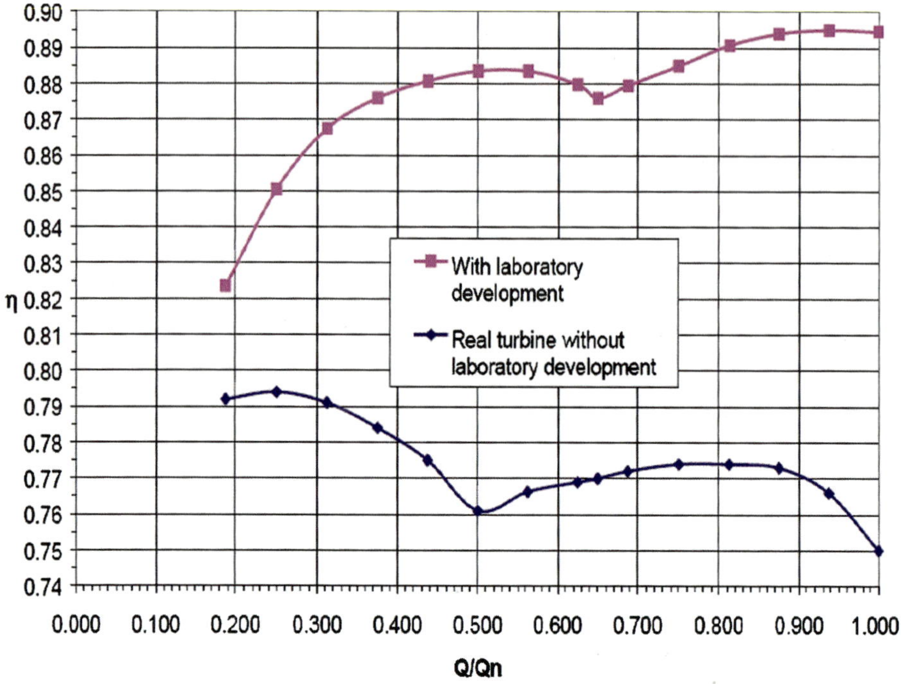

Fig. 4.35 Measured efficiency of a real turbine manufactured without laboratory research

Typical efficiency curves are shown in Fig. 4.36. It is important to note that the Pelton, Cross-flow and Kaplan turbines are highly efficient. On the contrary, the efficiency of most advanced Francis turbine does not exceed 80%.

There are two methods for those who want to determine the turbine power output. The first method envisages testing of installation after the turbine is put into operation. Generally it is necessary to use simple methods which results can be verifiable. If tests show that the guaranteed output power is not obtained then the installation is returned to the producer. Payment of penalties by the producer compensates the production losses of the operator.

The second method consists in carrying out laboratory tests on the turbines that are geometrically similar to industrial prototypes. In the case of small size hydraulic installations, test patterns are similar to those of industrial turbines. It is better to correct any malfunctions before the installation is manufactured. The efficiency guaranteed by the producer is the efficiency that can be verified accordingly [28, 29]. It is defined by the ratio between the turbine produced power and hydraulic power:

$$\eta = \frac{P_{mec.}}{P_h} \tag{4.29}$$

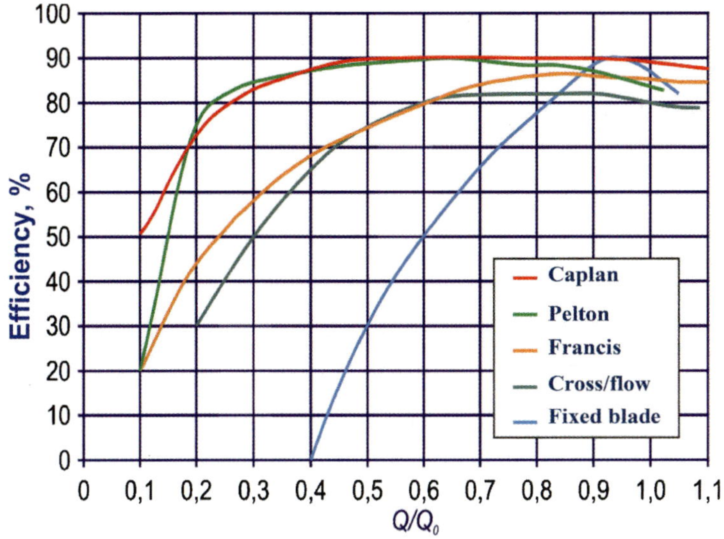

Fig. 4.36 Efficiency of various types of water turbines

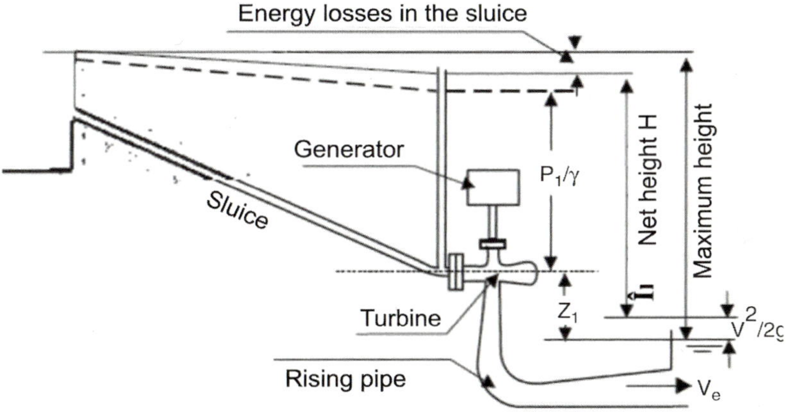

Fig. 4.37 General view of energy losses in the power installation diagram

As seen in Fig. 4.37, the turbine operation depends not only on the turbine efficiency but also on some other factors.

Modern Kaplan and Pelton turbines can function satisfactorily for a large range of streams – from approximately 1/5 of the discharge rate. Acceptable efficiency of common Kaplan turbines is from 1/3 and of Francis turbines – from 1/2 of the discharge rate. Propeller turbines with fixed blades can operate satisfactorily only in a very limited zone, depending on their discharge rate. Table 4.10 lists usual efficiencies of small hydraulic turbines (Kaplan, Francis, and Pelton) in two versions.

Table 4.10 Usual efficiency of small turbines

Turbine type	Optimal efficiency
Usual Kaplan turbine	0.91
Modern Kaplan turbine	0.93
Francis	0.94
Pelton with *n* nozzles	0.90
Pelton with one nozzle	0.89

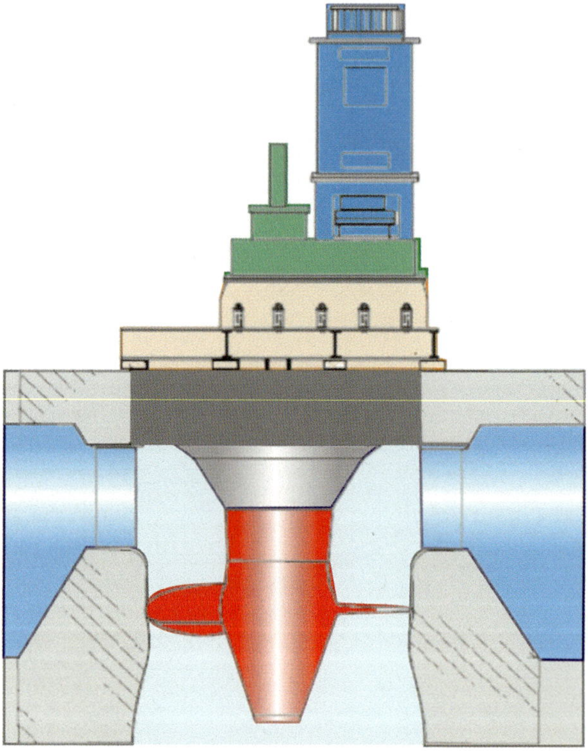

Fig. 4.38 Speed multiplier with parallel axes

Methods of speed multiplication. The speed multipliers, applied in micro hydroelectric power plants designs, can be classified as follows:

- Multiplier with cylindrical wheels which axes are parallel to those of the rotor (applied in micro hydroelectric power plants with vertical rotor). Figure 4.38 [22] shows the vertical configuration coupled to a Kaplan turbine;
- Conical planetary multiplier for shafts mounted at an angle of 90°. Figure 4.39 [22] shows a conical planetary speed multiplier (applied in hydraulic units with horizontal rotor of turbine);
- Belt multiplier (applied in small power hydraulic plants with horizontal axle turbine). It is convenient in maintenance (see Fig. 4.40) [22].

Fig. 4.39 Conical planetary speed multiplier with axes concurrent in space

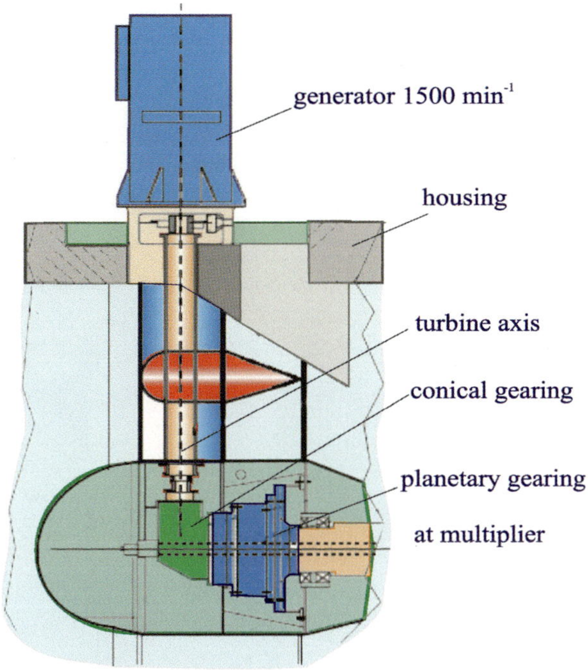

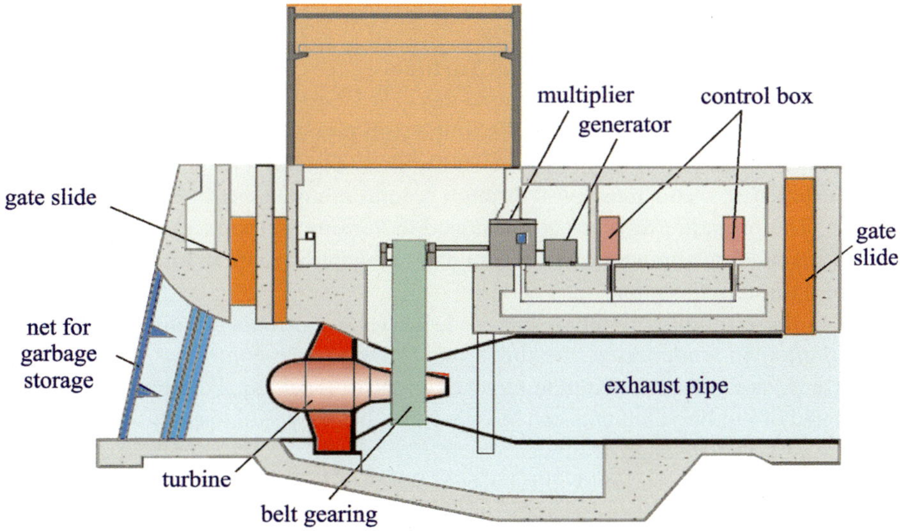

Fig. 4.40 Speed multiplier with belt transmission

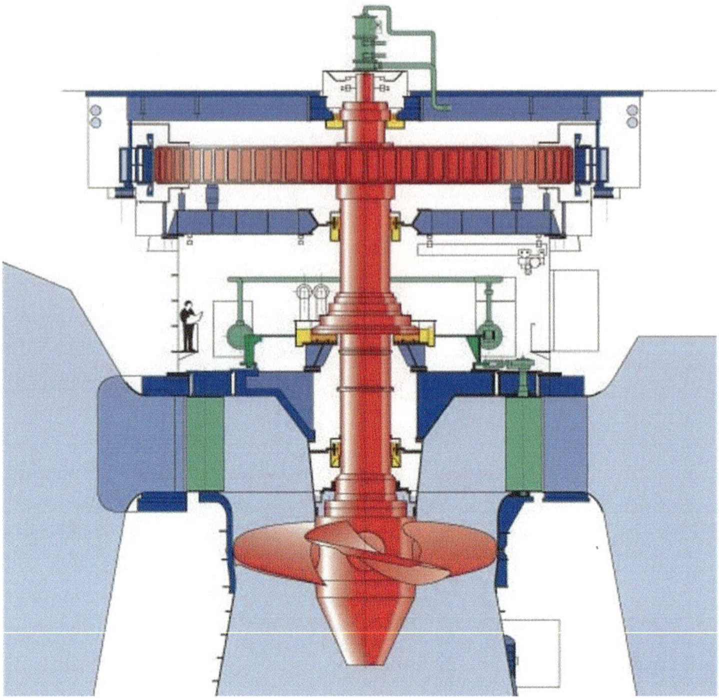

Fig. 4.41 Kaplan water turbine

4.2.2.3 Reaction Turbines: Kaplan Turbine

Propeller turbines are basically similar to the ship propellers, but operate in reverse mode. There is a wide variety of propeller turbines. The key-element is that for their better efficiency, water must pass through several components before entering the turbine. The propeller turbines are designed in various methods. One method is to form a special housing where the water enters tangentially and is forced to perform a spiral motion in the rotor.

When these guiding elements are used, the water currents are directed into the turbine inlet. In some cases, the directing blades can be controlled. In this case, the turbines are called the Kaplan turbines. Control mechanical systems of turbine blades and of guiding elements can be expensive, but they lead to essential increased efficiency of a wide range of flows.

The Kaplan turbine is an axial rotation water turbine with a rotor with adjustable blades, used commonly in plants. The turbine is the invention (1912) of the Austrian professor Viktor Kaplan (Fig. 4.41). This invention is, in fact, improving the Francis turbine (invented by the American engineer James B. Francis in 1849).

The Francis turbine has the problem of *cavitation* (air bubbles in the water current in the turbine), which produces pressure decrease by decreasing the turbine efficiency. This shortcoming is removed by the Kaplan turbine, which uses adjustable

Fig. 4.42 Operation principle of modern Kaplan water turbine

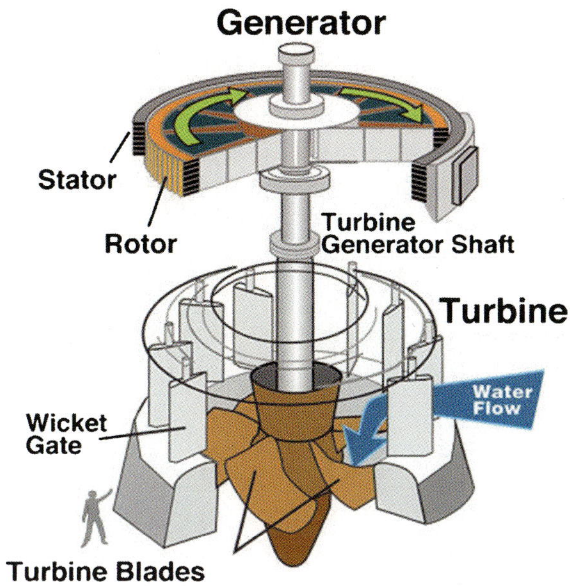

blades; for optimal operation, the turbine requires a water stream with constant flow rate. There are different design variants of the Kaplan turbine. Figure 4.42 shows a modern version of the Kaplan turbine; the turbine is supplemented with additional hydrodynamic profile blades for the orientation of water currents towards the main blades. The turbine operates through the effect of overpressure, the yield reaching 80–95%. A higher current flow rate and a small water fall are optimal for the Kaplan turbine.

Optimization of operational parameters of the Kaplan turbine is facilitated today by modern computers performances and related software. Figure 4.43 presents CFD numerical simulation of the turbine. At the basis of classical calculation of axial turbine is the classic method of calculation proposed in the nineteenth century by the Swiss mathematician Leonhard Euler:

$$D = m \cdot v \cdot r \tag{4.30}$$

Modification of the rotational impulse position in the centre of turbine:

$$M = \frac{dD}{dt} = \frac{m}{dt} \cdot r \cdot dc \tag{4.31}$$

With the help of the integral and taking into account the laws of levers, the below formula is obtained:

$$P = \dot{m} \cdot \int_1^2 u \cdot dc_u \tag{4.32}$$

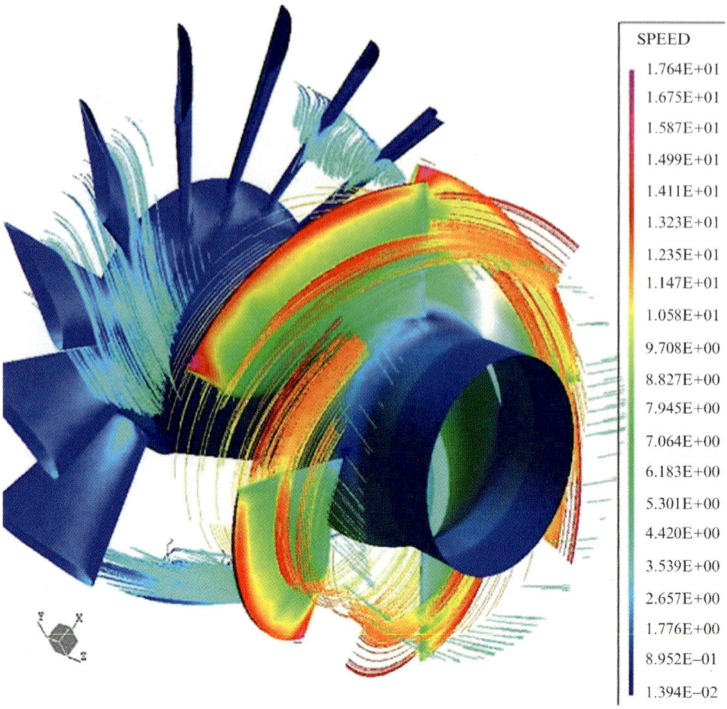

Fig. 4.43 CFD numerical simulation. Speed distribution for the Kaplan turbine

From where

$$\frac{P}{m} = \int_{1}^{2} u \cdot dc_u = Y \tag{4.33}$$

The last equation is called the Euler equation, which results in:

$$Y = u_2 \cdot c_{u2} - u_1 \cdot c_{u1} \tag{4.34}$$

where Y is the work performed by the turbine blade, u_1 – is the rotational speed of the blade point at inlet and at outlet – u_2; the fluid speed in the turbine is c_{u1} – at inlet and c_{u2} – at outlet.

4.2.2.4 Reaction Turbine: Francis Turbine

The Francis turbine is a type of water turbine designed by James B. Francis (Fig. 4.44). It is a flow reaction turbine that combines the radial and axial concepts.

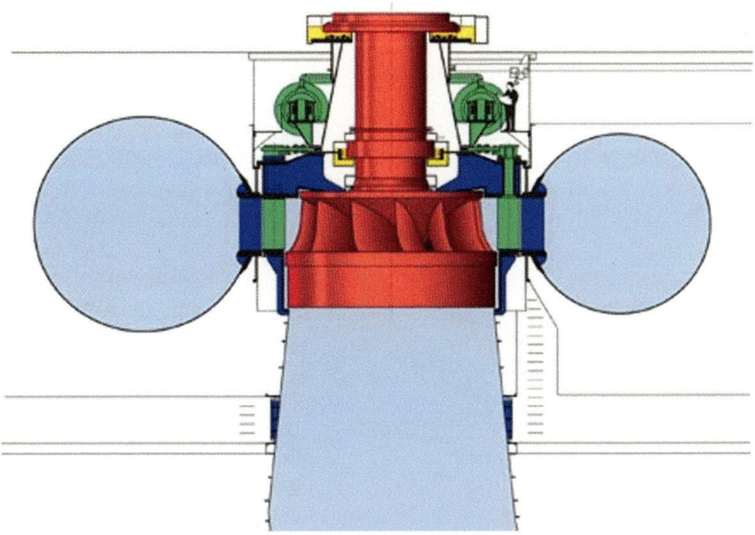

Fig. 4.44 Francis turbine

Nowadays, the Francis turbines are the most used water turbines. They operate in a range of heights from 10 to several hundred meters and are used in particular for electricity generation. The Francis turbine is a turbine with essentially modified propeller in which the water currents act radially on the blades at inlet and axially at outlet.

The Francis turbine was designed as a low height machine, installed in a chamber without spiral housing. The Francis turbines have nine or more blades mounted on the rotor. Water enters the turbine through the blades mounted in vertical position and leaves the turbine at outlet, changing their direction at 90°. The Francis turbines operate at water fall heights from 1 m to 500 m and can grow to 800 MW power output. Thousands of these turbines have been installed in the UK and elsewhere in Europe from 1920 to 1960. Figure 4.45 gives an example of computer simulation of water streams action on the blades, which allows optimization of the blade geometry. Figure 4.46 presents the core components, performances and industrial prototype of a hydraulic system based on the Francis turbine.

The power output of this system can exceed 2 kW for certain values of the water fall height and flow rate. The Francis turbines are improved continuously. The emergence of computers and powerful software has opened up new possibilities in this respect. Current research is carried out to upgrade the Francis turbines in terms of increasing their efficiency. The Francis turbines are used at the largest hydroelectric plants in the world [30]: Grand Culee (USA) – 611.5 MW; Geheyan (China) – 310 MW; Ertan (China) – 582 MW; Chirchill Falls (Canada) – 507 MW; Three Gorges (China) – 710 MW; Harspranget 4 (Sweden) – 469 MW.

Fig. 4.45 Computer simulation of water streams action on the blades of Francis turbine

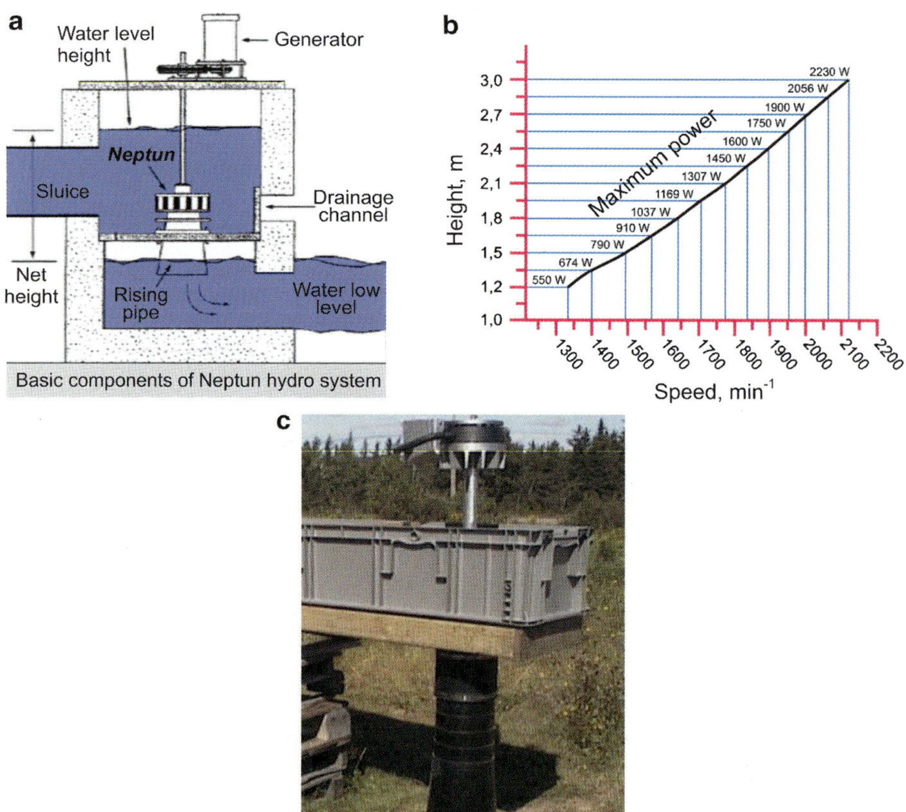

Fig. 4.46 Francis system

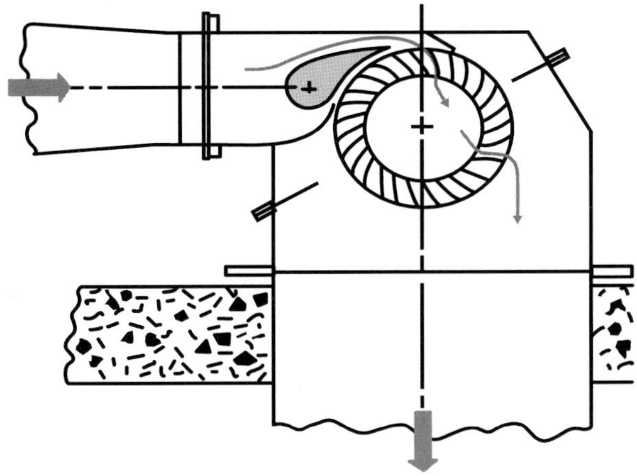

Fig. 4.47 Operation principle opf Pelton turbines with common buckets

4.2.2.5 Impulse Turbines: Pelton Turbine

Lester Pelton, the inventor of the turbine that bears his name, was born in 1829 in Vermillion, Ohio, USA. The Pelton turbine's operating principle is shown in Fig. 4.47 [31]. The Pelton turbine is a wheel with a series of buckets installed at the periphery. A water jet (Fig. 4.47) or more jets (Fig. 4.48), with a high speed, is directed tangentially to the wheel. Almost all water energy comes to spin the buckets, and the reflected water currents are discharged through a discharge pipe.

Figure 4.48 shows the operation principle of the Pelton turbine with double buckets – the figure is copied from the original patent of Pelton (1880). The Pelton water turbines are applied for big heights (30–1,000 m) and can develop power from 10 kW to 10 MW [32–34]. Generally speaking, the Pelton turbines of small dimensions need minimum volume of water streams, being ideal for small rivers. Figure 4.49 shows a model of the Pelton turbine with usual buckets (a) and with double buckets (b). The operation principle of mono- and double buckets is presented in Fig. 4.50 [15]. In the case of double buckets the water jet divides into two flows that interact with each of the two parts of the bucket.

Another type of impulse turbine is the Turgo turbine. The Turgo turbines are similar to Pelton turbines, but the water jet is designed to act on the blades plane at an angle (usually 20°) and the water has only one inlet and one exit. Consequently the Turgo turbines can have rotor diameters smaller than the Pelton turbines for the same power equivalents, but are more difficult to manufacture.

4.2.2.6 Cross-flow Turbine

A *cross-flow turbine*, *Banki-Michell turbine*, or *Ossberger turbine* is a water turbine developed by the Australian Anthony Michell, the Hungarian Donát Bánki and the German Fritz Ossberger. Ossberger patented (1922) and manufactured the turbine as a standard product becoming a world leader.

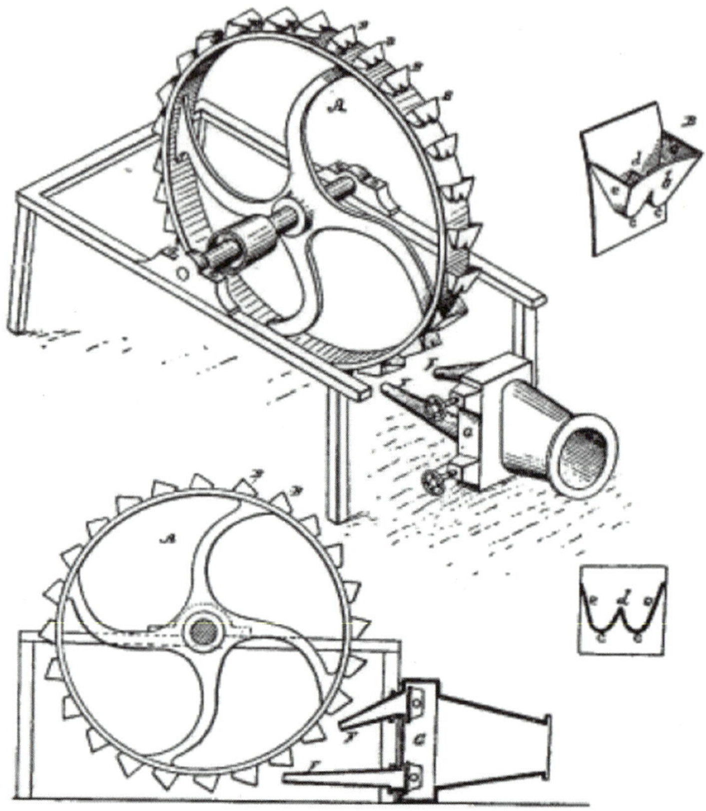

Fig. 4.48 Operation principle opf Pelton turbines with double buckets

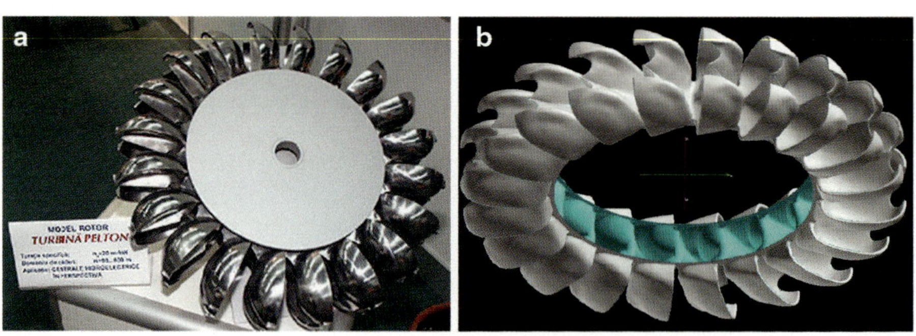

Fig. 4.49 Pelton turbine model

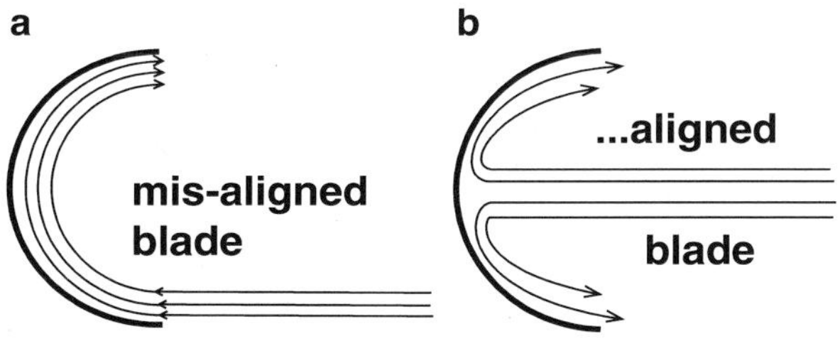

Fig. 4.50 Operation principle of mono- (**a**) and double (**b**) buckets

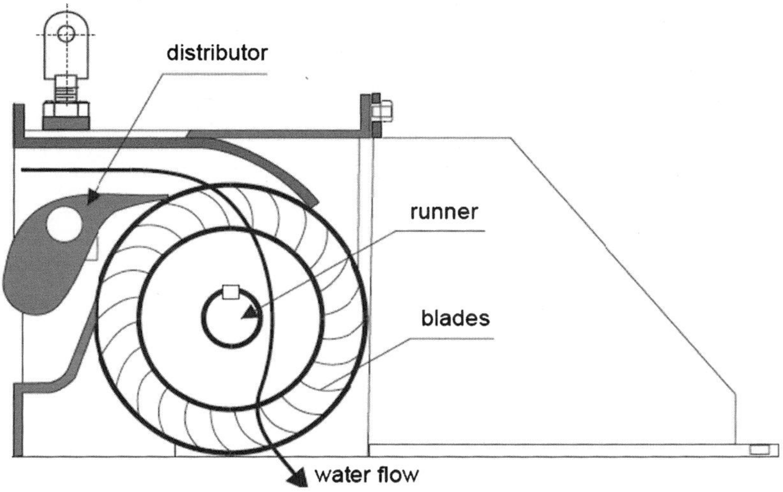

Fig. 4.51 Operation principle of Cross-flow turbine

Banki turbine (similar to Cross-flow and Ossberger turbines) is a hydro power system apparently similar to a water wheel. Unlike the water wheel, it uses a nozzle and bucket-shaped blades (Fig. 4.51) [22]. The middle of turbine is free and the blades which form the wheel has inclined and curved forms. The currents of water that enter the turbine, do not spin only the wheel by their weight, but because the currents pass through the blades and through the turbine middle, their direction is changed.

This allows extraction of additional energy according to the Newton's Third Law. Venturi element enhances this effect by high speed water connection to the blades. Crossflow turbines use both concepts in the construction of the impulse turbine and reaction turbine. This allows them to be more efficient in a wider range of heights. The conceptual system is similar to Pelton, but requires less engineering. The cross flow turbine has lower efficiency than other turbine designs, but earns the market due to the low cost and possibility of installation in the households. A wheel with reasonable efficiency (70–80% to about 4 m height) can be constructed from pipe sections. Venturi flow element and the associated ones can be made from metal sheets. It is more complicated as an open water wheel, but not difficult to manufacture.

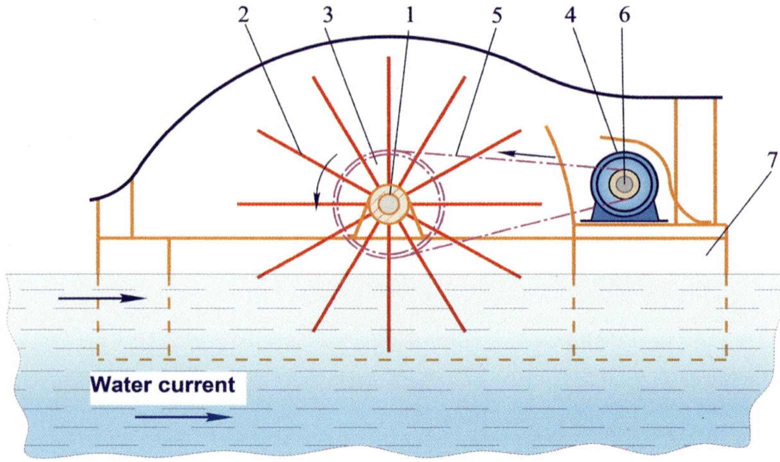

Fig. 4.52 Floating micro hydroelectric power plant with water wheel

4.2.3 Floating Micro Hydroelectric Power Plants for River Water Kinetic Energy Conversion

Floating micro hydro power plants are of special interest. In terms of costs, floating micro hydro power plants are efficient because they do not include essential costs related to civil engineering [34]. The conceptual scheme of these micro hydroelectric power plants is shown in Fig. 4.52. The water wheels or the open hydro turbines (usually, with propellers) are used as working body. Next the most representative examples of small flow hydropower plants with different working bodies will be considered.

4.2.3.1 Floating Micro Hydroelectric Power Plant with Water Wheel

The floating micro hydro power plant (Fig. 4.52) is the achievement of an ancient idea: the old water wheel installed on a floating platform, transforms the kinetic energy of flowing water from rivers into electricity or mechanical energy (for irrigation). The design of micro hydro power plant is simple. Rotational movement of the water wheel with the horizontal main axis 1 and blades 2 by belt pulley 3, is transmitted to the electric generator 4 through the belt transmission (belt pulley 3, 6 and belt 5). The water wheel and the electric generator are installed on the floating platform 7. The main advantages of the micro hydro power plant are: the lack of dams which excludes the negative environmental impact; the automatic control of the water wheel position depending on the water level and the running water course.

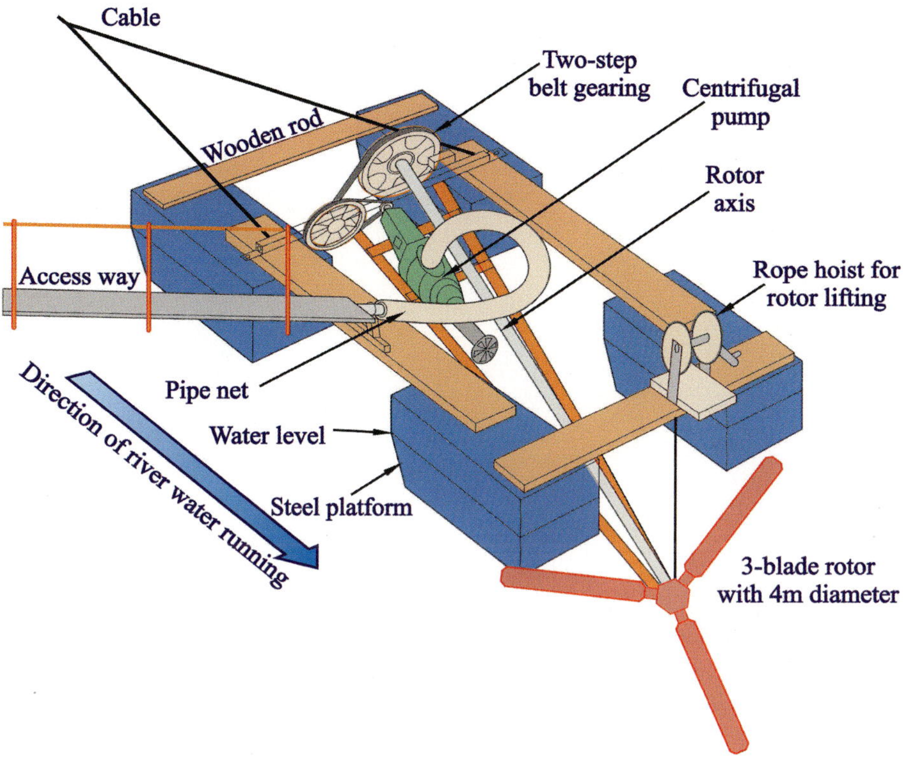

Fig. 4.53 Micro hydro power plant with 3-bladed rotor with inclined axis, Garman type

4.2.3.2 Floating Micro Hydroelectric Power Plant with Three Bladed Rotor

A design of floating micro hydropower plant, which can be made from the materials, available to a wide circle of amateur builders, is shown in Fig. 4.53 [35]. Micro hydro power plant includes a Garman-type three-bladed rotor. The three-bladed rotor converts kinetic energy of running water in rotational motion, that is transmitted via a system of belt transmissions to a hydraulic pump. The efficiency of this micro hydro power plant is 15, ..., 16%. Figure 4.54 shows the diagram of generated power dependence on the water flow velocity and rotor diameter. Despite its reduced efficiency the described micro hydro power plant gains by its design simplicity and reduced cost.

4.2.3.3 Floating Micro Hydroelectric Power Plant with Gorlov Turbine

A special interest presents the hydraulic turbine, invented by A. Gorlov (mechanical engineering professor at the Northeastern University in Boston (USA)). According

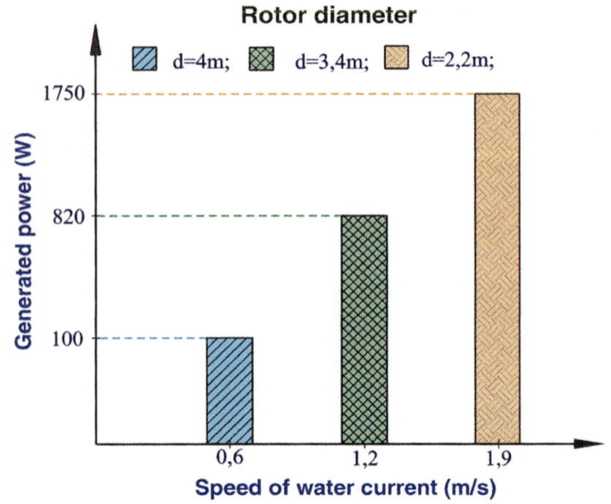

Fig. 4.54 Diagrams of generated power dependence on the water flow velocity

to A. Gorlov, 90% of the kinetic energy of water is in areas where dams cannot be built. He says his turbines are ideal for these areas with water and tidal currents, for which conventional hydropower is too expensive. The Gorlov turbines achieve a high conversion coefficient of hydraulic energy up to 35% (theoretically, this ratio is 0.59 – Betz coefficient). Turbines with 1 m diameter and1 m height can be installed in large numbers along the coast, without harming fish populations. The Gorlov turbine (Fig. 4.55) [36] was developed based on the Darrieus turbine, invented for the wind energy conversion in 1930. It has several advantages: simple construction, low cost, estimated at about $400–600 for 1 kW power – less than the construction of hydroelectric dams systems, and practically does not generate noise. These turbines can be connected in series. Figure 4.55b shows the construction of Gorlov turbine, and in Fig. 4.56 – the experimental model of Gorlov turbine. Four Gorlov turbines were tested at Cape Cod, and a 5 kW system has been installed, which generates sufficient electricity to supply a nearby motel with 14 rooms, in the zone of ebb in Vinalhaven, Maine. Another test took place in a region of the Amazon River, where turbines have been used to charge car batteries used for TVs. A. Gorlov proposes to use his turbines for the decomposition of seawater into hydrogen and oxygen by electrolysis, and the storage of hydrogen in pressure vessels. Hydrogen will be used in fuel cells. The turbines and equipment are tested, respectively, in the Uldomok Channel, southwestern coast of South Korea, and tidal energy farms are planned with the total power of 90 MW. If these tests are successful, Korea will develop near the island of Jindo a system of 3,500 MW [36].

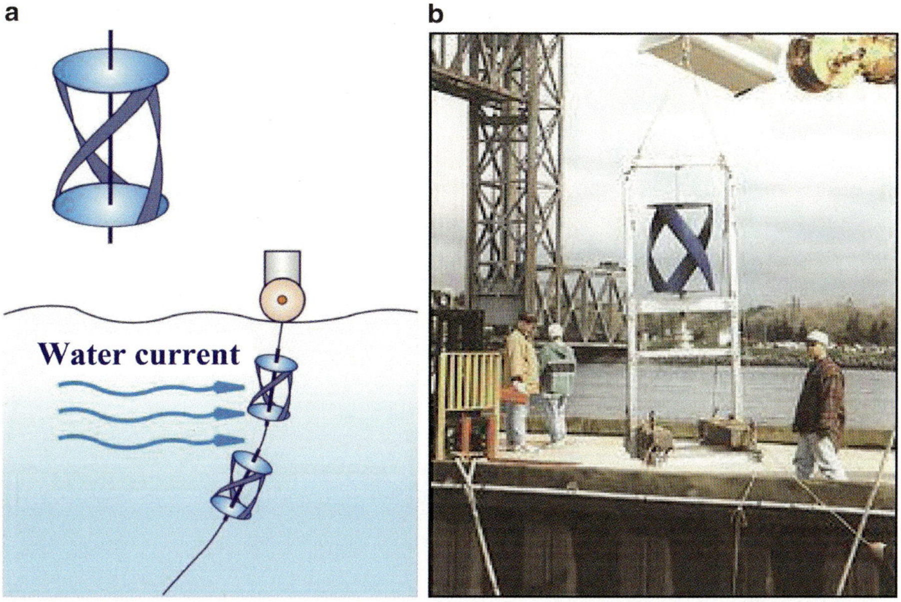

Fig. 4.55 Helical turbine, invention by de A. Gorlov

Fig. 4.56 Experimental model

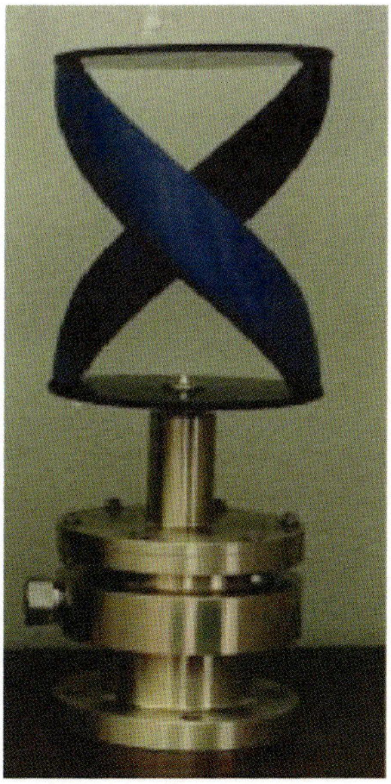

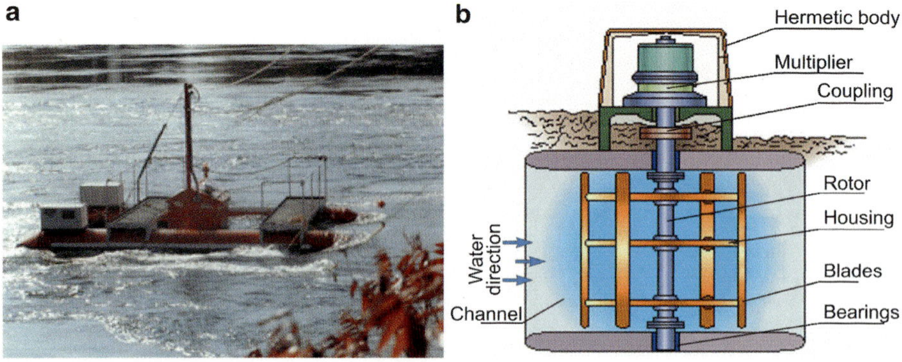

Fig. 4.57 Micro hydro power plant with Davis turbine

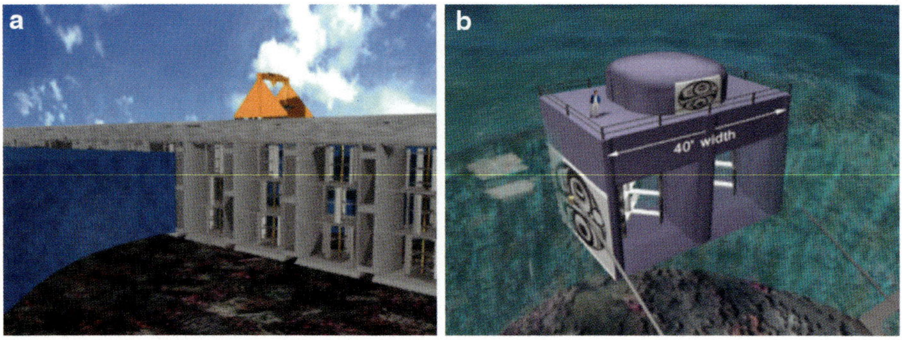

Fig. 4.58 Computer models of Davis rotor modules: (**a**) for oceans, (**b**) for rivers

4.2.3.4 Floating Micro Hydroelectric Power Plant with Davis Turbine

The Davis hydraulic turbine is of special interest [37]. In 1984 a small hydro power plant with Davis hydraulic turbine and vertical axis was designed and tested on the Harbour River, Nova Scotia, (Fig. 4.57a); it produced 100 kWh. The Davis turbine consists of four blades with hydrodynamic profile, fixed rigidly on the rotor (Fig. 4.57b).

The rotor is coupled through a coupling to the multiplier input shaft (planetary gear of speed multiplication). The power generator is mounted on the multiplier output shaft. All nodes are mounted on the platform, that is fixed on two pontoons. To obtain electricity from the kinetic energy of the running water a similar construction of the Davis turbine was built and tested in Florida Bay (Fig. 4.58) in 1985. The obtained power was 5 kW. The project was funded by *Nova Energy Ltd* company, Texas, and was called the Davis turbine VEGA–I.

The turbine was operating at a 65 m depth, in order to impede its collision with large cargo ships. To get more power several Davis turbines were joined

Fig. 4.59 Davis turbine mounting in Florida Bay

in modules (blocks) mounted vertically or horizontally. Thus it was possible to obtain a summary power range of the kinetic energy of running river water from 5 to 500 kW and a 200–8,000 MW power – from the kinetic energy of the ocean water. Figure 4.58a, b shows the computer models of the Davis turbine modules of 7–14 MW power – for oceans and of 250 kW power – for rivers. Figure 4.59 shows the installation of the Davis turbine in Florida, USA.

4.2.3.5 Floating Micro Hydroelectric Power Plant with Horizontal Axis and Multi-blade Turbine

Micro hydropower plants with horizontal axis turbines are more efficient for some shallow water rivers. Micro hydropower plant with horizontal axis (Fig. 4.60) [38] was developed at Krasnoyarsk State Technical University, Russia, and is designed to convert the kinetic energy of running river water into electricity. The micro hydropower plant can operate efficiently with other energy sources (e.g. diesel power plant), reducing drastically the consumption of fuel or replacing it by other in the case of emergency. The key hydrological parameters that ensure the efficient operation of micro hydropower plants are the following: the range of water flow velocity for river depth of, at least, 1.5 m; the segment width for this depth of rivers at least 10 m.

The outlet power of river water flow velocity between 1.8–2.5 m/s is from 10 to 30 kW, for three phase direct current voltage 220/380 V with 50 Hz frequency. The dimensions of submersible micro hydroelectric power plant: length – 5 m, width – 3 m, height – 2 m. The turbine type: horizontal sectional, each section having two blades with hydrodynamic profile rigidly fixed on the main shaft. Each section is

Fig. 4.60 Submersible micro hydro power plant with horizontal axis

fixed on the main shaft with a certain angular retardation. In front, two nozzles are fixed on both sides of rotor. The main shaft is installed on the structure of resistance.

4.2.3.6 Floating Micro Hydroelectric Power Plant with Horizontal Axis and Propeller Turbine

Another type of floating micro hydropower plants having a new principle of operation is the micro hydroelectric power plants with helical turbine. One of the first works is a report on a prototype of horizontal axis turbine developed by Harwood (1985), National Institute of Amazonian Research (INPA). It used two propeller turbines with a 4 m diameter, mounted in the nozzle. For micro hydro power plant equilibrium, the turbines are made with rotation in opposite directions. The rotational movement of the turbines is summed and transmitted through a multiplier system to the electric generator. The turbine housings are rigidly fixed on a metal construction, which in its turn is rigidly installed on two pontoons. The micro hydropower plant is anchored to the river. This equipment has been tested in areas of the Amazon River at water speeds between 0.7 to 1.5 m/s. Figure 4.61 shows a computer model of the floating micro hydropower plant with two multi-blade turbines mounted in the nozzles.

The rotation of turbines in different directions ensures stability to the micro hydropower plant. Based on this principle, the engineers from the Scientific Research Institute in Novosibirsk, Russia, manufactured a micro hydropower plant with two propeller turbines (Fig. 4.62) [38]. The micro hydropower plant contains a floating movable structure, which can be easily moved throughout the riverbed when the water level changes and, as well, improve the turbine efficiency on the basis of a more efficient use of the running water energy. The micro hydropower plant contains a platform on which the following components are rigidly fixed: the generator; the housings of the propeller turbines with the nozzles; the floating bodies in the form of pontoons mounted to the platform; the lifting and sinking mechanism of the propeller turbines. The testing of this construction was carried out on the rivers in Altai and Iakutia, Russia, during a whole season. The tests have shown good efficiency, technical characteristics, operation and installation simplicity.

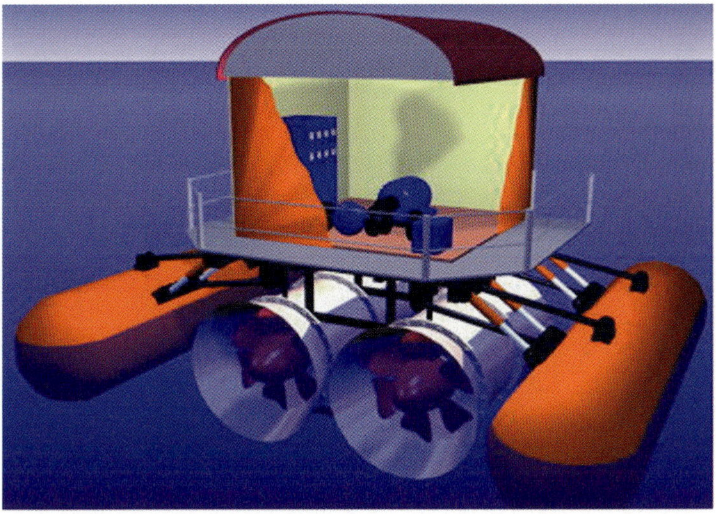

Fig. 4.61 Micro hydropower plant with two multi-blade turbines

Fig. 4.62 Floating micro hydropower plant

4.2.3.7 Micro Hydroelectric Power Plants Integrated with Other Renewable Energy Conversion Systems

One of the most common applications of alternative energy is power supply of isolated consumers: a vacation house or cottage, a motel or other social objects, located in an area without access to the public grid. To ensure fully the electricity needs, micro hydro power plants are often integrated into a complex energy system that includes both renewable energy conversion systems (wind, solar, thermal, biomass, hydrogen, etc.) and conventional energy systems (diesel and gas

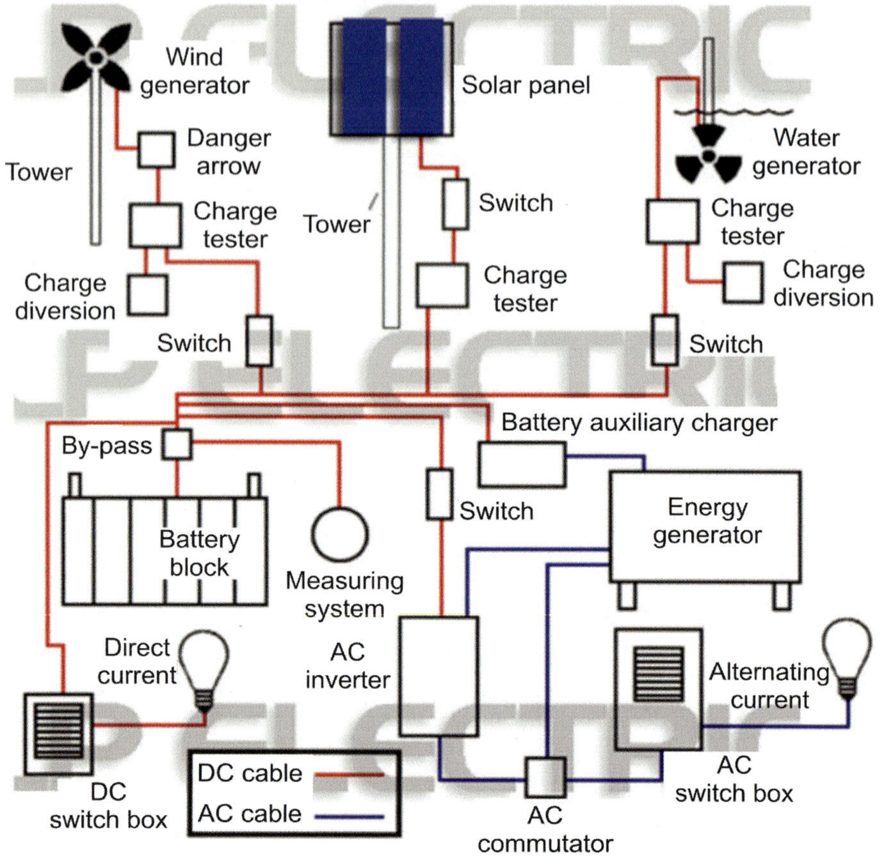

Fig. 4.63 Diagram of an individual power system for home, cottage, etc.

stations, etc.). Integration can be done in two ways: integration into a single power system of several renewable energy conversion systems (wind, hydro, solar, etc.); creation of joint operating energy facilities (e.g. the main shaft, which is linked to power generator, is driven by both a wind turbine and a hydraulic turbine). Their combined use is always possible. Figure 4.63 shows a diagram of an individual power system for home, cottage, motel, etc., based on the use of hydraulic, solar and wind energy. Figure 4.64 shows a modern cottage, which energy needs are satisfied by an integrated complex system, based on the use of hydraulic, solar and wind energy. It is difficult to imagine a modern cottage or a motel located in an inaccessible location without refrigerator, television, lighting, hot water, stereo, microwave and other essential elements of a comfort level requirements.

A schematic diagram of a mixed wind-hydraulic power installation is shown in Fig. 4.65 [39]. The installation includes a wind turbine and a hydraulic turbine for

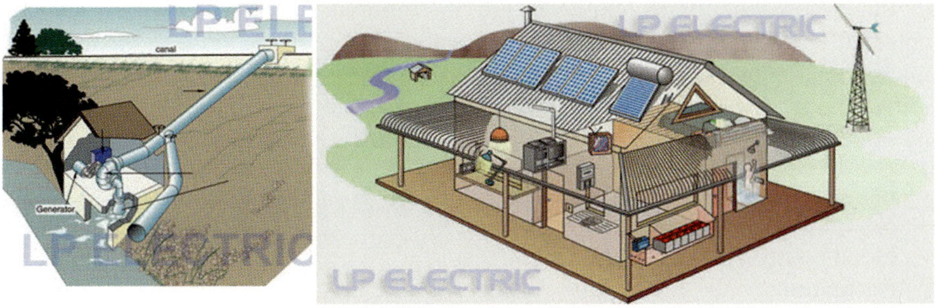

Fig. 4.64 Modern cottage with a power supply complex system

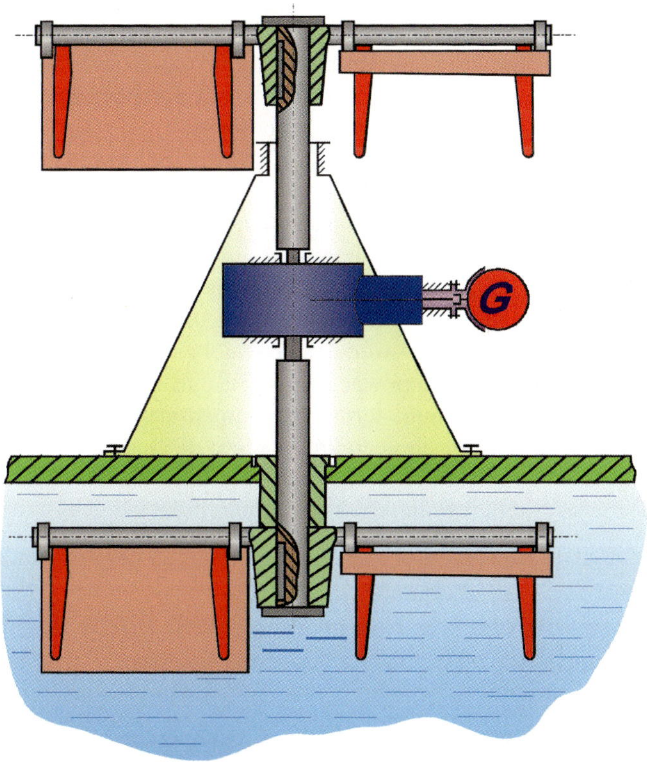

Fig. 4.65 Principle diagram of a wind-hydropower installation with mixed operation

the conversion of flowing rivers kinetic energy. Both working bodies are connected to a differential which, in its turn, sums rotational movements from both working bodies, multiplies and transmit them to the electric generator. This technical solution allows the conversion of both the hydraulic energy and wind energy, which in some areas of rivers with steep sides form a veritable wind tunnel.

Renewable energy technologies are intermittent in nature, so they are not constantly available. Solar energy is not available at night. Wind stations will stop in the lack of wind. Although micro hydro power plants are the most reliable renewable energy source, they also depend on the water flow rate, which is dictated by multicriteria optimization considering the following variables:

- Irrigation;
- Navigation;
- Flood control;
- Recreation;
- Energy demand.

It is therefore appropriate to integrate renewable energy conversion systems. Globally, some activities have been initiated with limited scope for the exploration of associated purposes with the operation of integrated hydro-wind power. The results obtained from the growing number of pilot studies and investigations, and he synthesis of best practices are being enriched in this respect. The lessons learned in the past can now be investigated for future projects.

The program will initiate case studies of hydro-wind power integration in the USA. The program will also involve international experts in activities related to the integration of hydroelectric power and its application on the US market. The main objective of the program is to establish a dialogue between energy producers and to accelerate immediate maintenance activities and market development of hydropower sustainable market under ecological, economic and political aspects.

In some cases, where hydropower is not widely applicable because of environmental reasons, micro hydro power plants serve as buffer systems in ensuring electricity for short and medium terms to compensate energy fluctuations and service costs, and to increase the economic value of electricity delivered. There have been discussed opportunities for hydropower large-scale integration with the wind and solar energy, but they have not been quantified, yet.

4.3 Micro Hydroelectric Power Plant with Pintle and Blades with Rectilinear Profile in Normal Section

The analysis of hydraulic energy conversion systems that was carried out in the previous parts of the hand book has demonstrated the opportunity for the development of water kinetic energy conversion systems compared to potential energy conversion systems. A number of advantages are observed obviously.

- *Technical advantages*: relatively simple hydraulic energy conversion systems.
- *Economic advantages*: the cost of civil engineering works is reduced considerably.
- *Ecological advantages*: absence of dams and storage basins.

The analysis of existing micro hydroelectric power plants for river water kinetic energy conversion has pointed that there are reserves to increase the efficiency of the utilized turbines. Betz coefficient, equal to 0.59, represents the maximum theoretical efficiency of the hydraulic energy conversion. The majority of the existing systems provides an output factor (coefficient) for water kinetic energy in the value range of 0.2. Only certain modern systems exceed the efficiency by over 30%. In this respect there are sufficient reserves to increase the efficiency of the flow hydraulic turbines, which become more and more attractive to the engineers and inventors in the field. Therefore, the Centre for Renewable Energy Conversion Systems Design was set up at the Technical University of Moldova. The Centre employs qualified staff and is endowed with advanced design and research equipment. Further, the authors describe several modern solutions increasing the efficiency of the river water kinetic energy conversion systems. Some research results were published in the monograph [40], exhibited at international fairs [41–43], and described in research reports [44].

4.3.1 Conceptual Diagrams

As a result of an ample study that was carried out, a conceptual diagram of the micro hydroelectric power plant with pintle (vertical axis) (Fig. 4.66a) was proposed; it comprises a 4-blade working element with pintle and horizontal blades [45]. The working element contains a shaft *1*, connected rigidly to the can *2*, and the rotating axles *4* and *5* are fixed in the cages *3* of the can. Blades *6* and *7,* positioned reciprocally under an acute angle (<90), are mounted on the ends of axles *4* and *5.* Axle *4* is incorruptible (Fig. 4.66b), and axle *5* is made of two separated parts *8* and *9*, interconnected rigidly through an intermediate support *10*. The intermediate support *10* allows alternative rotation of parts *8* and *9,* of axle *5,* in the boundaries of angle α (Fig. 4.66b) provided by axle *4*. Axle *4*, in its turn, has the possibility to rotate under angle α in the limit buffer *11* (Fig. 4.66c). The ailerons *12* and *13*, oriented under a certain angle towards the remnant part of the blade, are fixed rigidly on the ends of blades *6* and *7*. In its turn, the main shaft *1* is connected to the axles of multiplier *14* and, consequently, to the generator *15*. The platform (decking) *16* is mounted on a mainland base by a parallelogram mechanism *17*. The generator *15* and the multiplier *14* are fixed on the platform.

The hydraulic power plant operates, as follows: axles *4 and 5* with the blades *6* and *7* are located in the running river water. One blade reaches the water stream, and under its action it moves rotating the shaft under a certain angle until it gets out of the water stream action. At this moment, another blade is brought onto the way of the water stream by the shaft. The blade that escaped from the water stream influence is moved against the water stream by the shaft. In order to decrease the resistance that has occurred as the blade moves against the stream, the ailerons *12* and *13,* driven by the water stream, rotate the blades *6* and *7*, (while they are positioned against the water current) and get them out of the water current action, significantly reducing the resistance. At the same time, the blade that is fixed on the

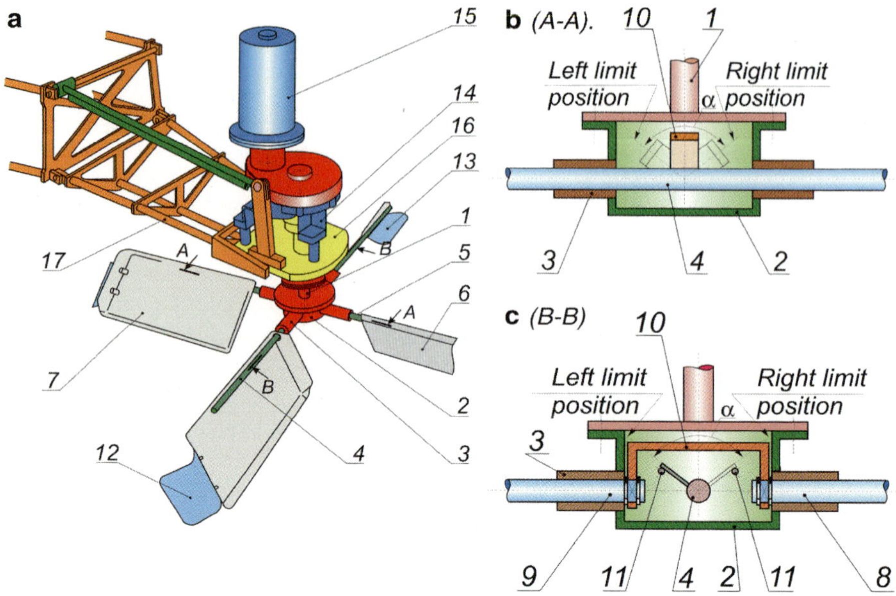

Fig. 4.66 Diagram of the micro hydropower plant with pintle (vertical axis) and horizontal blades with rectilinear profile

other end of the axle rotates under an angle α in the position of maximum resistance (in operating position). Thus, the position of blade maximum resistance (operating position) and the position of blade smallest resistance (the position of blade motion against the water current) will be provided. The rotary motion of the shaft *1* is multiplied by multiplier *14* and is further transmitted to generator *15* that produces electric current. The parallelogram mechanism *17* ensures vertical position of the shaft *1* for any level of water current. The shortcoming of this diagram consists in the relatively reduced conversion factor of water kinetic energy into electrical energy (simultaneously, on the average, only one blade, out of four, participates in the conversion process and it uses only the water pressure on the blade surface).

The increase in the number of blades of the micro hydroelectric power plant rotor (Fig. 4.67) [46] provides a higher uniformity of the generator shaft speed and a higher level of conversion. At the same time, its construction is more complex and demands a more sophisticated orientation mechanism for the blades. The curvilinear shape of blades *2* provides a higher conversion factor and a reduced degree of water turbulence.

4.3.2 Modeling of Blades Interaction with the Water

The efficiency of the micro hydroelectric power plant depends on the working element being the basic one. The design of the working element in the form of a

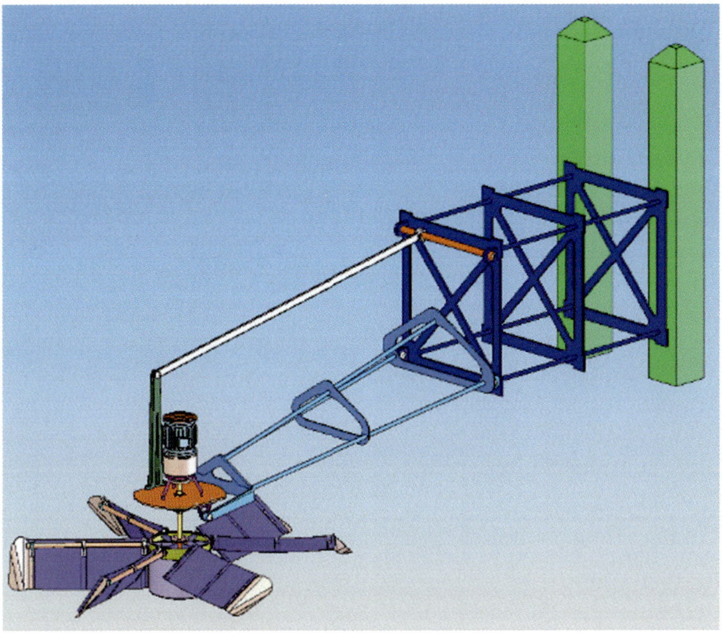

Fig. 4.67 Diagram of the micro hydropower plant with pintle (vertical axis) and horizontal blades

multiblade rotor with straight or curvilinear profile has demanded the determination of blades optimal sizes under resistance conditions, the stability of their form under the action of the water currents, their behaviour at various water velocities, the rational number of blades in terms of certain constructive or functional parameters. With this purpose it was necessary to carry out multicriteria research on the blades' interaction with the fluid (water). To justify their constructive and functional parameters, a modern design station and ANSYS CFX5.7 software, purchased within a grant funded by the Civilian Research and Development Foundation (CRDF) from the USA [47], was employed. High performance of CFX-5 media is based not only on one product, but presents a proper combination of advanced technologies integrated into all its modules. Computer simulation of the blades – fluid interaction for various shapes of blades was carried out in the digital analysis media CFD-ANSYS CFX5.7. As well, research was carried out on the distribution of pressure on the blades' surface and on the turbulence phenomena [41–43, 48]. The diagrams (Fig. 4.68) show some details of the comparative analysis of the effects generated by the blades with straight and curvilinear surfaces. The analysis of the obtained images has demonstrated a higher efficiency of the blades with curvilinear surface and a determined radius of curvature, expressed by

- More rational distribution of stress in the blade structure generated by the action of the water currents;
- More reduced degree of water turbulence in the blade rear.

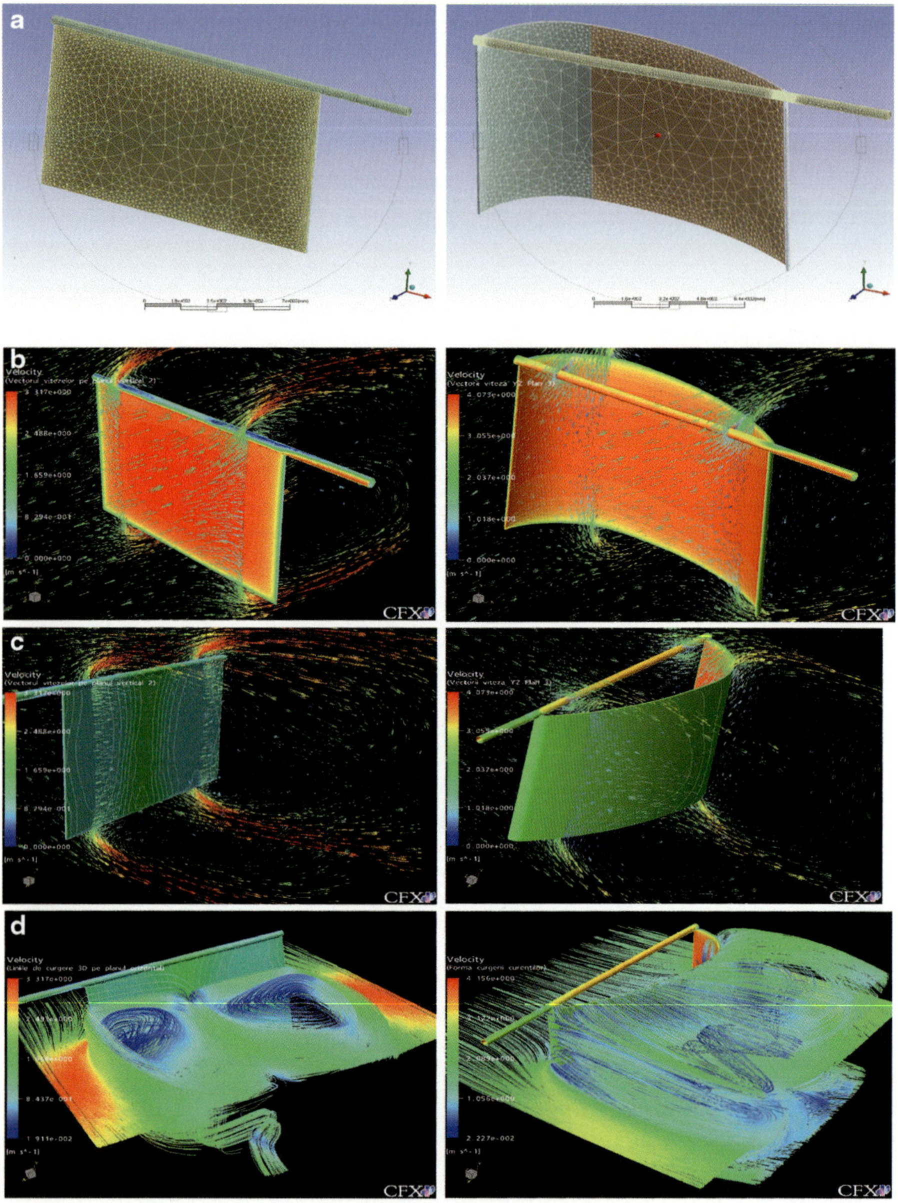

Fig. 4.68 (**a**) Division of multiblade rotor in finite element (tetrahedron), CFX CAD2Mesh. (**b**) Vector of flow rates on two vertical planes along the flow (upstream), CFX Post. (**c**) Vector of flow rates on two vertical planes along the flow (downstream),CFX Post. (**d**) 3D fluid flow lines in a horizontal plane along the flow (upstream), CFX Post

Fig. 4.69 The micro hydro power station laboratory model with vertical axis and horizontal straight blades

4.3.3 Laboratory Testing of the Micro Hydroelectric Power Plant with Pintle

As mentioned previously, it was necessary to manufacture and test a laboratory model of the plant in order to establish the efficiency of the micro hydroelectric power plant. On the basis of the theoretical research and computer modeling of water interaction with the rotor blades, a stand for experimental testing in laboratory conditions was developed, designed and manufactured (Fig. 4.69). The stand comprises a multifunctional experimental model of the micro hydroelectric power plant with *pintle* (made according to the conceptual diagram, Fig. 4.66), that provides water-pumping into the tank where water streaming has been simulated; two pumps (confined into the casing) that provide water pumping into the tank with determined pressures in order to simulate various water flow rates; and a computerized measuring and data processing system using LabVIEW operational system. The experimental model has ensured the simulation of water flow rate in a wide range of values, under various degrees of turbulence, height adjustment of the turbine depending on the water level, etc. The stand design has allowed the testing of a large number of working elements: with pintle and horizontal blades with straight (Fig. 4.66) and curvilinear (Fig. 4.67) surfaces.

4.4 Micro Hydroelectric Power Plant with Horizontal Axle and Helical Turbine

4.4.1 Conceptual Diagrams

Being in search of some constructive solutions to increase the conversion efficiency of the water kinetic energy by enlarging the contact surface of the blades with water, an old idea was addressed – the Archimedean screw – on which basis the helical turbine with two, three or four inceptions was designed. In order to increase the turbine energy efficiency several original ideas were considered:

- The utilization of the hydrodynamic profile of blades in normal section that allows an injection effect of the peripheral pool of water and the generation of a secondary buoyant force;
- The design of the helical turbines with exterior cylindrical, conical or parabolic surface allows the optimization of energy parameters of the turbine by reducing the power losses of water flow input, passage and outlet;
- The mounting of the helical turbines in casings endowed with convergent nozzle (at turbine input) and diffuser (at turbine output) with hydrodynamic profile in normal section allows the increase of energy efficiency of the helical turbine. Figure 4.70 [40, 49–51] shows the conceptual diagrams of the micro hydroelectric power plant with helical turbine that comprises two pontoons fixed in-line on a platform (decking). The helical turbine is laid out between the pontoons and is fixed in such a way as to ensure constant submergence of the turbine.

Fig. 4.70 Floating micro hydropower plant with helical turbine

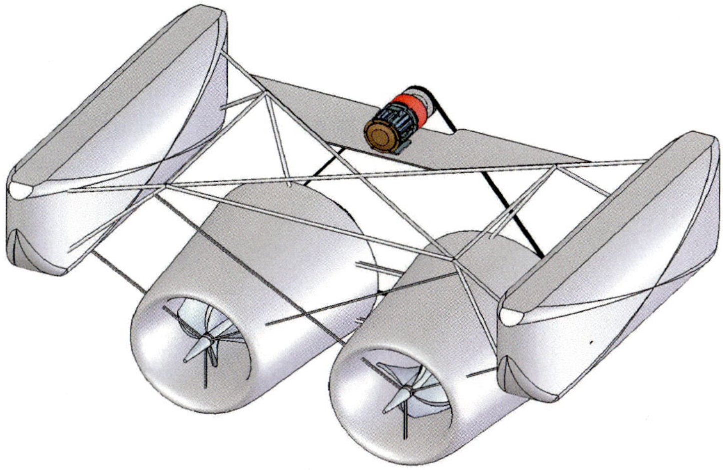

Fig. 4.71 Floatable micro hydropower plant with two helical turbines

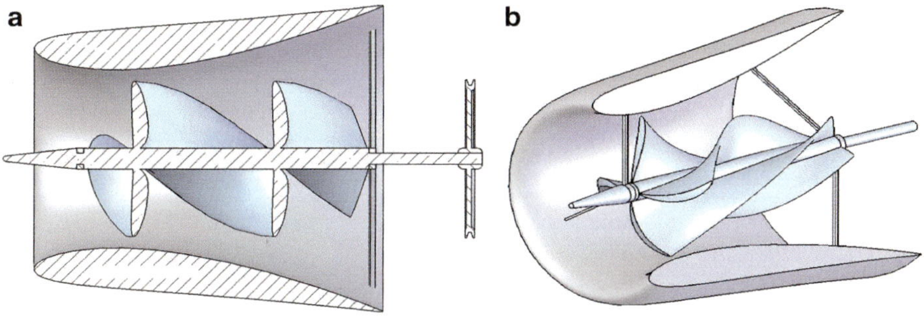

Fig. 4.72 Cylindrical and conical helical turbines with hydrodynamic profile convergent nozzle-diffuser

The helical turbine is spaced into a casing with convergent nozzle and diffuser, and its shaft is connected to the generator via a system of mechanical transmissions. To compensate the moment of rotation of the floatable micro hydroelectric power plant and to increase its energy efficiency, the micro hydroelectric power plant (Fig. 4.71) comprises two helical turbines interconnected via a system of mechanical transmissions with a single generator.

The reverse direction of the turbine helical lines provides the rotation in reverse directions of the rotors (impellers) of helical turbines, and the compensation of the moment of rotation of the micro hydroelectric power plant. The exterior surface of the helical turbine can be cylindrical (Fig. 4.72a) or conical (Fig. 4.72b).

The walls of the casing in the diffuser part (at input) are hydrodynamic in normal section (Fig. 4.72a). The floatable micro hydroelectric power plant with helical turbines is anchored on the bottom of the water source or on its banks (Fig. 4.73).

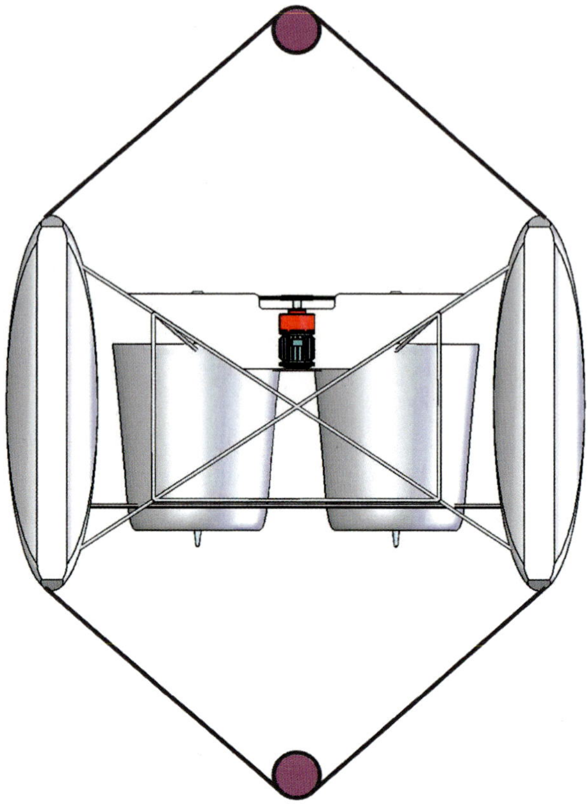

Fig. 4.73 Diagram of micro hydropower plant anchoring

4.4.2 *Some Aspects of Analytic Description of the Basic Geometrical Parameters*

The mathematic model of the helical turbine was developed with the purpose to carry out theoretical research, and for the optimization of its geometrical parameters. The description of the mathematic model is presented in [41–43, 50, 51]. Further on, some aspects of analytic description of the basic geometrical parameters will be described. According to Fig. 4.74, the helical turbine comprises the shaft 1 on which the blades 2 are fixed rigidly in helical line with constant pitch. The aerodynamic profile (Fig. 4.74b) is characterized by gibbous shape in its frontal part and by pointed shape in its rear part, respectively.

The pitch line of the aerodynamic profile is considered the geometrical position of the inscribed circle centers in the profile. The basic geometrical parameters of the profile are as follows:

- Relative thickness of profile $\vec{c}$ is the relation between the maximum thickness of the profile and the chord length b, $\vec{c} = c/b$;

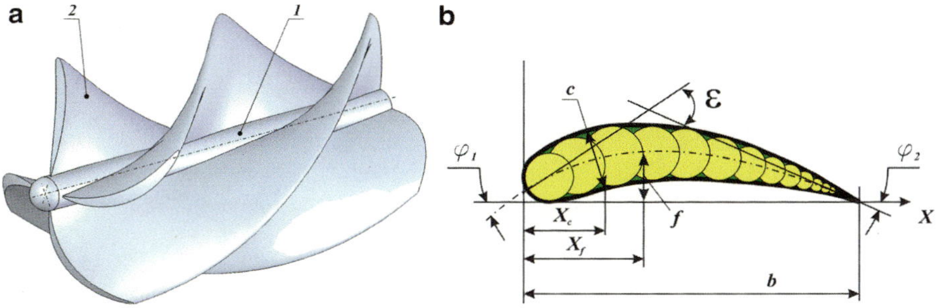

Fig. 4.74 (**a**) Helical turbine. (**b**) Aerodynamic profile of blade in normal section

- Relative hollow $\vec{f}$ is the relation between the maximum amount of deflection of the axial curve and the chord length b, $\vec{f} = f/b$;
- Profile curvature is characterized by the bending angle of the centre line $\boldsymbol{\varepsilon}$ that is the angle between the tangents drawn to the centre line of the profile in its frontal and rear parts.

Positions $\vec{c}$ and $\vec{f}$ are determined by the relative x-coordinate $\vec{x}_f = x/b$ and $\vec{x}_c = x/b$. Knowing the uplift (ascending) force that reacts on the profile lattice it is possible to determine the moments of force by which the water reacts on the turbine. The reciprocal position of the profiles in the lattice is characterized by the pitch t, the angle of sight Θ (the angle between the wing chord and the lattice flank) and angles φ_1 and φ_2 between the tangents drawn to the centre line of the profile in its points and the lattice flank. The pitch ratio of the lattice is the relation between the pitch t and the chord length b, $\vec{t} = t/b$. The reciprocal value of the pitch ratio is called lattice density $\varepsilon = 1/t = b/t$. The position of the lattice profile towards the water direction is characterized by the entering angle i, that is the angle between the velocity vector and the tangent drawn to the centre line of the profile.

To determine the forces developed by the water flows, the profile infinite lattice careening by a permanent parallel water flow plane was analyzed. Applying the equation of the kinetic (motional) quantity for the mentioned volume *ABCD* (Fig. 4.75):

$$F\Delta T = m\vec{\omega}_2 - m\vec{\omega}_1 \tag{4.35}$$

the forces that react on the profile have been determined finally as:

$$P = \sqrt{P_U^2 + P_Z^2} = \rho G\sqrt{\omega_Z^2 + [(\omega_{1U} + \alpha\omega_{2U})/2]^2} \tag{4.36}$$

The direction of the mean geometrical velocity is determined by:

$$ctg\beta_m = \frac{\omega_{mU}}{\omega_{mZ}} = \frac{\omega_{1U} + \omega_{2U}}{2\omega_Z} = \frac{1}{2}\left(\frac{\omega_{1U}}{\omega_Z} + \frac{\omega_{2U}}{\omega_Z}\right) = \frac{1}{2}(ctg\beta_1 + ctg\beta_2) \tag{4.37}$$

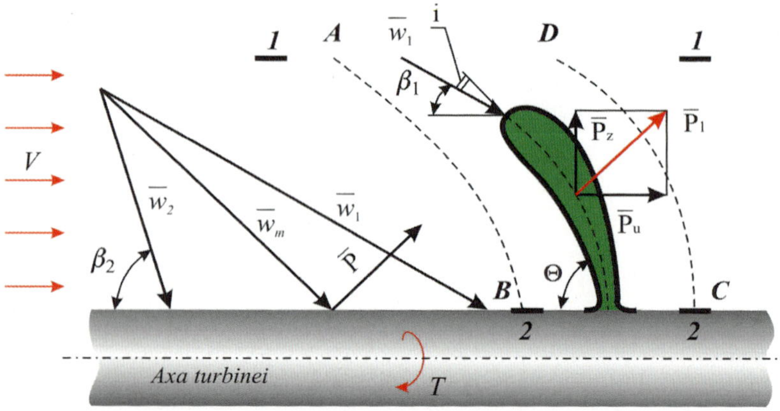

Fig. 4.75 Calculation diagram of the forces developed by water currents

Thus, the resultant of the forces that react on the lattice from the water side is equal to the product of density, mean geometrical velocity and velocity circulation round the profile. Its direction of action is perpendicular to the vector of water mean geometrical velocity.

4.4.3 Modeling of the Blades Interaction with Water

A series of conceptual diagrams of helical turbines was designed to undertake multicriteria research on the helical turbine. Figure 4.76 shows the possible diagrams of helical turbines:

- Helical turbine with exterior cylindrical surface and cylindrical core;
- Helical turbine with exterior conical surface and cylindrical core;
- Helical turbine with exterior parabolic surface and cylindrical core;
- Helical turbine with exterior cylindrical surface and conical core;
- Helical turbine with exterior conical surface and conical core;
- Helical turbine with exterior parabolic surface and conical core.

The diagrams of the propeller turbine can be also designed with constant and variable pitch of the helical lines. In this case the number of possible diagrams of helical turbines can be increased up to 12. As result of a comparative quantitative analysis, two of the most characteristic conceptual diagrams were taken as basis for subsequent multicriteria research on the helical turbine:

- Cylindrical helical turbine and constant core (Fig. 4.77a);
- Conical helical turbine and constant core (Fig. 4.77b).

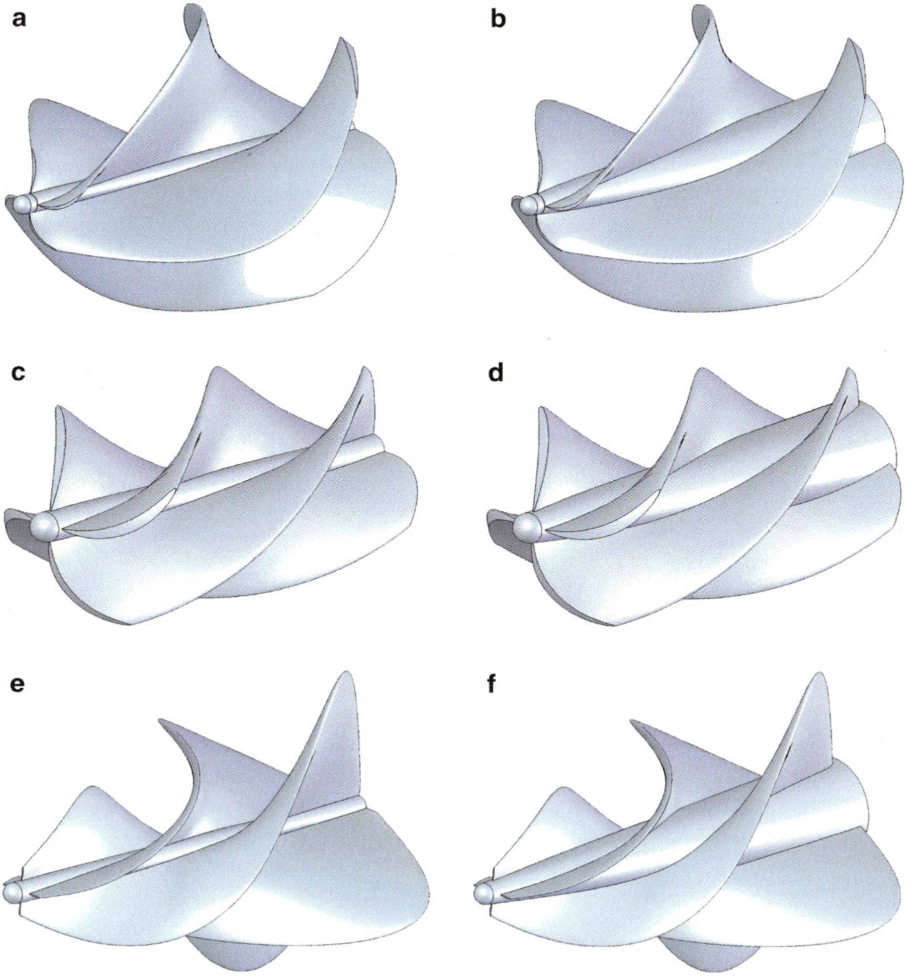

Fig. 4.76 Conceptual diagrams of helical turbines. (**a**) Conical turbine, cylindrical core. (**b**) Conical turbine, conical core. (**c**) Cylindrical turbine, cylindrical core. (**d**) Cylindrical turbine, conical core. (**e**) Parabolic turbine, cylindrical core. (**f**) Parabolic turbine, conical core

Two 3D models of helical turbine are presented in Fig. 4.77 [41–43, 51], modeled in SolidWorks software: cylindrical helical rotor with constant pitch of the helical line and cylindrical core (a), and conical helical rotor with variable pitch of the helical lines and cylindrical core (b).

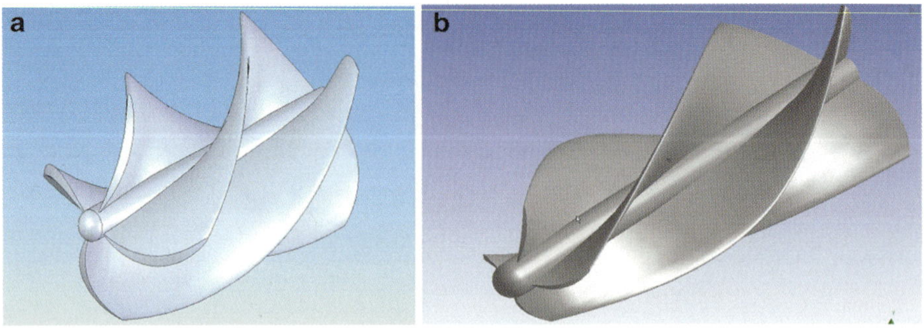

Fig. 4.77 3D models of the cylindrical and conical helical rotor, design software – SolidWorks

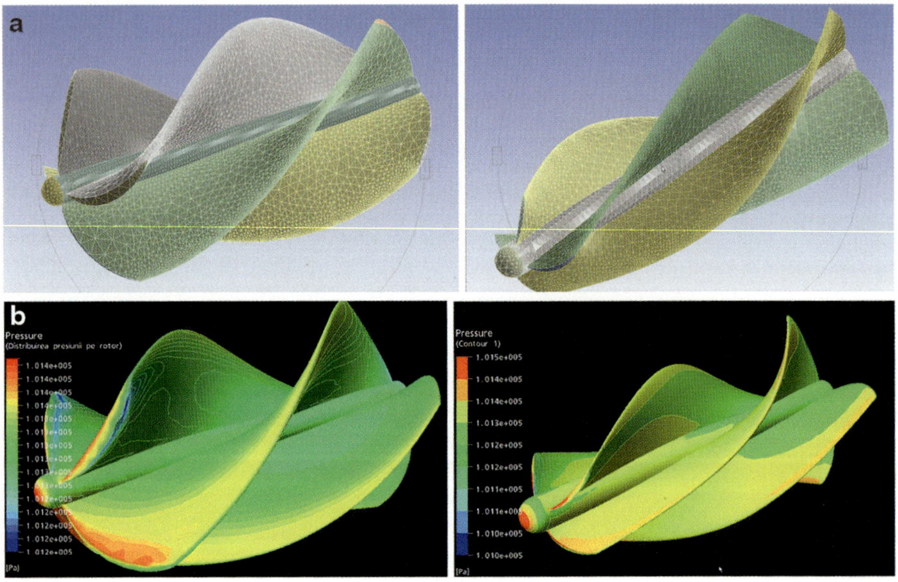

Fig. 4.78 (**a**) Splitting up of the helical turbine in finite elements (tetrahedron), CFX CAD2Mesh. (**b**) Total distribution of pressure on the helical rotor surface, CFX Post. (**c**) Velocity vector on a plane surface oriented to the length of the axis of rotation, CFX Post. (**d**) Velocity vector on a circular surface oriented to the length of axis of rotation, CFX Post. (**e**) Shape of the liquid flow around the rotor

The first model is the basic concept of the helical rotor, and the second model is the modification of the basic concept, designed in the result of studying literature and then developed as a worked out mathematical model.

The purpose of simulation of these two models was to carry out research on the helical rotor interaction with the fluid, which fact allowed the selection of the optimum version for its further investigation and constructive development. Further on, the simulation results of both versions of the helical rotors are presented (Fig. 4.78) in the succession of simulation phase development. Figure 4.78a shows

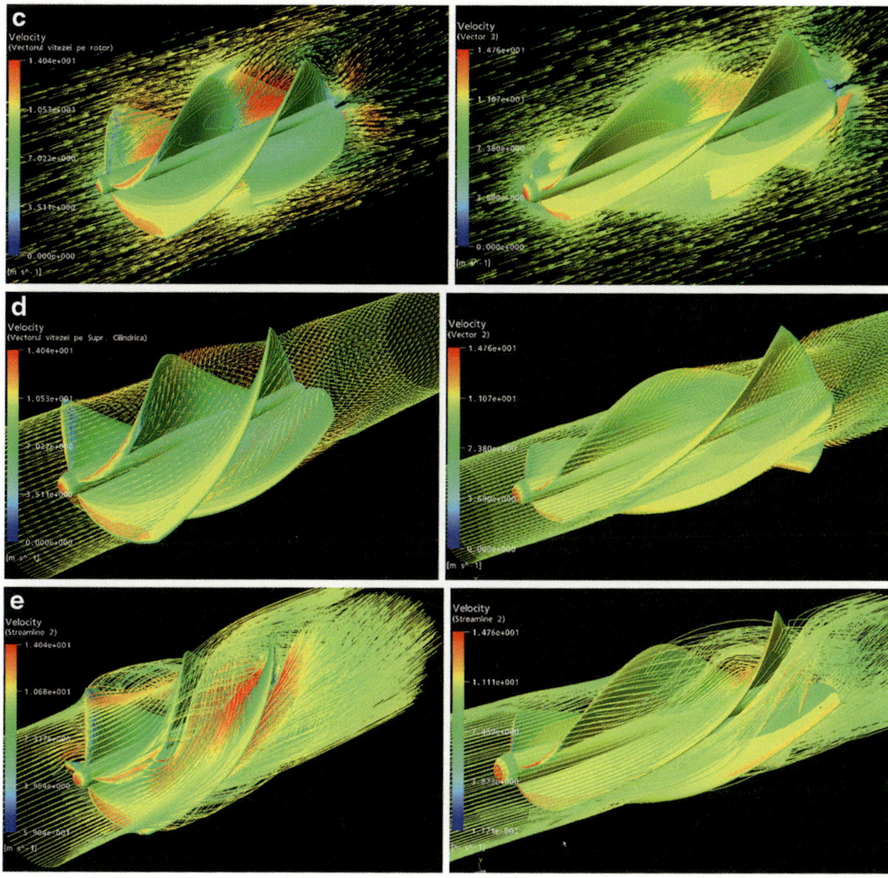

Fig. 4.78 (continued)

the splitting up of the rotor blades, designed as tetrahedron finite elements, and the determination of total pressure on the working surfaces of the helical rotor. The comparative analysis shows that pressure distribution is optimal in the conical helical rotor with variable pitch of the helical lines (Fig. 4.78b). The analysis of images (Fig. 4.78c) demonstrates the presence of certain more pronounced zones of the liquid turbulence at the rotor output.

The shape of fluid flow around the rotor is more efficient in the case of the conical turbine with variable pitch of the helical lines (Fig. 4.78e). Multicriteria analysis of the helical turbine has allowed the justification of geometrical and constructive parameters on which basis the technical documentation of the experimental models was designed. Figure 4.79 shows the experimental model of the cylindrical helical turbine with constant (cylindrical) core made of plastics (kaprolone-L fiber). The experimental model was executed by a digitally controlled tool employing the software developed on the basis of research. The experimental model was designed

Fig. 4.79 Experimental model of the cylindrical helical turbine

in sectional version in order to allow, by experimental research, the justification of the optimal relationship between the turbine length and its outside diameter with account of the conversion efficiency. Transition from the constructive parameters of the turbine experimental model to its real sizes was done by simulating the real values influence on the braking action during the water flow output from the turbine and, accordingly, on the conversion efficiency.

4.5 Micro Hydropower Plant with Rotor's Pintle and Hydrodynamic Profile of Blades

The content of this part has the goal to introduce the reader into the complexity and width of the issues under study on the segment "from the idea to the launch into production of the final product". This issue is quite important for the execution of the renewable energy conversion system, for instance, of the micro hydroelectric power plant for the conversion of the river water kinetic energy into electrical or mechanical energy using the hydrodynamic effects. The micro hydropower plant is a complex technical system that includes constructive components with distinct functions: rotor-turbine that draws off a part of the water kinetic energy at its interaction with the water flow; mechanical transmissions for the transformation of the converted energy; pumps and generators for useful power generation, etc. The conversion efficiency of the micro hydroelectric power plant depends on the performances of each component.

The main phases (in successive order) are as follows:

- Design of the functional concept of the micro hydroelectric power plant;
- Theoretical research of the factor of influence on the water kinetic energy conversion efficiency;
- Particular research and design of the working element for the water kinetic energy conversion efficiency;
- Research and design of the units participating in the transformation of converted energy into useful energy;
- Manufacturing and separate experimental research on the units;
- Design and manufacturing of the micro hydroelectric power pilot-plant;
- Experimental research on the units as integral technical system and the evaluation of the similarity of functional and constructive parameters that have been theoretically and experimentally determined;
- Introduction of partial modifications in the project documentation;
- Development of the execution technologies and manufacturing of the micro hydroelectric power plant, as a final industrial product.

The functional and constructive parameters of the hydrodynamic rotor, multiplier, generator and hydraulic pumps, adopted within the carried out research separately on each working element, demand experimental research of their functioning as an integral system, in real conditions. The experimental research on the units of the micro hydroelectric power plant as an integral system aims at the increase of the conversion efficiency of the water flow kinetic energy into useful energy by introducing the relevant constructive modifications in the project documentation of the final industrial product.

4.5.1 Conceptual Diagrams

To avoid the construction of dams, it is possible to use the river kinetic energy by utilizing water flow turbines. This type of turbines can be mounted easily and are simple in operation. Their maintenance costs are rather convenient. The stream velocity of 1 m/s represents an energy density of 500 W/m^2 of the flow passage. Still, only part of this energy can be extracted and converted into useful electrical or mechanical energy, depending on the type of rotor and blades. Velocity is important, in particular, because the doubling of water velocity leads to an eight times increase of the energy density. The section of e.g. Prut River is equivalent to 60 m^2 and its mean velocity in the zones of exploration is (1–1.3) m/s, which is equivalent to approximately (30–65) kW of theoretical energy. Taking into account the fact that the turbine can occupy only a part of the riverbed, the generated energy could be much smaller. There are various conceptual solutions, but the issue of increasing the conversion efficiency of the water kinetic energy stands in the attention of the researchers. The analysis of the constructive diversion of micro hydroelectric power

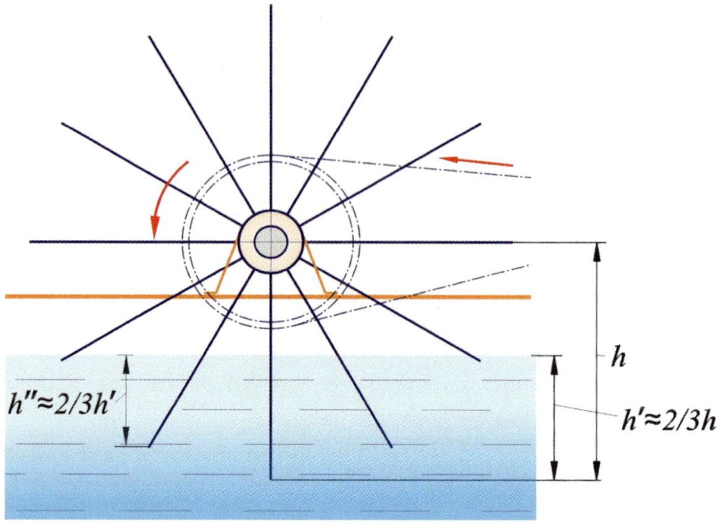

Fig. 4.80 Conceptual diagram of the water wheel with rectilinear profile of blades

plants, examined previously, does not satisfy completely from the point of view of water kinetic energy conversion efficiency.

The maximum depth of blade's immersion is about 2/3 of the blade height *h* in a classical hydraulic wheel with horizontal axle (Fig. 4.80). Thus, only this surface of the blade participates at the transformation of water kinetic energy into mechanical one. As well, the preceding blade covers approximately 2/3 of the blade surface plunged into the water to the utmost ($h'' \approx 2/3h'$), that reduces sensitively the water stream pressure on the blade. The blade, following the one that is plunged into the water to its utmost, is covered completely by it and practically does not participate in the water kinetic energy conversion. Therefore the efficiency of such hydraulic wheels is small.

Insistent searches of authors have lead to the design and licensing of some advanced technical solutions for outflow micro hydroelectric power plants. They are based on the hydrodynamic effect, generated by the hydrodynamic profile of blades and by the optimal blades' orientation towards water streams with account of energy conversion at each rotation phase of the turbine rotor (Fig. 4.81). To achieve this, it was necessary to carry out considerable multi-criteria theoretical research on the selection of the optimal hydrodynamic profile of blades and the design of the orientation mechanism of blades towards the water streams.

The main advantages of these types of micro hydroelectric power plants are:

- Reduced impact on the environment;
- Civil engineering works are not necessary;
- The river does not change its natural stream;
- Possibility to produce floating turbines by utilizing local knowledge.

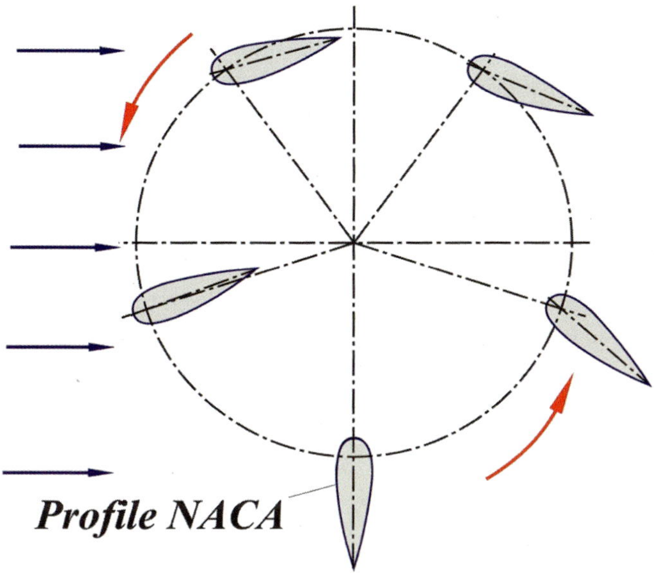

Fig. 4.81 Conceptual diagram of the water rotor with hydrodynamic profile of blades with its orientation towards the water streams

Another important advantage is the fact that it is possible to install a series of micro hydro power plants at small distances (about 30–50 m) along the river course. The influence of turbulence caused by the neighboring plants is excluded.

The results of investigations conducted by the authors (on the water flow velocity in the selected location for micro hydro power plant mounting, on the geological prospects of the river banks in the location of installing the anchor foundation and on the energy demands of the potential consumer) represent the initial data for the conceptual development of the micro hydro power plants and the working element.

The conceptual development of the plant structures with hydrodynamic profile of the blades was performed on the basis of three conceptual diagrams:

- Micro hydropower plant with pintle and blades fixed on the vertical axles anchored by steel structure;
- Floatable micro hydro power plant with pintle and blades fixed on the vertical axles;
- Floatable micro hydro power plant with horizontal axis and blades fixed on the horizontal axles.

In order to increase the conversion factor of water kinetic energy (Betz coefficient), a number of structural diagrams of floatable micro hydro power plants has been developed and patented [52–58]. The micro hydropower plants comprise a rotor with vertical axis and vertical blades with hydrodynamic profile in normal section. The blades are connected by an orientation mechanism towards the water streams direction. The rotational motion of the rotor with vertical axis is multiplied by a mechanical transmissions system and is transmitted to an electric generator or to a hydraulic pump. The mentioned nodes are fixed on a platform installed on floating bodies. The platform is connected to the shore by a hinged metal truss and by a stress relieving cable.

The selection of the optimal blades hydrodynamic profile is very important for functional optimization of micro hydro power plants. It will allow increasing the conversion factor (Betz coefficient) due to the hydrodynamic buoyant force. As well, conversion increase is achieved by ensuring the optimal position of blades towards the water streams at various phases of rotor revolution, employing an orientation mechanism of blades. Thus, practically all blades (even those blades which move against the water currents) participate in the generation of the summary torque. Moving in the water currents direction, for torque generation the blades use both the hydrodynamic forces and the water pressure exercised on the blade surfaces. Moving against the water currents direction the blades use only the hydrodynamic lift force for torque generation. Due to the fact that the relative velocity of blades concerning the water currents is twice bigger, practically, at their motion against the water currents, the hydrodynamic lift force is relatively big, and the generated torque is commensurable to the one generated by the water pressure. This effect makes the basis of all patented technical solutions. Next, six technical solutions of micro hydro power plants are presented, comprising various basic nodes and conversion principles that have been patented. These technical solutions allow essential increasing of the river water kinetic energy conversion coefficient. Full description of the most representative technical solutions and brief description of the conceptual diagrams of micro hydro power plants properties are given below.

In micro hydro power plant (Fig. 4.82) [53], the turbine 1 comprises blades 2, executed with the hydrodynamic profile and mounted on the axles 3, fixed by their upper part on the extreme ends of the bars 4, with the possibility to rotate around their axles. The position of the blades 2 at angle $\boldsymbol{\alpha}$ to the direction of water flow is ensured by the controlling mechanism 5. Platform 6 is consolidated additionally by a winch 7 fixed on the truss that is mounted unshiftable on the shore pillar 8. The turbine 1 and the blades 2 are placed in the river water flow. The floating bodies 9 and the hollow blades 2 themselves control the position of turbine 1 and blades 2 concerning the water level. The multi-blade rotor is connected cinematically and coaxially to the electric generator 11 by the multiplier 10. The winch 7 is used for turbine 1 maintenance which fact requires its removal from the water. The blade 2 (Fig. 4.83) is positioned under angle α towards the water flow; it changes depending on the blade position to the water flow direction.

The components of force $\boldsymbol{F}$, acting on the blade, are determined from the relationships:

$$F_x = C_x \cdot \frac{\rho \cdot v^2}{2} \cdot S$$

$$F_y = C_y \cdot \frac{\rho \cdot v^2}{2} \cdot S \tag{4.38}$$

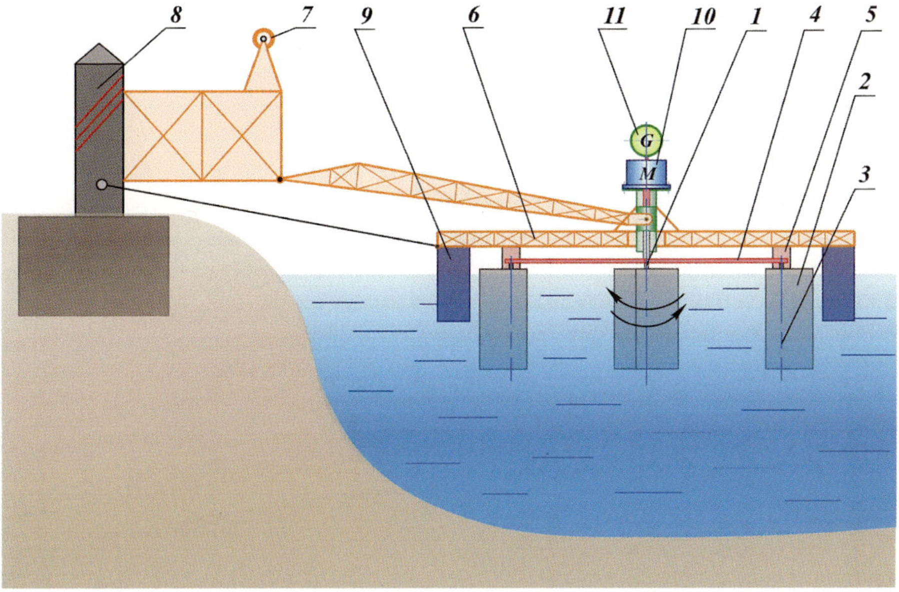

Fig. 4.82 Floatable micro hydropower plant with blades orientation mechanism

Fig. 4.83 Positioning of blades towards the water currents

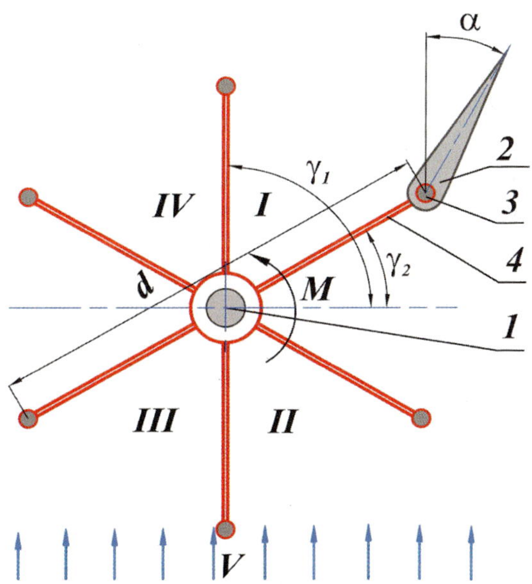

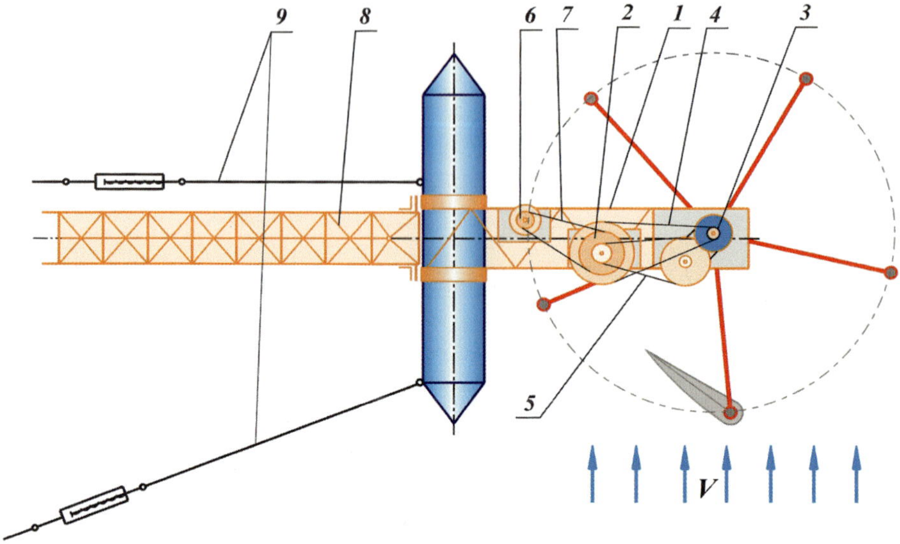

Fig. 4.84 Floating micro hydropower plant with electric generator and hydraulic pump

where: ρ is water density; v is the water flow linear velocity; s is the blade surface; C_x, C_y are lift and drag (resistance) coefficients of the blade profile. Coefficients C_x and C_y depend on the blade entering angle α (the angle between the blade and the water flow direction) and on the profile shape. The angle is determined either experimentally or by numerical calculations. The torque developed by one blade is described by the equation;

$$M = F_\tau \cdot \frac{d}{2} = (\cos\gamma \cdot F_y - \sin\gamma \cdot F_x)\frac{d}{2} \tag{4.39}$$

where F_τ is the projection of force F on the tangent drawn to the path of motion of the blade axis.

The summary torque includes the general component of the resistance force F_h. The torque moment generated by the turbine consists of the torques generated by each separate blade. Currently only one blade will not generate positive moment (it will generate a negative moment – the resistance one). Thus, the torque generated by the proposed turbine will be essentially bigger than the torque produced by the existing turbines for the same geometrical (blades dimensions) and kinematical parameters of water. The proposed micro hydro power plant allows the transformation of the water flow kinetic energy into mechanical or electrical energy with an increased utilization coefficient of water energy.

In the floating micro hydro power plant (Fig. 4.84) [54] an additional centrifugal pump 2 is mounted on the resistance structure 1. It is connected cinematically to the multi-blade rotor spindle 3 by belt transmissions 4 and 5. The electric generator

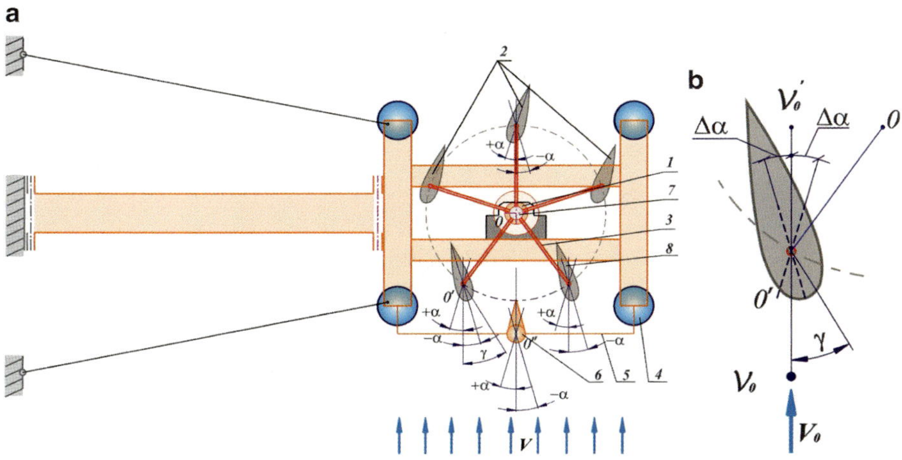

Fig. 4.85 Floating micro hydropower plant with influence compensation of water currents flow direction change

6 is connected to the multi-blade rotor spindle by belt transmissions 4 and 7. As well, the resistance structure 1 is connected to the shore by the metal truss 8 and supporting cables with cross-ties 9.

In the floating micro hydro power plant (Fig. 4.85a) [55] the rotor 1 contains an odd number of blades 2 that are fitted with the possibility of rotation on vertical axes O′–O′ (Fig. 4.85b) mounted on the extreme end of each horizontal bar 3. On frame 4, in the front part (through which water flows pass) a rigid bar 5 is installed on which, in front of the hydro turbine relative to the water flow direction, a sensor 6 is fixed that determines the water flow direction and connects to the rotation gear 7. The water flow moves in the direction of vector V_O (Fig. 4.85b). Angle γ is the entering angle of the blades formed by the hydrodynamic surface string and the working lines of the water flow vector $V_O V_O$. The angle depends on the form of the hydrodynamic surface and on the position in the plane surface. By changing the water flow direction due to the change of water discharge and river bed, the water currents will divert by angle $\Delta\alpha$ modifying the entering angle γ. To meet the angle of attack, optimal in terms of conversion, it is necessary to correct the position of all blades by angle $\pm\alpha$. When changing the water flow direction, the positioning of all blades 2 is corrected simultaneously by angle $\pm\Delta\alpha$ using the rotation mechanism 7:

$$\gamma_{1,2} = \gamma \pm \Delta\alpha. \tag{4.40}$$

In the floating micro hydropower plant (Fig. 4.86) [56] the steel structure 1 is connected to the shore masts 3 by a hinged metal truss 2. The cable 4, installed in line with the metal farm 2 and bound to the same mast 3, and to the lateral cables 5 and 6 connected to mast 7 and 8 are envisaged in order to relieve the stress of the swivel connections *truss 2-steel structure 1* and *truss 2-shore masts 3*.

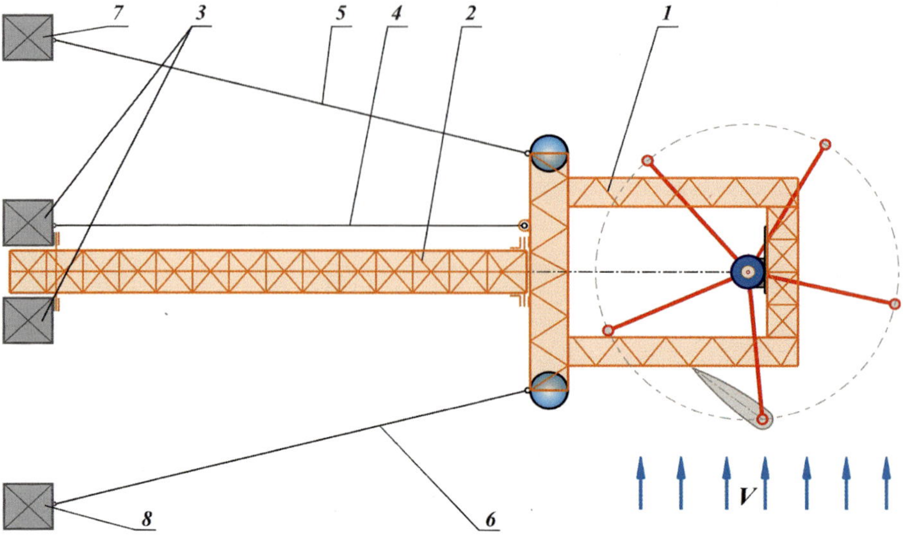

Fig. 4.86 Floating micro hydropower plant with stress cable for the link with the shore

In the floating micro hydro power plant (Fig. 4.87) [57, 58] a technical solution is proposed ensuring the transverse stability of platform 1 of the floating micro hydro power plant that is mounted on floating bodies 2 and 3, placed on the same side (shore side of the rotor spindle 4). Due to the fact that the rotor 4 blades 5 are hollow, the hydrostatic Archimedes force of the blades 5 fulfills the role of the floating bodies (Fig. 4.87a, b).

The analysis of the application points motion path of the Archimedes force F_a (point N in Fig. 4.87c) has shown that the distance from this point to the plane that crosses the rotor spindle 4 (O_1-O_1, Fig. 4.87b) will differentiate depending on the positioning angle of the rotor. Thus, these distances, for the blades that are placed in the upper semiplane defined by axis O_1O_1-OO differ from the distances of those blades placed in the lower semiplane. The migration of the points of application of the Archimedes force causes the pitching moment:

$$M_r = M_{\sum as} - M_{\sum ad} \tag{4.41}$$

where

$M_{\Sigma as}$ is the summary moment developed by the Archimedes forces that react on the blades currently located in the upper semiplane;

$M_{\Sigma ad}$ is the summary moment developed by the Archimedes forces that react on the blades currently located in the lower semiplane.

The summary moments developed by the Archimedes forces that react on the blades are determined by the relations:

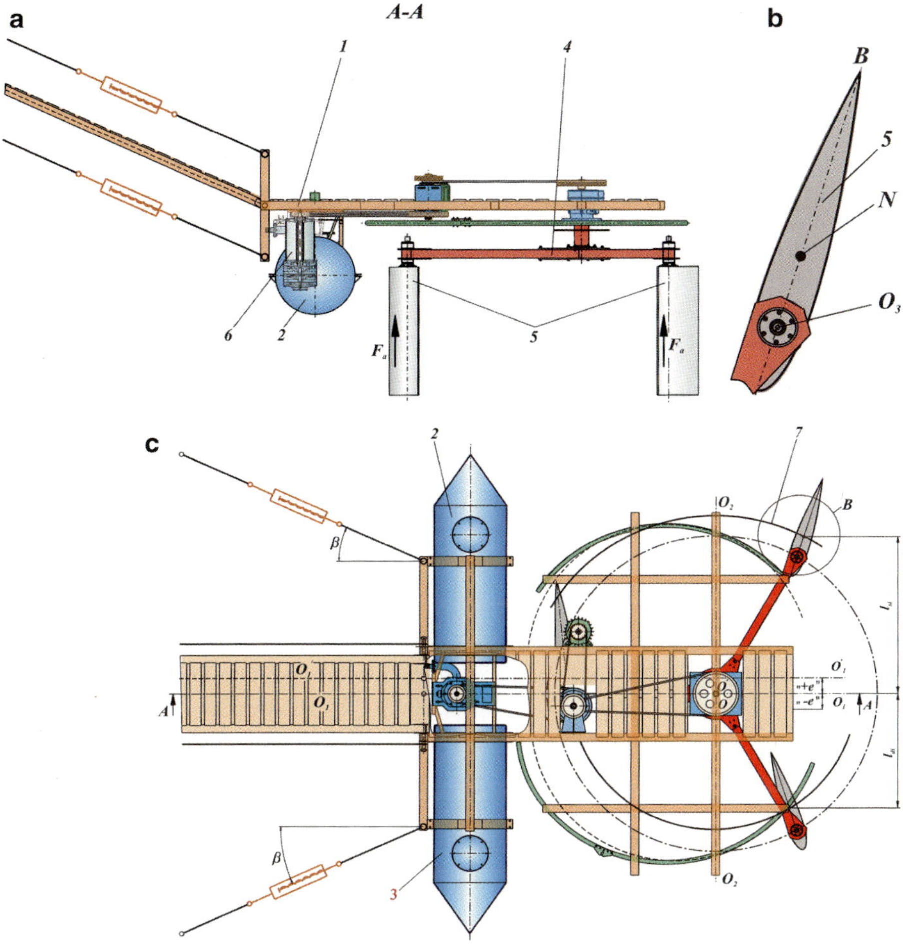

Fig. 4.87 Micro hydropower plant with increased transverse stability

$$M_{\sum as} = \sum F_{ai}..l_{si} \text{ and } M_{\sum ad} = \sum F_{ai}..l_{di} \tag{4.42}$$

where

F_{ai} is the Archimedes forces that react on the blades 5 currently located in the upper semi-plane;

l_{si} is the distance from the point of application of the Archimedes force that reacts on the blades 5 currently located in the upper semiplane;

l_{di} is the distance from the point of application of Archimedes force that reacts on the blades 5 currently located in the lower semiplane.

Distances l_{si} and l_{di} are calculated from the relation:

$$l^2 = R^2 + c_M^2 + 2Rc_M \cos(\alpha + \varphi) \tag{4.43}$$

where

R is the rotor radius 4;
C_M is the distance from the point of application of the Archimedes force and the blade fixing point to the turbine rotor;
α is the angle formed by the blade chord and the water flow direction;
φ is the angle formed by the rotor lever and the perpendicular direction on the watercourse.

To compensate the pitching moment M_r, the rotor spindle 10 is settled in plane $O_1' - O_1'$ at distance e compared to the longitudinal axial plane of the floating bodies $O_l - O_l$. Distance e is calculated from the relation:

$$e = \frac{\sum_{i=1}^{n} y_i}{n} \tag{4.44}$$

where n is the number of rotor blades, and y_i is the distance from the centre of application of Archimedes force on blade i up to the longitudinal axial plane (Fig. 4.87c). For each distance, y_i is calculated by the relation:

$$y_i = c_M \cos\alpha + R \sin\left(\varphi + (i-1)\frac{360^o}{n}\right) \tag{4.45}$$

where

R is the rotor radius;
C_M is the distance from the point of application of the Archimedes force and the blade fixing point to the turbine impeller, O_iN_i in Fig. 4.87c;
n is the number of rotor blades.

So, the distance e is calculated by the relation:

$$e = c_M \cdot \cos\alpha \tag{4.46}$$

where α is the angle formed by the blade chord and the water flow direction.

Conclusion: To ensure the floating stability of the micro hydro power plants the rotor is mounted on the main structure with displacement e against the water stream. Thus, the micro hydro power plants designed to be anchored on the left bank cannot be anchored on the right bank.

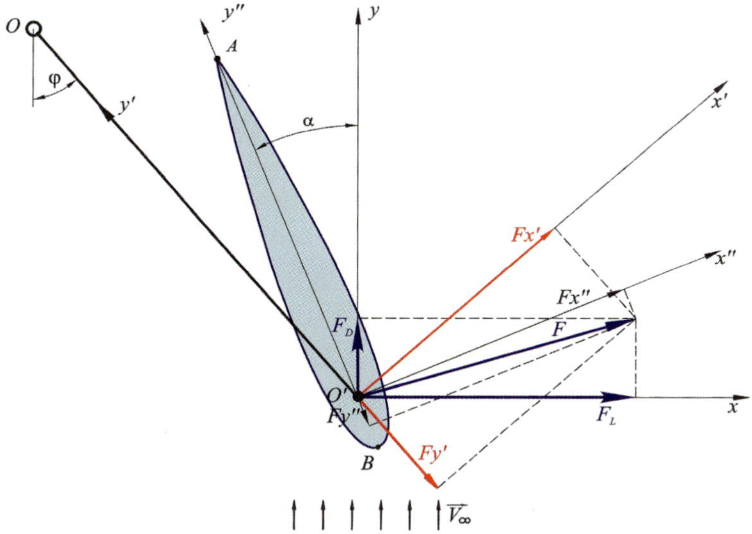

Fig. 4.88 Hydrodynamic profile blade

4.5.2 *Research on the Factors of Influence of Kinetic Energy Conversion Efficiency and Design of the Hydrodynamic Rotor*

4.5.2.1 Theoretical Justification of the Hydrodynamic Profile Selection of the Blade in Normal Section

Consider the symmetrical profile of the blade placed in a uniform water stream with velocity $\vec{V}_\infty$ (Fig. 4.88). In the fixing point O' of the symmetrical blade with lever OO' let consider three coordinate systems, namely: the $O'xy$ system with axis $O'y$ oriented in the direction of the velocity vector $\vec{V}_\infty$, and axis $O'x$ normal to this direction; the $O'x'y'$ system with axis $O'y'$ oriented along the lever direction $O'O$, and axis $O'x'$ normal to this direction, and finally the $O'x''y''$ system with axis $O'x''$ oriented along the profile's chord toward the trailing edge and axis $O'y''$ normal to this direction. Points A and B correspond to the trailing and the leading edges, respectively. The angle of attack α is the angle between the profile's chord AB and the direction of the velocity vector $\vec{V}_\infty$, and the positioning angle φ is the angle formed by the velocity vector direction and lever $O'O$.

The hydrodynamic force $\vec{F}$ has its components in directions $O'x$ and $O'y$, named the lift and drag forces, respectively given by:

$$F_L = \frac{1}{2} C_L \rho V_\infty^2 S_p \tag{4.47}$$

$$F_D = \frac{1}{2} C_D \rho V_\infty^2 S_p \tag{4.48}$$

where ρ is the fluid density, V_∞ is the flow velocity, $S_p = ch$ (c is the length of chord AB, and h is the blade height) represents the lateral surface area of the blade, and C_L and C_D are dimensionless hydrodynamic coefficients, called the lift coefficient and drag coefficient. The hydrodynamic coefficients C_L and C_D are functions of the angle of attack α, the Reynolds number Re and the hydrodynamic shape of the blade profile. The components of the hydrodynamic force in the coordinate system $O'x'y'$ are

$$\begin{aligned} F_{x'} &= -F_L \sin\varphi + F_D \cos\varphi \\ F_{y'} &= F_L \cos\varphi + F_D \sin\varphi \end{aligned} \tag{4.49}$$

The torque developed by blade i at the rotor spindle OO' is

$$T_{r,i} = F_{x'} \cdot |OO'| \tag{4.50}$$

and the total torque developed by all blades is

$$T_{r\Sigma} = \sum_{i=1}^{Npal} T_{ri} \tag{4.51}$$

where $Npal$ is the number of rotor blades.

Generally, the hydrodynamic force does not have application point in the origin of the blade axes system O' so that it produces a resultant moment. The produced moment is determined with respect to a certain reference point. As a reference point there is considered the point located at distance ¼ of the chord length measured from the leading edge B. The moment, called the pitching moment, is computed according to formula

$$M = \frac{1}{2} C_M \rho V_\infty^2 c S_p \tag{4.52}$$

where C_M is the hydrodynamic moment coefficient.

The shape of the hydrodynamic profile is chosen from the library of NACA 4 digits aerodynamic profiles. The standard NACA 4 digit profiles are characterized by three shape parameters measured in percents of the chord's length: maximum value of camber C_{max}, location of the maximum camber $x_{C,max}$ and maximum thickness G_{max}. For example, the NACA 5416 profile has a 5% maximum camber located at 40% from the leading edge and has a maximum thickness of 16%. The profile coordinates are obtained by combining the camber line and the distribution of thickness [41]. Since the considered blades will have a symmetric shape, the camber is null ($C_{max} = 0$, $x_{C,max} = 0$) and the camber line will coincide with x-axis.

4.5.2.2 Determination of the Hydrodynamic Coefficients C_L and C_M: Plane Potential (Cyclic) Motion

For simplicity, the profile chord length is considered unitary. Initially, the fluid is considered incompressible and inviscid, and its flow – plane and potential [59]. In the case of an incompressible plane potential flow the velocity components $\vec{V} = (u, v)$ in point $P(x,y)$ are given by the relations:

$$u(x,y) = \frac{\partial \Phi}{\partial x}$$
$$v(x,y) = \frac{\partial \Phi}{\partial y} \tag{4.53}$$

where is the potential obtained by superposition of a uniform velocity flow $\vec{V}_{\infty} = (V_{\infty}\cos\alpha, V_{\infty}\sin\alpha)$ and a distribution of sources and vortexes on the profile C. Thus, the potential is

$$\Phi = \Phi_{\infty} + \Phi_S + \Phi_V \tag{4.54}$$

where the uniform flow potential is

$$\Phi_{\infty} = V_{\infty}x\cos\alpha + V_{\infty}y\sin\alpha \tag{4.55}$$

the potential of the two-dimensional source distribution with strength q(s) is

$$\Phi_S = \oint_C \frac{q(s)}{2\pi}\ln(r)ds \tag{4.56}$$

and the potential of vortex distribution with strength γ(s) is given by formula:

$$\Phi_V = -\oint_C \frac{\gamma(s)}{2\pi}\theta ds \tag{4.57}$$

In relations (4.56, 4.57) variable s represents the arc length along profile C, and (r, θ) are the polar coordinates of point $P'(x,y)$ relative to the point on the contour corresponding to arc length s (Fig. 4.89).

Therefore the potential in point $P'(x,y)$ is computed as follows:

$$\Phi(P') = V_{\infty}x\cos\alpha + V_{\infty}y\sin\alpha + \oint_C \frac{q(s)}{2\pi}\ln(r)ds - \oint_C \frac{\gamma(s)}{2\pi}\theta ds \tag{4.58}$$

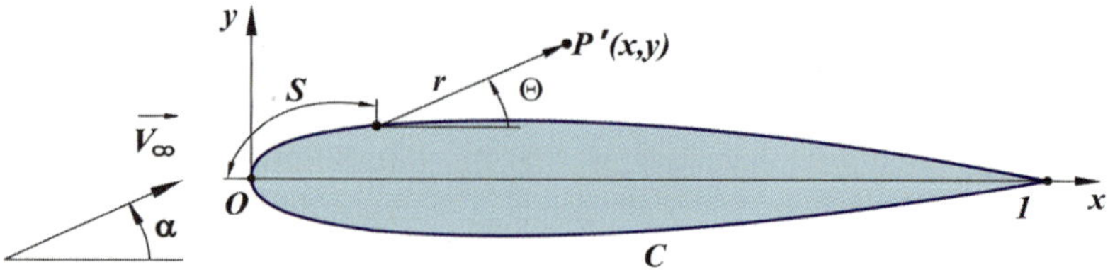

Fig. 4.89 Potential two-dimensional flow around profile C

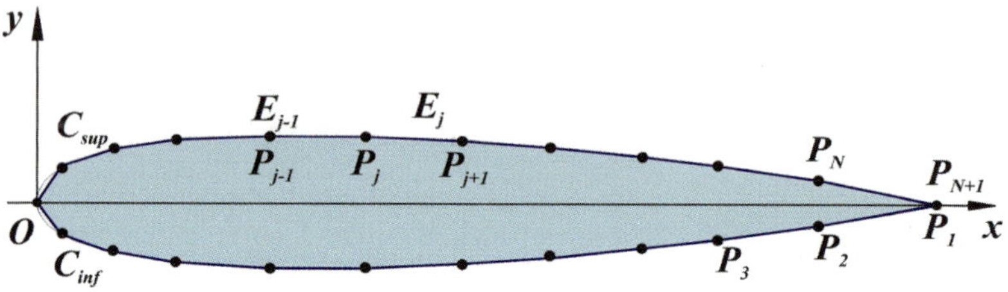

Fig. 4.90 Discretisation of profile C

In order to compute the plane flow potential the collocation method is used, that is: the boundary of profile C is approximated by a closed polygonal line

$$C \approx \bigcup_{j=1}^{N} E_j$$

with sides E_j having their vertices P_j and P_{j+1} on C. The vertices P_j are having a higher density near the leading and trailing edges. This is achieved by choosing Chebyshev points as their x-coordinates. Numbering of the vertices starts from the trailing edge on the lower side in the direction of leading edge, passing further to the upper side (Fig. 4.90). It is further assumed that the strength of vortexes $\gamma(s)$ distributed on profile C is constant on the boundary having value γ, and the strength of sources $q(s)$ distributed on the profile is piecewise constant on each boundary element E_j with value q_j, where $j = 1, \ldots, N$. Breaking the integrals in equation (4.58) along each panel gives:

$$\Phi = V_\infty x \cos\alpha + V_\infty y \sin\alpha + \sum_{j=1}^{N} \int_{E_j} \left(\frac{q_j}{2\pi} \ln(r) - \frac{\gamma}{2\pi}\theta \right) ds \tag{4.59}$$

with unknowns being γ and q_j, $j = 1, \ldots, N$.

Consider now the boundary element E_j with vertices P_j and P_{j+1} (Fig. 4.91). The unit normal and tangent vectors of the element E_j are given by:

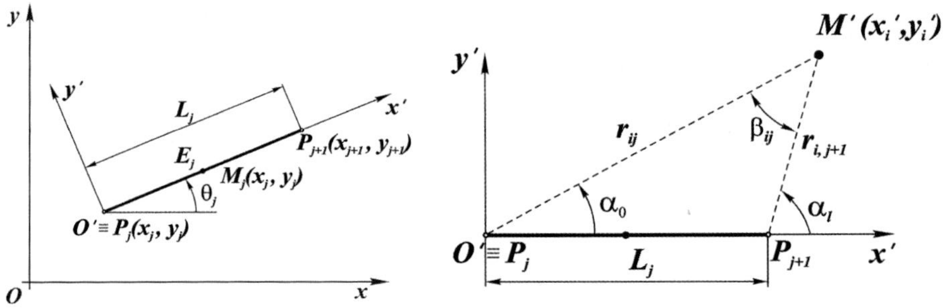

Fig. 4.91 Boundary element E_j

$$n_j = (-\sin\theta_j, \cos\theta_j),$$
$$\tau_j = (\cos\theta_j, \sin\theta_j), \tag{4.60}$$

where

$$\sin\theta_j = \frac{y_{j+1} - y_j}{L_j}$$
$$\cos\theta_j = \frac{x_{j+1} - x_j}{L_j}$$

The unknowns γ and q_j, $j = 1, \ldots, N$ from the relation (4.59) are determined from the boundary and Kutta conditions. The boundary condition is the flow tangency condition along the profile:

$$\vec{V} \cdot \vec{n} = 0 \tag{4.61}$$

where $\vec{n}$ is the profile normal direction. Imposing condition (4.61) on the collocation points $M_j\left(\overline{x}_j, \overline{y}_j\right)$ chosen to be the midpoints of E_j and denoting the velocity components in M_j by u_j and v_j, respectively, provides N algebraic relations:

$$-u_j \sin\theta_j + v_j \cos\theta_j = 0, \mathrm{j} = 1, \ldots, N \tag{4.62}$$

Relations (4.62) are used to determine the $N+1$ unknowns: γ and q_j, $j = 1, \ldots, N$. The last necessary relation is provided by Kutta condition:

$$\vec{V} \cdot \vec{\tau}\Big|_{E_1} = \vec{V} \cdot \vec{\tau}\Big|_{E_N} \tag{4.63}$$

where $\vec{\tau}$ denotes the unit tangent vector of the boundary element. Using introduced notations, Kutta condition (4.63) takes the form:

$$u_1 \cos\theta_1 + v_1 \sin\theta_1 = -u_N \cos\theta_N + v_N \sin\theta_N. \tag{4.64}$$

The velocity components in a collocation point M_i are determined by the contributions of velocities induced by the distribution of sources and vortexes on each boundary element E_j:

$$u_i = V_\infty \cos\alpha + \sum_{j=1}^{N} u_{ij}^s q_j + \sum_{j=1}^{N} u_{ij}^v \gamma$$

$$v_i = V_\infty \sin\alpha + \sum_{j=1}^{N} v_{ij}^s q_j + \sum_{j=1}^{N} v_{ij}^v \gamma \tag{4.65}$$

where $u_{ij}^s, v_{ij}^s, u_{ij}^v, v_{ij}^v$ are the influence coefficients. For instance, u_{ij}^s represents the velocity component at point M_i in x-direction, induced by a source distribution with unit strength located on element E_j.

Let β_{ij}, $i \neq j$, be the angle between $P_j M_i$ and $M_i P_{j+1}$, and set $\beta_{ii} = \pi$. Denote by r_{ij} the distance between M_i and P_j and let $D_{ij} = \frac{r_{i,j+1}}{r_{ij}}, i, j = 1, \ldots, N$. It can be shown that the influence coefficients can be computed from the following formulas:

$$u_{ij}^s = \frac{1}{2\pi}\left(-\ln D_{ij} \cos\theta_j - \beta_{ij} \sin\theta_j\right)$$

$$v_{ij}^s = \frac{1}{2\pi}\left(-\ln D_{ij} \sin\theta_j + \beta_{ij} \cos\theta_j\right)$$

$$u_{ij}^v = \frac{1}{2\pi}\left(-\ln D_{ij} \sin\theta_j + \beta_{ij} \cos\theta_j\right)$$

$$v_{ij}^v = \frac{1}{2\pi}\left(\ln D_{ij} \cos\theta_j + \beta_{ij} \sin\theta_j\right) \tag{4.66}$$

After substituting expressions (4.65) and (4.66) in the boundary conditions (4.62) and in Kutta condition (4.64) and some algebraic manipulations a linear system with equations $N + 1$ and $N + 1$ unknowns γ and q_j, $j = 1, \ldots, N$ is obtained:

$$\left[A_{ij}\right]_{i,j=1}^{N+1} \begin{bmatrix} q_1 \\ q_2 \\ \vdots \\ q_N \\ \gamma \end{bmatrix} = \begin{bmatrix} b_1 \\ b_2 \\ \vdots \\ b_N \\ b_{N+1} \end{bmatrix} \tag{4.67}$$

Coefficients A_{ij} and b_i, $i, j = 1, \ldots, N+1$ are given by formulas:

$$A_{ij} = \frac{1}{2\pi}\left(\sin\Delta_{ij}\ln D_{ij} + \beta_{ij}\cos\Delta_{ij}\right), i, j = 1, \ldots, N$$

$$A_{i,N+1} = \frac{1}{2\pi}\sum_{j=1}^{N}\left(\cos\Delta_{ij}\ln D_{ij} - \beta_{ij}\sin\Delta_{ij}\right), i = 1, \ldots, N$$

$$A_{N+1,j} = \frac{1}{2\pi}\left(\beta_{1,j}\sin\Delta_{1j} + \beta_{N,j}\sin\Delta_{Nj} - \cos\Delta_{1j}\ln D_{1j} - \cos\Delta_{Nj}\ln D_{N,j}\right)$$

$$A_{N+1,N+1} = \frac{1}{2\pi}\sum_{j=1}^{N}\left(\sin\Delta_{1j}\ln D_{1j} + \sin\Delta_{Nj}\ln D_{N,j} + \beta_{1j}\cos\Delta_{1j} + \beta_{Nj}\cos\Delta_{Nj}\right)$$

$$b_i = V_\infty\sin\left(\theta_i - \alpha\right), i = 1, \ldots, N$$

$$b_{N+1} = -V_\infty\cos(\theta_1 - \alpha) - V_\infty\sin(\theta_N - \alpha)$$

and $\Delta_{ij} = \theta_i - \theta_j$.

The solution of linear system (4.67) will provide the values of γ and q_j, $j = 1, \ldots, N$, using which the tangential components of the velocity in collocation points M_i, $i = 1, \ldots, N$ can be computed. Recall that the normal component of the velocity in collocation points is null. The tangential component of the velocity at point M_i is

$$u_{\tau i} = u_i\cos\theta_i + v_i\sin\theta_i.$$

Substitute (4.65) in the above relation to obtain:

$$u_{\tau i} = \left(V_\infty\cos\alpha + \sum_{j=1}^{N}u^s_{ij}q_j + \sum_{j=1}^{N}u^v_{ij}\gamma\right)\cos\theta_i + \left(V_\infty\sin\alpha + \sum_{j=1}^{N}v^s_{ij}q_j + \sum_{i=1}^{N}v^v_{ij}\gamma\right)\sin\theta_i.$$

By using relations (4.66) after some algebraic manipulations provide the following formulas for the tangential components of velocity:

$$u_{\tau i} = \cos\left(\theta_i - \alpha\right)V_\infty + \frac{1}{2\pi}\sum_{j=1}^{N}\left(q_i\left(\beta_{ij}\sin\Delta_{ij} - \cos\Delta_{ij}\ln D_{ij}\right) + \gamma\left(\sin\Delta_{ij}\ln D_{ij} - \beta_{ij}\cos\Delta_{ij}\right)\right) \quad (4.68)$$

Bernoulli equation $p + \frac{1}{2}\rho V^2 = p_\infty + \frac{1}{2}\rho_\infty {V_\infty}^2$ implies that

$$p = p_\infty + \frac{1}{2}\rho_\infty {V_\infty}^2 - \frac{1}{2}\rho V^2$$

Thus, the local pressure coefficient can be rewritten as follows:

$$C_p = \frac{p - p_\infty}{\frac{1}{2}\rho_\infty V_\infty^2} = 1 - \frac{V^2}{V_\infty^2}. \tag{4.69}$$

Therefore, the local pressure coefficient on the discretized profile can be computed from the relation

$$C_{p,i} = 1 - \left(\frac{u_{\tau i}}{V_\infty}\right)^2 \tag{4.70}$$

where $u_{\tau i}$ are the tangential components of the velocity given by (4.68).

The hydrodynamic forces acting on the boundary element E_j are given by:

$$\begin{aligned} f_{xj} &= C_{p,j}\left(y_{j+1} - y_j\right) \\ f_{yj} &= C_{p,j}\left(x_{j+1} - x_j\right) \end{aligned} \tag{4.71}$$

and the pitching moment coefficient reported to the point of reference $\left(x_{ref}, y_{ref}\right) = \left(\frac{c}{4}, 0\right)$, are calculated by formula:

$$c_{m,j} = -f_{xj}\left(\frac{y_{j+1} - y_j}{2}\right) + f_{yj}\left(\frac{x_{j+1} - x_j}{2} - \frac{c}{4}\right) \tag{4.72}$$

The total force is the sum of contributions of each boundary element:

$$F_x = \sum_{j=1}^{N} f_{xj}, \; F_y = \sum_{j=1}^{N} f_{yj} \tag{4.73}$$

and the lift coefficient and the moment coefficient are calculated as follows:

$$C_L = -F_x \sin\alpha + F_y \cos\alpha \tag{4.74}$$

$$C_M = \sum_{j=1}^{N} c_{m,j} \tag{4.75}$$

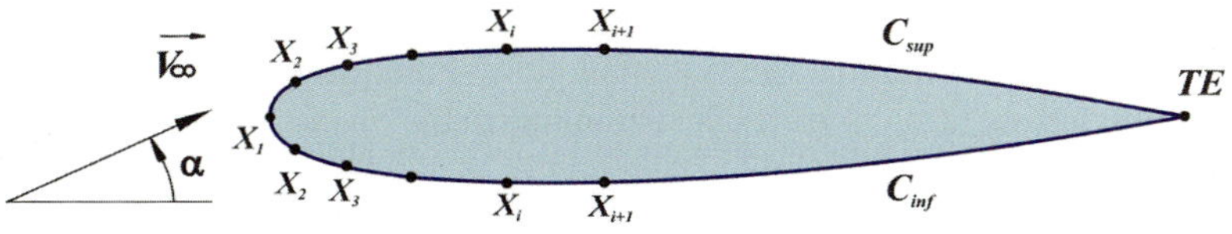

Fig. 4.92 Profile C discretization for the analysis of the boundary layer

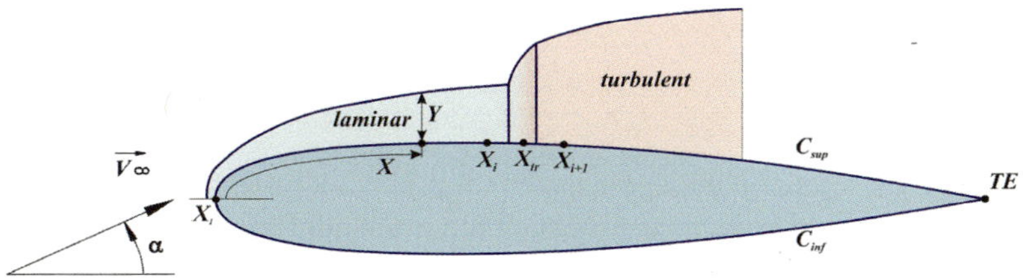

Fig. 4.93 Transition from laminar to turbulent boundary layer

4.5.2.3 Determination of the Hydrodynamic Coefficient C_D: Laminar and Turbulent Boundary Layer

In the case of fluid flow around the elongated bodies, Prandtl mentioned that the influence of fluid viscosity is limited to a rather thin layer in the immediate proximity of the body, in our case, of the profile. Due to adherence on the profile surface, the forces of viscosity influence the layer within which the flow velocity increases from zero value to the value corresponding to the flow without viscosity at the exterior part of the layer. Prandtl has called this area the boundary layer. Thus, it can be considered that the fluid rotational flow is concentrated in the boundary layer, but outside the layer, the flow is potential. Keeping this in mind, the area of fluid flow is divided into two zones: the boundary layer where the fluid behaves like a viscous one, and the zone outside the boundary layer where the fluid motion is potential [60, 61].

Therefore, the first phase of a standard method consists in the computation of the velocity distribution in the potential flow around the profile. The next phase consists in the computation of boundary layer parameters corresponding to the velocity distribution obtained in the first phase. In its turn, the boundary layer phase is divided into two sub-steps: laminar boundary layer and turbulent boundary layer.

The boundary layer starts at the stagnation point (the point on the profile's contour where velocity is zero, the distribution of velocities being provided by the previous phase) and follows the profile along the upper or lower surface in the direction of trailing edge. As soon as the stagnation point was determined (point x_1 in Fig. 4.92) the numbering of vertices starts in the direction of the trailing edge (Fig. 4.93). Profile discretisation is performed by taking into account a uniform partition of the upper and lower surfaces, respectively.

The computation of laminar boundary layer parameters is based on the integral momentum and kinetic energy equations. The boundary layer thickness is defined as the distance from the profile at which the flow velocity differs by 1% from the velocity corresponding to the potential flow. Prandtl laminar boundary layer equations are derived from the steady incompressible Navier-Stokes equations:

$$\frac{\partial u}{\partial x} + \frac{\partial v}{\partial y} = 0$$

$$\rho\left(u\frac{\partial u}{\partial x} + v\frac{\partial v}{\partial y}\right) = -\frac{\partial p}{\partial x} + \mu\frac{\partial^2 u}{\partial y^2}$$

$$\frac{\partial p}{\partial y} = 0 \tag{4.76}$$

In (4.76), x represents the measured distance along the contour from the stagnation point, and y is the measured distance in the normal to the surface direction (Fig. 4.93).

Introduce the displacement thickness δ^*:

$$\delta^* = \int_0^\infty \left(1 - \frac{u}{V}\right) dy \tag{4.77}$$

where V represents the velocity of the boundary layer exterior part in the considered point, and u is the tangential velocity in this point. Similarly, the momentum thickness θ is defined by

$$\theta = \int_0^\infty \left(1 - \frac{u}{V}\right) \frac{u}{V} dy \tag{4.78}$$

and the kinetic energy thickness θ^* is given by

$$\theta^* = \int_0^\infty \left(1 - \frac{u^2}{V^2}\right) \frac{u}{V} dy \tag{4.79}$$

By integrating Prandtl equations (4.76) and using relations (4.77—4.78) the Von Karman integral-differential momentum equation is obtained

$$\frac{d\theta}{dx} + \frac{\theta}{V}\left(2 + \frac{\delta^*}{\theta}\right)\frac{dV}{dx} = \frac{1}{2}C_f \tag{4.80}$$

where C_f denotes the local friction coefficient on the profile surface defined by

$$C_f = \frac{\tau_w}{\frac{1}{2}\rho V^2} \tag{4.81}$$

with τ_w denoting the tangential stress:

$$\tau_w = \mu \left. \frac{\partial u}{\partial y} \right|_{y=0} \tag{4.82}$$

Introduce the shape parameter

$$H = \frac{\delta^*}{\theta} \tag{4.83}$$

Then momentum equation (4.80) can be rewritten as follows:

$$\frac{d\theta}{dx} + (2 + H)\frac{\theta}{V}\frac{dV}{dx} = \frac{1}{2}C_f \tag{4.84}$$

On the other side, multiply equation (4.80) by u, and integrate it in order to obtain the integral the boundary layer kinetic energy equation:

$$\frac{d\theta^*}{dx} + 3\frac{\theta^*}{V}\frac{dV}{dx} = 2C_d \tag{4.85}$$

where the dissipation coefficient C_d is defined by the relation:

$$C_d = \frac{1}{\rho V^3}\int_0^\infty \mu \frac{\partial^2 u}{\partial y^2} dy \tag{4.86}$$

Introduce the second shape parameter:

$$H^* = \frac{\theta^*}{\theta} \tag{4.87}$$

Subtract equation (4.84) from equation (4.85) and rewrite the kinetic energy equation in the following form:

$$\theta\frac{dH^*}{dx} + H^*(H - 1)\frac{\theta}{V}\frac{dV}{dx} = 2C_d - \frac{1}{2}H^* C_f \tag{4.88}$$

The system of equations (4.84) and (4.88) is not sufficient to determine all unknowns. The additional relations are based on Falkner-Skan semi-empirical relations. Assume the following correlation between H^* and H:

$$H^* = \begin{cases} 0.076\dfrac{(H-4)^2}{H} + 1.515, \text{if } H < 4 \\ 0.04\dfrac{(H-4)^2}{H} + 1.515, \text{otherwise} \end{cases} \tag{4.89}$$

Also let $\mathrm{Re}_\theta = \mathrm{Re} \cdot \theta \cdot V$ and assume that

$$\frac{1}{2}\mathrm{Re}_\theta C_f = F_1(H) = \begin{cases} 0.01977\dfrac{(H-7.4)^2}{H-1} - 0.067, \text{if } H < 7.4 \\ 0.022\dfrac{(H-7.4)^2}{(H-6)^2} - 0.067, \text{otherwise} \end{cases} \tag{4.90}$$

$$2\mathrm{Re}_\theta \frac{C_d}{H*} = F_2(H) = \begin{cases} 0.00205(4-H)^{11/2} + 0.207, \text{if } H < 4 \\ -0.003\dfrac{(H-4)^2}{1+0.02(H-4)^2} + 0.207, \text{otherwise} \end{cases} \tag{4.91}$$

Multiply momentum equation (4.84) by Re_θ and denote $\omega = (\mathrm{Re}_\theta)^2$. The terms are re-arranged to obtain:

$$\frac{1}{2}V\frac{d\omega}{dx} + (2+H)\,\omega\frac{dV}{dx} = F_1(H) \tag{4.92}$$

Similarly, multiply equation (4.88) by $\frac{\mathrm{Re}_\theta}{H*}$ and re-arrange the terms to get:

$$\omega V\frac{d(\ln H^*)}{dH}\frac{dH}{dx} + (1-H)\,\omega\frac{dV}{dx} = F_2(H) - F_1(H) \tag{4.93}$$

Introduce the following notations:

$$A(x) = \frac{dV}{dx}, \mathrm{F}_3(H) = \frac{d(\ln H^*)}{dH}, \mathrm{F}_4(H) = F_2(H) - F_1(H)$$

Then equations (4.92) and (4.93) can be rewritten as follows

$$\begin{aligned} &\frac{1}{2}V(x)\frac{d\omega}{dx} + (2+H)\,\omega A(x) = F_1(H) \\ &\omega V(x)F_3(H)\frac{dH}{dx} + (1-H)\,\omega A(x) = F_4(H) \end{aligned} \tag{4.94}$$

The unknowns in the system of differential equations (4.94) are H and ω. The initial values are chosen in order to ensure that

$$\frac{dw}{dx}(0) = 0 \text{ and } \frac{dH}{dx}(0) = 0$$

Therefore $H(0)$ is the solution of the following equation:

$$\frac{1-H}{2+H} = \frac{F_4(H)}{F_1(H)}$$

Solving the above equation provides us the root $H(0) \approx 2.24$. Consequently,

$$\omega(0) = \frac{F_1(H(0))}{A(0)(2 + H(0))}$$

Therefore the initial conditions are

$$\begin{aligned} w(0) &= \frac{F_1(2.24)}{4.24A(0)} \\ H(0) &= 2.24 \end{aligned} \tag{4.95}$$

The system of nonlinear ODE (4.94) with initial conditions (4.95) is solved by the Backward Euler method, but in order to avoid the implicit iterations at the transition from step j to step $j+1$ the functions F_1 and F_4 are linearized in the neighborhood of H_j, while F_3 takes the value $F_3(H_j)$. Thus, there is obtained a system of two bilinear equations with the unknowns H_{j+1} and ω_{j+1}:

$$\begin{aligned} A_j^1 \omega_{j+1} + B_j^1 H_{j+1} \omega_{j+1} + C_j^1 H_{j+1} &= D_j^1 \\ A_j^2 \omega_{j+1} + B_j^2 H_{j+1} \omega_{j+1} + C_j^2 H_{j+1} &= D_j^2 \end{aligned} \tag{4.96}$$

where

$$\begin{aligned} A_j^1 &= \frac{V\left(x_{j+1}\right)}{2\Delta x} + 2A\left(x_{j+1}\right) \\ B_j^1 &= A\left(x_{j+1}\right), C_j^1 = -F_1{}'\left(x_j\right) \\ D_j^1 &= \frac{V\left(x_{j+1}\right)}{2\Delta x}\omega_j + F_1\left(x_j\right) - F_1{}'\left(x_j\right) H_j = 0 \\ A_j^2 &= A\left(x_{j+1}\right) - \frac{F_3(x_j) H_j V\left(x_{j+1}\right)}{\Delta x} \\ B_j^2 &= \frac{F_3(x_j) V\left(x_{j+1}\right)}{\Delta x} - A\left(x_{j+1}\right) \\ C_j^2 &= -F_4{}'\left(x_j\right), D_j^1 = F_4\left(x_j\right) - F_4{}'\left(x_j\right) H_j \end{aligned}$$

This method is iterated till either the transition from the laminar layer to the turbulent layer is predicted (transition point x_{tr}, Fig. 4.98) or till the trailing edge is reached (point *TE*, Fig. 4.98).

In order to locate the transition from laminar boundary layer to turbulent boundary layer the Michel criterion is being used. Let $\text{Re}_x = \text{Re} V x$ and

$$\text{Re}_{\theta\,\text{max}} = 1.174 \left(1 + \frac{22.4}{\text{Re}_x}\right) \text{Re}_x^{0.46} \tag{4.97}$$

Then, transition takes place if $\mathrm{Re}_\theta > \mathrm{Re}_{\theta\,\mathrm{max}}$. The transition point is the root of the linear interpolation of $\mathrm{Re}_\theta(x) - \mathrm{Re}_{\theta\,\mathrm{max}}(x)$ The other criterion that might be used is the so-called ISES e9 method.

To analyze the turbulent boundary layer introduce the mean values and fluctuations:

$$\overline{q}(x, y) = \lim_{T\to\infty} \frac{1}{T} \int_{t_0}^{t_0+T} q(x, y, t)dt \tag{4.98}$$

$$q'(x, y, t) = q(x, y, t) - \overline{q}(x, y) \tag{4.99}$$

The equations of the turbulent boundary layer are obtained from the Navier-Stokes equations:

$$\begin{aligned} \frac{\partial \overline{u}}{\partial x} + \frac{\partial \overline{v}}{\partial y} &= 0 \\ \rho\left(\overline{u}\frac{\partial \overline{u}}{\partial x} + \overline{v}\frac{\partial \overline{v}}{\partial y}\right) &= -\frac{\partial \overline{p}}{\partial x} + \frac{\partial}{\partial y}\left(\mu\frac{\partial \overline{u}}{\partial y} - \rho\overline{u'v'}\right) \\ \frac{\partial \overline{p}}{\partial y} &= \frac{\partial}{\partial y}\left(\mu\frac{\partial \overline{v}}{\partial y} - \rho\overline{v'^2}\right) \end{aligned} \tag{4.100}$$

Similarly to the case of laminar boundary layer, the Von Karman integral equations are obtained. The computations of the turbulent boundary layer parameters are done by applying the Head's model. Consider the flow volume in the boundary layer at an arbitrary point x:

$$Q(x) = \int_0^{\delta(x)} u dy \tag{4.101}$$

Then the displacement thickness is given by:

$$\delta^* = \delta - \frac{Q}{V} \tag{4.102}$$

Introduce the entrainment velocity:

$$E = \frac{dQ}{dx} \tag{4.103}$$

that can be rewritten as

$$E = \frac{d}{dx}(V\theta H_1) \tag{4.103}$$

where

$$H_1 = \frac{\delta - \delta^*}{\theta}$$

According to the Head's model the dimensionless velocity *E/V* depends only on H_I, and H_I, in its turn, depends only on *H*. Cebeci and Bradshow have considered the empirical relations

$$\frac{E}{V} = 0.0306(H_1 - 3)^{-0.6169} \tag{4.104}$$

And consequently

$$H_1 = \begin{cases} 0.8234(H - 1.1)^{-1.287} + 3.3, H \leq 1.6 \\ 1.5501(H - 0.6778)^{-3.064} + 3.3, \text{otherwise} \end{cases} \tag{4.105}$$

Finally, the last equation to determine the unknowns θ, H, H_1 and C_f is the Ludwieg-Tillman wall friction law:

$$C_f = \frac{0.246}{10^{0.678H}\mathrm{Re}_\theta^{0.268}} \tag{4.106}$$

Combine the Von Karman integral equation and the relations (4.104)–(4.106) **to obtain the following system of ODE**:

$$\frac{dY}{dx} = g(Y, x) \tag{4.107}$$

where

$$Y = \begin{bmatrix} \theta \\ H_1 \end{bmatrix} \tag{4.108}$$

and

$$g(Y, x) = \begin{bmatrix} -\frac{\theta\,(H_1 + 2)}{V}\frac{dV}{dx} + \frac{1}{2}C_f \\ -\frac{H_1}{V}\frac{dV}{dx} - \frac{H_1}{\theta}\frac{d\theta}{dx} + \frac{0.0306}{\theta(H_1 - 3)^{0.6169}} \end{bmatrix} \tag{4.109}$$

The initial values are the final values supplied by the laminar boundary layer step. The numerical integration of system (4.107) is done by Runge-Kutta method of order 2, namely:

$$\begin{aligned} Y^* &= Y_j + h_j g(Y_j, x_j) \\ Y_{j+1} &= Y_j + h_j\left(\frac{1}{2}g(Y_j, x_j) + \frac{1}{2}g(Y^*, x_j)\right) \end{aligned} \tag{4.110}$$

The calculation is done either till the rear edge is reached or till the separation of the turbulent layer occurs.

Squire-Young formula is applied to compute the drag coefficient C_D. Suppose that the following values are known: the momentum thickness θ, the form parameter H and the velocity V, all of them being computed in the point corresponding to the trailing edge TE on the upper and lower surfaces of contour profile C. Then, the drag coefficient is given by the relation:

$$C_D = 2\left(\theta V^{\lambda}|_{XTE,C_{upper}} + \theta V^{\lambda}|_{XTE,C_{lower}}\right) \tag{4.111}$$

where $\lambda = (H|_{XTE} + 5)/2$.

4.5.2.4 Selection of the Optimal Hydrodynamic Profile of Blades

The optimization of the performance of the turbine with hydrodynamic blades demands an optimal hydrodynamic profile of the blade. The numerical methods, described previously, are used to compute the hydrodynamic coefficients $C_{L,ref}$ and $C_{D,ref}$ for the symmetrical profiles selected from the NACA library of aerodynamic profiles with a reference chord length $c_{ref} = 1$ m. It should be noted that the numerical procedure converges for the angles of attack α that do not exceed 20° – 25° depending on the selected profile and the corresponding Reynolds number (Re = 1,300,000). For the angles of attack exceeding this critical value, the values corresponding to a flat (plane) profile are considered. Some of the considered profiles are shown in Fig. 4.94: NACA 0012, 0016, 63018 and 67015. Figure 4.95 shows the hydrodynamic lift $C_{L,ref}$ and drag $C_{D,ref}$ coefficients as function of the angle of attack. Taking into account the data from Fig. 4.99, the NACA 0016 hydrodynamic profile is being selected as the reference profile. Subsequently, this profile will be optimized in order to increase the turbine performance.

4.5.2.5 The Torque and the Forces Acting on the Multi-blade Hydrodynamic Rotor

The numerical methods, described previously, are used to compute the hydrodynamic coefficients corresponding to NACA 0016 profile with the chord length $c_{ref} = 1$m: $C_{L,ref}$, $C_{M,ref}$ and $C_{D,ref}$ by formulas (4.85), (4.86) and (4.111). The coefficients corresponding to the profile with the chord length 1.3 m are calculated from the relations:

$$\begin{aligned} C_L &= 1.3C_{L,ref} \\ C_M &= (1.3)^2 C_{M,ref} \\ C_D &= 1.3C_{D,ref} \end{aligned} \tag{4.112}$$

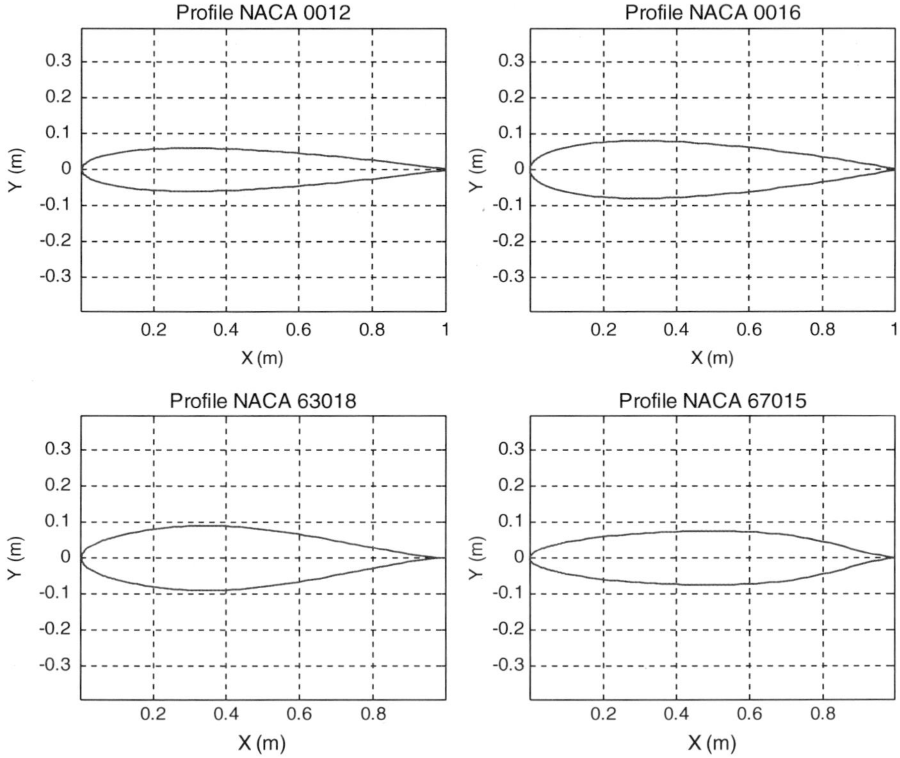

Fig. 4.94 Symmetric hydrodynamic profiles: NACA 0012, 0016, 63018 and 67015

The values of the lift and drag coefficients versus the angle of attack α are shown in Fig. 4.96. Taking into account these values, the angle $\alpha = 18^{\circ}$ is selected as the working angle of attack.

During a full revolution around the rotor's axis the blade changes its attack angle depending on its position (Fig 4.97). Thus, in sector I the angle of attack (angle formed by the blade's chord and water flow direction) is 18°; in sector II the angle of attack shifts from 18° up to −18°, but the blade does not contribute to the total torque developed at the rotor shaft. In this sector, extended up to approximately 60°, the blade is carried freely by the water flow and its re-positioning ends with an angle of attack of −18° at the end of sector III. The angle of attack is −18° in sector III. In sectors IV–VI the hydrodynamic effect is minimal and the blade has to be re-positioned from angle −18° to angle 18. In order to use the kinetic energy in the sectors IV–VI it is proposed to re-position the blade from −18° to 90° in sector IV; in sector V the blade remains under an angle of 90°, and in sector VI the angle of attack returns to 18°. Knowing the values of the hydrodynamic coefficients C_L and C_D, the lift force F_L and drag force F_D are computed using formulas (4.47) and (4.48), and the formula (4.49) supplies the normal and tangential components of the hydrodynamic force.

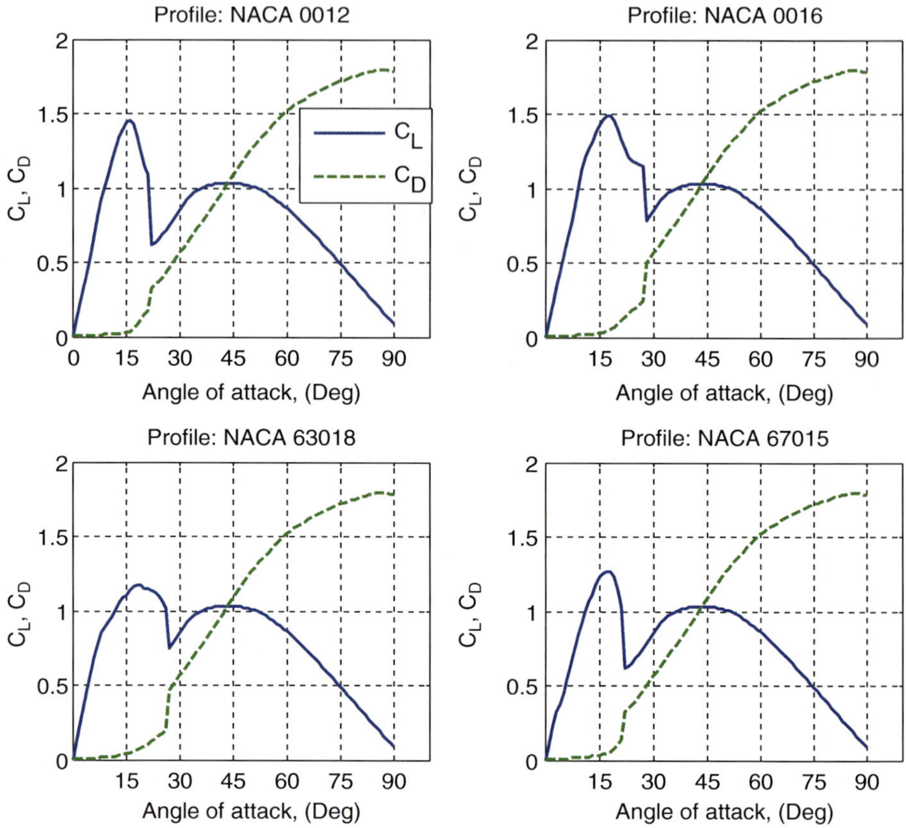

Fig. 4.95 Hydrodynamic lift C_L and drag C_D coefficients versus the angle of attack for NACA 0012, 0016, 63018 and 67015 profiles

The magnitude of the hydrodynamic force $\overrightarrow{F}$ acting on the blade, and its tangential and normal components F_x and F_y, versus the positioning angle are presented in Fig. 4.98. The following constructive parameters of the rotor were considered:

Rotor radius $R = 2\,\text{m}$;
Height of the submersible blade $H = 1.4\,\text{m}$;
Blade chord length $c = 1.3\,\text{m}$;
Main angle of attack $\alpha = 18^{\text{o}}$;
Number of blades $N_{pal} = 5$.

Figure 4.99 shows the torque $T_{r,i}$ developed by the blade versus the positioning angle; the torque is computed using the formula (4.50). Figure 4.100 shows the total torque at the rotor shaft $T_{r\Sigma}$ developed by all blades versus the positioning angle. The total torque is determined by the formula (4.51). Figure 4.101 shows the total

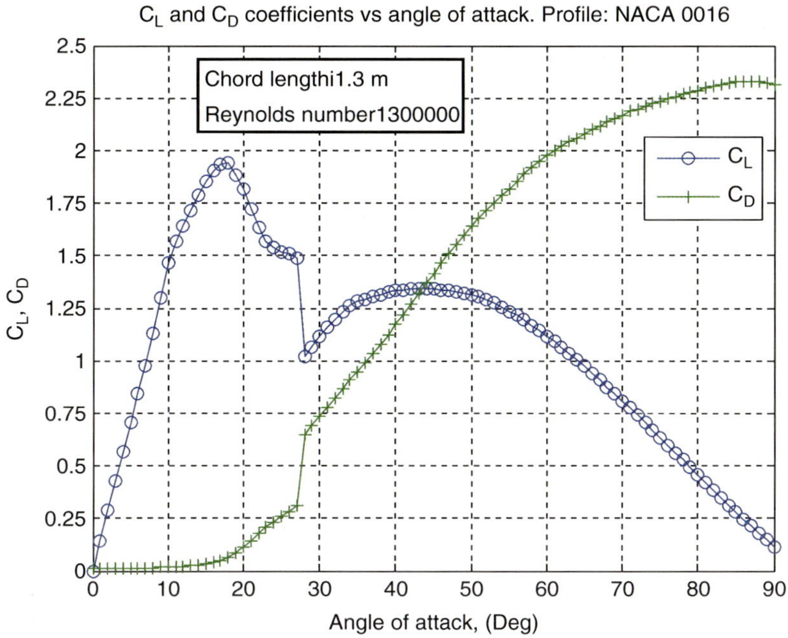

Fig. 4.96 Hydrodynamic lift C_L and drag C_D coefficients versus the angle of attack for NACA 0016 profile

torque $T_{r\Sigma}$ versus the positioning angle for three water flow velocities V_∞: 1 m/s, 1.3 m/s and 1.6 m/s. The graph of the moment coefficient $C_{M,ref}$ versus the angle of attack α is shown in Fig. 4.102.

Taking into account the fact that the hydrodynamic force is not applied in the blade fixing point O' (Fig. 4.103) this force produces the so-called pitching moment. This moment is determined with respect to the reference point ***P*** situated at ¼ distance of the chord from the leading edge B (Fig. 4.103). For the angle of attack $\alpha = 18°$ the moment coefficient $C_{M,ref} = -0.026$. Thus, from the relation (4.112) results that $C_M = 0.0439$. The moment with respect to the reference point ***P*** is

$$M = \frac{1}{2} C_M \rho V_\infty^2 c S_p = -39.92\,\text{Nm} \tag{4.113}$$

where $V_\infty = 1$ m/s, $c = 1.3$ m and $H = 1.4$ m. In the $O'x''y''$coordinate system, the components of the hydrodynamic force are given by the relation (4.49).

Applying the previously obtained values F_L and F_D gives:

$$\begin{aligned} F_{x''} &= 1601.2\,N \\ F_{y''} &= -413.8\,N \end{aligned} \tag{4.114}$$

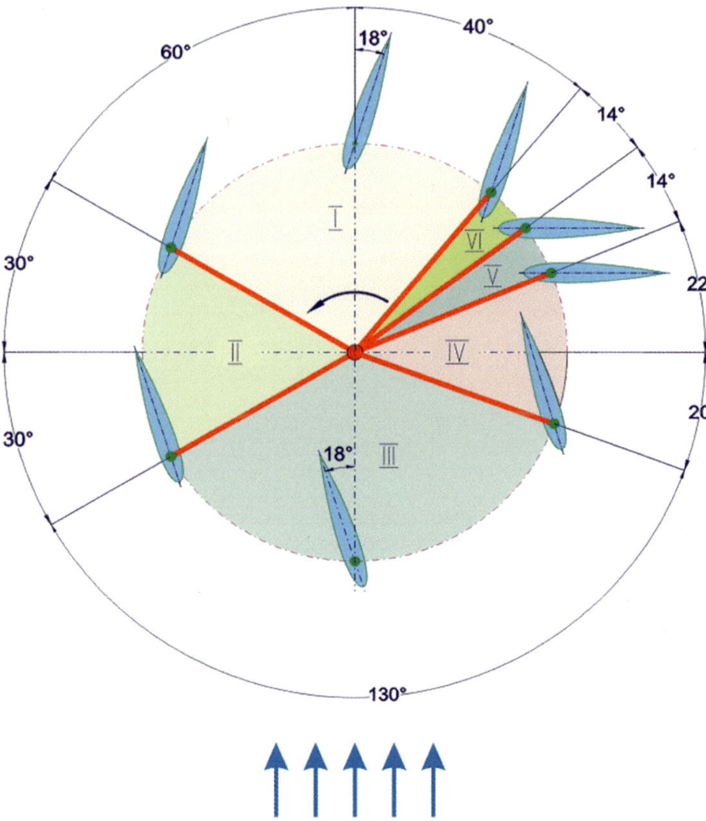

Fig. 4.97 Blade position and working areas

Then

$$\left|O'P\right| = |M|/|F_{x''}| = 0.0249 \text{ m} \approx 25 \text{ mm}. \tag{4.115}$$

In order to ensure the stability of the blade motion, the fixing point W should be selected in the interval $25\text{mm} \leq |O'W| \leq H$, where $H_{\min} \leq H \leq H_{\max}$. Values H_{min} and H_{max} are chosen in order to ensure that the friction force in the kinematical couples of the orientation mechanism is minimal.

To determine the optimal working angle of attack it is necessary to calculate the value of the torque developed by one blade and the total torque for several values of the angle of attack, namely: $\alpha = 15°$, $17°$, $18°$, $20°$, (Figs. 4.104, 4.105). In this context the angle of attack for the blade with hydrodynamic profile NACA 0016 was chosen $\alpha = 18°$.

Also, the performance of 3-, 4- and 5-blades rotor was analyzed. The total moment developed by the rotor shaft was computed and the results are presented in Fig. 4.106.

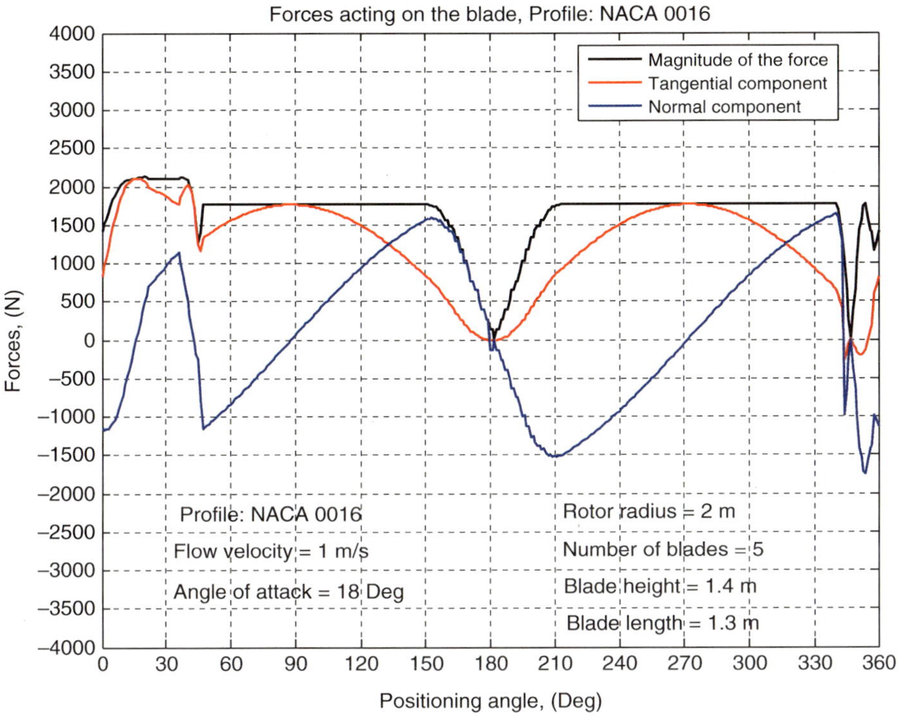

Fig. 4.98 Magnitude, tangential component and normal component of the hydrodynamic force of a rotor blade versus the positioning angle

4.5.2.6 Optimisation of NACA 0016 Hydrodynamic Profile

In order to maximize the torque produced by the micro hydro power plant rotor, the optimization of the hydrodynamic profile will be considered. The torque depends on the lift and drag hydrodynamic forces given by formulas (4.47) and (4.48). The hydrodynamic forces through the hydrodynamic coefficients depend on the angle of attack α, Reynolds number Re and the shape of the hydrodynamic profile. The hydrodynamic shape of the profile was selected from the NACA 4 and 5 digit library of aerodynamic profiles having as parameter (considering the profile symmetry) only the maximal thickness. The angle of attack constitutes the second parameter. The optimization aims at maximizing the lift force and, at the same time, does not allow the pitching moment and the drag force to take very large values. The following optimization problem is considered:

$$\text{Maximize } C_L = C_L(\theta, \alpha) \text{ subject to constraints on } C_D \text{ and } C_M \qquad (4.116)$$

where θ is the maximum thickness and α is the angle of attack.

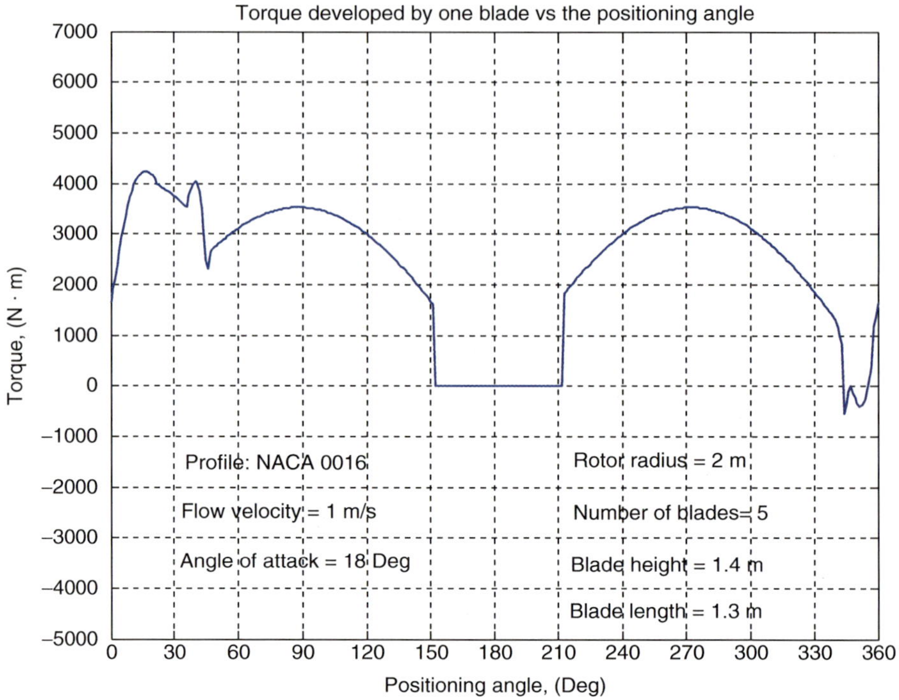

Fig. 4.99 Torque $T_{r,i}$ developed by the rotor blade versus the positioning angle

The values of the lower and upper bounds are determined, as follows: the negative maximum value for the moment coefficient will correspond to the solution for the angle of attack 0°. The maximum value for the drag coefficient will correspond to the solution for the angle of attack $\alpha = 18°$. Also, restrictions have been added to the optimization parameters $10\% \leqslant \theta \leqslant 20\%$ and $0^o \leqslant \alpha \leqslant 20°$. To find the optimal values of function $f = f(x_1, \ldots, x_n)$ an iterative method is used:

While given accuracy is not attained

$$\begin{aligned}
&\text{do}\\
&\text{Solve } B_i s_i = -\nabla f(x_i),\\
&x_{i+1} = x_i + \alpha_i s_i,\\
&\text{Enddo}
\end{aligned} \tag{4.117}$$

where α_i are the multipliers and B_i are the positive definite approximations of the Hessian of function f. The partial derivatives of f are approximated with the finite difference formulas:

$$\frac{\partial f}{\partial x_i}(x) = \frac{f(x + he_i) - f(x - he_i)}{2h} \tag{4.118}$$

where e_i denotes the basis vector.

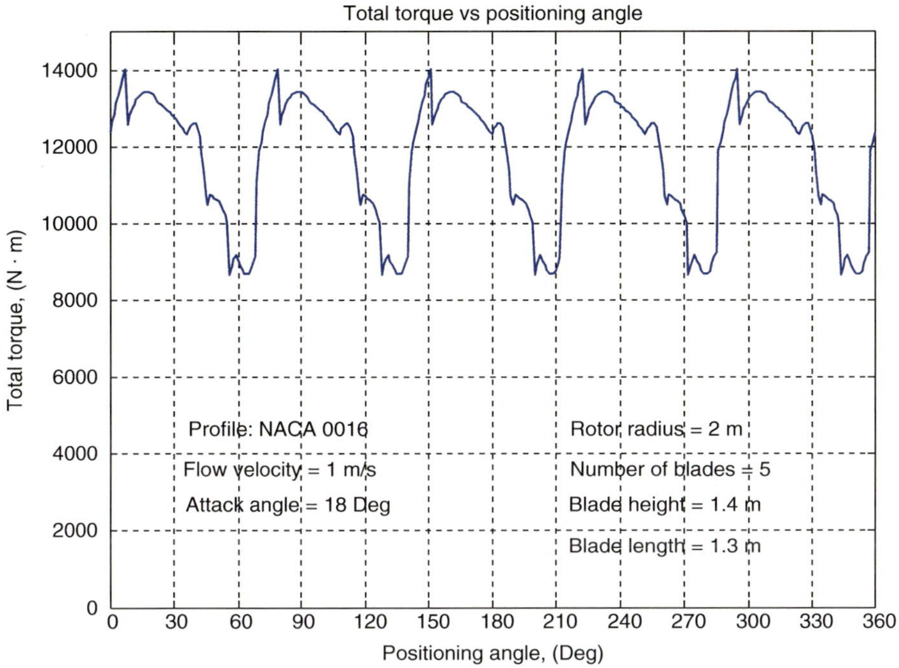

Fig. 4.100 Total torque $T_{r\Sigma}$ developed by 5 blades at rotor shaft versus the positioning angle

The optimization is implemented in MATLAB with a sequential quadratic programming algorithm with a line search and a BFGS Hessian update. The quadratic sub-tasks are solved by modified projection method. The gradient of function $C_L = C_L(\theta, \alpha)$ is approximated with finite differences with uniform step $h = 10^{-4}$. NACA 0016 profile was considered as initial profile (Fig. 4.107). The result of the shape optimization procedure is presented in Fig. 4.108. The results of the carried out research were published in [41, 45, 58, 62, 63].

4.5.2.7 Floating Stability of the Micro Hydropower Plant

Consider the micro hydropower plant placed in the river water flow. The position of the blades with respect to the water level is ensured by the Archimedes forces acting on the floating blades. The blade cavity generates the Archimedes force determined by the relation

$$F_A = \rho V g \tag{4.119}$$

where ρ is the water density, V is the volume of the blade interior and g is the gravitational acceleration. The analysis of the trajectory of the application points

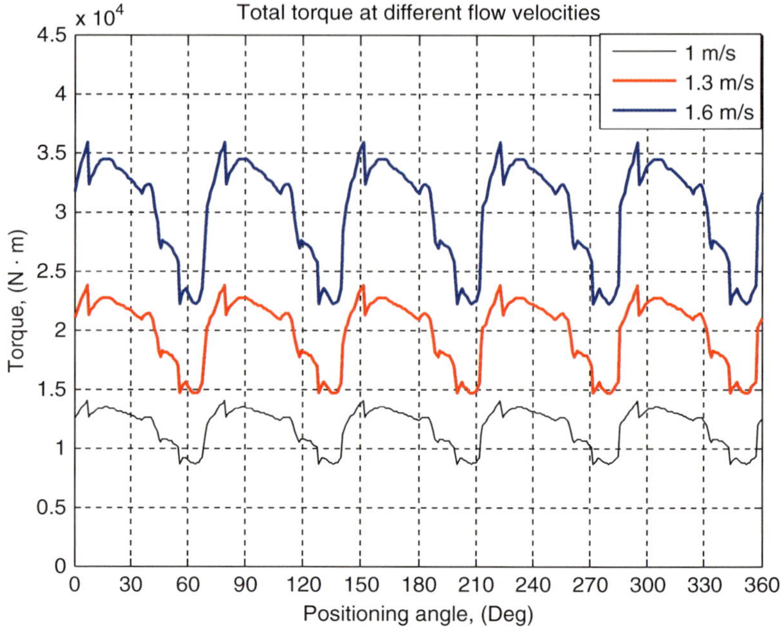

Fig. 4.101 Total torque $T_{r\Sigma}$ at rotor shaft versus the positioning angle for various water flow velocities

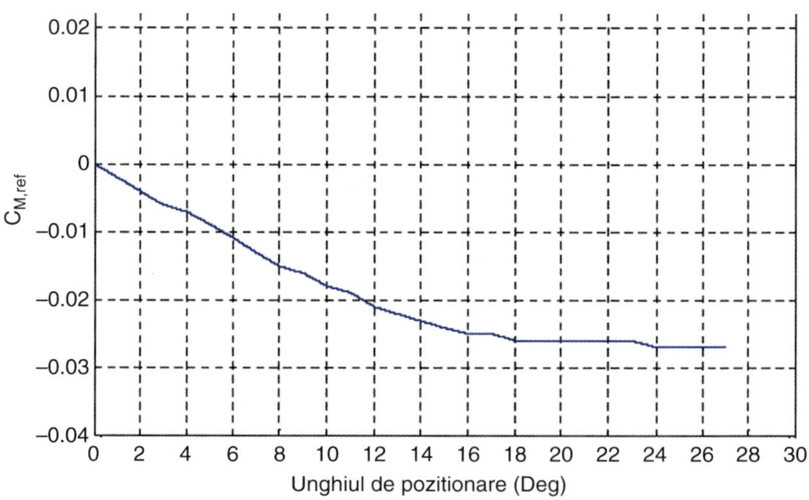

Fig. 4.102 Moment coefficient $C_{M,ref}$ versus the angle of attack for NACA 0016 profile

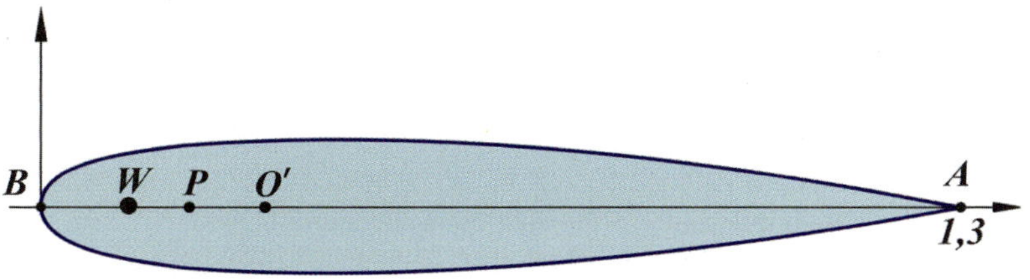

Fig. 4.103 Location of the blade fixing point

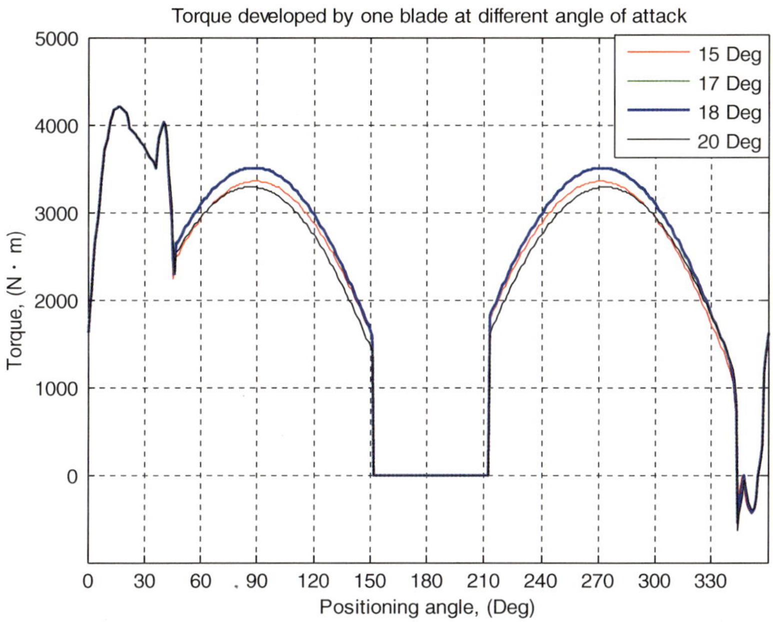

Fig. 4.104 Torque $T_{r,i}$ developed by one blade versus the positioning angle for various values of the attack angle $\alpha = 15°, 17°, 18°, 20°$

of the Archimedes force F_A (points N_i, $i = 1, 2, 3$, Fig. 4.109) has shown that the distance from these points to the rotor axis O will oscillate depending on the positioning angle φ. Thus, for the blades located in the upper half-plane defined by line OO' these distances are different from the respective distances of the blades located in the lower half-plane.

This fact leads to the appearance of the pitching moment with respect to the axis of longitudinal symmetry of the floating bodies:

$$M_r = M_{\Sigma,S} - M_{\Sigma,I} \tag{4.120}$$

where $M_{\Sigma,S}$ is the total moment developed by the Archimedes forces that act on the blades, currently located in the upper half-plane, and $M_{\Sigma,I}$ is the total moment

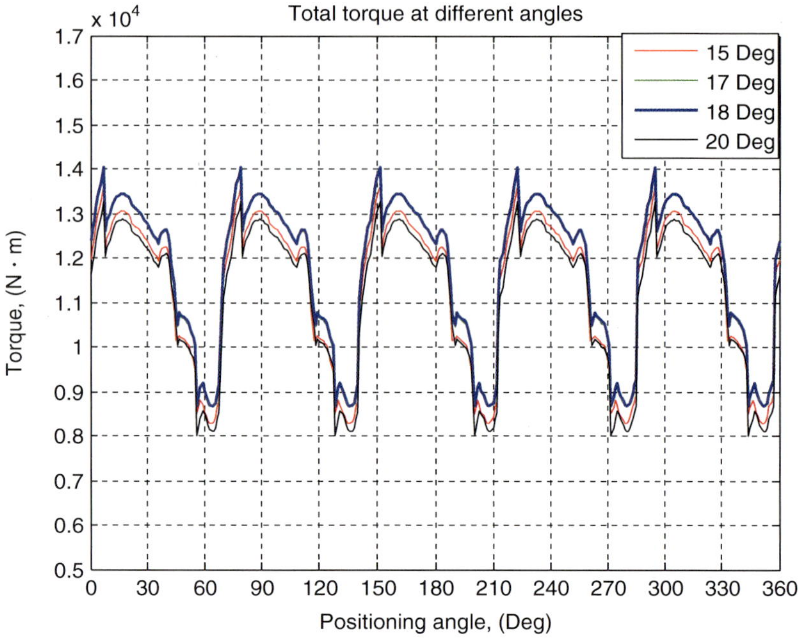

Fig. 4.105 Total torque $T_{r\Sigma}$ versus the positioning angle for various values of the angle of attack $\alpha = 15°, 17°, 18°, 20°$

developed by the Archimedes forces that act on the blades currently located in the lower half-plane. The total moments developed by the Archimedes forces that act on the blades currently located in the upper half-plane are determined by the relation

$$M_{\Sigma,S} = \sum F_{A,i} \cdot D_{A,i} \tag{4.121}$$

where $F_{A,i}$ is the Archimedes force that acts on the blade i, $D_{A,i}$ is the distance from the application point of the Archimedes forces to the rotor axis, and the summation is done for all blades located in the upper half-plane. Similarly,

$$M_{\Sigma,I} = \sum F_{A,i} \cdot D_{A,i} \tag{4.122}$$

Distances $D_{A,i}$ are given by:

$$D_{A,i}^2 = R^2 + c_A^2 + 2Rc_A \cos(\alpha + \varphi_i) \tag{4.123}$$

where R is the rotor radius, c_A is the distance between the application point of the Archimedes forces and the fixing point of the blade to the rotor lever, α is the angle of attack between the blade chord AB and the direction of water flow, and φ_i is the angle between the rotor lever and direction OO'.

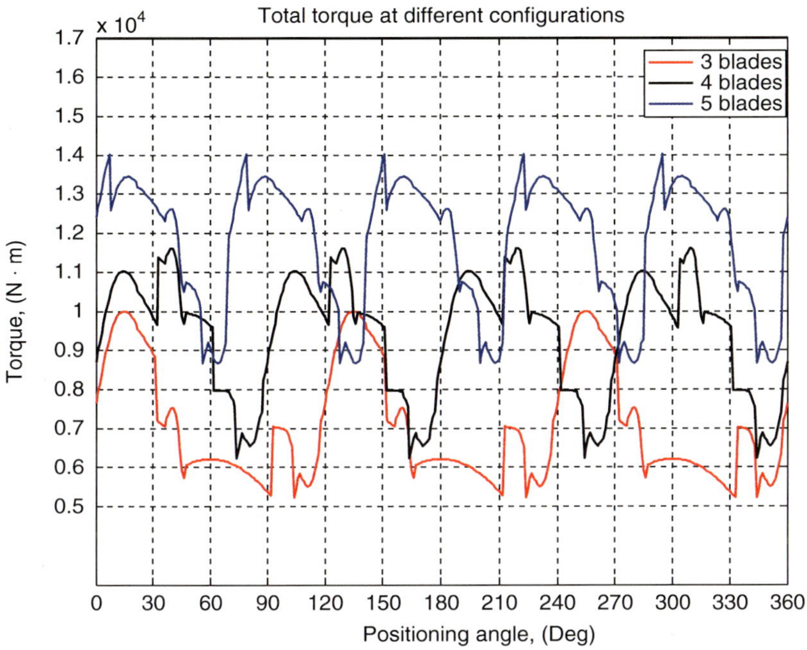

Fig. 4.106 Total torque $T_{r\Sigma}$ developed at the 3-, 4- and 5-blade rotor shaft versus the positioning angle

In order to compensate for the resulting pitching moment M_r it is proposed to move the rotor axis position with distance e with respect to the plane of longitudinal symmetry of the floating bodies. Distance e is given by:

$$e = \frac{\sum_{i=1}^{N_{pal}} y_i}{N_{pal}} \tag{4.124}$$

where N_{pal} is the number of rotor blades and y_i is the distance from the application point of the Archimedes force acting on blade i till the plane of longitudinal symmetry (Fig. 4.109). For each blade, distance y_i is computed by:

$$y_i = c_A \cos\alpha + R \sin\left(\varphi + (i-1)\frac{360^o}{n}\right) \tag{4.125}$$

Substitute (4.125) into (4.124) to get:

$$e = c_A \cos\alpha \tag{4.126}$$

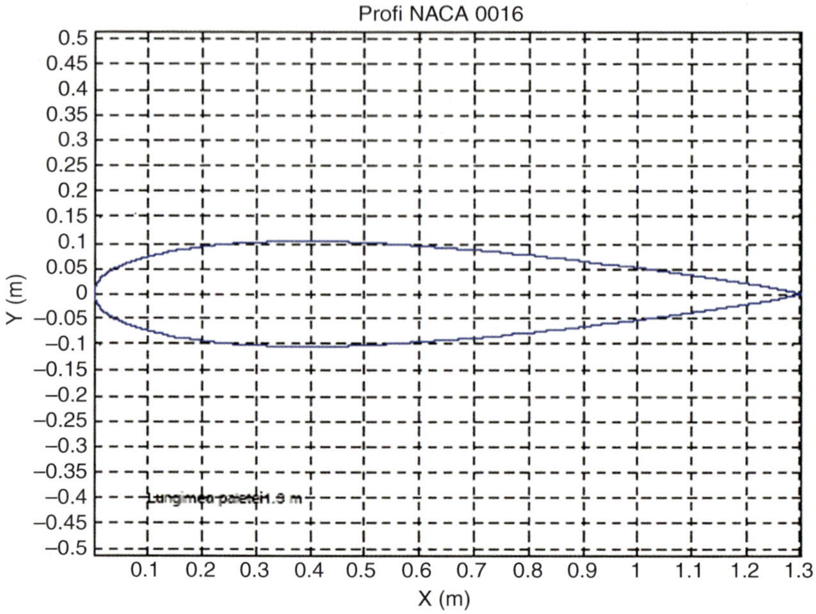

Fig. 4.107 Standard NACA 0016 hydrodynamic profile

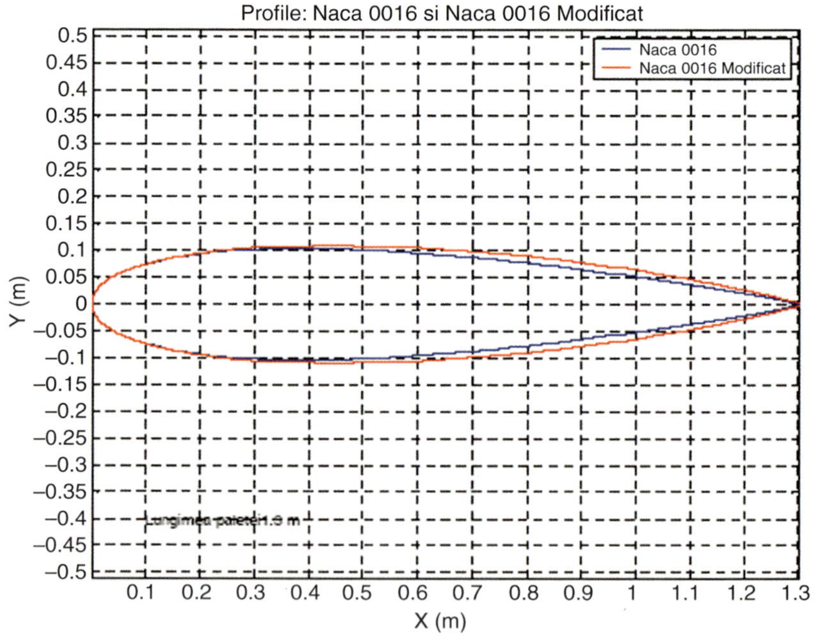

Fig. 4.108 Standard and optimised NACA 0016 hydrodynamic profile

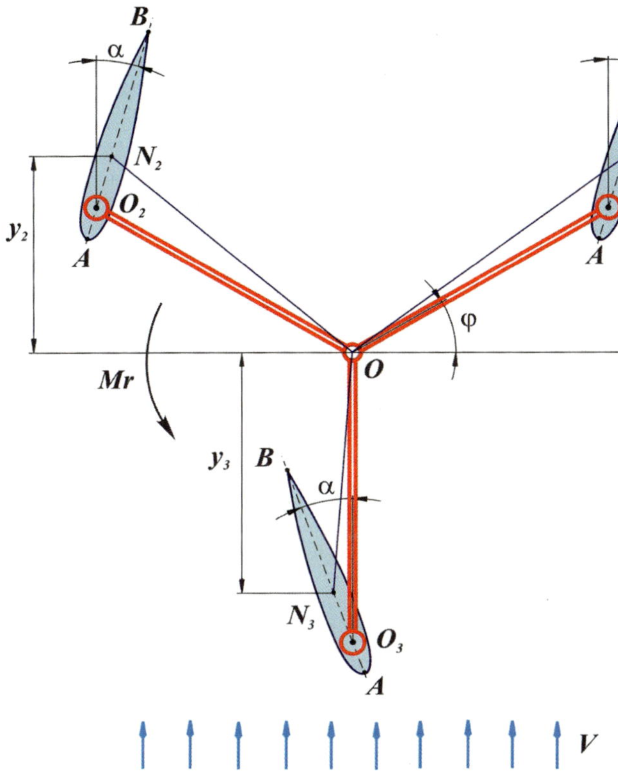

Fig. 4.109 Floating stability analysis

The application point of the Archimedes force on each blade is the centroid of hydrodynamic profile, in this case NACA 0016 profile. The centre of application points of the Archimedes forces that act on a number N_{pal} of submersed blades will migrate during rotor revolution (Fig. 4.110).

The migration trajectory generated by a full rotor revolution of 3- and 5-blade rotor represents closed curves shown in Fig. 4.110a and Fig. 4.110b. A point on the closed curve corresponds to the position of the centre of application points of the Archimedes force system in a fixed angular position of the rotor.

To identify the technical solution ensuring the floating stability of the micro hydropower plant it is necessary to estimate the values of the distance between the centre of application point of the Archimedes force system and the longitudinal symmetry axis of the floating bodies. Figure 4.111a, b shows the distance e versus the positioning angle φ of the 3-blade (a) and 5-blade (b) rotor. It has been established that in the case of a 3-blade rotor, distance e takes values bound by $e_{min} = 0.238\text{m}$ and $e_{\max} = 0.363\text{m}$. And in the case of a 5-blade rotor e is bounded by $e_{min} = 0.289\text{m}$, and $e_{\max} = 0.363\text{m}$. The average value of distance e for both 3- and 5-blade rotors is $e_{ave} = 0.33\text{m}$.

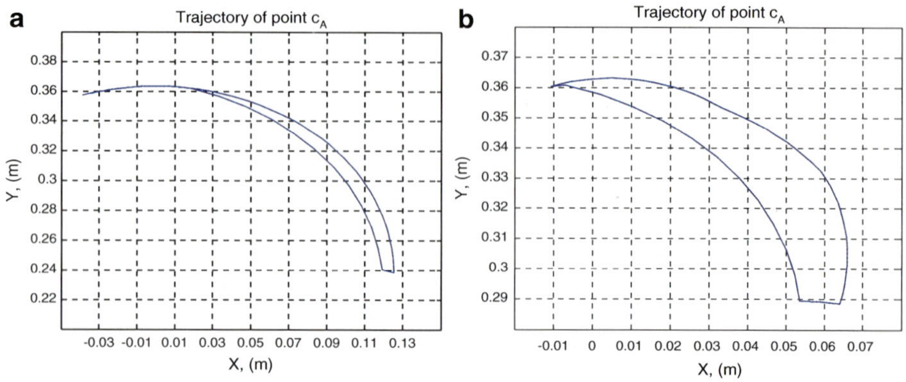

Fig. 4.110 Migration trajectory of the central point of application of the Archimedes forces for the 3-blade (**a**) and 5-blade rotor (**b**)

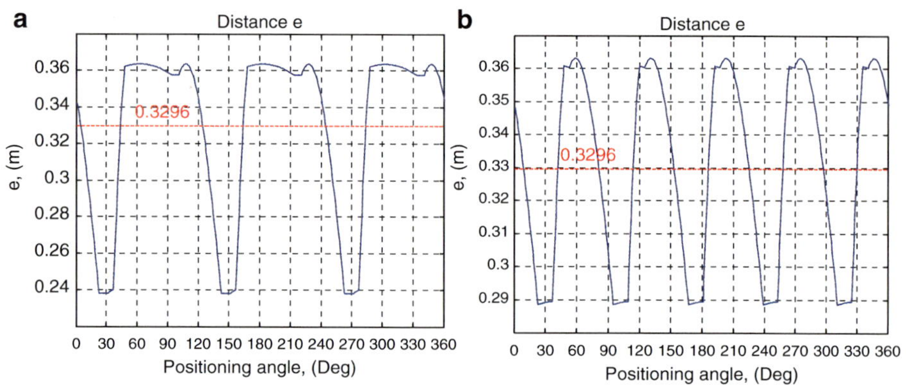

Fig. 4.111 Distance e versus the positioning angle φ for the 3-blade rotor (**a**) and 5-blade rotor (**b**)

In conclusion, in order to ensure the floating stability of the micro hydropower station the axis of 3- and 5-blade rotor must be shifted from the axis of longitudinal symmetry of the floating bodies at a distance of $e_{med} = 0.33\,\text{m}$ in the upriver direction. Also, micro hydropower stations anchored on the left bank must differ from those anchored on the right bank by the spatial truss constructions, in particular, by the constructional elements of the hydrodynamic rotor being shifted with distance e_{ave}.

4.5.2.8 Turbulence and Stability of the Hydrodynamic Rotor

Figure 4.112 shows the distribution of flow velocities around NACA 0016 profile corresponding to an attack angle of 18° and Reynolds number given by:

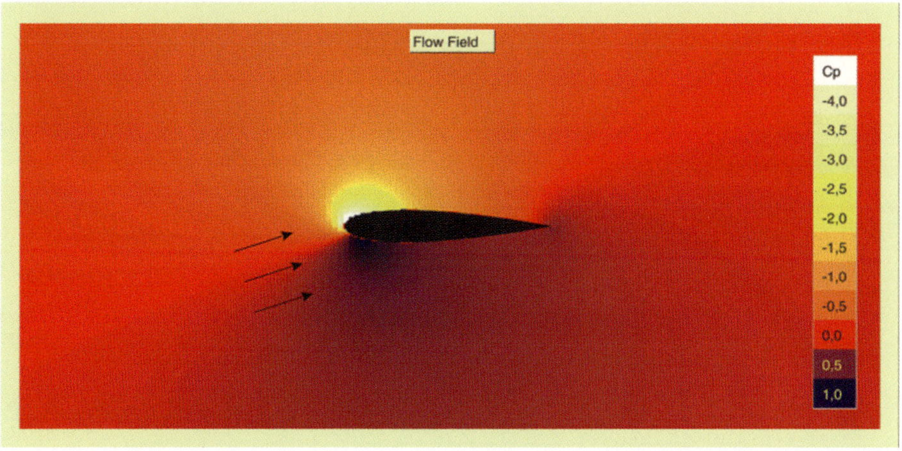

Fig. 4.112 Flow velocity around NACA 0016 profile at angle of attack 18°

$$\mathrm{Re} = \frac{\rho c \overrightarrow{V}}{\eta} = \frac{c \overrightarrow{V}}{\nu} \tag{4.127}$$

where the water density value is $\rho = 998.4\,\mathrm{kg}/\mathrm{m}^3$, the kinematical viscosity is $\nu = 1.012 \cdot 10^{-6}\,\mathrm{m}^2/\mathrm{s}$, and the length of the profile chord is $c = 1.3\,\mathrm{m}$. For the fluid flow velocities $V_\infty = 1\,\mathrm{m}/\mathrm{s}, 1.5\,\mathrm{m}/\mathrm{s}, 2\,\mathrm{m}/\mathrm{s}$ the corresponding values of the Reynolds number Re = 1,284,600, 1,798,400 and 2,312,300 are obtained.

The transition and separation of the boundary layer on the lower and upper surfaces of the blade profile are shown in Fig. 4.113. Points T.U. and T.L. stand for the points of transition from the laminar flow to the turbulent flow on the lower and upper surfaces of the blade (Fig. 4.93), respectively. Points S.U. and S.L. represent the boundary layer separation points on lower and upper surfaces. It was observed that on the upper surface the transition from the laminar flow to the turbulent flow takes place in the proximity of the stagnation point, while the separation of the boundary layer is predicted at approximately 40–50% of the blade chord length. On the lower surface the transition from laminar to turbulent flow as well as the separation of the boundary layer will take place in the vicinity of the trailing edge.

4.5.2.9 Estimation of the Energy Potential of the Multi-blade Hydrodynamic Converted at the Rotor Shaft

The energy potential of the water flow and the power generated by the micro hydro power plant rotor with vertical axis are estimated for the following constructive and functional parameters:

- 5-blade rotor with NACA 0016 hydrodynamic profile (Fig. 4.114);
- Effective immersion of blade in the water flow h = 1.4 m;
- Length of the blade chord l = 1.3 m;

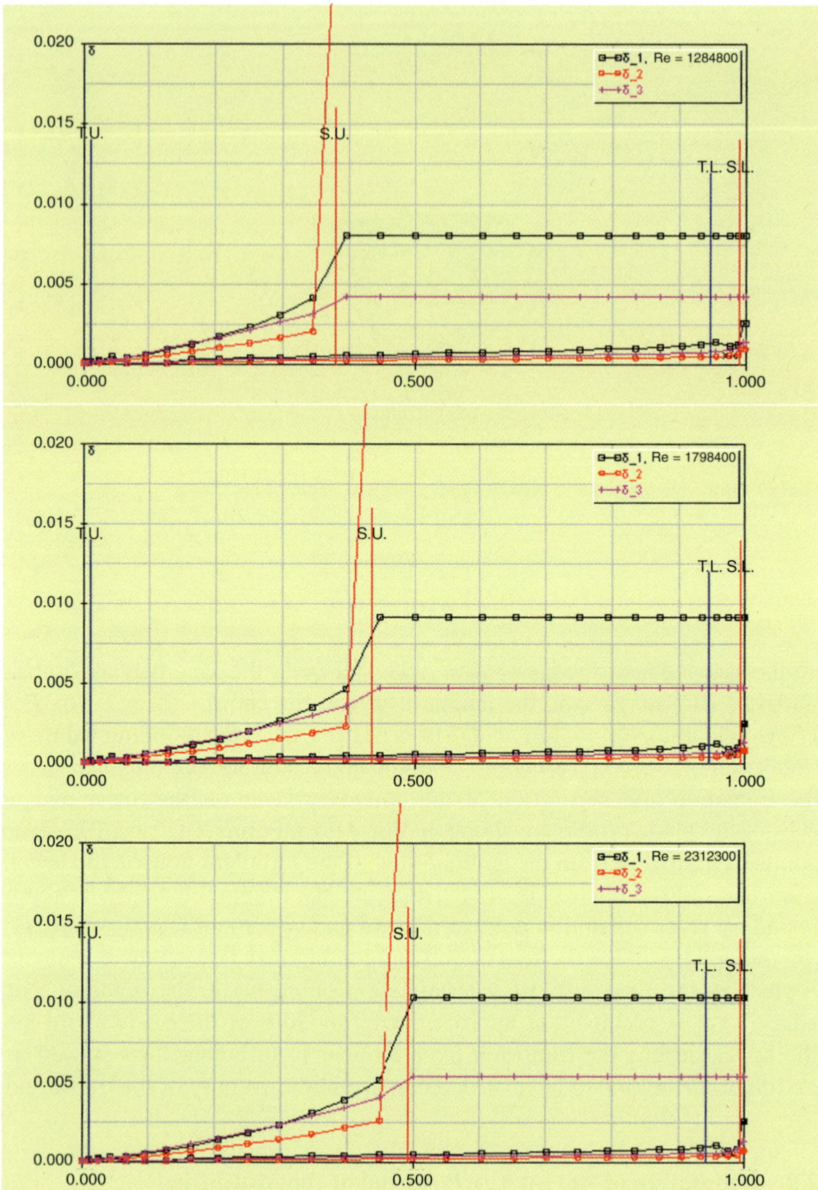

Fig. 4.113 Boundary layer analysis for the flow velocities 1 m/s, 1.5 m/s and 2 m/s

- Effective length l of the blade chord $l = 1$ m;
- Rotor diameter $D = 4$ m;
- Blade working angle of attack $\alpha = 18°$;
- Blade angle of attack in the neutral zone (zone V, Fig. 4.97) $\alpha = 90°$.

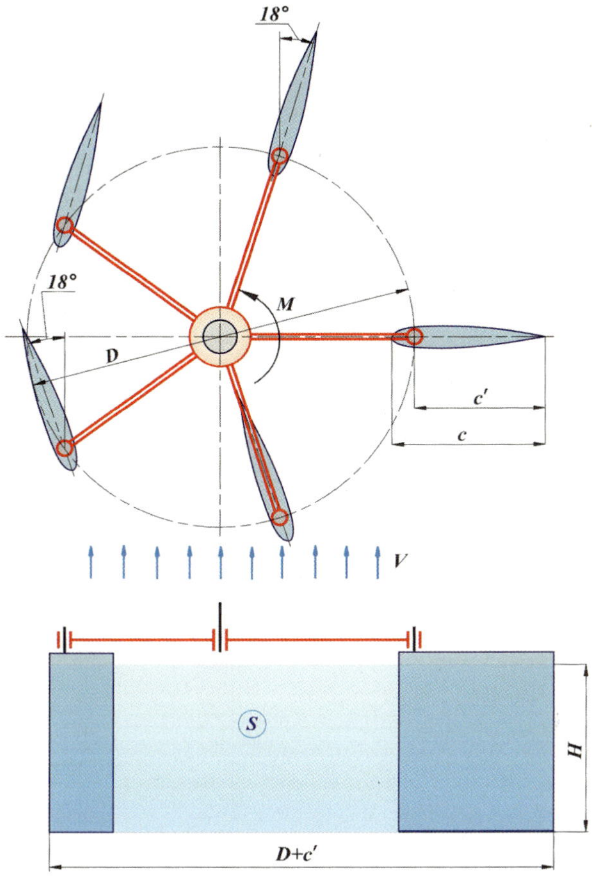

Fig. 4.114 Estimation of the energy potential of the water flow

Energy potential of the water flow estimated for the section corresponding to the rotor dimensions is given by:

$$P = \frac{1}{2}\rho {V_\infty}^3 S \tag{4.128}$$

where ρ is the fluid density, V_∞ is the water flow velocity and $S = H(D + c')$ is the area of the rotor cross section (Fig. 4.114). The power generated by the water flow at the rotor axis is compute by the formula:

$$P_a = PK \tag{4.129}$$

where P is given by relation (4.128) and K is the efficiency factor of energy conversion. Table 4.11 shows the energy potential of the water flow that corresponds to the effective surface of the dimension cross section of the rotor $S = 7\ \text{m}^2$, $S = 9.15\ \text{m}^2$, $S = 11.5\ \text{m}^2$ depending on the respective velocities of the water flow.

Table 4.11 Energy potential of the water flow versus the water flow velocity

Water flow velocity, V (m/s)	P (kW) ($S = 7$ m^2)	P (kW) ($S = 9.15$ m^2)	P (kW) ($S = 11.5$ m^2)
1.0	3.5	4.6	5.75
1.1	4.65	6.09	7.653
1.2	6.05	7.9	9.94
1.3	7.68	10.05	12.63
1.4	9.6	12.55	15.78
1.5	11.8	15.44	19.41
1.6	14.3	18.74	23.55
1.7	17.19	22.47	28.25
1.8	20.4	26.68	33.53
1.9	24	31.38	39.44
2.0	28	36.6	46.00

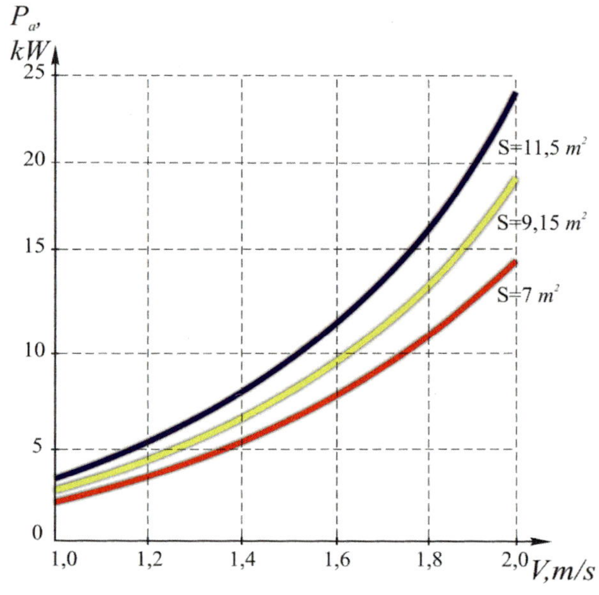

Fig. 4.115 Generated power at rotor axis

Figure 4.115 shows the converted power P_a at the rotor shaft versus the water flow velocity with a conversion efficiency coefficient 50% ($K = 0.5$) for various dimensions of the rotor cross section area S.

The generated power P_a of the rotor under the action of the water flow is provided by the formula:

$$P_a = T_{r\Sigma}\omega \tag{4.130}$$

where $T_{r\Sigma}$ is the total torque developed at the rotor axis given by formula (4.51), and ω is the angular velocity of the rotor. Figures 4.116 and 4.117 show the total torque $T_{r\Sigma}$ applied at the rotor (with 3- and 5 blades, respectively) axis and its average

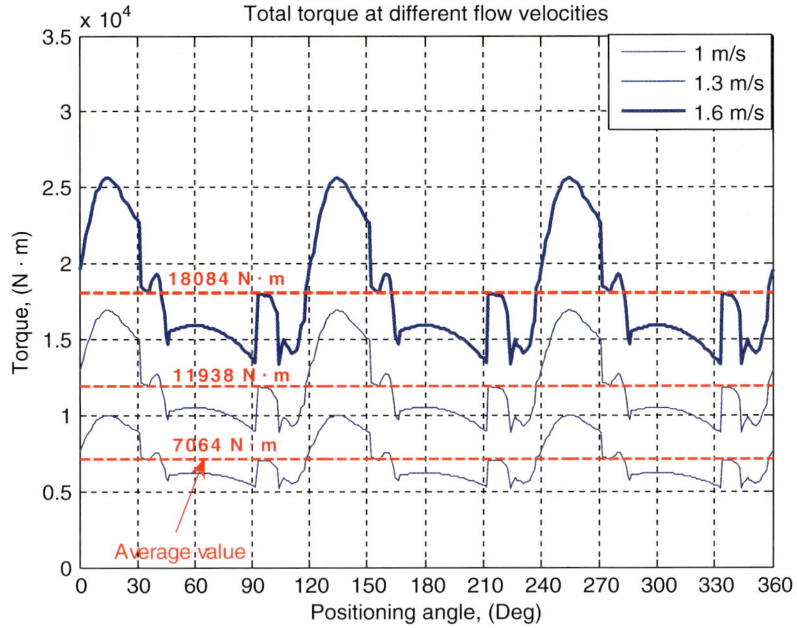

Fig. 4.116 Total torque of 3-blade rotor versus the positioning angle for flow velocities 1 m/s, 1.3 m/s and 1.6 m/s

value versus the positioning angle for various flow velocities. For instance, for the flow velocity $V_\infty = 1.3\,\mathrm{m/s}$, $T_{r\Sigma} = 19893\,\mathrm{Nm}$. Thus, the power of the rotor shaft generated by the energy potential of the water flow with velocity $V_\infty = 1.3\,\mathrm{m/s}$ is $P = 3.84\,\mathrm{kW}$. The angular velocity and the number of rotations per minute of the rotor are:

$$\omega = \frac{\mathrm{P}}{T_{r\Sigma}} = \frac{3840}{19893} = 0.1913\,\mathrm{s}^{-1} \tag{4.131}$$

$$n = \frac{30\omega}{\pi} = 1.843\,\mathrm{min}^{-1} \tag{4.132}$$

4.5.2.10 Computational Simulation of Water Flow Interaction with the Hydrodynamic Blades

In order to perform the deformation and stress analysis of a submersed blade with NACA 0016 hydrodynamic profile, the finite element ANSYS software is being used. Consider a submersed blade in the water flow subjected to hydrodynamic forces corresponding to the flow velocity $V_\infty = 2m/s$ and hydrostatic pressure [64, 65]. The maximum value of the forces acting on the blade is 11 KN. The

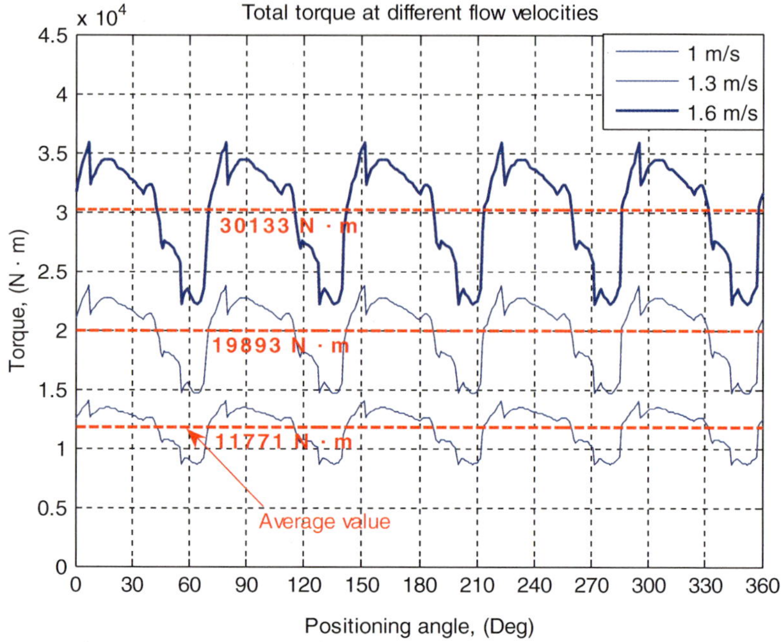

Fig. 4.117 Total torque of 5-blade rotor versus the positioning angle for flow velocities 1 m/s, 1.3 m/s and 1.6 m/s

cover of the hydrodynamic blade is made of aluminium alloy H37 with Young module $E = 1.97 \cdot 10^{11}\,\mathrm{N}/\mathrm{m}^2$ and Poisson coefficient $\nu = 0.27$ and the blade's interior is hollow. Two values for the lateral cover thickness are proposed: 1 mm and 1.5 mm. Also, three proposals for inner resistance structure of the blades have been examined: with 3, 4 and 5 transversal stiffening ribs (Fig. 4.118a, b, c). Since the estimated deformations can be considered small, the linear elasticity framework is adopted. The blade lateral cover is modelled with Kirchhoff-Love thin shell theory. The adaptive finite element discretization uses *shell63* elements form ANSYS library. The discretization with 18,248 elements and 18,683 nodes is shown in Fig. 4.119a. Forces applied on the discretized blade cover are shown in Fig. 4.119b.

The displacements in the blade lateral cover with thicknesses 1 mm and 1.5 mm and 3, 4, and 5 transversal stiffening ribs are presented in Figs. 4.120–4.122. For the blades with lateral cover thickness of 1 mm (Figs. 4.120a, 4.121a and 4.122b) the maximal displacements in the lower submersed area are 7.8 mm, 5.1 mm and 3.5 mm, respectively. These local displacements in the hydrodynamic profile can negatively influence the fluid flow regime in the adjacent zone and, consequently, the conversion efficiency of the water flow kinetic energy. For this reason, the profiles with the cover thickness of 1 mm have been abandoned.

On the other hand, the maximal displacements in the blades with cover thickness 1.5 mm and 3, 4 and 5 transversal ribs were reduced by 2.1, 2.4 and 2.6 times,

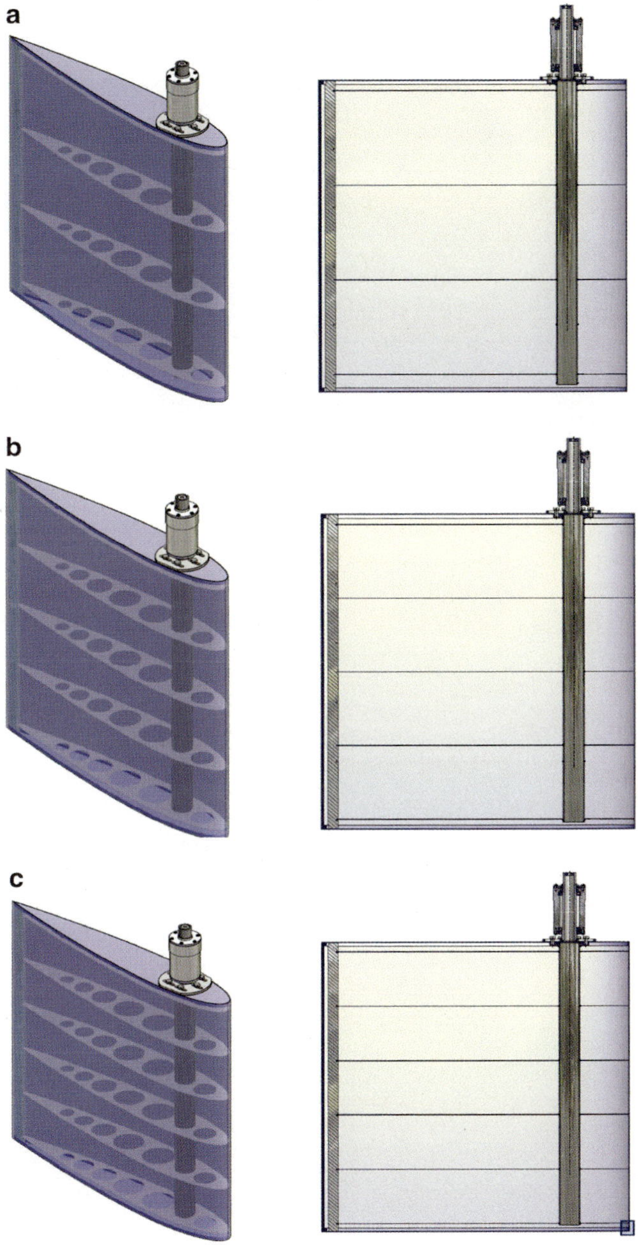

Fig. 4.118 Resistance inner structure of the blade with: 3(**a**), 4(**b**) and 5(**c**) transversal stiffening ribs

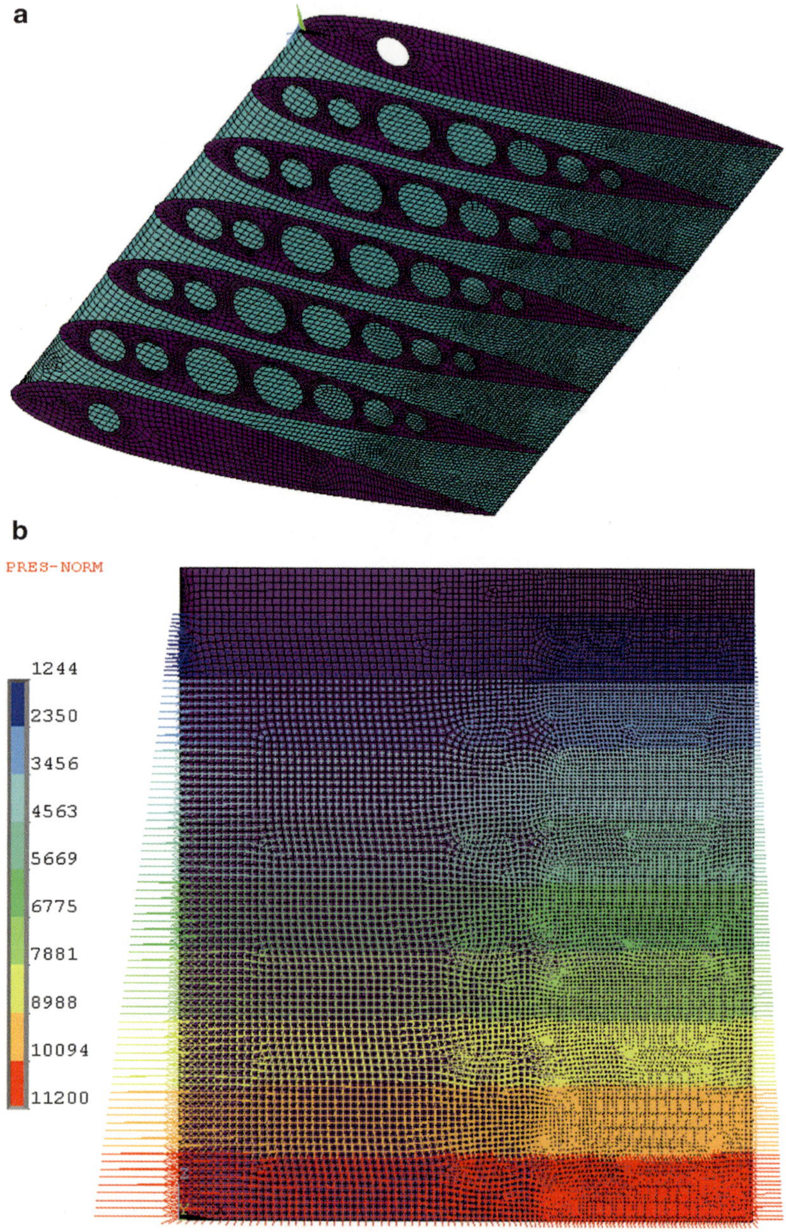

Fig. 4.119 Discretization of the blade lateral cover into finite elements (**a**) and the applied forces (N) (**b**)

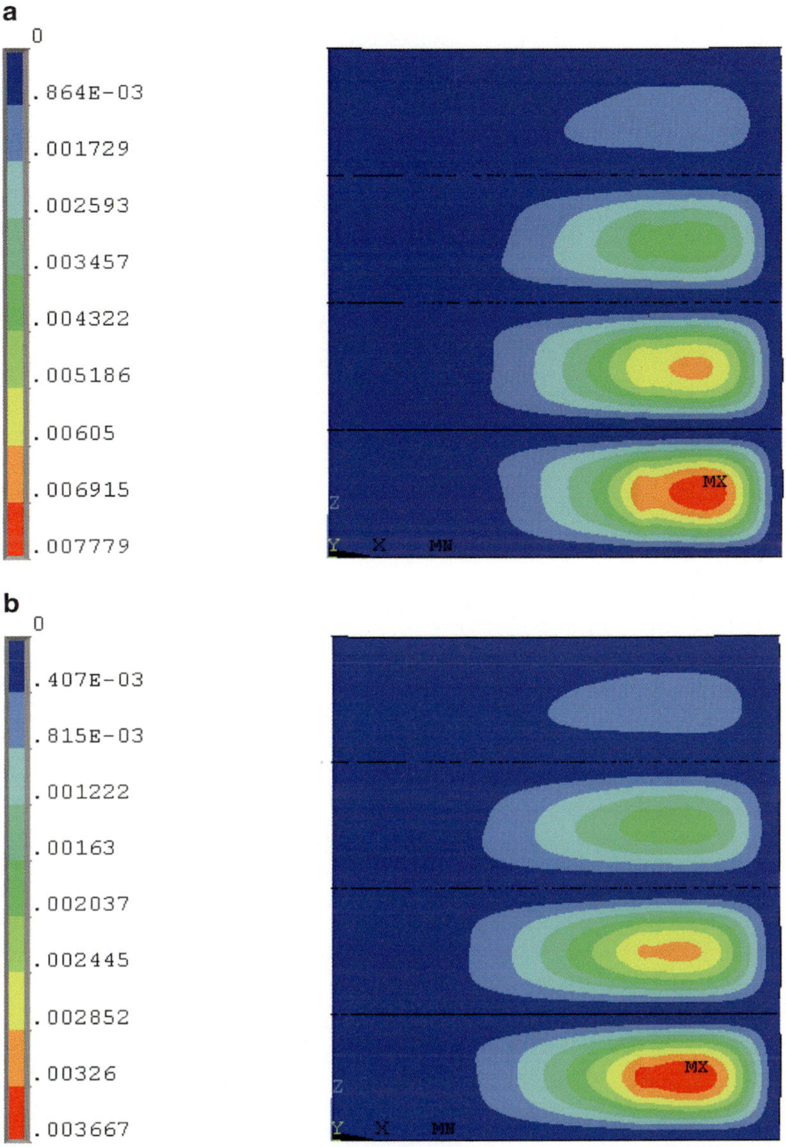

Fig. 4.120 Displacements (mm) in the NACA 0016 blade cover with three transversal stiffening ribs and thickness $S = 1$ mm (**a**) and $S = 1.5$ mm (**b**)

respectively, totalling 3.7 mm, 2.1 mm and 1.3 mm, respectively. The blades with the cover thickness 1.5 mm and 5 transversal ribs have a maximal displacement of 1.33 mm, acceptable from the point of view of decreasing the negative impact on the conversion efficiency. Therefore, the resistance structure with 5 transversal stiffening ribs and a lateral cover with thickness of 1.5 mm, made of aluminium alloy H37 will be examined further.

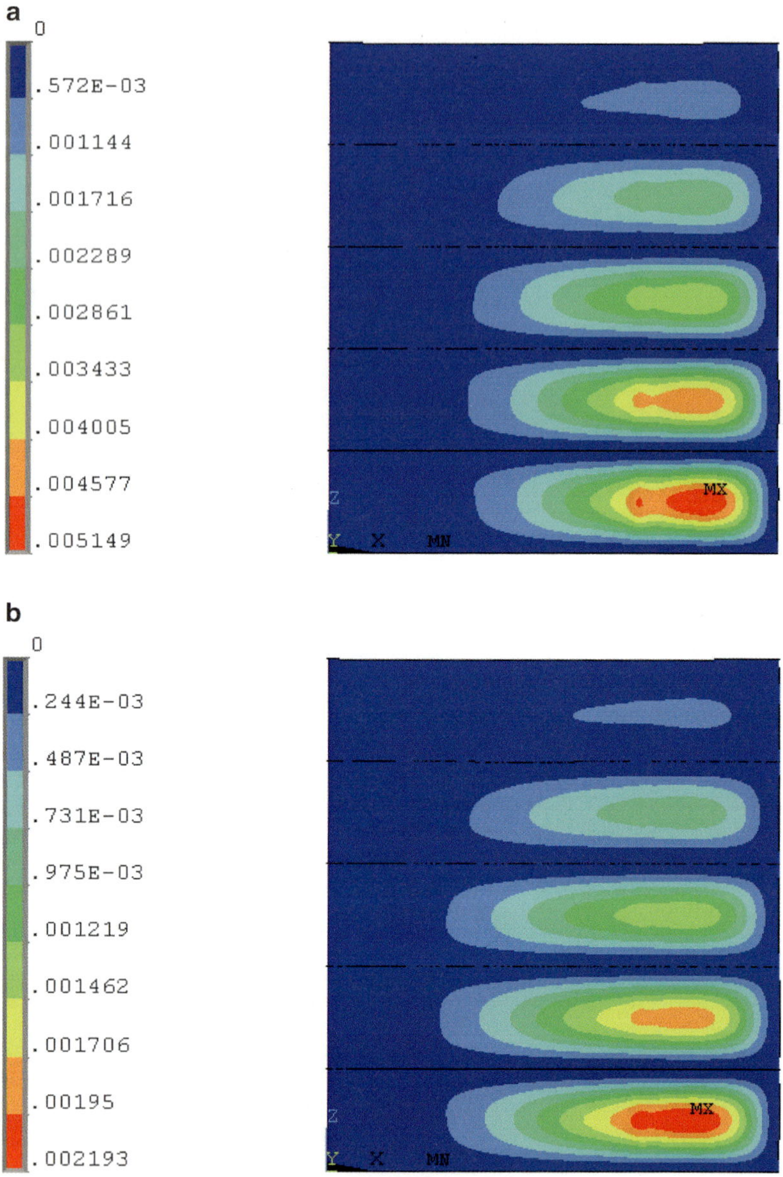

Fig. 4.121 Displacements (mm) in the NACA 0016 blade cover with four transversal stiffening ribs and thickness $S = 1$ mm (**a**) and $S = 1.5$ mm (**b**)

Clearly, a larger number of transversal stiffening ribs and/or a thicker cover induces a higher blade weight and manufacturing cost. The conversion efficiency highly depends on the compliance with the prescribed design shape parameters, along both the blade length and height. Therefore, it is necessary to estimate the

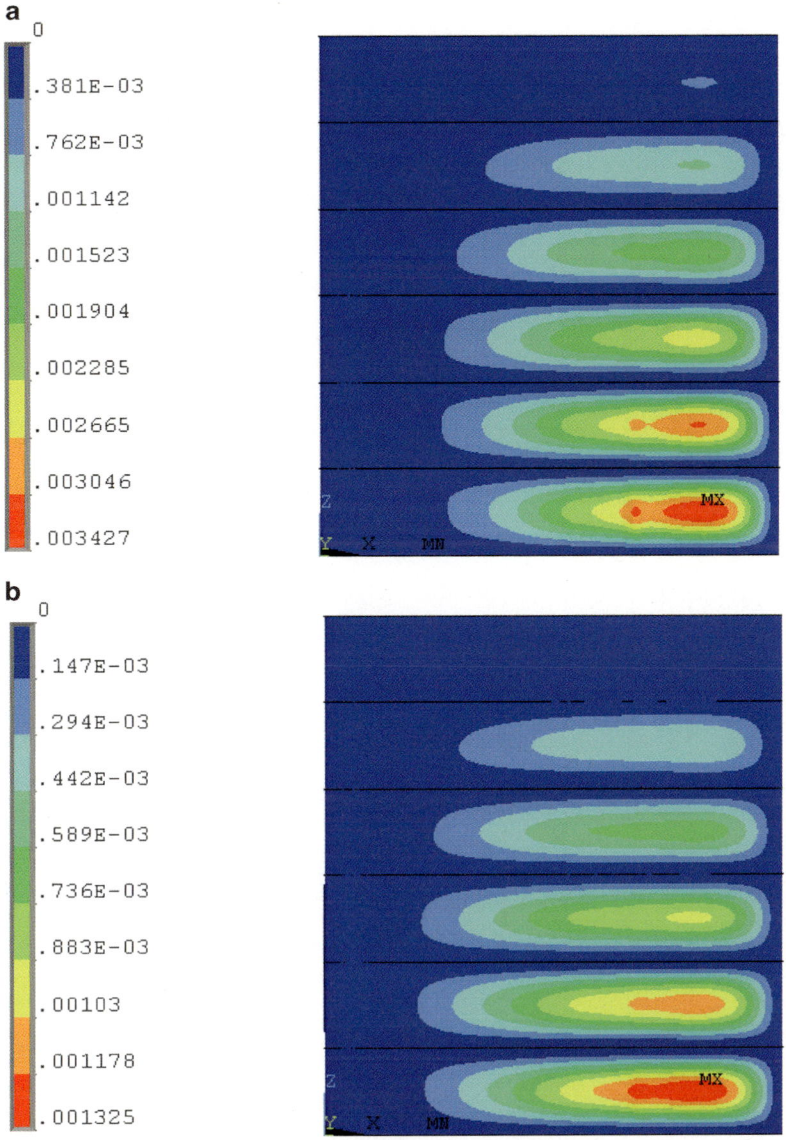

Fig. 4.122 Displacements (mm) in the NACA 0016 blade cover with five transversal stiffening ribs and thickness $S = 1$ mm (**a**) and $S = 1.5$ mm (**b**)

displacements along the blade length and height in the areas of maximal loading by hydrostatic and hydrodynamic forces. The displacements (mm) in the blade cover along the blade length in cross section A, located between two transversal ribs are shown in Fig. 4.123a. Thus, the maximum displacement of 1.3 mm is located at a distance of 235 mm from the trailing edge. The area with displacements larger

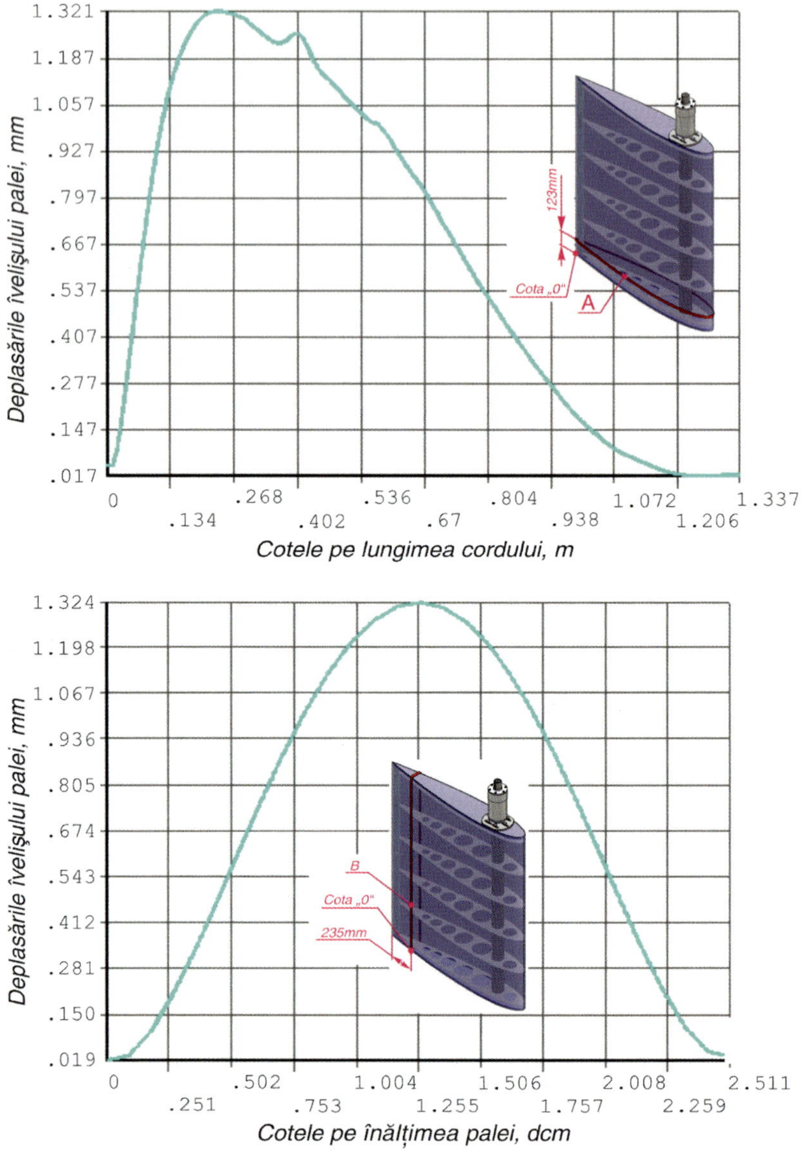

Fig. 4.123 Displacements (mm) in the blade cover in cross sections A (**a**) and B (**b**)

than 1 mm extends from a distance of 130 mm to 530 mm from trailing edge that consequently implies a local modification of the attack angle by ±0.29°, a change hydro dynamically admissible. Figure 4.123b shows the displacements from cross section B (comprising the area between the last two transversal ribs) placed at a distance of 235 mm from the blade trailing edge.

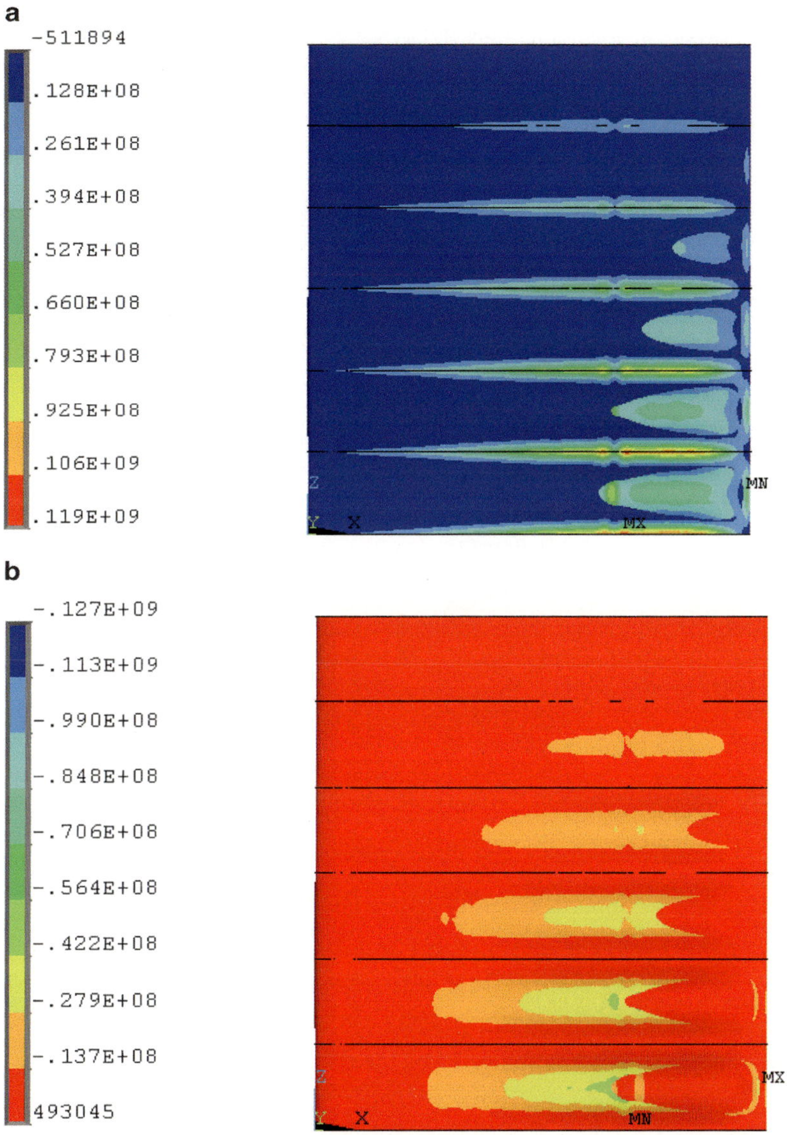

Fig. 4.124 Main stresses (N/m^2) σ_1 (**a**) and σ_3 (**b**) in the NACA 0016 hydrodynamic profile blade cover with five transversal stiffening ribs and cover thickness $S = 1.5$ mm

In order to quantify the stresses appearing in the blade cover with thickness 1.5 mm consider the eigenvalues of stress tensor arranged in decreasing order, called the main stresses, σ_1, σ_2 and σ_3. Figure 4.124 shows the distribution of σ_1 (a) and σ_3 (b).

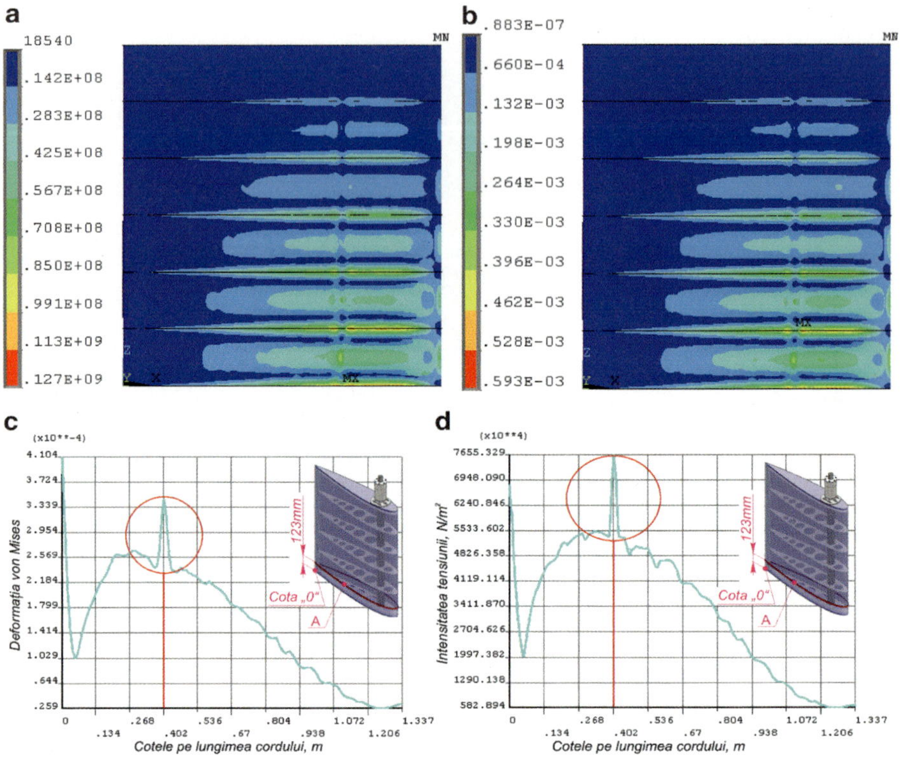

Fig. 4.125 Stress intensity (N/m^2) σ_I (**a**, **c**) and Von Mises deformation criterion ε_e (**b**, **d**) in the blade cover

Also, consider the stress intensity, defined by relation

$$\sigma_I = \frac{1}{\sqrt{2}}\left[(\sigma_1 - \sigma_2)^2 + (\sigma_2 - \sigma_3)^2 + (\sigma_1 - \sigma_3)^2\right]^{1/2} \tag{4.133}$$

and the Von Mises deformation given by the relation:

$$\varepsilon_e = \frac{1}{1+\nu}\left(\frac{1}{2}\left((\varepsilon_1 - \varepsilon_2)^2 + (\varepsilon_2 - \varepsilon_3)^2 + (\varepsilon_1 - \varepsilon_3)^2\right)\right)^{1/2} \tag{4.134}$$

where ν is the Poisson coefficient, and $\varepsilon_1, \varepsilon_2, \varepsilon_3$ are the eigenvalues of the deformation tensor.

Figure 4.125a, b shows the stress intensity σ_I (N/m^2) and the Von Mises deformation in the blade cover. From the analysis of the variation of stress intensity σ_I and Von Mises deformation in cross section A along the blade length (Fig. 4.125c, d), the concentration of stress intensity and Von Mises deformations was observed at a distance of 402 mm from the trailing edge. This distance corresponds to the

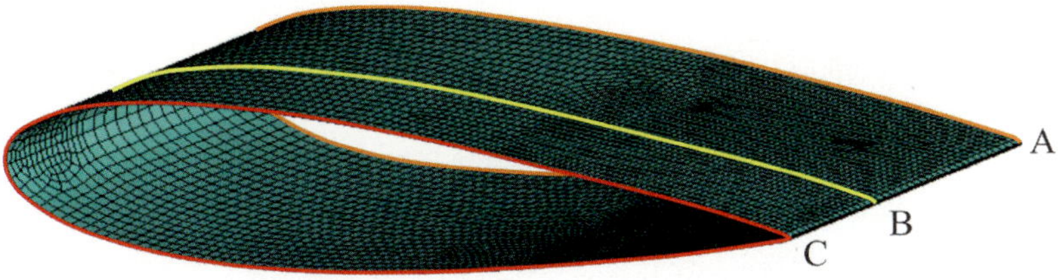

Fig. 4.126 Discretization of the blade cover with Solsh190 finite elements

transition of NACA 0016 profile boundary from a region with lower curvature radius (distances 1,137 to 402 mm) to the region with a larger curvature radius (distances 402 to 0 mm).

Due to the fact that the micro hydropower station blades will be acting in aggressive conditions (water and possible debris), an important issue consists in the assurance of the blades impermeability. For this purpose a hydrofoil injected with polyurethane foam should be also considered.

To this end consider a blade with cover made of aluminium alloy H37 with Young modulus $E = 1.97 \times 10^{11}$ N/m^2 and Poisson coefficient $\nu = 0.27$ injected with polyurethane foam with density 0.6 kg/cm^3, Young modulus $E = 0.95$ Gpa and Poisson coefficient $\nu = 0.24$. Since maximal deformations and stress occur between the last two stiffening ribs, only that portion of the blade was modelled. Discretization of the blade cover with 13,340 elements Solsh190 is presented in Fig. 4.126. The stiffening ribs are not modelled since their deformations are negligible, but instead the constraints on the degrees of freedom are imposed: on line *A* (Fig. 4.126) $u_x = 0$, $u_y = 0$ and $u_z = 0$; on line B, $u_x = 0$ and $u_y = 0$; on line *C*, $u_z = 0$. The polyurethane foam is modelled with 20,785 elements of type Solid45.

The displacements in a hydrofoil injected with polyurethane foam and alluminum alloy cover with thickness 1.5 mm are presented in Fig. 4.127, u_x (a), u_y (b) and u_z (c), maximal displacement being 0.01 mm. In Fig. 4.128, there are shown the main stresses σ_1 (a), σ_2 (b) and σ_3 (c) (Pa). The maximal stress is of approximately 0.64 MPa.

On the other hand, since the manufacturing technology of the blades with aluminium cover needs expensive welding processes, special machines and qualified workers, it was proposed to use composite materials for lateral cover. Consider a laminated composite material shell composed of the following layers as shown in Fig. 4.132; first layer bidirectional lining of type E fiberglass and polyetheric resin matrix; second layer has two sublayers consisted of chopped fiberglass linings with an armoured polypropylene lining between them; third layer is again a chopped fiberglass lining in a polyetheric matrix; and the fourth is a gelcoat covering layer.

The computation of the material constants was performed using the manufacturer recommendations. The Young modulus for the composite layer is computed by relation

$$E = E_m V_m + E_f V_f \tag{4.135}$$

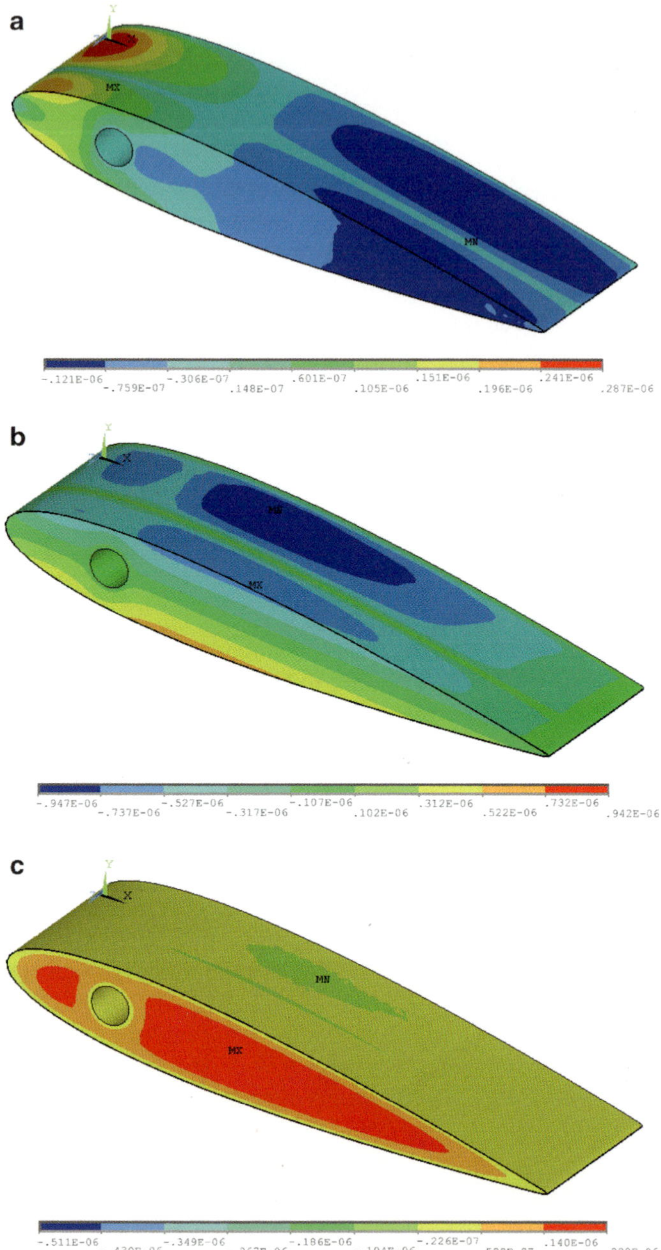

Fig. 4.127 Displacements in the blade injected with polyurethane foam and aluminium cover with thickness 1.5 mm: u_x (**a**), u_y (**b**) and u_z (**c**)

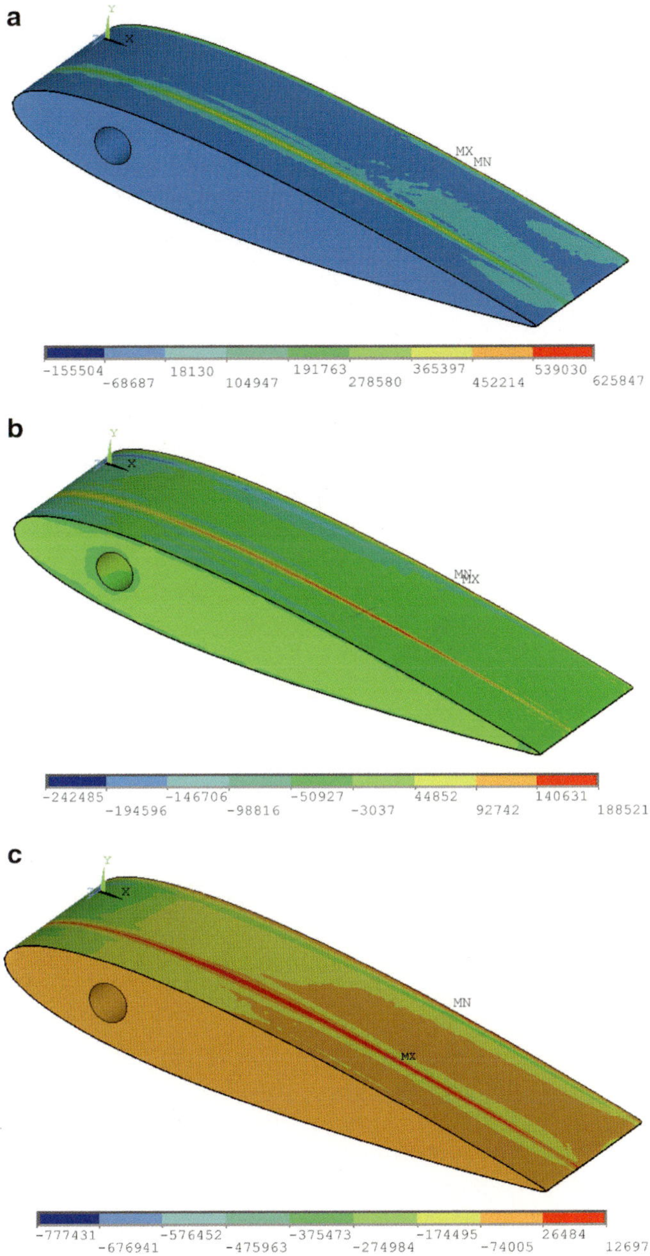

Fig. 4.128 Main stresses in the blade injected with polyurethane foam and alluminum cover with thickness 1.5 mm: σ_1 (**a**), σ_2 (**b**) and σ_3 (**c**) (Pa)

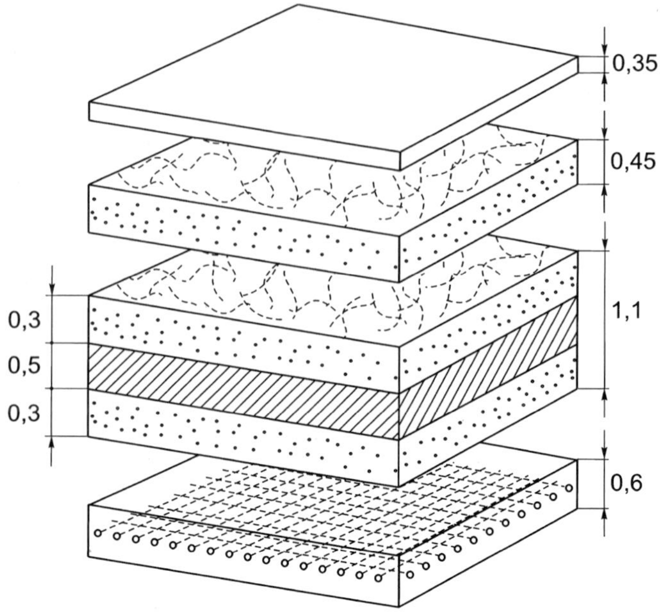

Fig. 4.129 The structure of the composite material blade cover

where E_m and E_f are Young modulus for matrix and fiberglass, respectively, and V_m and V_f are the volume fractions of the matrix and fiberglass. Evidently, $V_m + V_f = 1$. Poisson coefficient is computed by formula

$$\nu_{12} = \nu_m V_m + \nu_f V_f \tag{4.136}$$

where ν_m and ν_f are matrix and fibreglass Poisson coefficients. For the considered laminar plate (Fig. 4.129) the volume fractions are $V_m = 0.5$ and $V_f = 0.5$. Thus, the following values have been obtained: Young modulus $E = 13.2$ GPa and Poisson coefficient $\nu_{12} = 0.3$.

Two models have been considered: a portion of the hollow blade discretised with finite elements Solsh190 and a portion of the blade injected with polyurethane foam discretised with Solsh190 elements (lateral cover) and Solid45 elements (interior). In both cases the lateral cover is made of laminated composite materials with thickness 2.6 mm, polyurethane foam material constants, hydrodynamic forces, hydrostatic pressure, constraints and boundary conditions. In the hollow hydrodynamic blade with composite material cover with thickness 2.6 mm the maximal displacement is 4.3 mm and the maximal composite material cover with thickness 2.6 mm: σ_1 (a), σ_2 (b) and σ_3 (c) (Pa).

Stress value is approximately 38 MPa. The stresses and displacement distributions are similar with those of hollow blade with aluminium cover. In Fig. 4.130 there are presented displacements in the injected with polyurethane foam blade

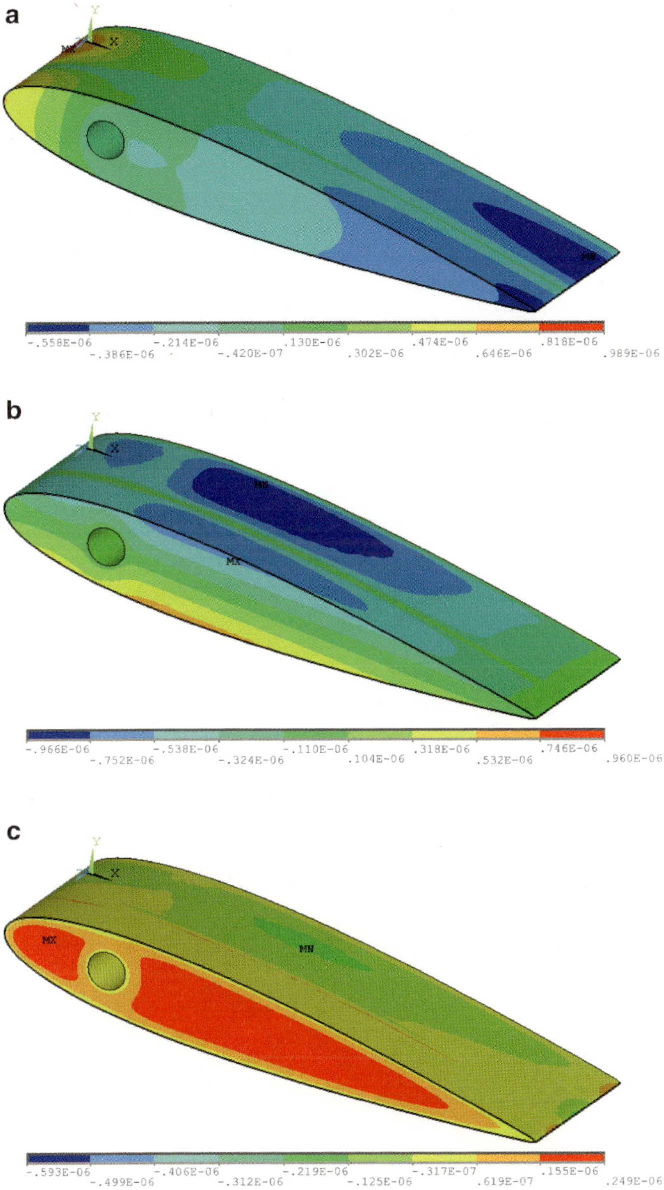

Fig. 4.130 Displacements in the blade injected with polyurethane foam and composite material cover with thickness 2.6 mm: u_x (**a**), u_y (**b**) and u_z (**c**)

with the same thickness of lateral cover: u_x (a), u_y (b) and u_z (c), with maximal displacement 0.01 mm. Figure 4.131 shows the main stresses in a blade injected with polyurethane foam: σ_1 (a), σ_2 (b) and σ_3 (c) (Pa).

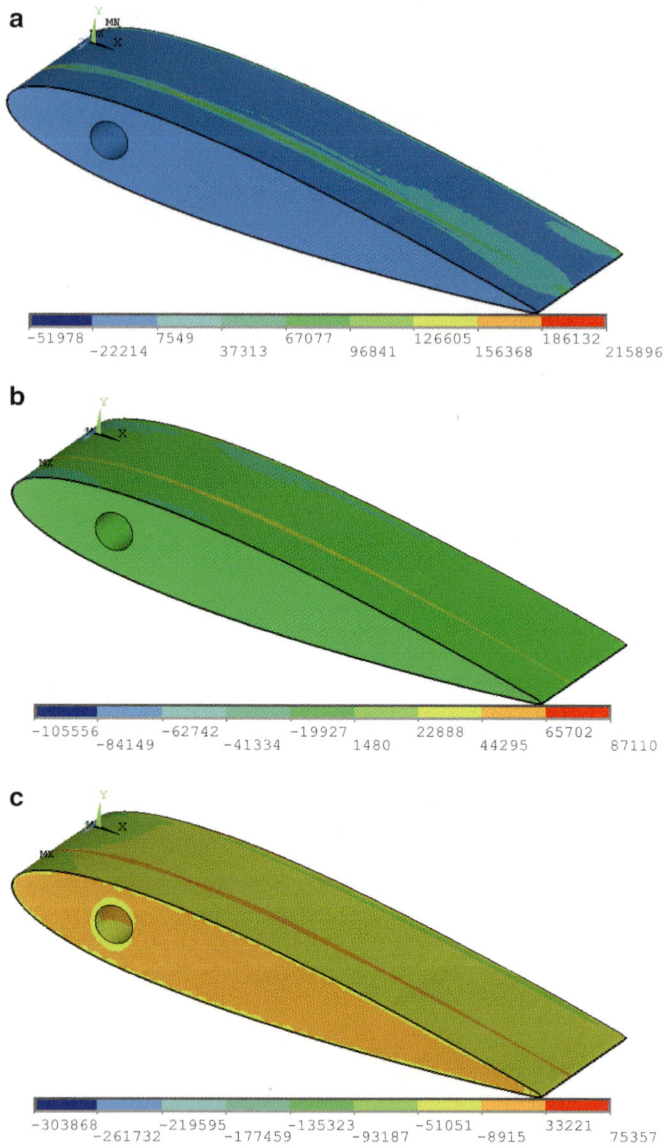

Fig. 4.131 Main stresses in the blade injected with polyurethane foam

4.5.2.11 Manufacturing Technologies for Hydrodynamic Profile Blades

The blade with hydrodynamic profile is the basic element that influences the efficiency of water kinetic energy conversion. To ensure the conversion efficiency of the kinetic energy it is important that the blade preserves its geometrical shape and

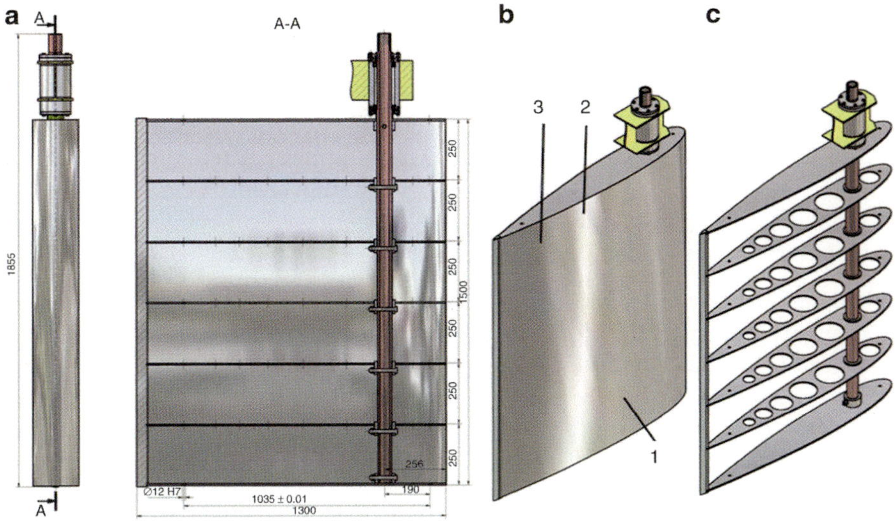

Fig. 4.132 Blade with metal resistance structure and metal coating

resistance parameters determined during the theoretical research and design phase. At the phase of manufacturing technology development the main goal was to ensure the reduction of the produced "green energy" cost aiming at its competitiveness on the market. This is possible by simplifying the construction of blades with hydrodynamic profile and by updating the manufacturing technologies.

Manufacturing of hydrodynamic profile blades with metal coating. Initially the manufacturing technology of blades was developed. It consisted in the design of the resistance metal structure and in the utilization of the coating made of aluminium alloy H37 sheets with its thickness of 1.5 mm (Fig. 4.132).

Numerical analysis of the deformation state carried out in Sect. 4.5.2.10 has shown that under the action of hydrostatic pressure and of water flows the maximum submersed end of the blade will resist to a deformation of approximately 1.3 mm. This deformation does not influence considerably the conversion efficiency as result of the blade profile deviation from the theoretical one. But, the production of blades with aluminium alloy H37 coating demands expensive manufacturing technologies, such as single fillet weld, stamping of the resistance plates with hydrodynamic profile and relatively high accuracy, etc. These technologies also demand expensive equipment and qualified labour that leads to an increase in the cost of blades manufacturing.

Manufacturing of hydrodynamic profile blades made of composite materials. In order to increase the energy conversion performance and to reduce the production cost, modern technology for blade manufacturing from composite materials was developed, such as plastics reinforced with glass fibre.

The developed technology comprises the following basic steps:

- Development, design and manufacturing of the prototype model of the hydrodynamic profile blade coating;
- Development, design and manufacturing of blade resistance structure;
- Step-by-step layer shaping of the coating made of plastics reinforced with glass fibre;
- Moulding of composite material coating, of resistance structure and filling the mould cavity with expandant (polyurethane).

The construction of the blade resistance structure should comply with the manufacturing technology of the blade hydrodynamic profile coating. In this respect more options of blade manufacturing have been tested. Finally, an optimal manufacturing option was selected with account of resistance and cost efficiency, namely, the division of hydrodynamic profile coating on the symmetry axis. This allows the use of a single mould for the manufacturing of both parts of the coating. This technology envisages separate manufacturing of the frontal ends and the blade rear edge (Fig. 4.132b). The metal resistance structure is made as a separate unit (Fig. 4.132c). This constructive configuration of the blades ensures material savings and demands reduced labour force ensuring the manufacturing quality of the blade and complying with the geometrical parameters obtained from the theoretical calculation.

Materials used to manufacture the moulding of the blade coating. The following materials have been used to produce the moulding of the blade coating:

Nr. crt	Manufacturing materials
1.	Plates MDF, thickness 30 mm, 16 mm, 4 mm
2.	Cleiberit D3 (modification of PVA glue)
3.	Lute NOVAL APP
4.	Polyurethane primer; Polyurethane paint
	Polishing paste OSKAR'S M100, M50
5.	Separation agent OSKAR'S W50
	Meriguards MirorGlass nr. 88
	NORSLIP 9600

The following materials have been used to manufacture the blade coating:

Nr.crt	Manufacturing materials
1.	Solidifier MEKP (methyl-ethyl-ketone-pirexit)
2.	Luperox K1 Standard, ATOFINA, France
3.	Polyethers resin CRYSTIC
4.	Gelcoat/white and black colours
5.	Glass fibre:
	Chapped Strand Mat M440 300/1250E
	Chapped Strand Mat M440 450/1250E
6.	Scrint Gobain Vetrotex
7.	Woven Rowind
8.	Spray Up Rowind
9.	Aluminium Hydroxide ATH
10.	Polyurethanes glue IMFI (France)

The blades have been manufactured using the technology of vacuum infusion. Technological operations for blade coating (working surface) manufacturing (Figs. 4.133 and 4.134):

Nr.crt	Order of technological operations
1.	Mould preparation (deposition of the wax separation layers)
2.	Technological maintenance (four layers for 6 h)
3.	Application of the polyvinyl material (Norslipp 9600)
4.	Technological maintenance – 15 min
5.	Application of the gelcoat substance, obtained from two components: solidifier and gelcoat
6.	Technological maintenance – drying
7.	Fabrication and dimension cutting of the glass fibre reinforced material
8.	Laying into the mould:
	Layer I – Woven Rowind
	Layer II – Multimatelite 300/600/300
	Layer III – Shaped Strand Mat M440/450
	Layer IV – Phil ply (single use separation material)
9.	Tube setting for vacuum creation and resin supply
10.	Mould gluing, pressurization with rubber strip Intafoil SP
11.	Membrane dimension cutting (multilayer polyamides cover (skin))
12.	Mould pressurization by: membrane gluing; creation of the vacuum *bag*; membrane laying into the mould; creation of working pressure; check of leaking via weld or membrane
13.	Moulding (resin casting)
14.	Technological maintenance (till complete polymerization of resin (2–3 h depending on the temperature and other factors))
15.	Technological maintenance for the elimination of bottoming stress
16.	Removal from the mould
17.	Overplus elimination

The same technological process is used to manufacture the blade end plates.

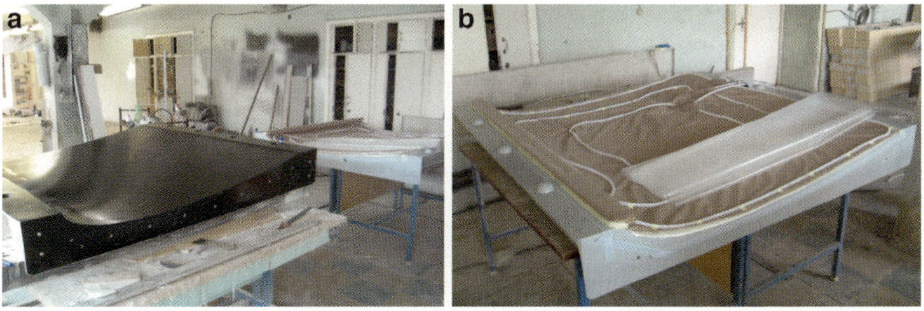

Fig. 4.133 Preparation of the mould with inverted profile: (**a**) mould covered with wax separation layer, (**b**) laying the membrane and the dumping tubes into the mould

Fig. 4.134 Blade coating manufacturing: (**a**) resin infusion into the vacuum bag; (**b**) removal of the coating semi product from the mould

Technological operations for the manufacturing of the blade rear edge:

Nr.crt.	Order of technological operations
1.	Mould preparation (deposition of wax separation layers)
2.	Technological maintenance (four layers for 6 h)
3.	Application of polyvinyl material (Norslipp 9600)
4.	Technological maintenance – 15 min
5.	Application of gelcoat substance, obtained from two components: solidifier and gelcoat
6.	Technological maintenance – drying
7.	Fabrication and dimension cutting of the glass fibre reinforced material
8.	Laying into the mould:
	Layer II – Multimatelite 300/600/300
	Layer III – Shaped Strand Mat M440/450
9.	Moulding (resin casting)
10.	Technological maintenance (till complete polymerization of resin (2–3 h depending on the temperature and other factors))
11.	Technological maintenance for the elimination of bottoming stress
12.	Removal from the mould
13.	Overplus elimination

Blade assembly. The components of the blade coating – two symmetric halves with hydrodynamic profile of the blade working surface, two end plates and the rear edge, as well as the metal resistance structure with the main shaft, are assembled within the mould (Fig. 4.134a):

Injection of expandant material into the blade cavity. To decrease coating displacements and ensure blade floatability, in case of coating deterioration, the inner space of the blades is filled with expandant material. Polyurethane was selected as expandant material for its optimal properties demanded by the operation conditions of blades. The injection of expandant material is done by using the mould within which all blade components have been installed (Fig. 4.134a), in the following sequence:

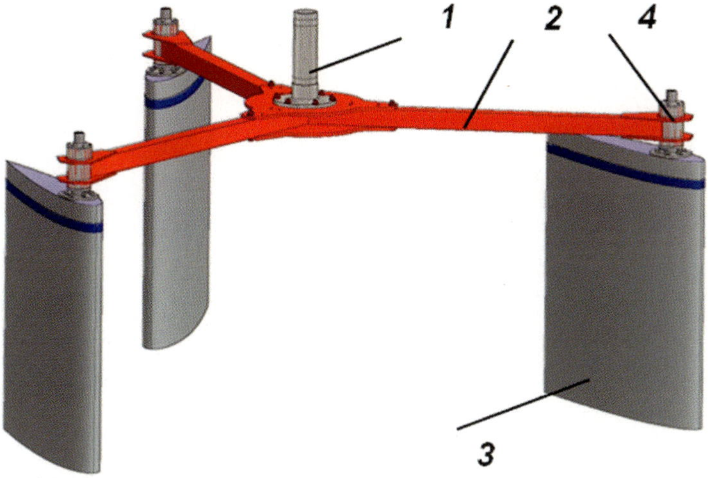

Fig. 4.135 3-blade hydrodynamic rotor

- An elastic tube is introduced into the blade inner space through special technological holes through which liquid polyurethane is injected;
- The injection is done gradually beginning with the blade last section separated by the resistance structure plates;
- After the injection of polyurethane into the whole inner space of the blade through special technological holes the remaining air is eliminated;
- The technological holes are covered and a 30 min technological break is taken for reinforcement;
- The blade is removed from the mould in 24 h, surpluses are eliminated and the blade external surfaces are polished.

The blade manufactured according to the described technology is shown in Fig. 4.134b.

1.	Cleaning of technological joints
2.	Laying of lower half of the coating in the semi-mould
3.	Appropriate setting of the resistance structure by centring the main shaft
4.	Laying of upper half of the coating
5.	Setting the blade lateral caps (covers) and the blade rear edge
6.	Laying the second half of the mould and its rigid fixing
7.	Technological holes done in the upper lateral cap

4.5.2.12 Design and Manufacturing of the Hydrodynamic Rotor

The hydrodynamic rotor has been designed in the Autodesk MotionInventor software (Fig. 4.135 – 3 blades rotor, and Fig. 4.136 – 5 blades rotor). The

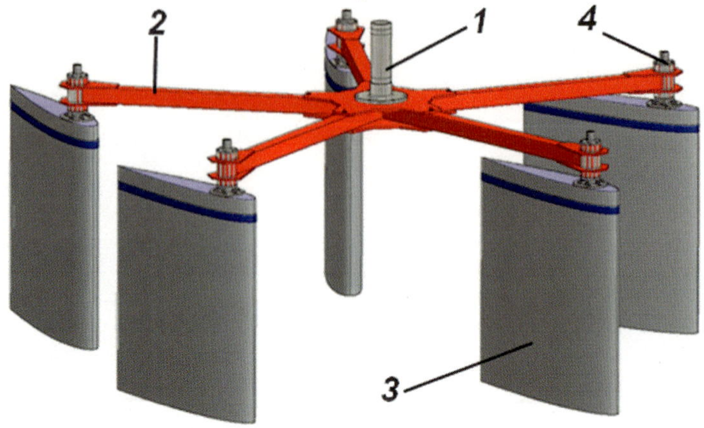

Fig. 4.136 5-blade hydrodynamic rotor

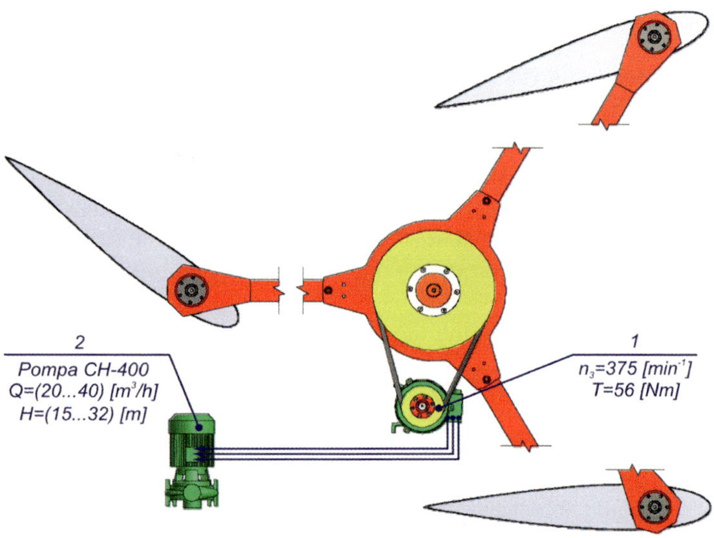

Fig. 4.137 Multiblade rotor connected kinematically to the electric (generator *1*) energy or mechanical hydraulic (pump *2*) energy production units

hydrodynamic rotor is the main working element of the micro hydropower plant, which converts kinetic energy of the water flow and transmit it via the kinematical linkage to the production units of electrical (generator 1) energy or mechanical (hydraulic pump 2) energy (Fig. 4.137).

From the point of view of its design, the rotor comprises the main shaft 1 (Fig. 4.133), the casing with radial bars 2, on which ends the hydrodynamic profile blades 3 are mounted with the help of node 4. The main shaft 1 and the casing with

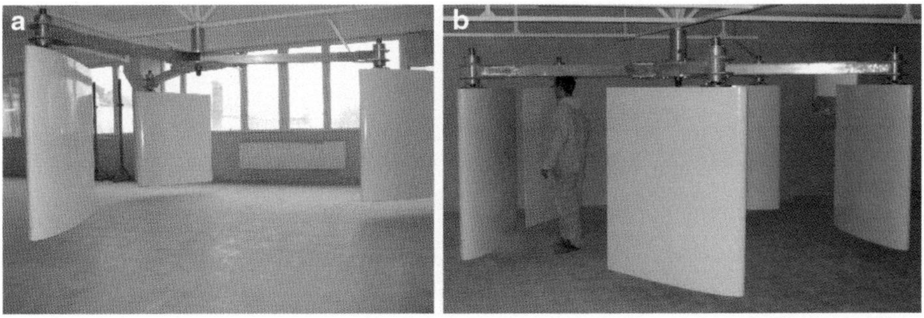

Fig. 4.138 Rotors with 3- (**a**) and 5-blades (**b**) with hydrodynamic profile, manufactured in the laboratory of the Centre for Renewable Energy Conversion Systems Design, TUM

the bars 2 are mounted removable. The hydrodynamic rotor is a spatial structure, strained complex with the bending and twisting moments. The casing with radial bars are made of aluminium alloy plates with calculated dimensions able to ensure design positioning (calculated) of blades with minimal deviations (the deflexion of blade axles till 5 mm, the angle of twist of the radial bars $\pm 1°$).

The technical solutions adopted in the final construction of the hydrodynamic rotor have resulted into computer simulated research, by utilising ANSYS CFX 5.7, sub programmes in the package of mathematic modelling MathCAD, etc., by applying possible loading in real conditions. The node 4 allows the changing of blade 3 positioning towards their axis of rotation, aiming at the insurance of the optimal pitching moment (the technical solution contains Know-How elements). Figure 4.138 shows 3-blade (a) and, respectively, 5-blade (b) rotors, the diameter of blade location is $D = 4$ m. The blades 3 with the height $h = 1.5$ m and the length $l = 1.3$ m are made of composite materials in conformity with the above described technology (Sect. 4.5.2.11). The shape and dimensions of the hydrodynamic profile of blades have been justified within the carried out research (Sects. 4.5.2.1–4.5.2.10) [41–43, 63–65].

4.5.3 Precessional Multiplier as Main Component in the Kinematical Structure of the Micro Hydropower Plant

4.5.3.1 Study of Performances and Selection of Multiplier Kinematical Structure

The multiplier is an indispensable part of the micro hydropower plant. It helps to increase rotor low speeds limited by the water flow small velocity and by the relative big placement diameter of the blades that participate in the energy conversion. For example, the rotor speed is (2–3) min^{-1} for water flow velocity $V = (1, \ldots, 1.6)$ m/s and for blade placement diameter $D = 4$ m.

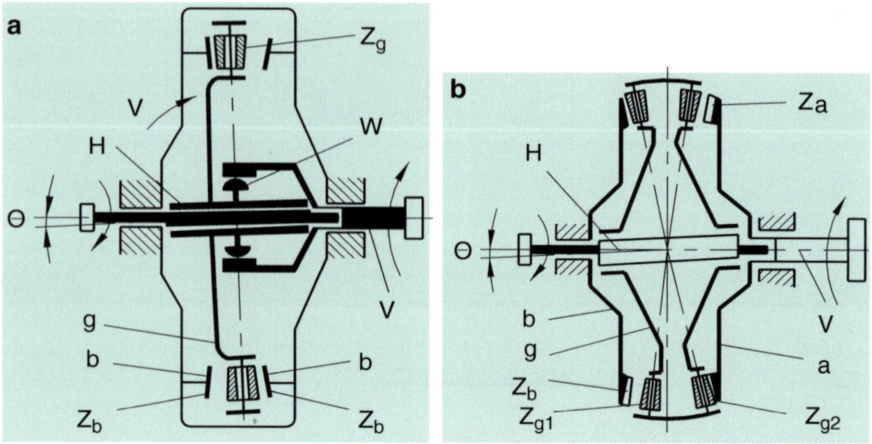

Fig. 4.139 Conceptual diagrams of precesional transmissions that operates efficiently in the multiplication regime

Diversity of requirements forwarded by the beneficiaries of mechanical transmissions consists, in particular, in increasing reliability, efficiency and lifting capacity, and in reducing the mass and dimensions. It becomes more and more difficult to satisfy the mentioned demands by partial updating of traditional transmissions. The target problem can be solved with special effects by developing new types of multipliers based on precessional planetary transmissions with multiple gear, that were developed by the authors. Absolute multiplicity of precessional gear (up to 100% pairs of teeth simultaneously involved in gearing, compared to 5–7% – in classical gearings) provides increased lifting capacity and small mass and dimensions. To mention that until now precessional planetary transmissions have been researched and applied mainly in reducers. Therefore it was necessary to carry out theoretical research to determine the geometrical parameters of the precessional gear that operates in multiplier mode. Also, it was necessary to develop new conceptual diagrams of precessional transmissions that function under multiplier regime.

The majority of precessional planetary transmissions diagrams developed previously operate efficiently in reducer's regime [85]. Depending on the structural diagram, precessional transmissions fall into two main types – *K-H-V* and *2 K-H*, from which a wide range of constructive solutions with wide kinematical and functional options that operate in multiplier regime. The kinematical diagram of the precessional transmission *K-H-V* (Fig. 4.139a), comprises five basic elements: planet career *H*, satellite gear *g*, two central wheels *b* with the same number of teeth, controlling mechanism *W* and the body (frame). The roller rim of the satellite gear *g* gears internally with the sun wheels *b*, and their teeth generators cross in a point, so-called the centre of precession. The satellite gear *g* is mounted on the planet (wheel) career *H,* designed in the form of a sloped crank, which axis forms some angle with the central wheel axis θ.

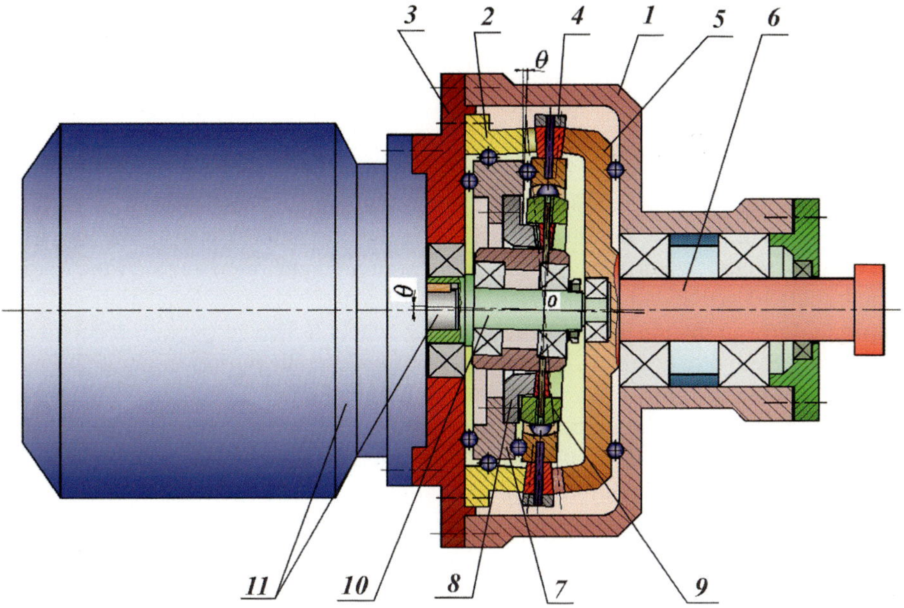

Fig. 4.140 Planetary precessional multiplier with satellite gear mounted radially

Revolving, the sloped crank *H* transmits sphero-spatial motion to the satellite wheel regarding the ball hinge installed in the centre of precession. For the transmission with the controlling mechanism designed as clutch coupling (Fig. 4.139a), the gear ratio (gear reduction rate) varies in the limits:

$$i_{HV}^{g} = -\frac{z_g cos\Theta - z_b}{z_b}\ i_{HV}^{g} = -\frac{z_g cos\Theta - z_b}{z_b cos\Theta} \tag{4.137}$$

Reaching the extreme values of four times for each revolution of the crank *H*. If necessary this shortcoming can be eliminated using as a controlling mechanism the constant cardan joint (Hooke's joint), the ball synchronous couplings, etc.

$$i_{HVmed}^{g} = -\frac{z_g - z_b}{z_b} \tag{4.138}$$

For $z_g = z_b + 1$, $i_{HV}^{g} = -\frac{1}{z_b}$, the driving and driven shafts have opposite directions. For $z_g = z_b - 1$, $i_{HV}^{g} = \frac{1}{z_b}$, the shafts revolve in the same direction.

This kinematical diagram of the precessional transmission ensures a range of gear ratios $i = 8, \ldots, 60$, but in the multiplication regime it operates efficiently only for the range of gear ratios $i = 8, \ldots, 25$. As well, in the controlling mechanism *W*, that operates with pitch angles of the semi couplings up to 3°, power losses occur reducing the efficiency of the multiplier on the whole.

To avoid power losses in the multiplier and to widen the kinematical options, (Fig. 4.140) [83] the conceptual diagram of the precessional multiplier with wide kinematical options was designed. The planetary precessional multiplier comprises the following units: the housing 1, inside which the fixed sun wheel 2 is placed and connected rigidly to the housing cover 3, exterior satellite wheel 4 with the teeth in the shape of rollers, movable sun wheel 5, linked rigidly to the input shaft 6. The satellite wheel 3 is connected kinematically with the sloped flange of the disk 7, connected rigidly with the sun wheel 8, that gears with the interior satellite wheel 9 mounted unbound on the output crank shaft 10, and linked rigidly to the rotor generator 11. The exterior satellite wheel 4 is mounted void on gear bodies on exterior spherical surfaces of the interior wheel 9. The pitch angle of the output crank shaft 10 axis and of the sloped flange is equal to θ. The rotational motion of the input shaft 6 is transmitted to the movable sun wheel 5. Due to the difference in the number of teeth of wheel 5 and of the exterior satellite wheel (gear) 4 ($Z_6 = Z_5 \pm 1$), the last will have to carry out a precession motion around the fixed point O (centre of precession). Precessional motion around its axis is excluded as the number of teeth of the sun wheel 2 is equal to the number of rollers of the satellite gear 4 ($Z_2 = Z_4$). The precessional motion of the exterior satellite gear 4 is transformed, by means of an inclined flange of the disc 7, into rotational motion around the disc axis 7 that will revolve by the degree of multiplication

$$i_7 = -\frac{z_5}{z_4 - z_5} \tag{4.139}$$

where Z_4 is the number of rollers of the exterior satellite gear 4;

Z_5 – is the number of teeth of the movable sun wheel 5.

The multiplied rotational motion of the disc 7 is transformed into multiplied precessional motion of the interior satellite gear 9, due to the difference in the number of teeth $Z_8 = Z_9 \pm 1$. At the rotation of disc 7 of the sun wheel 8 at an angle equal to the teeth angular pitch, the interior satellite gear will perform a complete precession cycle around point *O*. The precessional motion of the interior satellite gear 9 is transformed by means of output crank shaft 10 into rotational motion, multiplied to the degree of multiplication

$$i_{10} = -\frac{z_8}{z_9 - z_8} \tag{4.140}$$

where Z_8 is the number of teeth of the sun wheel 8,

Z_9 – is the number of rollers of the satellite gear 9.

The multiplied rotational motion of the output crank shaft 10 is transmitted to the electrical generator rotor 11.

The precessional transmission *2 K-H* has higher performances, including the kinetostatic one as well (Figs. 4.139b and 4.141). The transmission comprises a satellite gear *g* with two crown gears Z_{g1} and Z_{g2}, that gears with the unshiftable *b* and movable *a* central wheels.

$$i = -\frac{z_{g_1} z_a}{z_b z_{g_2} - z_{g_1} z_a} \tag{4.141}$$

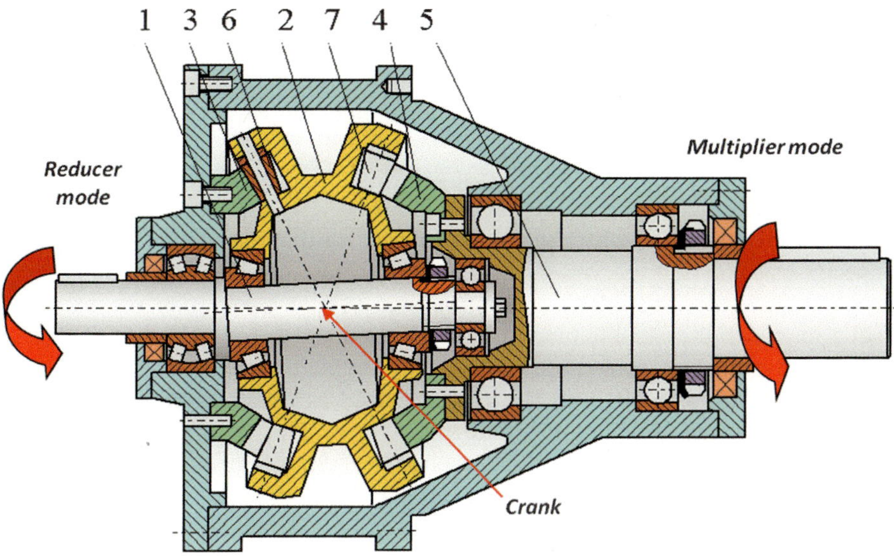

Fig. 4.141 Constructive diagram of the planetary precessional transmission 2 K-H

The analysis of this relation demonstrates that precessional transmissions *2 K-H* provide the fulfilment of a large range of transmission ratios $i = \pm(12, \ldots, 3{,}599)$. But in multiplier regime the transmission operates efficiently only in the limits $i = \pm(12, \ldots, 30)$. The process of self-braking (self-stopping) occurs at bigger transmission ratios. It is necessary to point out the series of peculiarities of the precessional transmissions *2 K-H* that ensure higher performances compared to similar planetary transmissions with cylindrical gears: precessional transmissions do not demand conditions of distance equality between the axis. This factor widens the area of their optimal design; precessional transmission kinematics does not limit the selection of the gear couples modules or of the rollers placement pitch. This factor increases the possibilities of shaping teeth pairs and of the transmission ratios interval; the peculiarities of the designed precessional gears allow increasing in the number of teeth that transmit the load simultaneously and this fact reduces significantly the dimensions and mass for the same loads compared to the traditional involute gearings.

Based on the carried out analysis a constructive diagram of the precessional planetary transmission was designed, taken as the base of precessional multipliers design. The precessional planetary transmission (Fig. 4.141) comprises the crank shaft 1 on which the satellite block 2, and the fixed and the movable wheels 3 (the movable wheel is connected to the shaft 5) are installed. The satellite block 2 has two crown gears (6 and 7) with the teeth executed as conical rollers mounted on the axle with the possibility of revolving around them. The transmission operates in the multiplier mode, as follows: at the rotation of the input shaft 5 with the gear 4, due to the difference in the number of geared teeth ($Z_4 = Z_7 - 1$, $Z_3 = Z_6 - 1$), the satellite block 2 will perform a spherical-spatial motion around the point – centre of precession (the point of intersection of the crown gear roller axes and of the crank shaft axes 1), producing a complete precessional cycle at the rotation of the gear 4 at an angle equal to the angular pitch. Due to its mounting on the sloped side of the crank shaft 1, the precessional motion of the satellite block 2 is transformed into rotational motion of the crank shaft 1 that will produce a complete rotation during a complete precessional cycle of the satellite block.

Based on the carried out analysis a constructive diagram of the precessional planetary transmission was designed, taken as the base of precessional multipliers design. The precessional planetary transmission (Fig. 4.141) comprises the crank shaft 1 on which the satellite block 2, and the fixed and the movable wheels 3 (the movable wheel is connected to the shaft 5) are installed. The satellite block 2 has two crown gears (6 and 7) with the teeth executed as conical rollers mounted on the axle with the possibility of revolving around them. The transmission operates in the multiplier mode, as follows: at the rotation of the input shaft 5 with the gear 4, due to the difference in the number of geared teeth ($Z_4 = Z_7 - 1$, $Z_3 = Z_6 - 1$), the satellite block 2 will perform a spherical-spatial motion around the point – centre of precession (the point of intersection of the crown gear roller axes and of the crank shaft axes 1), producing a complete precessional cycle at the rotation of the gear 4 at an angle equal to the angular pitch. Due to its mounting on the sloped side of the crank shaft 1, the precessional motion of the satellite block 2 is transformed into rotational motion of the crank shaft 1 that will produce a complete rotation during a complete precessional cycle of the satellite block.

4.5.3.2 Analytic Description of Teeth Profile and Justification of Precessional Gear Parameters Selection

Teeth profiles have an important role in the efficient transformation of motion in the precessional transmissions that operate as multiplier. Multiple precessional gear theory, previously developed, did not take into consideration the influence of the diagram error of the linking mechanism in the processing device for gear wheel on the teeth profile. Functioning under the multiplication regime, these errors have major influence, which can lead to instant blocking of gear and to power losses. With this purpose, a thorough analysis was conducted on the motion development mechanism under multiplication, and on the teeth profile error generating source. On the basis of fundamental theory of multiple precessional gear, previously developed, a new gear with modified teeth profile and the technology for its industrial manufacturing was proposed and patented [84].

Kinematically, the link between the semi product and the tool, in which one of them (the tool) makes spherical-spatial motion being, at the same time, limited from rotating around the axis of the main shaft of the teething machine tool, is similar to the *satellite-driven shaft* link from the precessional planetary transmission of the K-H-V type. The kinematical link between the tool and the stationary part of the device represents a Hooke articulation that generates the variability of transfer

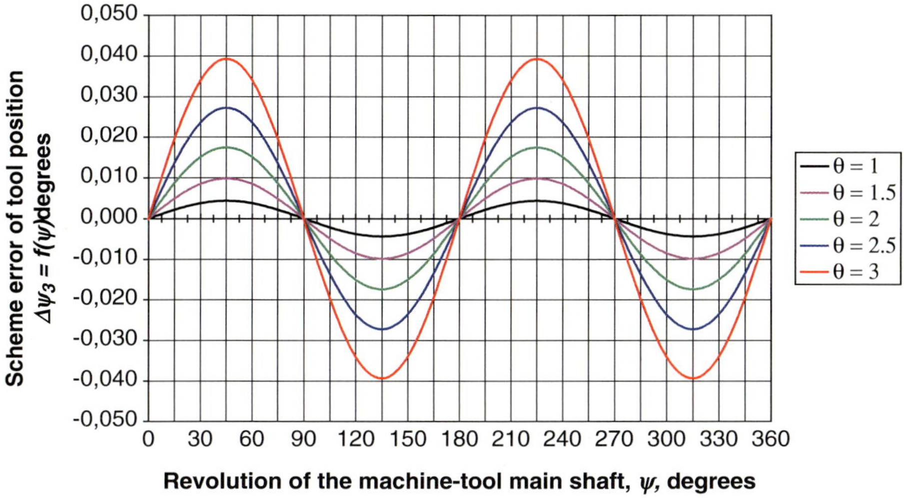

Fig. 4.142 Dependence of the scheme error of tool position $\Delta\psi_3$ at a revolution of the machine-tool main shaft ψ

function in the kinematical link *tool-semi product*. This variation will influence the teeth profile. Thus, the connection of tool with the housing registers a certain diagram error $\Delta\psi_3$ (to understand the deviation of the semi product angle of rotation ψ_3 from the angle of rotation of the semi product itself ψ_3^m at its uniform rotation):

$$u_{31}^m = -\frac{z_2 - z_3}{z_3}; \Delta\psi_3 = \psi_3 - u_{31}^m = \frac{z_2}{z_3}\left(\psi - arctg(\cos\theta \cdot tg\psi)\right). \qquad (4.142)$$

Figure 4.142 shows the dependence of the tool position diagram error $\boldsymbol{\Delta\psi_3}$ at a revolution of the machine tool main shaft ψ. This error is transmitted to the tool that shapes the teeth profile with the same error. To ensure continuity of the transfer function and to improve the performances of precessional transmission under multiplication it is necessary to modify teeth profile with the diagram error value $\Delta\psi_3$ by communicating supplementary motion to the tool. In this case the momentary transmission ratio of the manufactured gear will be constant. Usually, in theoretical mechanics the position of the body making spherical-spatial motion is described by Euler angles. The mobile coordinate system $OX_1Y_1Z_1$ is connected rigidly with the satellite wheel, which origin coincides with the centre of precession 0 (Fig. 4.143) and performs spherical-spatial motion together with the satellite wheel relative to the motionless coordinate system *OXYZ*.

The elaboration of the mathematic model of the modified teeth profile is based integrally on the mathematic model of teeth profile, previously developed by the authors. With this purpose it is necessary to present the detailed description of teeth profile without modification and, then, to present of the description of modified profile peculiarities.

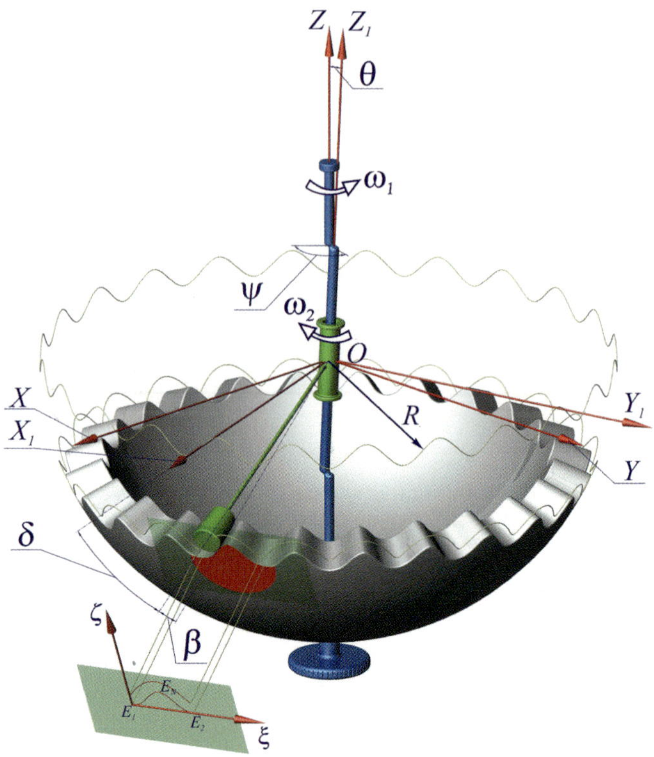

Fig. 4.143 Tooth profile in normal section

Description of teeth profile designed on sphere. An arbitrary point ***D*** of the tool axis describes a trajectory relative to the fixed system according to the equations:

$$
\begin{aligned}
X_D^m &= -\sin\delta\sin\left[Y_C^m\sin\theta + Z_C^m(1-\cos\theta)\cos\psi\right]\\
Y_D^m &= -Y_C^m\cos\delta + Z_C^m\sin\delta\left[\cos^2\psi + \cos\theta\sin^2\psi\right]\\
Z_D^m &= -Y_C^m\sin\delta\left(\cos^2\psi + \cos\theta\sin^2\psi\right) - Z_C^m\cos\delta \qquad (4.143)
\end{aligned}
$$

Index m means *modified*.

The motion of point D^m compared to the movable system connected rigidly to the semi product is described by formulas:

$$
\begin{aligned}
X_{1D}^m &= X_D^m\cos\frac{\psi}{Z_1} - Y_D^m\sin\frac{\psi}{Z_1};\\
Y_{1D}^m &= X_D^m\sin\frac{\psi}{Z_1} + Y_D^m\cos\frac{\psi}{Z_1};\\
Z_{1D}^m &= Z_D^m. \qquad (4.144)
\end{aligned}
$$

The projections of point D^m velocities is expressed by formulas:

$$\begin{aligned}
\dot{X}_D^m &= -\sin\delta\cos\psi\left[Y_C^m\sin\theta + Z_C^m(1-\cos\theta)\cos\psi\right]\dot{\psi} \\
&\quad -\sin\delta\sin\psi\left[\dot{Y}_C^m\sin\theta + \dot{Z}_C^m(1-\cos\theta)\cos\psi - Z_C^m(1-\cos\theta)\sin\psi\cdot\dot{\psi}\right]; \\
\dot{Y}_D^m &= -\dot{Y}_C^m\cos\delta + \dot{Z}_C^m\sin\delta\left[\cos^2\psi + \cos\theta\sin^2\psi\right] \\
&\quad + Z_C^m\sin\delta\left[-2\cos\psi\sin\psi + 2\cos\theta\sin\psi\cos\psi\right]\dot{\psi}; \\
\dot{X}_{1D}^m &= \dot{X}_D^m\cos\frac{\psi}{Z_1} - \frac{\dot{\psi}}{Z_1}X_D^m\sin\frac{\psi}{Z_1} - \dot{Y}_D^m\sin\frac{\psi}{Z_1} - \frac{\dot{\psi}}{Z_1}Y_D^m\cos\frac{\psi}{Z_1}; \\
\dot{Y}_{1D}^m &= \dot{X}_D^m\sin\frac{\psi}{Z_1} + \frac{\dot{\psi}}{Z_1}X_D^m\cos\frac{\psi}{Z_1} + \dot{Y}_D^m\cos\frac{\psi}{Z_1} - \frac{\dot{\psi}}{Z_1}Y_D^m\sin\frac{\psi}{Z_1}.
\end{aligned} \tag{4.145}$$

The coordinates of point E^m on the sphere is calculated by formulas:

$$X_{1E}^m = k_2^m Z_{1E}^m + d_2^m$$

$$Y_{1E}^m = k_1^m Z_{1E}^m - d_1^m$$

$$Z_{1E}^m = \frac{(k_1^m d_1^m - k_2^m d_2^m) - \sqrt{(k_1^m d_1^m - k_2^m d_2^m)^2 + (k_1^{m2} + k_2^{m2} + 1)\cdot(R_D^2 - d_1^{m2} - d_2^{m2})}}{k_1^{m2} + k_2^{m2} + 1} \tag{4.146}$$

where:

$$k_1^m = \frac{X_{1D}^m\left(X_{1D}^m\dot{X}_{1D}^m + Y_{1D}^m\dot{Y}_{1D}^m\right) + Z_{1D}^{m\,2}\dot{X}_{1D}^m}{Z_{1D}^m\left(X_{1D}^m\dot{Y}_{1D}^m - Y_{1D}^m\dot{X}_{1D}^m\right)};\quad k_2^m = -\frac{\left(k_1^m Y_{1D}^m + Z_{1D}^m\right)}{X_{1D}^m};$$

$$d_1^m = \frac{R_D^2\cos\beta\,\dot{X}_{1D}^m}{\left(X_{1D}^m\dot{Y}_{1D}^m - \dot{X}_{1D}^m Y_{1D}^m\right)};\quad d_2^m = \frac{\left(R_D^2\cos\beta + d_1^m Y_{1D}^m\right)}{X_{1D}^m}.$$

According to the obtained analytical relations a soft for the calculation and generation of teeth was developed in CATIA V5R7 modelling system that allowed obtaining the modified trajectories of points $E^m{}_e$ and $E^m{}_i$ on the spherical front surfaces, both exterior and interior ones, by which the teeth surface was generated (Fig. 4.144).

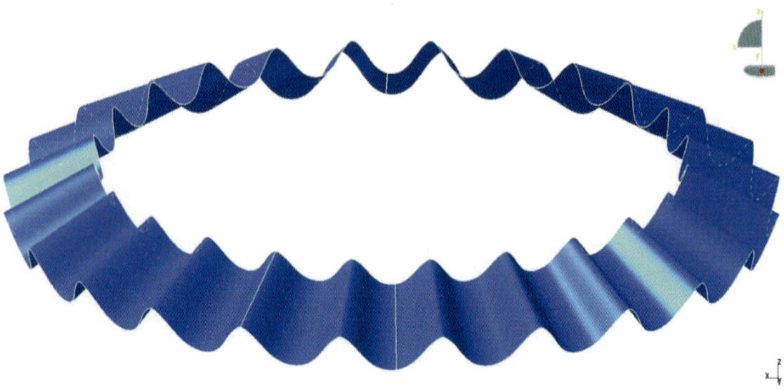

Fig. 4.144 Teeth generating surface

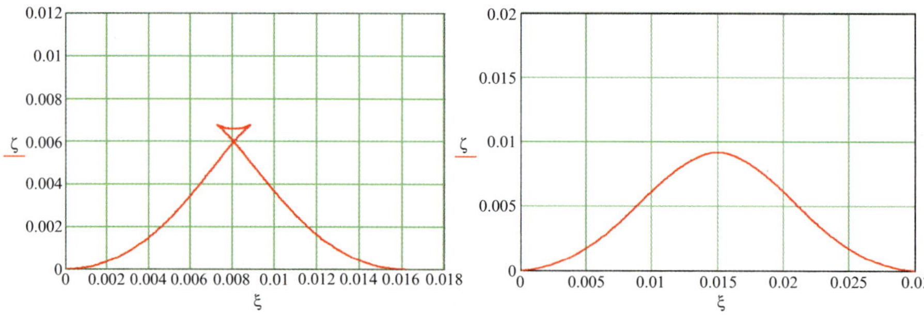

Fig. 4.145 Teeth profiles for multipliers

Description of modified teeth profile projected on a transversal surface. Projection of point E^m on the tooth transversal plane has the following coordinates:

$$\mathrm{X}''^{m}_{E} = \varepsilon^m \cdot X^m_{1E}, \mathrm{Y}''{}_E{}^m = \varepsilon^m \cdot Y''_{1E}{}^m, \mathrm{Z}''^{m}_{E} = \varepsilon^m \cdot Z''^{m}_{1E} \tag{4.147}$$

where $\varepsilon^m = -\frac{D}{AX^m_{1E}+BY^m_{1E}+CZ^m_{1E}}$.

The modified teeth profile in plane is described by the equations:

$$\begin{aligned}
\xi^m &= X''^{m}_{E} \cos \frac{\pi}{Z_1} + \left[R_D \cos(\delta + \theta + \beta) + Y''^{m}_{E}\right] \sin \frac{\pi}{Z_1};\\
\zeta^m &= X''^{m}_{E} \sin\gamma \sin \frac{\pi}{Z_1} - \left[R_D \cos(\delta + \theta + \beta) + Y''^{m}_{E}\right] \sin\gamma \cos \frac{\pi}{Z_1} + \\
&\quad + \left[R_D \sin(\delta + \theta + \beta) + Z''^{m}_{E}\right] \cos\gamma.
\end{aligned} \tag{4.148}$$

A wide range of modified teeth profiles with different geometrical parameters were generated in MathCAD 2001 Professional software (Fig. 4.145). The solid model of a gear wheel is shown in Fig. 4.146.

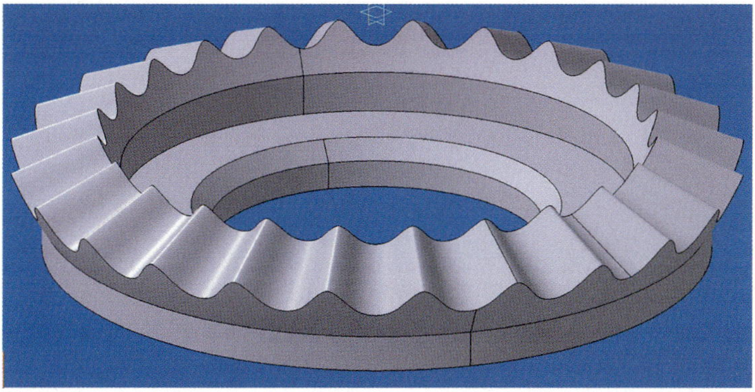

Fig. 4.146 Computerised model of the sun gear

Based on the carried out research it was established that from the point of view of decreasing energy losses in gearing, in the multiplication mode of operation, the gearing angle should be $\alpha > 45°$, and the nutation angle (the pitch angle of the crank shaft) should be $-\theta \leq 2.5°$. This is dictated by the reverse principle of movement in the multipliers compared to the reducers: the axial component of the normal force in gear must be maximal to drive the crank shaft in the rotation movement through the satellite wheel.

4.5.3.3 Calculation of Precessional Gear at Contact Pressure

Calculation is done for the wheel with the smaller number of teeth Z_4, as it supports higher loads. The Anthology of Inventions [82], written by the authors, describes the calculation methodology and the nomographic charts for the selection of values of five basic parameters, that influence the teeth profile. The crank shaft pitch angle θ is recommended in the boundaries 1.5–3°. Roller taper angle β, generating angle (pressure angle) α_w, the coefficient of proportionality of the associated teeth radiuses v and the pitch angle of the roller axis δ are selected in conformity with the nomographic charts [82].

Preliminary calculation of the sun gear effective diameter. The basic parameters that need calculation are shown in Fig. 4.147. The calculation is done in the following sequence.

Multiplicity gear ε is selected from the recommendations [82].

The number of teeth Z_ε that carry simultaneously the load is calculated:

$$Z_\varepsilon = \frac{Z_4 - 1}{2} \cdot \frac{\varepsilon}{100} \qquad (4.149)$$

(as Z_ε is an integer, the decreasing integral value is selected).

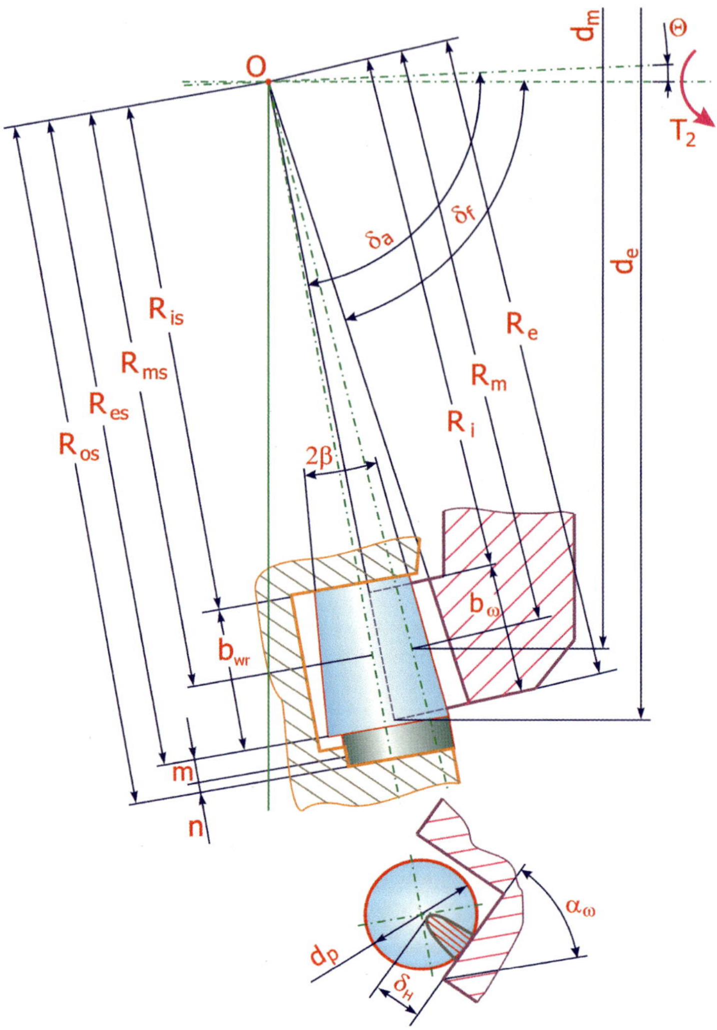

Fig. 4.147 Precessional gear geometry

The effective diameter of gear wheels d_m is calculated from the relationship:

$$d_m = 53 \cdot \sqrt[3]{\frac{T_2\,(1-\nu) \cdot K_{HP} \cdot K_{H\beta} \cdot K_{HV} \cdot \cos{(\delta + \theta)}}{(\sigma_{HO})^2 \cdot \psi_{bd} \cdot Z_\varepsilon \cdot tg\beta \cdot \cos\alpha_w}}, \text{ mm} \qquad (4.150)$$

where: K_H is the experimental coefficient [82], that characterises the irregularity of load distribution between the teeth;

$K_{H\beta}$ is the experimental coefficient [82], that characterises the irregularity of load distribution on teeth length;

K_H is the experimental coefficient [82], that characterises load dynamics;
$\psi_{bd} = 0, 11$ is the coefficient of tooth length [82] compared to diameter d_m.
According to the effective diameter d_m the following can be calculated:

- Length of gear wheel tooth b_w:

$$b_w = \psi_{bd} \cdot d_m, \text{ mm} \tag{4.151}$$

- Length of rollers b_{wr}:

$$b_{wr} = b_w + 2, \text{ mm} \tag{4.152}$$

- Rollers' diameter in medium and abut section d_{mr}, d_{rr}:

$$d_{mr} = d_m \frac{tg\beta}{\cos(\delta + \theta)}, \text{ mm} \tag{4.153}$$

$$d_{rr} = d_{mr} + b_{wr} \cdot tg\beta, \text{ mm} \tag{4.154}$$

- Diameter of rollers' axes d_a:

$$d_a = d_{mr} - b_{wr} \cdot tg\beta - 2\Delta, \text{ mm} \tag{4.155}$$

where Δ, mm is the thickness of roller plate (wall) in minimal section selected in the boundaries 1, ..., 5 mm.

Determination of the allowable contact pressures considering the rolling friction. Initially, the tooth-roller friction velocity V_{gl} is determined:

$$V_{gl} = K_1 \cdot \ln\left[\frac{f_{\max}}{f_{\max} - \frac{K}{d_{mr}} - \frac{f \cdot d_a + f \cdot \left(\frac{d_{rr}+d_a}{2}\right) \sin\alpha_w \cdot \sin(\delta \cdot \beta_i) \cdot \cos\delta + K}{d_{mr}}}\right] \tag{4.156}$$

where K_1 – is the coefficient depending on the working conditions of the upper *kinematical coupling* [82];

f_{max} – *maximal value of the friction coefficient;*
K – *rolling friction coefficient;*
f – *friction coefficient;*

Taking into account $\boldsymbol{V_{gl}}$, the allowable contact pressure σ_{HOi} is calculated from the relationship:

$$\sigma_{HOi} = \sigma_{HO}^{\left(0,28+0,72 \cdot e^{-\frac{V_{gl}}{K_3}}\right)} \tag{4.157}$$

where: K_3 – is the coefficient, selected from the recommended ones [82].

Determination of gear wheel effective diameter and calculation of the geometrical parameters. The average diameter d_m of the gear wheels is defined by the relationship:

$$d_m = 53 \cdot \sqrt[3]{\frac{T_2 (1 - \nu) \cdot K_{HP} \cdot K_{H\beta} \cdot K_{HV} \cdot \cos(\delta + \theta)}{(\sigma_{HOi})^2 \cdot \psi_{bd} \cdot Z_\varepsilon \cdot tg\beta \cdot \cos\alpha_w}}, \text{ mm} \tag{4.158}$$

And the medium conical distance of the satellite rim with rollers R_{ms} is defined by the relationship:

$$R_{ms} = \frac{d_m}{2\cos(\delta + \theta)(1 - tg\beta \cdot \sin\alpha_w \cdot tg(\delta + \theta))}, \text{ mm} \tag{4.159}$$

Depending on the medium diameter d_m the following can be calculated:

- Length of gear wheel tooth b_w:

$$b_w = \psi_{bd} \cdot d_m, \text{ mm} \tag{4.160}$$

- Length of rollers b_{wr}:

$$b_{wr} = b_w + 2, \text{ mm} \tag{4.161}$$

- Rollers' diameter in medium and abut section d_{mr}, d_{rr}:

$$d_{mr} = d_m \cdot \frac{tg\beta}{\cos(\delta + \theta)}, \text{ mm} \tag{4.162}$$

$$d_{rr} = d_{mr} + b_{wr} \cdot tg\beta, \text{ mm} \tag{4.163}$$

- Diameter of rollers' axes d_a:

$$d_a = d_{mr} - b_{wr} \cdot tg\beta - 2\Delta, \text{ mm} \tag{4.164}$$

where Δ is the thickness of roller plate (wall) in minimal section selected in the boundaries 1, ..., 5 mm (calculation is done similarly for another couple of rollers).

Further on, the following can be calculated:

- The medium conical distance of the gear wheel:

$$R_m = \frac{d_m}{2\cos\left(\delta + \theta + ctg\left(\frac{d_{mr} \cdot \sin\alpha_w}{2R_{ms}}\right)\right)}, \text{ mm} \tag{4.165}$$

- Exterior and interior conic distances of the roller rim R_{es}, R_{is}:

$$R_{es} = R_{ms} + \frac{b_{wr}}{2}, \text{ mm} \quad R_{is} = R_{ms} - \frac{b_{wr}}{2}, \text{ mm} \tag{4.166}$$

– Exterior and interior conic distances of the gear wheel rim R_{er}, R_{ir}:

$$R_{es} = R_{mr} + \frac{b_w}{2}, \text{ mm} \quad R_{is} = R_{mr} - \frac{b_w}{2}, \text{ mm} \tag{4.167}$$

– Crest and foot cone angles of the teeth δ_f, δ_v

$$\delta_f = 90^o - (\delta + \theta + \beta)\,,^o;\ \delta_v = \delta_f + 2\theta,^o. \tag{4.168}$$

Verification of teeth resistance at contact pressure. Allowable contact pressures σ_H are calculated by formula:

$$\sigma_H = 275\sqrt{\frac{2T_2\,(1-\nu)\,K_{HP}K_{H\beta}K_{HV}}{d_m d_{mp} b_w Z_\varepsilon \cos\alpha_w}} < [\sigma, \text{MPa}] \tag{4.169}$$

Calculation is repeated with the constructive parameters modification, in particular of the *tooth-roller* couple to the condition $\sigma_H \geq \sigma_H$.

Given the specific operation of the micro hydropower plant (24 h out of 24) and taking into account the fact that the multiplier is overloaded with dynamic tasks, it is strictly necessary that the gear is subject to verification of teeth resistance at contact pressure for each operation system.

4.5.3.4 Design of Precessional Multiplier Structure

On the basis of the undertaken study, diagram 2 K-H was selected for the development of precessional multiplier of the micro hydropower plant. As a result of analysis of a wide range of tooth profiles with different geometrical parameters of gear by using the mathematical modelling package MathCAD 2001 Professional, the optimum tooth profiles were selected with account of their functioning in conditions of multiplication. Also, in MathCAD 2001 Professional software the calculation of geometrical parameters of precessional gear was done. The structures of two versions of precessional multipliers for micro hydropower plants were designed by use of the SolidWorks software. The precessional multiplier is connected by flange with an electric generator, which allows obtaining a compact module, coaxial with the micro hydropower plant rotor. The structure from Figs. 4.148 and 4.149 is proposed for multiplier functioning in conditions of lower temperatures. To simulate the multiplier assembly and functioning, the dynamic computerised model of the precessional multiplier multiplier was developed in AutoDesk MotionInventor (see the annexed CD). The precessional multiplier has reduced dimensions and mass, high lifting capacity and multiplication ratio up to i = 144 (based on a two stage diagram) with satisfactory mechanical efficiency.

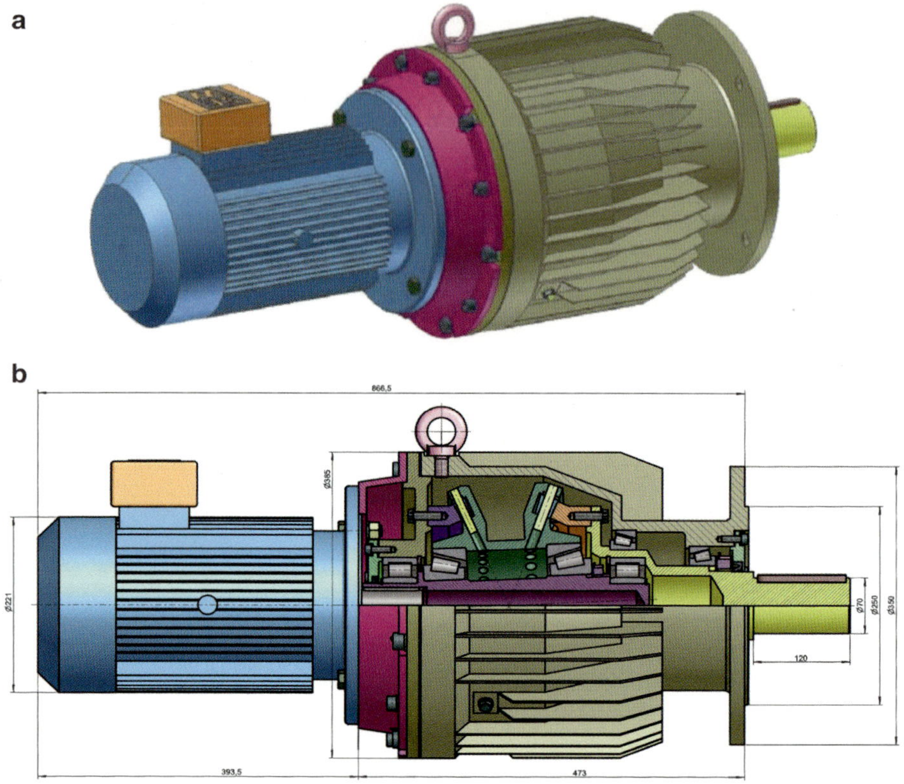

Fig. 4.148 Planetary precessional multiplier: (**a**) general view, (**b**) section view

4.5.4 *Design, Manufacturing and Testing of Low Speed Centrifugal Pump for Micro Hydropower Plants*

4.5.4.1 Study of Functional Characteristics and Selection of Pump Type

The rotational speed of the rotor of micro hydropower plant for the conversion of water kinetic energy into mechanical energy depends on the water stream flow rate and on the hydrodynamic rotor diameter, and makes up (2, . . . , 4) min^{-1} only. To transform and transmit the converted energy from the hydrodynamic rotor to the hydraulic pump it is necessary to multiply the number of revolutions at its shaft by:

- (500–250) times – for the pumps designed to operate at the number of revolutions $n = 1{,}000\ \text{min}^{-1}$;
- (750–375) times – for the number of revolutions $n = 1{,}500\ \text{min}^{-1}$;
- (1,500–750) times – for the number of revolutions $n = 3{,}000\ \text{min}^{-1}$.

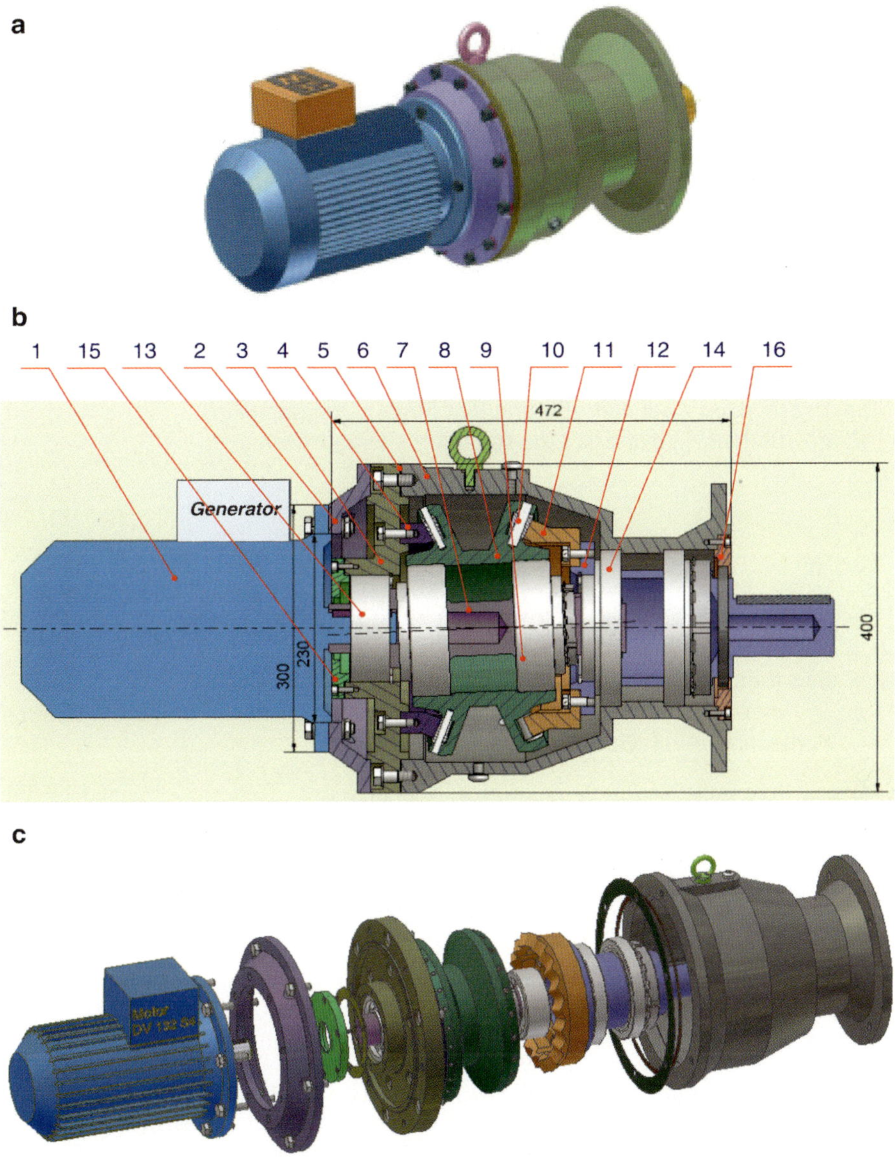

Fig. 4.149 Planetary precessional multiplier for operation in conditions of low temperatures (river water): (**a**) General view. (**b**) Section view *1* – permanent magnet generator; *2* – flange: generator – multiplier; *3* – cover; *4* – fixed gear wheel; *5* – adjustment pitch (washer); *6* – reducer housing; *7* – crank shaft; *8* – satellite gear; *9* – bearings for the satellite abutment on the crank shaft; *10* – unit (subsystem): conical roller – axes – disk; *11* – mobile wheel gear; *12* – input shaft; *13* – bearings for the central abutment of the crank shaft; *14* – bearings for the input shaft abutment; *15*, *16* – fittings heads. (**c**) Components in progress.

The multiplication of revolutions in the interval i = 250, ..., 1,500 to ensure rotational frequency (1,000, ..., 3,000) min^{-1}, for which the hydraulic pumps are designed traditionally, is not convenient because of the mechanical losses increase in transmissions and from the point of view of structural compactness of the micro hydropower plant, on the whole. To realise various opportunities of micro hydropower plant utilisation related to water pumping, it is proposed to study the functional characteristics and the constructive and functional peculiarities of hydraulic pumps, in production at global level.

The basic parameters that characterise pumps functioning are pressure, flowrate, power, efficiency, number of revolutions, and admissible absorption height:

- Pump flow rate Q is the volume of liquid pumped in a unit of time, m^3/s;
- Pump pressure H is the useful specific work created on the account of hydraulic energy increase for a unit of liquid that flows in the pump, m;
- Pump efficiency η expresses the summary energy losses:

$$\eta = \eta_h \cdot \eta_v \cdot \eta_{imec} \tag{4.170}$$

where

η_h determines the hydraulic losses related to the losses caused by friction and whirpools generation;
η_v determines the quantity of flow rates through pump packing;
$\eta_{i.mec}$ determines the interior mechanical losses as result of pump rotating elements friction with the liquid;

- Useful hydraulic power N determines the increase of energy obtained by the liquid that flows through the pump, kW:

$$N = \frac{\rho \cdot g \cdot Q \cdot H}{1000\eta} \tag{4.171}$$

- Admissible absorption height H_{vid} characterises the value of the drop pressure at pump input, m.

Among the dynamic pumps the following can be mentioned: vortex pumps, labirinto-helical pumps and disk (plate)-pumps. Maximum achievable efficiency of the vortex pump makes up no more than 45% of the calculation scheme. But it is difficult to use these pumps in many cases because of friction areas wearing and gripping in narrow packing gaps.

Maze-screw pumps, which work with low-viscosity liquids, are characterized by simple construction and lack of direct contact and friction between its components, and are able to create high pressure at low flow rates. But the overall efficiency of maze pumps usually does not exceed 25%. In the process of design and production of these pumps it is necessary to take into account high accuracy and installing rigidity of the screw, which provides guaranteed gap between the screw and

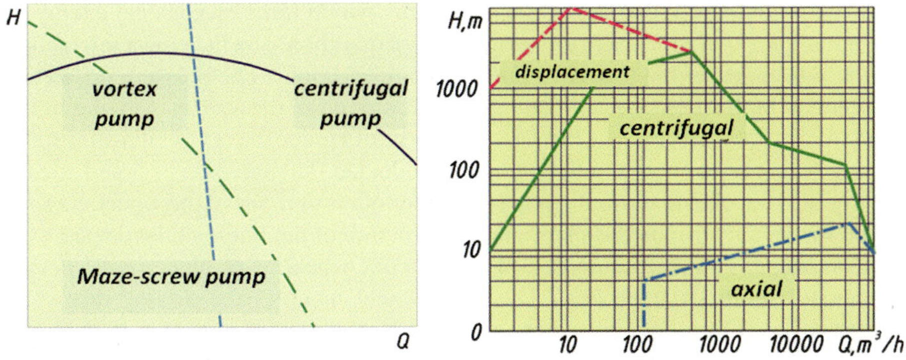

Fig. 4.150 (**a**) Fields of application of different types of pumps according to their Q (m^3/h) and H (m) characteristics, (**b**) pressure comparative chart for various types of pumps

bush labyrinths. Failure to meet these requirements has an impact on the pump efficiency. Disk (plate, drive) pumps can achieve maximum efficiency 45–55%, and are characterized by small pressure, comparatively. Disk pumps are more effective for pumping viscous liquids and liquids which do not permit their *rolling* in hydraulic channels, including the main pump anticavitation properties increasing, at pumping abrasive mixtures. At the same time it may be mentioned that the operating characteristics of the mentioned pumps differ from the vane pumps characteristics and are characterized by a sudden drop in pressure and efficiency at flow rate small changes (sudden pressure drop characteristics). Types of pumps by their fields of application are shown in Fig. 4.150. Figure 4.150b shows the comparative chart of the pressure characteristics for several types of pumps [85].

Currently, dynamic vane pumps have a much wider spread in all areas of husbandry. The choice of blades pump (centrifugal, diagonal or axial) types is determined by their concrete parameters. In general, all pumps are characterized by the following criterion – high speed coefficient:

$$n_s = 3.65 \cdot n \cdot \frac{\sqrt{Q}}{H^{0.75}} \tag{4.172}$$

where:

n is the number of revolutions at pump shaft, min^{-1};
Q – pump flow rate, m^3/s;
H – pump pressure, *m*.

Therefore, centrifugal pumps occupy the niche within $n_s = (20–300)$ min^{-1}, diagonal pumps – within (300–600) min^{-1} and axial pumps – within (600–1,800) min^{-1}.

Centrifugal pump itself is a power machine, which turns mechanical energy of the liquid into hydraulic energy. Energy conversion is based on the power interaction of the power wheel blade system and the pumped liquid. The pressure difference appears on both sides of each blade at wheel rotation in the stream. Moving in the field of centrifugal forces, the fluid gets a reserve of kinetic energy, which afterwards turns into pressure energy in the pump outlet device. The increase of liquid stream energy will depend on current velocity, number of revolutions of the wheel, its size, blade shape, and also on the shape and construction of the liquid outlet device.

Centrifugal pump is composed of spiral work wheel or of blade outlet device in case of multi-stage exhaust. The work wheel includes the driving disks and driven disks with a certain number of shaped blades, in particular cylindrical or shaped blades. The liquid stream that enters the wheel axial direction moves radially towards the wheel periphery under the action of centrifugal forces and due to the special shape of the meridian section.

The outlet device of the centrifugal pump is composed of spiral elements and the diffuser in which the current of velocity transforms into potential.

As a result of the undertaken study, the following opportunities were determined with account of the specific kinematic and functional peculiarities of the micro hydropower plant for river water kinetic energy conversion, and also the diverse consumer requirements for hydro power plants:

1. In the construction of micro hydropower plant for river water kinetic energy conversion into electrical and mechanical energy for water pumping purposes, high speed revolution centrifugal pumps can be applied (n = 1,500–3,000) min^{-1}; the pumps are supplied with electric energy produced by a low speed generator that is coupled kinematically to the hydrodynamic impeller (of the micro hydropower plant).
2. It is necessary to design, research and manufacture low speed centrifugal hydraulic pumps for those micro hydropower plants that will be used to convert river water kinetic energy into mechanical energy and to pump water.

As result of study of the catalogue of centrifugal pumps that are produced nowadays, two types of centrifugal pumps with rotational frequency n = 3,000 min^{-1} have been selected, namely:

- Centrifugal pump, series PC model KM, produced by *Mezon* SA enterprise in the Republic of Moldova [86]
- Centrifugal pump, model CHT–400, produced by the Italian company *Pentax Water Pumps* [87]

Table 4.12 shows the general view of the centrifugal pump (series PC, model KM) and describes its functional characteristics. Table 4.13 presents the general view and fucntional characteristics of pump CHT–400.

Further on the design, research and manufacturing of two typo dimensions of low speed hydraulic pumps are presented, that are destined, in particular, for micro hydropower plants.

Table 4.12 Characteristics of centrifugal pump, series PC, model KM

Parameter	Measure unit	Rated value
Flow rate	m^3/h	20
Pumping height	m	20
Mode of operation	–	Permanent
Temperature of pumped water, max	°C	
Entrance pressure	kg/cm^2	1
Engine rated output	kW	2.2
Alternating current stress	V	380
Unit mass, max	kg	28.8

Table 4.13 Characteristics of centrifugal pump, model CHT – 400

Type	Flow rate (m^3/h)						
3~	6	12	18	24	30	36	42
230/400 V −50 Hz	Pumping height (m)						
CHT 550	37	37	35.5	33	29.4	25	19
CHT 150	24.5	22.5	18.5	12.5	4	–	–
CHT 200	28	25	21	15	5.5	–	–
CHT 300	33.5	31.5	28	22.5	13.5	–	–
CHT 350	26	25	23	20	16	10.5	–
CHT 400	31.8	31.6	29.7	27.8	23	18.6	12.3

4.5.4.2 Preliminary Calculation of Centrifugal Pump and Optimisation of Centrifugal Wheel Geometry at Low Rotation Speeds

As result of preliminary calculations, the main geometrical dimensions and energy values of the pump have been determined depending on the:

- Nominal flow $Q = 40\ m^3/h$;
- Nominal advance pressure $H = 10$ m;
- Number of revolutions of the connecting shaft $n = 750, 500, 300\ min^{-1}$;
- Operation position – vertical.

Limiting the maximum torque of the driving mechanism shaft is an additional condition for the calculation. Therefore, one of the variable parameters in the calculations will be the number of pump stages. In this case, it is possible to choose the working wheel with allowable geometric dimensions and maximum satisfactory efficiency, corresponding to the determined field of the pump high speed coefficient n_s.

The calculation is based on the classical methodology of calculation of centrifugal pumps [88–90]. It is accepted that the cavitation capacity of the pump wheel is "normal." Mechanical losses, related to the friction of bearings and the compression node, are not taken into account in the hydraulic calculation, but the value of interior mechanical efficiency $\eta_{mec} = 0.95, \ldots, 0.99$ can be taken on the basis of the statistical data. Shaft diameter for all previously supported versions is unique and determined with account of simplified calculation at torque, taking into consideration the multistage construction:

$$d_a = \sqrt[3]{\frac{9554.14 \cdot N_{\max}}{0.2 \cdot \tau_f \cdot n}} = \sqrt[3]{\frac{9554.14 \cdot 4}{0.2 \cdot 12 \cdot 10^6 \cdot 750}} = 0.028, \text{ m} \quad (4.173)$$

where:

$N_{max} = 4\,\text{kW}$ is the highest capacity on the shaft;
$\tau_f = 12\,\text{MPa}$ – admissible stress of shaft material at torque.
It is accepted $d_a = 34$ mm.

Calculation data are presented in Table 4.14. Aiming at a comparative analysis, the table includes the values of the centrifugal pumps with the number of revolutions of shaft 3,000 and 1,500 min^{-1}.

To select the pump execution options, the hydraulic calculation has been carried out in order to determine the working wheel geometrical dimensions and the connection of the pump stage with the blade. The optimisation calculation is done by utilisation of specialised software on the basis of classical calculation methodology for centrifugal pumps. The obtained shapes of the blade profile and the meridian section is checked by numerical FEMLIMS media environment, which allows three-dimensional analysis and visualization of viscous fluid flow in the wheel sharp cavity. Determination of the working wheel size is done with the purpose to exclude the occurrence of nominal regimes in the flow between the blades of low power areas and the opposite direction flows, which affects negatively the hydraulic characteristics of the pump. Final correction of the centrifugal pump dimensions and parameters is made with consideration of the experimental model of pump test results. It is necessary to note that changing the shaft speeds towards increasing or decreasing would influence pump parameters. In accordance with the laws of hydraulic similarity this influence is determined from the following relations:

$$Q_2 = Q_1 \cdot \frac{n_2}{n_1};\ H_2 = H_1 \cdot \left(\frac{n_2}{n_1}\right)^2;\ N_2 = N_1 \cdot \left(\frac{n_2}{n_1}\right)^3 \quad (4.174)$$

Research and optimisation of the geometrical parameters of centrifugal pump are based on the ensurance of the following pump functional parameters:

	Pump PSS 40-10/50	Pump PSS 30-10/50
Speed frequency	500 min^{-1}	300 min^{-1}
Efficiency	40 m^3/h	30 m^3/h
Pumping height	10 m	10 m

Table 4.14 Calculated data of centrifugal pump for various revolutions

Nr.var.	Number of shaft revolutions n (min^{-1})	Number of pump stages, i	Pump efficiency, η	Pump effective power N (kW)	Necessary moment of the pump shaft M (Nm)	Pump working wheel diameter D_2 (mm)	Wheel input diameter D_0 (mm)	High speed coefficient of pump stage, n_s
1	3,000	1	0.8	1.363	4.34	110	70	205.3
2	1,500	1	0.76	1.434	9.13	185	90	102.6
3	750	1	0.63	1.73	22.03	325	108	51.3
4	500	1	0.55	1.982	37.85	470	120	34.2
5		2	0.68	1.602	30.6	345		57.5
6		3	0.73	1.491	28.49	300		78
7	300	1	0.42	2.595	82.6	760	140	20.5
8		2	0.58	1.88	59.84	550		34.5
9		3	0.65	1.674	53.28	460		46.8
10		4	0.69	1.58	50.28	405		58.1
11		5	0.72	1.515	48.19	370		68.6

4.5.4.3 Manufacturing and Experimental Research of Centrifugal Pump Characteristics at Low Speeds

The 3D model of three stage centrifugal pump (Fig. 4.151a) and the technical documentation were developed in SolidWorks design software. In cooperation with *Hidrotehnica SA enterprise* (Chişinău), for the first time in the Republic of Moldova, two industrial prototypes of hydraulic pump PSS 40-10/50 were manufactured (Fig. 4.152). Complex experimental research on low speed centrifugal pumps was carried out on the test stand (Fig. 4.153) for various functional regimes, specific for their operation conditions in the structure of the micro hydropower plant for water kinetic energy conversion. The pumps were manufactured according to climatic (Y, GOST 15150-69) and reliability (category 1, GOST 27.003-90) requirements. The basic power parameters, including noise emission and vibroactivity, have been determined within the carried out experimental research. The results of experimental research are presented in Fig. 4.154a, b, c, Fig. 4.155a, b, c and in Table 4.15 [85, 91]. Overall efficiency of pump PSS 40-10/50 is $\eta_p = 0.72$. The overall level of noise emission (in the air) at the revolutions n = 520 min^{-1} is 53/48 dB/dBA, and maximum vibroactivity is 75 dB.

As result of experimental research that was carried out the following conclusions can be made. For the revolution frequency of the hydrodynamic rotor of the micro hydropower plant, multiplied up to $n \approx 500$ min^{-1} at the hydraulic pump shaft, the functional parameters of the last that have been obtained by experimental research are satisfactory, namely:

Mechanical efficiency of the pump $\eta = 0.72$;
Flow rate $Q = 40$ m^3/h;
Pumping height $h = 10$ m,

and the deviation from the obtained theoretical parameters is not significant (see Table 4.16).

4.5.5 Development of the Design Concept and Manufacturing of Pilot Station of Polyfunctional Floating Micro Hydro Power Plant for Experimental Research in Real Conditions

4.5.5.1 Pilot Station Destination

The micro hydro power plant is a complex system, which includes the main working element – the hydrodynamic rotor, destined for the conversion of a part of the river water kinetic energy potential into useful energy, and the units with distinct functions, such as the multiplier, generator, hydraulic pump etc. The results of

a

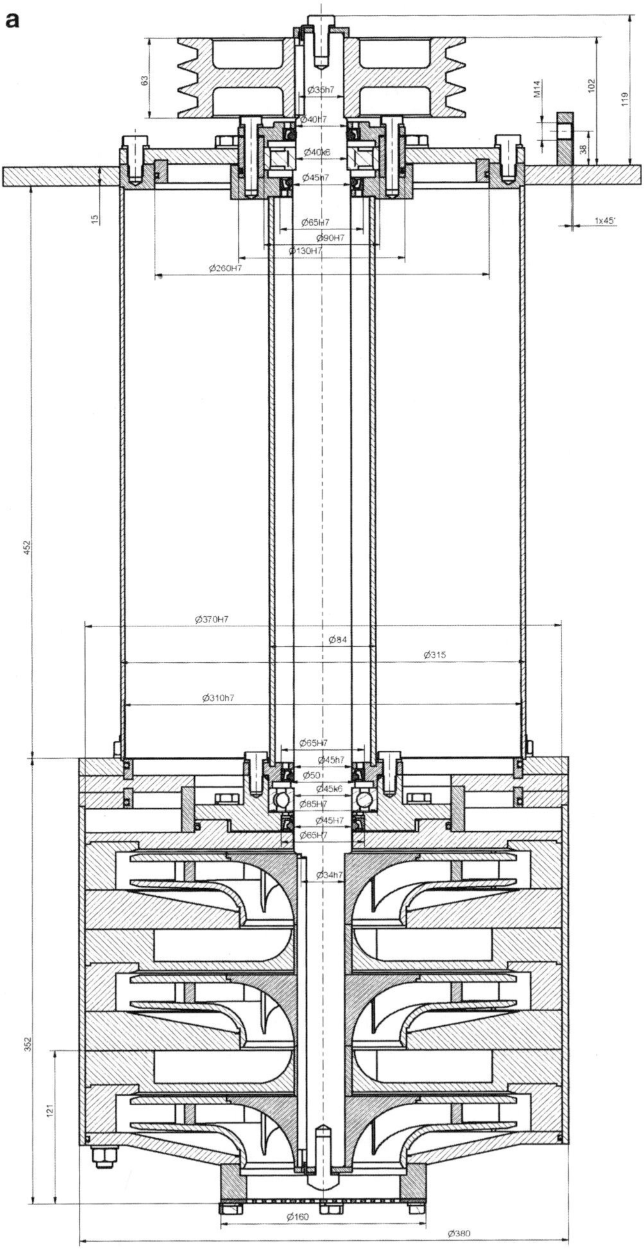

Fig. 4.151 3D model of centrifugal pump PSS 40–10/50 (**a**), general view (**b**) and working bodies (**c**)

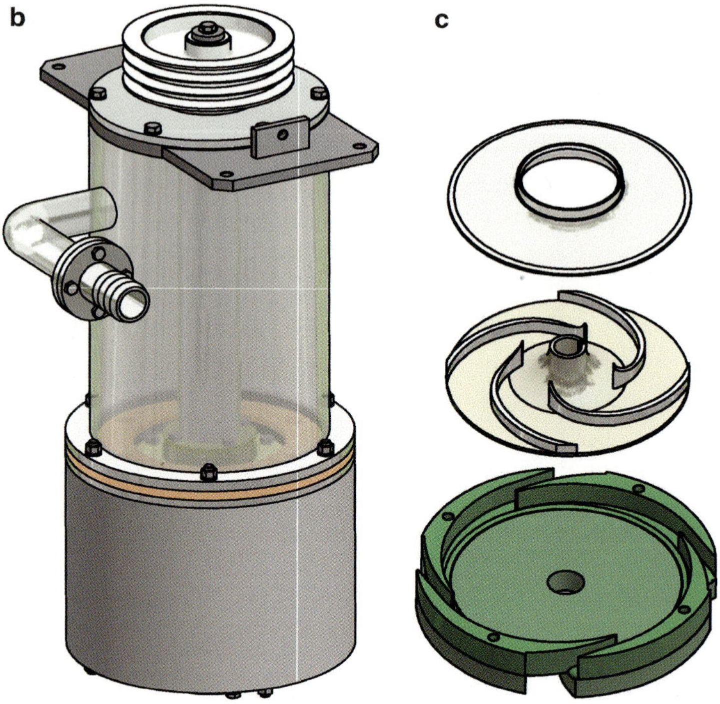

Fig. 4.151 (continued)

research on the construction and functional parameters of the mentioned units, shown separately from each other regardless of their functional interdependence as an integral system, are presented in Sects. 4.5.2–4.5.4.

To carry out research on the functional parameters, in field conditions, of all units participating in the process of conversion of flowing water kinetic energy into useful energy as an integral technical system, the pilot plant design was developed in MotionInventor AutoDesk design environment, using the developed conceptual diagrams [53–58] and the results of theoretical research findings, of computer modeling and simulation of hydrodynamic profile blade interaction with the fluid (water) [40–43, 45, 58, 62, 63, 64–66]. The pilot station is intended for research and verification of the technical solutions adopted at each stage of unit development, in real conditions, and, if necessary, changes will be made at the design stage of the industrial prototypes of micro hydro power plants.

4.5.5.2 Description of the Pilot Station Construction and Manufacturing

The pilot station (Fig. 4.156) is an integral technical system and performs conversion of the river water kinetic energy into electrical, mechanical or combined energy.

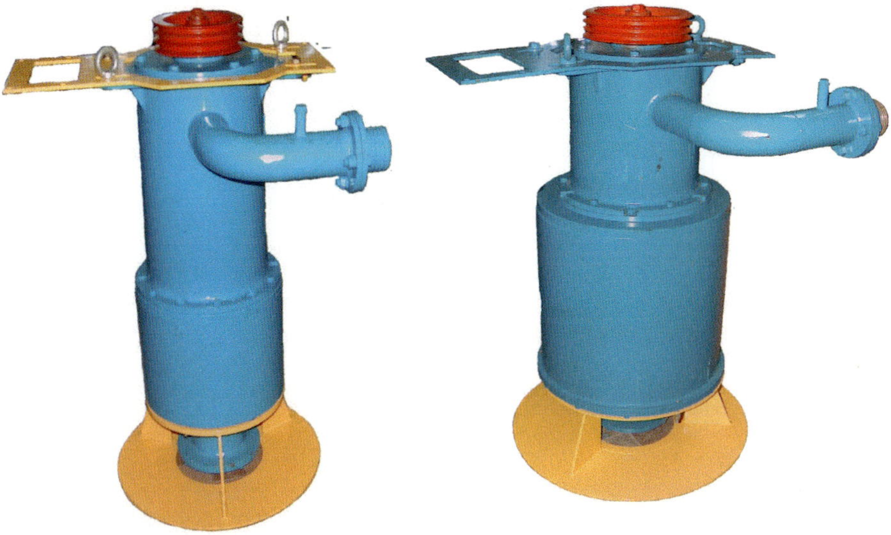

Technical characteristics:	Technical characteristics:
Number of revolutions $n=500min^{-1}$;	Number of revolutions $n=300min^{-1}$;
Pump mechanical efficiency $\eta=0,7$;	Pump mechanical efficiency $\eta=0,7$;
Flow rate $Q=40m^3/h$.	Flow rate $Q=30m^3/h$.
Pumping height – $(10\text{-}15)\ m$.	Pumping height – $(10\text{-}15)\ m$.

Fig. 4.152 Two industrial prototypes of hydraulic centrifugal pump manufactured by *Hidrotehnica –SA enterprise*, Chişinău

Fig. 4.153 Experimental test stand for hydraulic pumps PSS 40–10/50

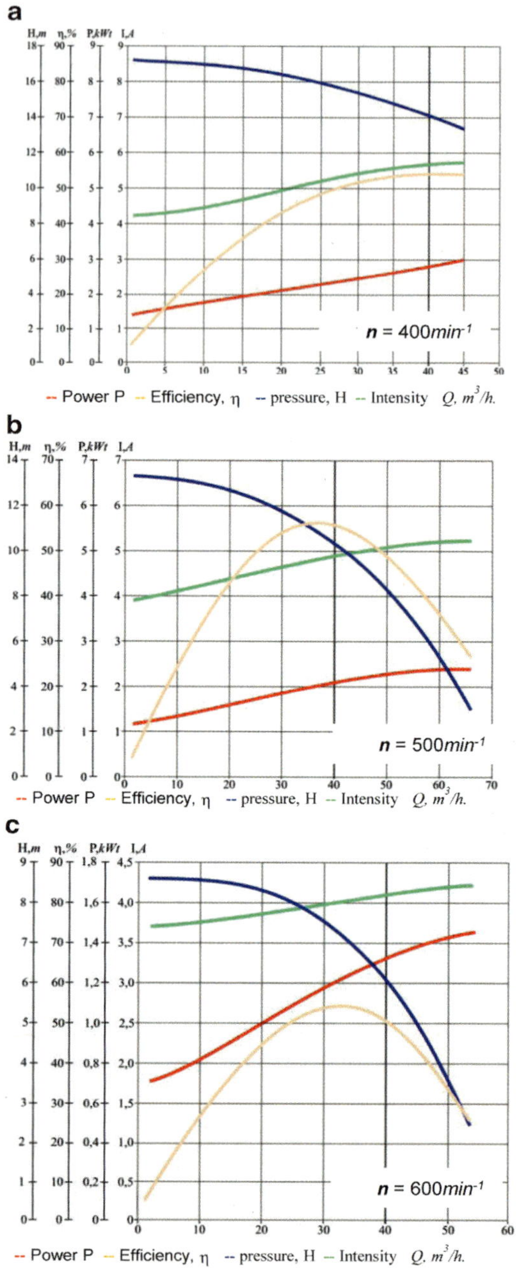

Fig. 4.154 Pump performance for different operating regimes

The pilot station includes the following nodes and units: working body as rotor 1 with three or five blades with hydrodynamic profile 2, precessional multiplier 3, permanent magnet generator 4 and hydraulic pump 5. The mentioned nodes and units are mounted on a spatial resistance housing 6, installed on four floating bodies

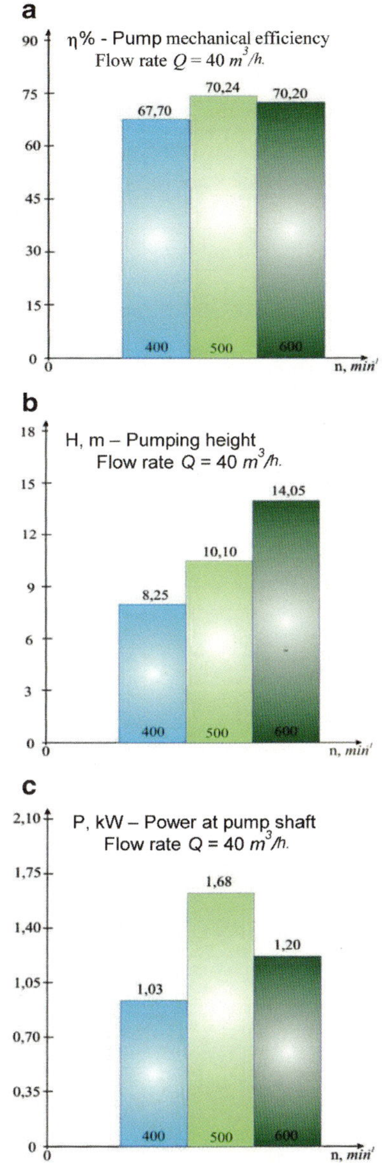

Fig. 4.155 Comparative analysis of the basic power parameters of the produced pumps

7 and anchored to the shore by a deck 8 (communication bridge). To ensure the pilot station buoyancy at various levels of river water, the bridge 8 is assembled with the spatial housing 6 through two kinematic couplings with a degree of freedom, and with the pillars 9 on the shore – through a kinematic coupling with three degrees of freedom. To protect the blades 2, when water levels in the river become lower, four level spring bars 10 are mounted on the spatial housing 6. To transmit the reactive

Table 4.15 Experimental test results

Nr. crt.	Q_{op} (m^3/h)	H_{op} (m)	I (A)	P_1 (kW)	Efficiency η (%)	$n_{tur.mot.}$ (tur/min)	Q_{pr} (m^3/h)	H_{pr} (m)	P_{1pr} (kW)
1	1.35	13.269	3.917	1.060	4.598	534.6	1.35	13.269	1.060
2	13.50	12.596	4.200	1.400	33.049	531.8	13.50	12.596	1.400
3	22.50	12.246	4.383	1.620	46.278	529.8	22.50	12.246	1.620
4	27.00	11.980	4.517	1.740	50.582	528.8	27.00	11.980	1.740
5	31.50	11.420	4.633	1.860	52.625	527.9	31.50	11.420	1.860
6	36.00	10.666	4.767	1.980	52.768	526.9	36.00	10.666	1.980
7	40.05	10.113	4.817	2.040	54.025	526.1	40.05	10.113	2.040
8	45.00	9.077	4.967	2.160	51.457	525.0	45.00	9.077	2.160
9	49.50	8.042	5.067	2.240	48.355	522.3	49.50	8.042	2.240
10	54.00	6.812	5.117	2.300	43.521	519.6	54.00	6.812	2.300
11	64.80	3.007	5.267	2.420	21.906	513.1	64.80	3.007	2.420

Note: The table shows the driving mechanism efficiency for the integral stand

Table 4.16 Theoretical functional parameters of the centrifugal pump PSS40-10/50 for different rotation frequencies

Nr. var.	Input shaft rotation speed n (rot/min)	Pump stages i	Pump efficiency, η	Pump power N (kW)	Required moment applied to the pump shaft T (Nm)	Diameter of the pump working wheel D_2 (mm)	Diameter of the wheel input D_0 (mm)	Pump swiftness, n_s
1	3,000	1	0.8	1.363	4.34	110	70	205.3
2	1,500	1	0.76	1.434	9.13	185	90	102.6
3	750	1	0.63	1.73	22.03	325	108	51.3
4	500	1	0.55	1.982	37.85	470	120	34.2
5		1	0.68	1.602	30.6	345		57.5
6		3	0.73	1.491	28.49	295		78
7	300	1	0.42	2.595	82.6	760	140	20.5
8		2	0.58	1.88	59.84	550		34.5
9		3	0.65	1.674	53.28	460		46.8
10		4	0.69	1.58	50.28	405		58.1
11		5	0.72	1.515	48.19	370		68.6

Note: 1. Shaft diameter under the working wheel hub for all options is $d_v = 40$ mm
2. Pump efficiency is calculated without account of losses in the bearings
3. Cavitation capacities of stages are ordinary

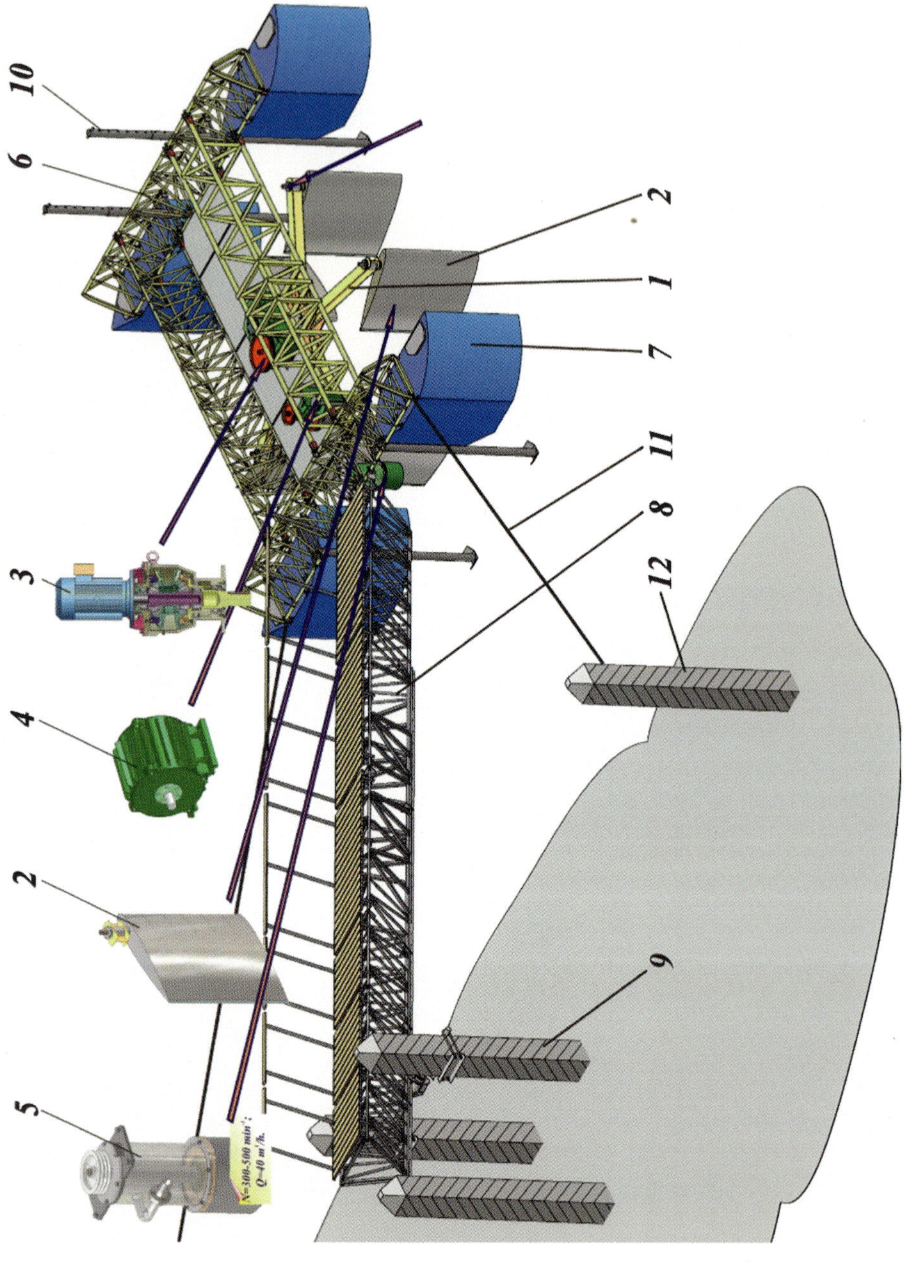

Fig. 4.156 Pilot station of the polyfunctional floating micro hydropower plant for experimental research in field conditions

Spatial housing with four floating bodies at the stage of manufacturing, „Incomaş" SA enterprise, Chişinău.

Spatial housing with four floating bodies at the stage of assembling, CITA Etalon, TUM.

Fig. 4.157 Spatial housing with four floating bodies

torque moment from the hydrodynamic rotor, the spatial housing 6 is anchored to the shore through cable 11, attached to the pillars 12. The unit maintenance decking is installed inside the spatial housing 6. The orientation of hydrodynamic profile blades 2, at the optimal entering angle towards the water flow velocity vector, is carried out in automatic mode via a special device, which contains Know-how elements and is not described here. The units and core nodes of the pilot station developed by the team of authors are shown in Fig. 4.157 and Fig. 4.158a, b, c, d. The overall view of the spatial resistance housing with four floating bodies and the deck, manufactured by "Incomas" SA enterprise (Chisinau) and assembled at the Center for Advanced Technology Implementation "Etalon" (Technical University of Moldova), are presented in Fig. 4.155.

Figure 4.158a, b, c, d shows the general view of the: (a) low speed centrifugal pumps manufactured by *Hidrotehnica* SA enterprise, Chişinău; (b) centrifugal

Fig. 4.158 General view of the main nodes and units – pilot station components: low speed centrifugal pumps (**a**); centrifugal pumps, series PC, model KM (**b**); low speed permanent magnet electric generator (**c**); precessional multiplier (**d**); 5-blade hydrodynamic rotor (**e**); flowmeter CD-O5 (**f**)

pumps, (series PC, model KM) manufactured by *Mezon* Ltd enterprise, Chişinău; (c) permanent magnet low speed electrical generator produced at *Moldovahidromaş* SA; (d) precessional multipliers produced by *Topaz* JS company and Reupies Ltd, Chişinău; (e) flowmeter CD-05 manufactured by the Research&Production Centre *TIS*, Chişinău; (f) 5-blade hydrodynamic rotor, manufactured by the Centre for Renewable Energy Conversion Systems, Technical University of Moldova. The functioning principle of the pilot station is presented on CD support (see Fig. 4.159).

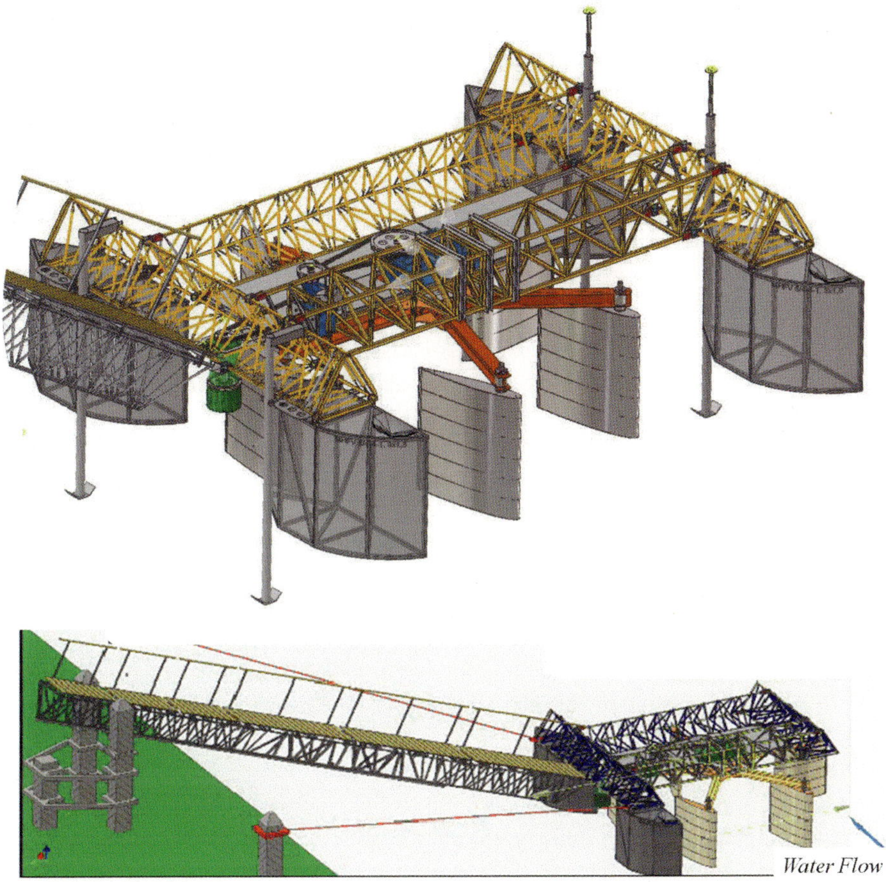

Fig. 4.159 Pilot station micro hydropower plant in dynamics (see the attached CD)

4.5.5.3 Schedule and Methodology of Experimental Research on the Pilot Station in Field Conditions

Pilot station positioning. The pilot station is installed on the river under the following positioning conditions:

- Rotor axis is perpendicular to the water mirror;
- Longitudinal axis of spatial housing is perpendicular to the water flow velocity vector;
- Level of blades submersion meets the project rate (h = 1.4 m). Perpendicularity of the hydrodynamic rotor shaft to the water mirror is ensured by changing the buoyancy of four floating bodies, and the perpendicularity of the longitudinal axis of the spatial housing to the water flow stream vector is ensured through

supporting cables secured to the anchoring rods. The submersion level of blades with hydrodynamic profile h = 1.4 m is maintained by changing the buoyancy of the four equal floating bodies.

Schedule of experimental research on the pilot station. Structural and functional parameters of hydrodynamic rotor, planetary multiplier, generator and low speed centrifugal pumps, determined separately for each working body, needs to be checked by experimental investigations in real conditions of their operation in an integral technical system.

In this context, the time table of experimental research on the pilot station in field conditions includes:

- Study of diversity of water flow speed cadastre within the boundaries of rotor effective section (the width of the rotor blades and level of submersion) and assessment of the water flow energy potential;
- Study of the influence of force factors on the stability of the hydrodynamic profile blades positioning (angle of attack α) and the kinetostatic analysis of the blades positioning mechanism;
- Study of energy and kinetic conversion efficiency parameters (for the electric generator terminals and the input shaft of hydraulic pumps);
- Study of kinematic parameters of hydrodynamic rotor and mechanical losses in the kinematic chain (linkage);
- Setting the influence of structural and functional parameters of hydrodynamic rotor on the hydrodynamic effects and water turbulence flow mode under field conditions;
- Study of functional characteristics of electrical generator and centrifugal pumps.

The experimental research of the integral technical system – hydrodynamic rotor coupled kinematically with component units of micro hydro power plants aims at increasing efficiency of water flow kinetic energy conversion into useful energy by identifying and, where necessary, introducing in the technical documentation of partial structural changes and, sometimes, of conceptual and technical solutions adopted previously.

4.5.6 Floating Micro Hydropower Plants for River Water Kinetic Energy Conversion into Electrical and Mechanical Energy

4.5.6.1 General Overview

When developing industrial prototypes of micro hydropower plants for river water kinetic energy conversion the following criteria and requirements were taken into account:

– Exclusion of dam construction and of the negative impact on the environment, implicitly;

- Lowest cost;
- Simplicity of construction and operation;
- Increased reliability at dynamic overload in operating conditions;
- Resistant composite materials, including conditions of high humidity;
- Automatic adjustment of the micro hydropower plant platform position in conditions of water level changing.

The technical solutions adopted in the process of micro hydropower plants design result from theoretical and experimental research presented in Sects. 4.5.2–4.5.5. To justify the constructive and functional parameters, additional numerical modeling and simulations were performed, using ANSYS software CFX5.7, and the sub-software developed by the authors for MathCAD, Autodesk MotionInventor, etc., namely, simulation of the *fluid–blade* interaction and floating stability, hydrodynamic optimization of the blade profile in order to increase kinetic energy conversion efficiency of river water at its different velocities using 3-, 4- and 5-blades rotor. The experience gained at the research – design – manufacture stages of the pilot station of micro hydropower plants is described in Sect. 4.5.5.

The efficient operation of micro hydropower plants by individual customers for particular destination depends on their constructive configuration choices and on the functional characteristics of component units participating in the conversion of flowing water kinetic energy into useful energy. Pre-feasibility study shows the economic efficiency of micro hydropower plants in producing electrical and mechanical energy by means of river water kinetic energy conversion.

To meet the objectives and consumer demands for micro hydro power plants, and also to increase the efficiency of conversion of flowing water kinetic potential in the certain river area, the authors have developed the following structural and functional concepts based on modular assembly:

- Micro hydropower plant with hydrodynamic rotor for river water kinetic energy conversion into mechanical energy for water pumping (MHCF D4 $\times$ 1.5 M);
- Micro hydropower plant with hydrodynamic rotor for river water kinetic energy conversion into electric and mechanical energy (MHCF D4 $\times$ 1.5 ME);
- Micro hydropower plant with hydrodynamic rotor for river water kinetic energy conversion into mechanical energy at low speeds (MHCF D4 $\times$ 1.5 ME);
- Micro hydropower plant with hydrodynamic rotor for river water kinetic energy conversion into electric energy (MHCF D4 $\times$ 1.5 E).

The micro hydropower plants designed modularly, allow change of destination and of functional characteristics by replacing some units with others (generator, pump, blades with other hydrodynamic profile, 3- and 5-blade rotor).

The micro hydropower plants possess similar resistance in structure construction calculated for strength and rigidity to dynamic applications.

Buoyancy and maintenance of hydropower plant rotor axis perpendicularity at variable river water level is ensured by patented technical solutions [55–60].

Continuous orientation mechanism of the blades at the constant entering angle relative to the direction of fluid stream contains Know-how elements and is not described here. The main working body, which depends mainly on the amount of kinetic energy converted into useful energy, is the blade with hydrodynamic NACA 0016 profile, developed on the basis of carried research and presented in Sect. 4.5.2.

Two types of 3- and 5-blade rotors were designed for the described micro hydropower plants. Installed capacity of micro hydropower plants with diameter $D = 4$ m, submerged height of blades $h = 1.4$ m and the length of blade chord $l = 1.3$ m at water flow velocity $V = 1, \ldots, 2$ m/s can be in the boundaries $P = 2$–19 kW (see Fig. 4.115).

4.5.6.2 Micro Hydropower Plant with Hydrodynamic Rotor for River Water Kinetic Energy Conversion into Mechanical Energy

This model of hydro power plant is designed to convert river water kinetic energy into mechanical energy used to pump water into irrigation and sewerage systems, and supply industrial water etc. with the flow rate $Q = 40\ \mathrm{m^3/h}$ at the height pumping, $H = (10–15)$ m.

Static description of the microhydro power plant. The blades 1 (Fig. 4.160) are connected to the hydrodynamic rotor 2 by roller friction bearings to ensure their orientation under a certain entering angle $\boldsymbol{\alpha}$. The hydrodynamic rotor 2 is mounted on the input shaft of the planetary multiplier 3 through an auxiliary shaft, which is fixed on the bearings. The belt pulleys of the transmission 4 are mounted on the output shaft of the planetary multiplier – the big one, and the small one – on the input shaft of the centrifugal pump 5. The hydrodynamic rotor 2 and blades 1, the multiplier 3, the centrifugal pump 5 and guides 6 are mounted on the spatial housing 7, installed on the pontoons 8.

Functioning principle. The river flowing water with the energy potential dependent on the flow velocity drives the hydrodynamic profile blades 1 (Fig. 4.161), oriented continuously by the entering angle α, and revolving in their relative movement in relation to the rotor through the bearings mounted in body 5. The micro hydropower rotor 2 comprises three blades oriented at an entering angle α, which is dependent on the water flow velocity. In the areas of blades 1 location, inefficient from the point of view of river kinetic energy conversion, under hydrodynamic forces the blades 1 are repositioned at an angle of 90° to the currents of water or are carried by the water unhampered to the angle $\alpha = 0$. Thus, the respective positioning of blades allows the increase of water kinetic energy rate converted into useful energy. As result, the water currents transmit a part of their kinetic energy to the blades 1, stressing them under the hydrodynamic forces and reporting rotational motion with angular frequency ω_l and torque T_l to the rotor 2. The summary torque T_l, developed by the hydrodynamic forces and applied to the 3-blade rotor shaft at water flow velocities 1.3, 1.6 and 1.8 m/s and at the entering angle of blades $\alpha = 18°$, is presented in Fig. 4.162.

For rotor diameter $D = 4$ m, the submerged height of blades $h = 1.4$ m and length of blade chord $l = 1.3$ m, the torque is: $T_l = 11{,}938$ Nm for water flow velocity

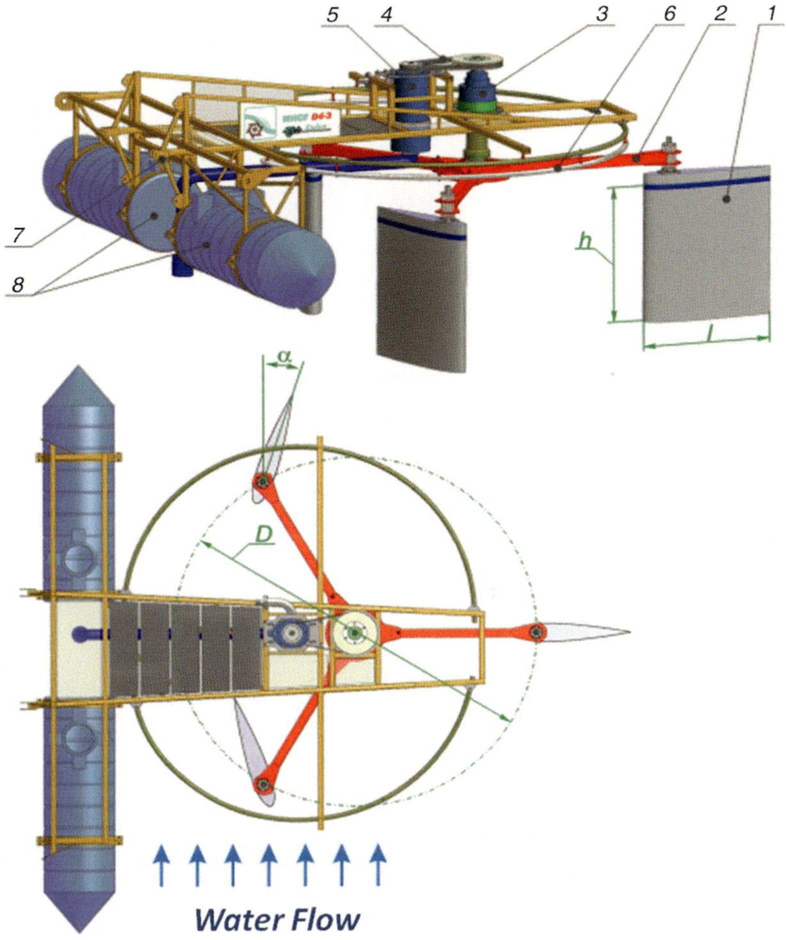

Fig. 4.160 Micro hydropower plant with hydrodynamic rotor for river kinetic energy conversion into mechanical energy for water pumping (flow rate Q = 40 m^3/h, pumping height H = 10, . . . , 15 m) (MHCF D4 x 1.5 M). *1* – Blade with hydrodynamic 0016 NACA profile; *2* – 3-blade rotor; *3* – planetary multiplier with multiplying ratio $i = 112$; *4* – belt drive with multiplying ratio $i = 1.9$; *5* – permanent magnet generator (characteristics – see Sect. 4.5.4); *6* – centrifugal pump *PSS40-10/50* (characteristics – pump flow rate $Q = 40$ m^3/h at pumping height (10–15) m); *7* – plastic pontoons; *8* – guide; *9* – spatial housing

$V = 1.3$ m/s; $T_1 = 18{,}084$ Nm for $V = 1.3$ m/s; $T_1 = 22{,}887$ Nm for $V = 1.8$ m/s. The calculations of kinematics and lifting capacity of all constructive elements as well as of all functional and energy parameters of micro hydropower plant have been carried out for the torque value $T_1 = 18{,}084$ Nm.

Rotor 2, rigidly coupled by means of auxiliary shaft with the input shaft of the multiplier 3, transmits rotational motion to the last with angular frequency $\boldsymbol{\omega_1}$ and torque $\boldsymbol{T_1}$. The multiplier reproduces the rotor 2 revolutions up to

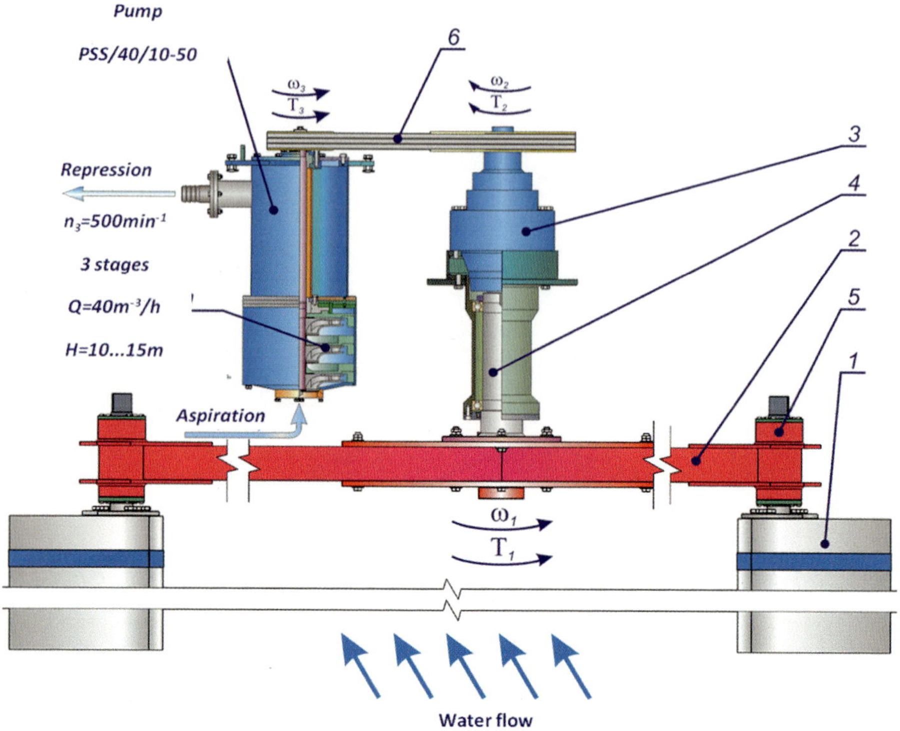

Fig. 4.161 Kinematics of micro hydropower plant MHCF D4 × 1.5 M

$n_2 = \frac{30\omega_1}{\pi} i_1 (\text{min}^{-1})$ where $\boldsymbol{i_1}$ represents the multiplying ratio of the multiplier ($i_1 = 112$). Rotational motion at angular frequency $\omega_2 = \frac{\pi n_2}{30} (\text{s}^{-1})$ is transmitted from the multiplier input shaft via a transmission belt 4 of the centrifugal pump input shaft with multiplying ratio $i_1 = 2.25$. As result, the input shaft of the centrifugal pump swivels with angular frequency $\omega_3 = \omega_1 \cdot i_1 \cdot i_2\ (\text{s}^{-1})$ and is stressed at torque:

$$T_3 = \frac{T_1 \cdot \eta_1 \cdot \eta_2 \eta_r}{i_1 \cdot i_2}, (\text{Nm})$$

where:

η_1 is the multiplier mechanical efficiency ($\eta_1 = 0.9$);
η_2 – is belt transmission mechanical efficiency ($\eta_1 = 0.95$);
η_r – mechanical efficiency of hydrodynamic rotor bearings ($\eta_1 = 0.99$).

According to the experimental research presented in Sect. 4.5.5.3, the mechanical efficiency of centrifugal pump is $\eta_1 = 0.72$ at rated speed frequency

$$n_3 = \frac{30\omega_3}{\pi} = 500\ \text{min}^{-1}.$$

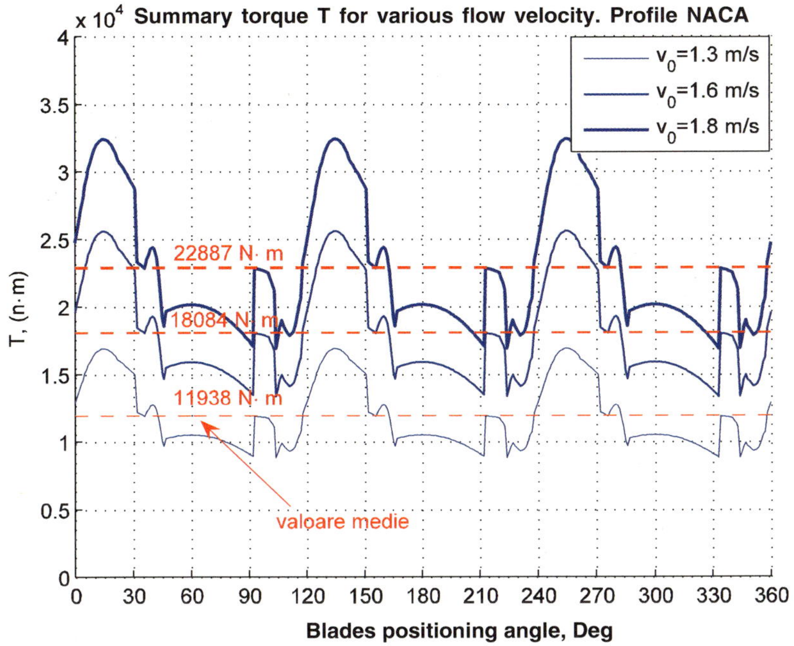

Fig. 4.162 Torque T_I at the hydrodynamic rotor shaft with NACA 0016 profile blades

The mechanical efficiency of the micro hydropower plant with hydrodynamic rotor for river water kinetic energy conversion into mechanical energy, with account of all mechanical losses in the linkage (at the hydraulic pump shaft) is:

$$\eta_{\sum} = \eta_1\eta_2\eta_3\eta_r = 0.9 \cdot 0.95 \cdot 0.99 = 0.846$$

Accordingly, the micro hydropower plant (MHCF D4 × 1.5 M) ensures the transformation into useful energy of 84.6% of the kinetic energy potential of the flowing water transmitted to the hydrodynamic rotor.

4.5.6.3 Micro Hydropower Plant with Hydrodynamic Rotor for River Water Kinetic Energy Conversion into Electrical and Mechanical Energy

The micro hydropower plant MHCF D4 × 1.5 ME for river water kinetic energy conversion into electrical and mechanical energy (Fig. 4.163) is polyfunctional and can be utilised for electrical lighting of streets, heating, water pumping in drip irrigation systems, and also for draining agricultural lands adjacent to rivers.

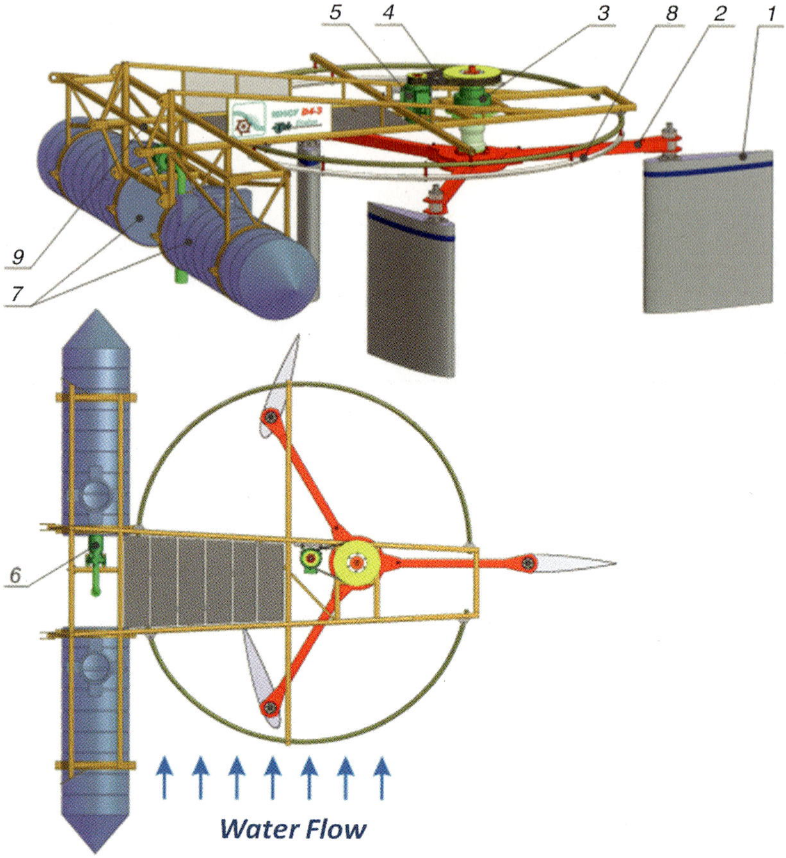

Fig. 4.163 Micro hydropower plant with hydrodynamic rotor for river kinetic energy conversion into electrical and mechanical energy (rotor diameter $D=4$ m, submerged height of blade $h=1.4$ m, length of blade chord $l=1.3$ m) (MHCF D4 × 1.5 ME). *1* – Blade with hydrodynamic NACA 0016 profile; *2* – 3-blade rotor; *3* – planetary multiplier with multiplying ratio $i=112$; *4* – belt drive with multiplying ratio $i=1.9$; *5* – permanent magnet generator (characteristics – see Sect. 4.5.4); *6* – centrifugal pump *CH* – 400 (characteristics – pump flow rate $Q=(20\text{–}40)\text{m}^3/\text{h}$ at pumping height 15–32 m); *7* –plastic pontoons; *8* – guide; *10* – space housing

Rigging the NACA 0016 profile blades 1 in the hydrodynamic rotor 2 and its attachment to the multiplier input shaft 3 is done similar to micro hydropower plant MHCF D4 × 1.5 M. Kinematic and construction peculiarities of MHCF D4 × 1.5 ME are as follows: rotational movement of the hydrodynamic rotor 2 (Fig. 4.164) with an angular frequency (velocity) $\omega_{1,}$ by means of multiplier 3 and belt transmission 4 with effective multiplying ratio $i=212.8$, is multiplied to the operating angular frequency of the permanent magnet low speed generator 5:

$$\omega_3 = \omega_1 \cdot i_1 \cdot (s^{-1}).$$

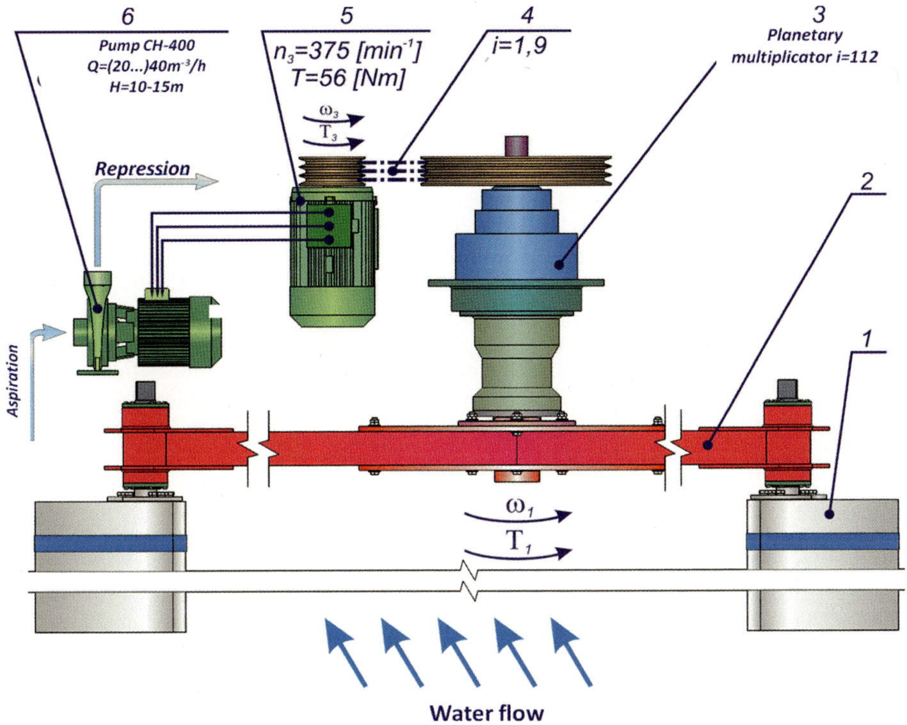

Fig. 4.164 Kinematics of micro hydropower plant MHCF D4 × 1.5 ME

Torque $\boldsymbol{T_3}$, applied to rotor 5, is:

$$\eta_{\sum} = \eta_1 \eta_2 \eta_r \eta_g = 0.9 \cdot 0.95 \cdot 0.99 \cdot 0.87 = 0.736$$

$$T_3 = \frac{T_1 \cdot \eta_1 \cdot \eta_2 \eta_r}{i} (\text{Nm}),$$

where:

η_1 is multiplier mechanical efficiency ($\eta_1 = 0.9$);
η_2 – is belt transmission mechanical efficiency ($\eta_1 = 0.95$);
η_r – mechanical efficiency of hydrodynamic rotor bearings ($\eta_1 = 0.99$).
i – effective multiplying ratio equal to the multiplying ratios product of the planetary multiplier and belt transmission.

The electric energy produced by the permanent magnet generator 5 (Fig. 4.164) can be utilised to satisfy the energy needs of the private consumers and, as well, to supply the centrifugal pump 6 (model CH 400) with electrical energy in order to pump water in drip irrigation systems or for drainage of agricultural land adjacent to river (with relocation of the centrifugal pump 6).

In the case of electric energy production, with account of mechanical losses both in the micro hydropower plant linkage and in the permanent magnet generator, the efficiency of energy utilisation at generator's terminals is, and in the case of water pumping (at the centrifugal pump shaft) the efficiency is:

$$\eta_{\sum} = \eta_1 \eta_2 \eta_r \eta_g \eta_{me} = 0.9 \cdot 0.95 \cdot 0.99 \cdot 0.87 \cdot 0.91 = 0.67$$

where:

η_g is the generator efficiency;
η_{me} is the efficiency of the hydraulic pump electromotor.

Thus, the micro hydropower plant MHCF D4 × 1.5 ME ensures the transformation into useful energy of 73.6% and 67% of the energy potential of flowing water, picked up by the hydrodynamic rotor, in producing electrical energy and, respectively, in water pumping.

4.5.6.4 Micro Hydropower Plant with Hydrodynamic Rotor for River Water Kinetic Energy Conversion into Electrical and Mechanical Energy at Small Speeds

Micro hydropower plant MHCF D4 × 1.5 ME (Fig. 4.165) is designed to convert river water kinetic energy into electrical and mechanical energy, by utilising low speed permanent magnet generator 5 (n = 375 min^{-1}) and three-stage low speed centrifugal pump PSS 40-10/50 (n = 500 min^{-1}) 7 designed, in particular, for the micro hydropower plant and manufactured at "Hidrotehnica" S.A., Chişinău. Research results and functional characteristics of the low speed pump are presented in Sect. 4.5.5.3.

Kinematics and functional principle of the micro hydropower plant are analogic to the micro hydropower plant presented in Sect. 4.5.6.3 (Fig. 4.165). Constructive peculiarities of this micro hydropower plant refer, in particular, to the driving mechanism unit of the centrifugal pump (Fig. 4.166) and to the supply of the pump low speed electromotor 2 from the permanent magnet low speed generator 5 (Fig. 4.165). This design configuration can be used both to meet the needs of irrigation by pumping water at relatively low heights (10–15) m (e.g. over the river dam) and to perform the draining works of the land adjacent to the river. When the micro hydropower plant is used for draining works, the centrifugal pump driving mechanism (Fig. 4.166) is relocated from the spatial housing of the micro hydropower plant to the floating platform in the flooded area of agricultural land adjacent to the river.

Three-stage centrifugal pump PSS 40-10/50 1 (Fig. 4.166) is coupled to the electromotor 2 through the toroidal coupling 3 and the housing 4 for the transmission of the reactive torque.

Taking into account the mechanical losses both in the linkage and in the electric generator, the kinetic energy efficiency transmitted by the water flow to the hydrodynamic rotor at the permanent magnet generator terminals makes up:

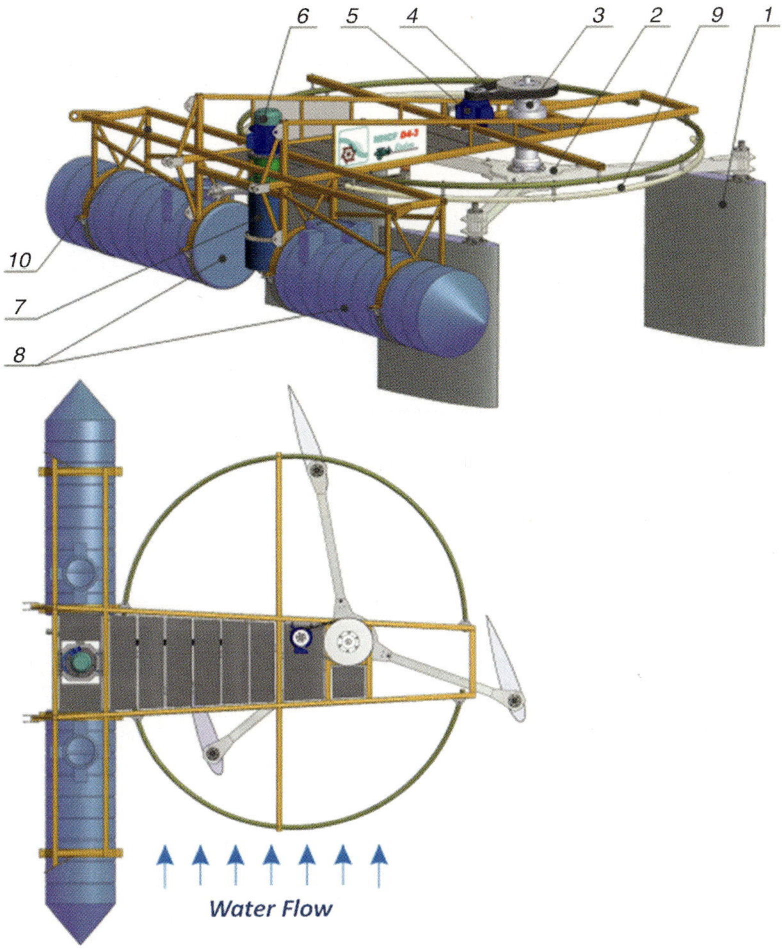

Fig. 4.165 Micro hydropower plant with hydrodynamic rotor for river kinetic energy conversion into electrical and mechanical energy used for water pumping (rotor diameter $D = 4$ m, submerged height of blade $h = 1.4$ m, length of blade chord $l = 1.3$ m). *1* – Blade with hydrodynamic NACA 0016 profile; *2* – 3-blade rotor; *3* – planetary multiplier with multiplying ratio $i = 112$; *4* – belt drive with multiplying ratio $i = 1.9$; *5* – permanent magnet generator (characteristics – see Sect. 4.5.4); *6* – asynchronous electromotor; *7* – centrifugal pump *PSS-40-10/50* (characteristics Q and H – see Sect. 4.5.4.3); *8* – plastic pontoons; *9* – guide; *10* – spatial housing

$$\eta_{\sum} = \eta_1 \cdot \eta_2 \cdot \eta_r \cdot \eta_g = 0.9 \cdot 0.95 \cdot 0.99 \cdot 0.87 = 0.736$$

and at the inlet shaft of the hydraulic pump PSS40-10/50 it is:

$$\eta_{\sum} = \eta_1 \cdot \eta_2 \cdot \eta_r \cdot \eta_g \cdot \eta_{m.e.} = 0.9 \cdot 0.95 \cdot 0.99 \cdot 0.87 \cdot 0.81 = 0.596$$

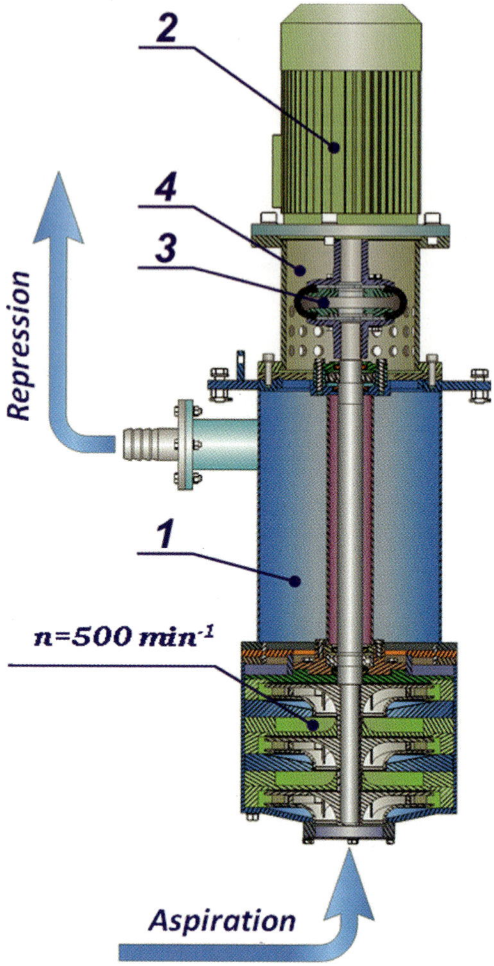

Fig. 4.166 Unit of three-stage hydraulic pump driving mechanism PSS 40-10/50

The micro hydropower plant (MHCF D4 × 1.5 ME) with the given configuration ensures the transformation of 73.6% of river water energy potential into useful energy and only 59.6% – at water pumping. Relatively small efficiency at water pumping is due to a more reduced efficiency of the low speed electromotor.

4.5.6.5 Micro Hydropower Plant with Hydrodynamic Rotor for River Water Kinetic Energy Conversion into Electrical Energy

The micro hydro power plant (Fig. 4.167) is designed to convert river water kinetic energy into electrical energy only. The construction peculiarities are as follows: the hydraulic rotor comprises five blades *1*; the permanent magnet

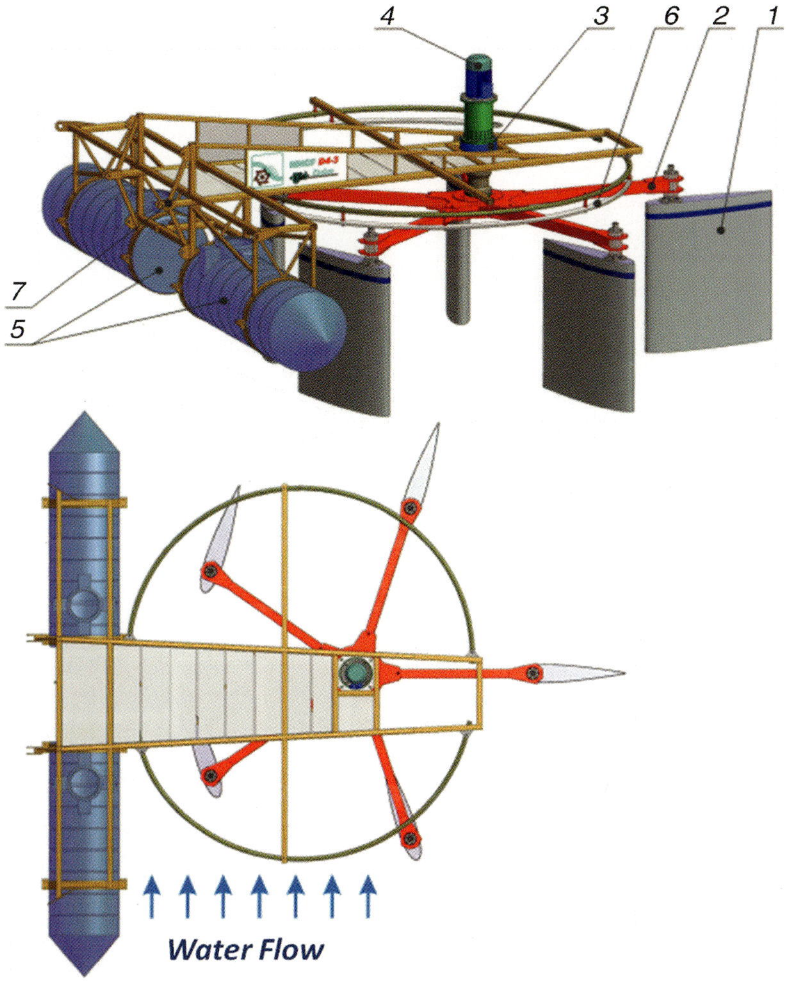

Fig. 4.167 Micro hydro power plant with hydrodynamic rotor for river water kinetic energy conversion into electrical energy (5-blade rotor diameter $D = 4$ m, submerged height of blade $h = 1.4$ m, length of blade chord $l = 1.3$ m). *1* – blade with hydrodynamic NACA 0016 profile; *2* – 5 blade rotor; *3* – planetary multiplier with multiplying ratio $i = 112$; *4* – permanent magnet generator (characteristics, see Sect. 4.5.3); *5* – plastic pontoons; *6* – guide; *7* – spatial housing

generator *3* (Fig. 4.168) is assembled coaxially with a planetary multiplier *1* through toroidal coupling *2* and housing *4* for taking over the reactiv moment of torsion.

Dependence of summary torque T_I applied to the 5-blade rotor shaft depending on water flow velocity ($V = 1.3$–1.8) m/s is presented in Fig. 4.169. Kinematic and lifting capacity calculations of all structural elements, including functional parameters and technical characteristics of micro hydropower plants have been

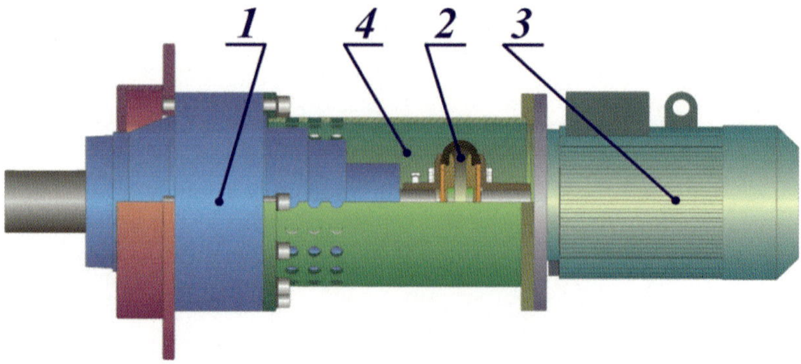

Fig. 4.168 Unit of low speed electric generator driving mechanism (MCHF D4 × 1.5E)

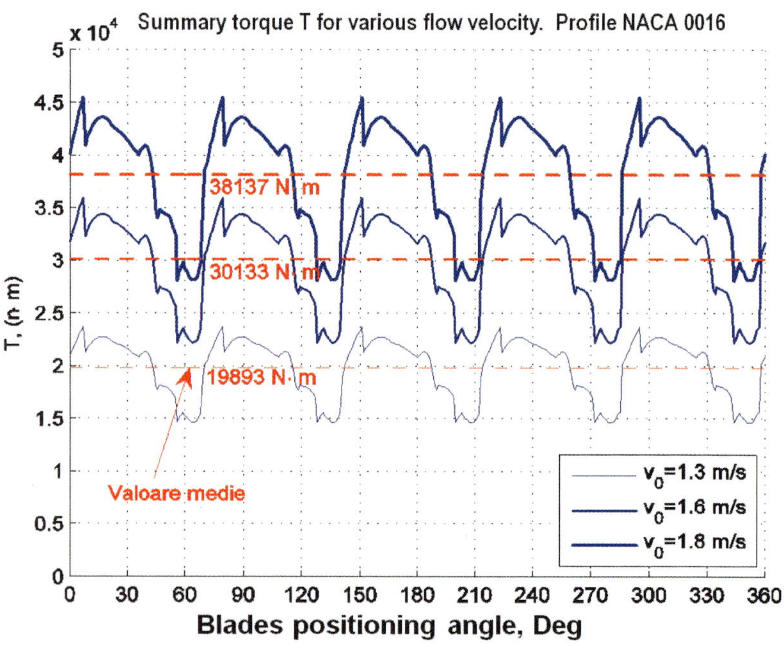

Fig. 4.169 Torque T_l at the shaft of 5-blade hydrodynamic rotor with NACA 0016 profile

carried out for the torque value $T_l = 19{,}893$ Nm, corresponding to the water flow velocity $V = 1.3$ m/s (maximum velocity specified for Prut, Dniester and Raut rivers).

Efficiency of the kinetic energy transmitted by the water flow to the hydraulic rotor can be considered (at the permanent magnet generator terminals):

$$\eta_{\sum} = \eta_1 \cdot \eta_r \cdot \eta_g = 0.9 \cdot 0.99 \cdot 0.87 = 0.775.$$

Fig. 4.170 Industrial prototypes of the micro hydropower plant installed on the Prut River

In conclusion, we state that micro hydropower plants ensures the transformation of 70–86% of the flowing water potential energy into useful electrical energy transmitted to the hydrodynamic rotor.

The industrial prototypes of the micro hydropower plants, developed by authors (Fig. 4.170), was demonstrated at the various International exhibitions [71–84, 97–105], where was appreciated by special prices and gold and silver medals.

References

1. Nakicénovic N, Grübler A, MacDonald A (eds) (1998) Global energy perspectives. IIASA, WEC, Cambridge University Press, Cambridge
2. WEC (1998) Survey of energy resources (18th Edition). World Energy Council, London
3. World Atlas & Industry Guide (2001) Int J Hydropower Dams April 2001
4. World Atlas and Industry Guide (2003) Int J Hydropower Dams. Aqua-Media International, Surrey
5. International Water Power and Dam Construction Venezuela country profile
6. International Water Power and Dam Construction Canada country profile
7. Egré D, Gagnon L, Milewski JC (1999) Are large hydro projects renewable and green? Int J Hydropower Dams Issue One
8. Einhorn B (2006) Blinding science: China's race to innovate. Business Week, 31 March 2006. Accessed 16 April 2006.
9. BBC News Online (2006) Three Gorges dam wall completed. 20 May 2006. Retrieved on 21 May

10. New Scientist report on greenhouse gas production by hydroelectric dams
11. ISO 1100-1 (1996) Measurement of liquid flow in open channels. Part 1: Establishment and operation of a gauging station
12. ISO/DIS 110-2 (1982) Measurement of liquid flow in open channels. Part 2: Determination of the stage-discharge relation (revision of ISO 1100-2: 1982)
13. www.nwl.ac.uk/ih/nrfa
14. Harvey A, Brown A (1993) Micro-hydro design manual. ITDG Publishing, London
15. Fraenkel P, Paish O, Bokalders V, Harvey A, Brown A (2003) Micro-hydro power: a guide for development workers. ITDG Publishing, IT Power, Stockholm Environment Institute, London
16. Siddiqui IH (1986) Water works and irrigation system in India during pre-Mughal times. J Econ Social Hist Orient 29(1):52–77
17. Pacey A (1991) Technology in world civilization: A thousand-year history. The MIT Press, Cambridge. Reprint edition (July 1, 1991). ISBN 0262660725.
18. Ingpen R, Wilkenson P (1968) Encyclopedia of ideas that changed the world. Viking Studio Books, New York
19. Notice sur la vie et les ouvrages du général J. V. Poncelet, par M. le général Didion. in Mémoires de l'Académie Nationale de Metz 1870 (50e année/1868–1869; 2e série) pp. 101–159
20. www.waterhistory.org
21. Norman J (2009) Production of electricity by micro hydro power plants. Booklet. ABS Alascan, Inc. www.scribd.com. Accessed 15 May 2010
22. European Small Hydropower Association (2004) Guide on how to develop a small hydropower plant. Thematic network on small hydropower (TNSHP). European small hydropower association – ESHA. Celso Penche 1998
23. de Siervo F, Lugaresi A (1978) Modern trends in selecting and designing Pelton turbines. Water Power & Dam Construction, December 1978
24. Lugaresi A, Massa A (1987) Designing Francis turbines: trends in the last decade. Water Power & Dam Construction, November 1987
25. Lugaresi A, Massa A (1988) Kaplan turbines: design trends in the last decade. Water Power & Dam Construction, May 1988
26. Schweiger F, Gregori J (1987) Developments in the design of Kaplan turbines. Water Power & Dam Construction, November 1987
27. Schweiger F, Gregory J (1989) Developments in the design of water turbines. Water Power & Dam Construction, May 1989
28. International Code for the field acceptance tests of hydraulic turbines (publication IEC 60041)
29. International Code for model acceptance tests (publication IEC 60193)
30. http://www.gepower.com/prod_serv/products/hydro/en/turbines/francis.htm
31. Durrand WF (1939) The Pelton water wheel. Stanford University, Mechanical Engineering, San Francisco
32. Thake J (2000) The micro-hydro Pelton turbine manual: design, manufacture and installation for smallscale hydropower. ITDG Publishing, London
33. www.smallhydropower.com/manual3.htm etc.
34. Curtis D (1999) Going with the flow: small-scale water power. CAT
35. Garman P (1986) Water current turbines: a fieldworker's guide. Intermediate Technology Publications, London, 144 pp
36. Gorlov AM (1995) The helical turbine: a new idea for low-head hydropower. Hydro Rev 14(5):44–50
37. http://www.bluenergy.com/public/technology/turbine.html, Davis Hydro Turbine Prototypes, 10 January 2007
38. Golovin et al. Patent Request No. 2247859 (RU). Submersible free flow micro power plant. I.Cl.: F03B13/00, 2005.03.10
39. Bostan I, Dulgheru V, Moroz N, Roşcovan G. Patent No. 1179 MD. Power unit. I.Cl.:F03 D 3/00. Publ. 30.01.1999, BOPI nr. 3/99

40. Jula A, Mogan Gh, Bostan I, Dulgheru V et al. (2006) ECOMECA – ECO-mechanical engineering (monograph). Braşov, Publ. House of "Transilvania" University, Braşov, p. 324.
41. Bostan I, Dulgheru V, Sobor I, Bostan V, Sochirean A (2007) Renewable energy conversion systems. Tehnica-Info, 592 pp.
42. Bostan I, Dulgheru V, Bostan V, Ciupercă R (2009) Anthology of inventions: renewable energy conversion systems Vol. 4. Ch.: Bons Offices SRL, 458 pp.
43. Bostan I, Gheorghe A, Dulgheru V, Bostan V, Sochireanu A, Dicusară I (2011) Conversion of renewable kinetic energy of water: synthesis, theoretical modeling, and experimental evaluation. Energy security: international and local issues, theoretical perspectives, and critical energy infrastructures (NATO science for peace and security series – C: Environmental security). Springer, pp. 125–177. ISBN 978-94-007-0718-4
44. Bostan I, Ţopa M, Sochireanu A, Vengher D, Ciupercă R, Trifan N, Dicusară I, Ciobanu O (2004) Micro hydro power station to use water kinetic energy. Phase I: "Estimation of river power potential (Nistru, Prut and Răut) and conceptual development of micro hydro power plant. Theoretical justification of rotor functional parameters and development of technical task". Report. Research supervisor: Bostan I, T.U.M., Chişinău, 62 p.
45. Bostan I, Bogdan V, Dulgheru V, Bostan N, Ciupercă R Patent 2288 (MD), CIB F03B7/00. Hydraulic Plant. T.U.M. – Nr. 2001 - 0301; Decl. 13. 09. 2001; Publ. BOPI - 2003 - Nr. 10
46. Bostan I, Dulgheru V, Ciupercă R, Ciobanu O, Ciobanu R Patent 2888 (MD), CIB F 03B7/00. Hydraulic Plant. T.U.M. – Nr. 2005 0067; Decl. 04 March 2005; Publ. BOPI - 2005 - Nr. 10
47. Bostan I, Sochireanu A, Ciupercă R, Poştaru Gh, Trifan N (2004) Helical turbine system for wind and hydraulic energy recovery. Grant CRDF MP2-3023, 2003. Research Report. Coordinators: Dulikravich George Stevo, professor, University of Arlington, Texas, USA; acad. Bostan I., T.U.M., Chişinău
48. Bostan I, Dulgheru V, Ciupercă R, Ciobanu R (2005) Simulation of fluid interaction with the working element blades for renewable energy conversion systems, using ANSYS CFX–5.7. ECODESIGN: National Workshop (2005; Braşov): National Workshop "Machine parts": Ecodesign: XXV edition: Braşov, 9–10 June 2005. "Transilvania" University Publ. House, Braşov, pp 119–126
49. Bostan I, Dulgheru V, Cartofeanu V, Ciupercă R, Ciobanu O Patent No. 2916 (MD), CIB F 03 B 7/00. Floating hydro power plant. T.U.M. – No. 2005 0065; Decl. 04.03. Publ. BOPI No. 11/2005
50. Bostan I, Dulgheru V, Ciupercă R, Ciobanu O Patent No. 2994 (MD), CIB F03 D 1/06. Helical wind turbine. T.U.M. – No. 2006 006; Decl. 30 December 2004; Publ. BOPI - 2006 - No. 2
51. Bostan I, Dulgheru V, Ciupercă R (2004) Helical turbine for wind systems and micro-hydro plant. Product engineering (Eco-design, technologies and green energy). Springer, pp 519–527
52. Bostan I, Dulgheru V, Bostan V, Ciobanu O, Sochireanu A Patent No. 2991(MD), CIB F03 B 7/00. Hydro power plant. T.U.M. – No. 2005 0136; Decl. 16. 05. 2005; Publ. BOPI - 2006 - No. 2
53. Bostan I, Dulgheru V, Sochireanu A, Bostan V, Ciobanu O, Ciobanu R Patent No. 2992 (MD), CIB F03 B 7/00. Hydraulic plant. U.T.M. – No. 2005 0270; Decl 15 September 2005; Publ. BOPI - 2006 - No. 2
54. Bostan I, Dulgheru V, Bostan V, Sochireanu A, Trifan N Patent No. 2993 (MD), CIB F03 B 7/00: F 03 B 13/00. Hydraulic turbine. T.U.M. – No. 2005 0272; Decl. 15 September 2005; Publ. BOPI - 2006 - No. 2
55. Bostan I, Dulgheru V, Bostan V Sochireanu A, Ciobanu O, Ciobanu R, Dicusară I Hydraulic plant. Patent No. 3104 (MD). BOPI No. 7/2006
56. Bostan I, Dulgheru V, Bostan V, Sochireanu A, Ciobanu O, Ciobanu R, Patent 3845 (MD), CIB F 03 B 13/00; F 03 B 7/00; F 03 B 13/10; F 03 B 13/22; F 03 B 17/06. Hydraulic station. U.T.M. Publ. BOPI - 2009 - Nr. 2
57. Bostan I, Dulgheru V, Bostan V, Ciupercă R Patent No. 2981 (MD), CIB B63 B 35/44: E02 B 17/00. Hydraulic plant. U.T.M. – No. 2005 0274; Decl. 15. 09. 2005; Publ. BOPI - 2006 - No. 2

58. Bostan I, Gheorghe A, Dulgheru V, Bostan V, Sochireanu A, Ciobanu O, Ciobanu R Patent 3846 (MD), CIB F 03 B 13/00; F 03 B 7/00; F 03 B 13/18; F 03 B 13/22; F 03 B 17/06. Hydraulic station with horizontal axle. U.T.M. Publ. BOPI - 2009 - Nr. 2
59. Moran I (1984) An introduction to theoretical and computational aerodynamics. Wiley, New York
60. Versteeg HK, Malalasekera W (1995) An introduction to computational fluid dynamics, the finite volume method. Longman, UK
61. Batcelor GK (1970) An introduction to fluid dynamics. Cambridge University Press, Cambridge
62. Bostan I, Dulgheru V, Bostan V, Sochireanu A (2005) Micro hydro power plant for river water kinetic energy conversion. Phase II: Research on working element functional parameters based on the experimental model. Calculation of bearing elements and constructive design of micro hydro power plant. (Experimental Processing). Report. Research supervisor: V. Dulgheru, Chişinău, 112 p
63. Bostan I, Bostan V, Ciobanu O Aspects of constructive and functional optimization of rotors with multiple blades with hydrodynamic profile from high power plants for energy conversion.Acta Technica Napocensis. Series: Applied mathematics and mechanics 50, vol. II. ISSN 1221-5872, pp 255–258
64. Bostan I, Bostan V, Dulgheru V (2006) Numerical modelling and simulation of the fluid flow action on rotor blades of the micro-hydropower station. Ovidius University Annual Scientific J. Mechanical engineering series, Volume VIII, Nr. 1, 19–21 May 2006, Ovidius University Press, Constanţa. ISSN 1223-7221, pp 70–78.
65. Bostan I, Bostan V, Dulgheru V (2007) Numerical simulation of the fluid flow interaction with hydrodynamic profile blades of the rotor of micro-hydropower station for river kinetic energy conversion. Pro-active partnership in creativity for the next generation. Proceedings of the 31st Annual Congress of the American Romanian Academy of Arts and Science. ARA Doval E. (Coord.) – Quebec, Canada. Presses Internationales Polytechnique, pp 59–62. ISBN 978-2-553-01412-3
66. Bostan I, Dulgheru V, Bostan V, Sochireanu A (2005) Conversion of renewable kinetic energy of water: synthesis, theoretical modelling and experimental evaluation. Phase I. Grant SCOPES IB7320 – 110902/1. Coordinators: Prof. Dr. Adrian Gheorghe (University ETHZ, Zurich, Switserland); acad. Bostan I, T.U.M. Intermediary Research Report. Phase II, T.U.M., Chişinău, 22 p
67. Bostan I, Dulgheru V, Bostan V, Sochireanu A, Ciobanu O, Ciobanu R, Dicusară I, Trifan N (2006) Floatable micro-hydro power station with adjustable hydrodynamic Blades. Geneve'2006. 4–8 April. Official Catalogue, p 47
68. Bostan I, Dulgheru V, Bostan V, Sochireanu A, Ciupercă R, Ciobanu O, Ciobanu R, Trifan N (2006) Micro-hydro power plant with multi-blade rotor with vertical axis and blades with hydro dynamic profile. International Fair of Inventions, Research and Technological Transfer INVENTICA 2006, Iaşi, 5–9 July, p 570
69. Bostan I, Dulgheru V, Bostan V, Sochireanu A, Ciobanu O (2006) Micro-hydro power plant with vertical rotor. International Fair of Inventions and New Technologies "Novii Ceas", 27–29.09 2006. Sevastopol. Catalogue, p 32
70. Bostan I, Bostan V, Sochireanu A, Trifan N, Ciobanu O, Ciobanu, R (2006) Micro-hydro power plant with vertical rotor. International Fair of Inventions and New Technologies INVENTIKA 2006, Bucharest, 3–7 October, 2006, p 94.
71. Bostan I, Bostan V, Sochireanu A, Trifan N, Ciobanu O, Ciobanu R (2006) Floatable micro-hydropower plant. Salon International des Inventions, Bruxelles' 2006. 22–26 November, p 57
72. Bostan I, Dulgheru V, Bostan V, Sochireanu A, Dicusară I, Ciupercă R, Ciobanu O, Ciobanu R, Cozma T, Trifan N, Vengher D (2005) Floatable micro-hydropower plant with adjustable hydrodynamic blades. Technological transfer in electronics, multifunctional materials and fine mechanics. International Workshop 17, IX, 2005. Î.E.P. Ştiinţa, Chişinău, pp 180, 188–193
73. Bostan I, Dulgheru V, Bostan V, Sochireanu A, Dicusară I, Ciupercă R, Ciobanu O, Ciobanu R, Cozma T, Trifan N, Vengher D (2005) Floatable micro-hydropower plant with adjustable hydrodynamic blades. Energetics of Moldova–2005. International Conference "Energetics of

Moldova-2005" 21–24 September, 2005. Republic of Moldova, Chişinău: Presentations. Publ. House, Moldovan Academy of Sciences, Chişinău, pp 604–608
74. Bostan I, Dulgheru V, Bostan V, Sochireanu A, Dicusară I, Ciupercă R, Ciobanu O, Ciobanu R, Cozma T, Trifan N, Vengher D (2005) Micro hydro power plant for river water kinetic energy conversion. Register of research developments in the field of electronics, multifunctional materials and fine mechanics. Moldovan Academy of Sciences; Agency for Innovations and Technological Transfer. E.P. Ştiinţa, Chişinău, p 54
75. Bostan I, Dulgheru V, Ciobanu O (2006) Constructive-technological aspects of multiblade rotor of micro hydro power plant. Technical University of Iasi, Vol. LII(LVI) Fasc. 5D. Section of Machine Building, Iaşi, p 1275–1279. ISSN: 1011-285
76. Bostan I, Dulgheru V, Bostan V, Sochireanu A, Ciupercă R, Ciobanu O, Ciobanu R, Trifan N (2006) Micro hydro power plant with multiblade rotor with vertical axis and hydrodynamic blades. XVIII International Conference of Inventions "Research and Advanced Technologies", Iaşi, 5–9 July 2006, p 529
77. Bostan I, Dulgheru V, Bostan V, Ciobanu O, Sochireanu A, Trifan Bostan I, Dulgheru V, Bostan V, Ciobanu O, Sochireanu A, Trifan N (2006) Hydroelectric station. The second meeting of the National Club "Science & Business" "Energy-saving Technologies and Alternative Energy Sources", 19 June 2006. Catalogue of Innovation in Energetics,Vol 2, Chişinău, pp 32–33
78. Bostan I, Moraru Gh, Ţopa M, Sochireanu A, Dicusară I, Ciupercă R, Ciobanu O (2006) Micro hydro power plant for river water kinetic energy conversion. Phase III: "Development of Manufacturing Technology for Working Element. Experimental Processing of Micro Hydro Power Plant and Testing in Real Conditions". Research Report. Supervisor: Dulgheru V, Chişinău, 65 p
79. Bostan I, Bostan V, Dulgheru V, Ciobanu O (2006) Design of micro hydro power plant with multiblade rotor. National Seminar on Machine Parts "Ioan Drăghici", University "Oil and Gas", Ploieşti, Romania,13–14 July 2006, pp 67–70. ISBN (10) 973-719-110-0
80. Bostan I, Dulgheru V, Bostan V, Ciobanu O (2006) Design of micro hydro power plant with multiblade rotor. VIII National Workshop "Computer Assisted Design PRASIC'06". BRAŞOV, November 9–10, 2006, Volume "Mechanisms". Tribology, pp 67–70
81. Bostan I, Dulgheru V, Bostan V, Sochireanu A, Ciobanu O, Trifan N (2006) Micro hydro power plants for river water kinetic energy conversion. International Fair "Vinahodi Innovacii", 10–13 April 2006, Kiev. Official Catalogue, p 68
82. Bostan I, Dulgheru V, Sochireanu A, Babaian I (2011) Anthology of inventions: Vol. 1. Planetary precessional transmissions, 593 p. ISBN 978-9975-4100-9-0.
83. Bostan I, Dulgheru V, Ciobanu R, Ciobanu O Patent No. 3153. Planetary precessional multiplier. Publ. BOPI - 2006 - No. 9
84. Bostan I, Ţopa M, Dulgheru V, Vaculenco M Patent No. 1886 B1 (MD), CIB F16H1/32; B23F9/06. Precessional gearing and procedure for its fulfilment. T.U.M. – No. 2000 – 071; Decl. 13. 10. 2000; Publ. BOPI - 2002 - No. 3
85. Bostan I, Cebotari D, Donţu V, Dulgheru V, Ciobanu O (2006) Elaboration of the low speed centrifugal pump. J "Meridian Ingineresc", Nr. 2, p 11–15
86. Electrical pump unit, PC series, KM type. Chişinău, 2006. Pentax Water Pumps. http://www.pentax-pumps.it/ita/html/index.html
87. Lomakin A (1966) Centrifugal and axial pumps, M-L, Mašinostroenie, Moscow
88. Mihajlov A, Malûšenko V (1997) Vane Pumps. M, Mašinostroenie, Moscow
89. Pflejderer K (1990). Vane-type engines for liquids and gases. Under the reduction of. V. I. Polikovckoo. M., Mašiz
90. Bostan I, Cebotari D, Donţu V, Dulgheru V, Ciobanu O, Kokidico A (2006) Low speed centrifugal pump. Nasosy & Oborudovanie, VNIIAEN, Ukraine, Nr. 2, pp 32–34
91. Bostan I, Dulgheru V, Bostan V (2006) Micro hydro power plant for river water kinetic energy conversion without dam construction: ecologically clean method to meet rural consumers demands. International Workshop on Socio-Economic Integration after Adhesion of Romania to European Union. 3–4 November 2006, Braşov, Romania, pp 34–38

92. Bostan I, Dulgheru V, Poştaru Gh, Ciupercă R, Trifan N (2008) Conversion of renewable kinetic energy of water: synthesis, theoretical modeling and experimental evaluation. Grant SCOPES IB7320 – 110902/1. Coordinator: prof.dr. A. Gheorghe (University ETHZ, Zurich, Elveţia); acad. Ion Bostan (TUM). Scientific reports, Chişinău, 2006–2008
93. Bostan I, Dulgheru V (2006) Some aspects of market study of renewable energy obtained from river water kinetic energy. Monthly J "Convorbiri economice". Braşov. ISSN1582-3555. No. 8, pp 32–34
94. Bostan I, Bostan V, Sochireanu A, Trifan N, Ciobanu O (2007) Micro hydro power plant for river water kinetic energy conversion without dam construction. Xth Moscow International Fair on Industrial Property. Catalogue, 27–30 March 2007, p 6
95. Bostan I, Dulgheru V, Bostan V, Ciobanu O (2007) Micro hydro power plant for river water kinetic energy conversion. International Fair of Inventions, Research and technological transfer ECOINVENT 2007. 30 May–02 June 2007. Official Catalogue, pp 27–28
96. Bostan I, Dulgheru V, Sochireanu A, Ciobanu O, Ciobanu R, Ciupercă R, Dicusară I (2005) Floatable micro hydro power plant with adjustable hydrodynamic profile blades: poster, industrial prototype of the hydraulic pump and generator. International Fair INFOINVENT 2005, 9–12 November 2005, Chişinău, p 38 (Gold Medal)
97. Bostan I, Greorghe A, Dulgheru V, Bostan V, Sochireanu A, Cartofeanu V, Ciobanu O, Ciobanu R, Dicusară I, Trifan N (2011) Floatable micro-hydropower station with self oriented hydrodynamic blades. Cluj-Napoca, 2011. The International Salon of Research and Innovations "PROINVENT' 2011". 14–18 March 2011. The DIPLOMA of EXCELLENCE and Gold medal
98. Bostan I, Dulgheru V, Sobor I, Bostan V, Sochireanu A, Crudu R, Guţu M, Ciobanu O, Ciobanu R, Trifan N (2010) Industrial prototype of mini hidropower station for flow water kinetic energy conversion. Salon des Inventions, Geneva, Palexpo, 6 au 10 Avril 2010 (Silver medal)
99. Bostan I, Dulgheru V, Sobor I, Bostan V, Sochireanu A, Crudu R, Guţu M, Ciobanu O, Ciobanu R, Trifan N (2010) Industrial prototype of mini hidropower station for flow water kinetic energy conversion. XIIIth Moskow International Salon of Research and Innovations ARHIMED-2010. 30 March–02 April 2010 (Gold medal)
100. Bostan I, Dulgheru V, Bostan V, Ciobanu O, Ciobanu R, Sochireanu A, Dicusară I, Trifan N (2010) Industrial prototype of mini hidropower station for flow water kinetic energy conversion into electrica land mechanical energy. EUROINVENT-European Exhibition of Creativity and Innovation, Iaşi, România. 07–09 May 2010 (Gold medal, Special prise of Inventors Association of Zagreb)
101. Bostan I, Dulgheru V, Bostan V, Sochireanu A, Ciobanu O, Ciobanu R, Dicusară I, Trifan N (2010) Industrial prototype of mini hidropower station for flow water kinetic energy conversion into electrica land mechanical energy. International Salon of Research and Innovations, INVENTICA 2010, XIVth edition, 9–11 June 2010 (Gold medal)
102. Bostan I, Dulgheru V, Bostan V, Sochireanu A, Ciobanu O, Ciobanu R, Dicusară I, Trifan N (2009) Industrial prototype of mini-hydropower station for flow water kinetic energy conversion. EURECA 2009, Bruxel (Gold medal)
103. Bostan I, Dulgheru V, Bostan V, Sochirean A, Ciobanu O, Ciobanu R, Trifan N (2009) Floatable micro-hydropower station. PRIZE ENVIRONMENT PROTECTION. "EUROINVENT'2009", Iaşi, 9/05/2009
104. Bostan I, Dulgheru V, Bostan V, Sochirean A, Ciobanu O, Ciobanu R, Trifan N (2009) Floatable micro-hydropower station with adjustable blades "EUROINVENT'2010", Iaşi, 8/05/2009 (Gold medal)
105. Bostan I, Vişa I, Dulgheru V, Bostan V, Sochireanu A, Ciobanu O, Trifan N (2009) Micro-hydropower station for the rivers water kinetic energy conversion. Cluj-Napoca, 2009. The International Salon of Research and Innovations "PROINVENT' 2009". The DIPLOMA of EXCELLENCE and PROINVENT medal

Chapter 5
Wind Energy

5.1 Wind as Energy Source

Wind energy has been used by mankind over thousands of years. For over 3,000 years the windmills have been used for pumping water or grinding (milling). And nowadays, in the century of information technologies, nuclear energy and electricity, thousands of windmills are used for pumping water and oil, for irrigation and production of mechanical energy to drive low-power mechanisms on different continents.

Electricity can be obtained using different methods, but absolutely all require fuel, in most cases fossil fuels – coal, natural gas, oil or uranium 235 and plutonium 239 in thermonuclear plants. By burning or nuclear fission the primary energy embedded in the fuel is converted into thermal energy (heat). The turbine, designed specifically for each type of fuel, drives the generator that produces electricity. In this context, the electricity produced from the wind is not differing from the electricity produced from fossil or nuclear fuels. The wind as a fuel is essentially different – it is free of charge and does not pollute the environment.

Nowadays, the phrase *use of wind energy* means, primarily non-pollutant electrical energy produced at a significant scale by modern *windmills* called *wind turbines*, a term that attempts to outline their similarity to steam or gas turbines, which are used for producing electricity, and also to make a distinction between their old and new destination.

The attempts to obtain electricity from the wind date back over a hundred years since the late nineteenth century. But a true flourishing of this technology is registered only after the 1973 oil crisis. An unexpected increase in oil prices has forced the governments of developed countries to allocate substantial financial resources for research, development and demonstration programmes. Over 20 years, worldwide, a new technology, a new industry and, in fact, a new market – the market of the Wind Energy Conversion Systems (WECS), have been developed.

If in 1973 the main incentive for the development of WECS was the oil price, today another incentive is added – the tendency of mankind to produce "clean" or

I. Bostan et al., *Resilient Energy Systems*, Topics in Safety, Risk, Reliability and Quality 19, DOI 10.1007/978-94-007-4189-8_5,

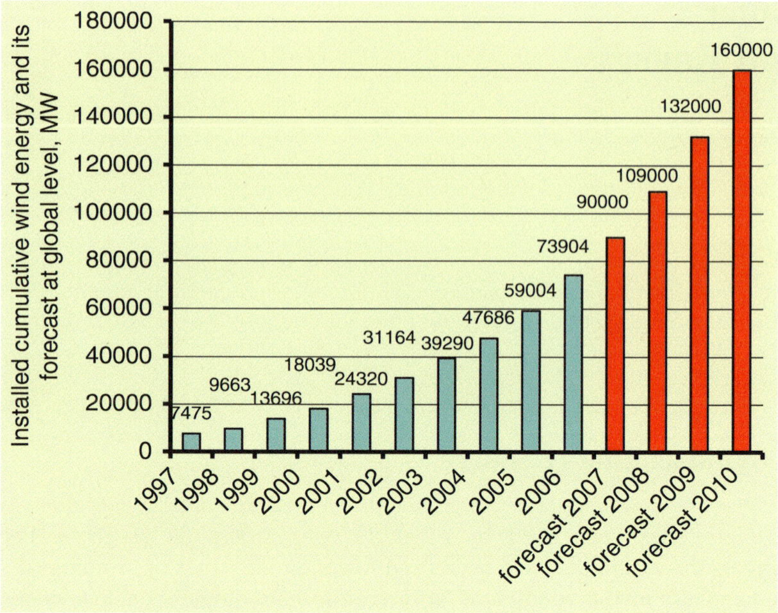

Fig. 5.1 Installed cumulative wind energy and its forecast at global level

"green" electricity with little or no carbon monoxide emissions. The year 1993 was marked as the beginning of a wind boom characterized by an annual increase of over 20% of installed power capacity. Thus, in 1999 the global capacity increased by 4,033 MW, which was a record for the wind energy sector. It is rather significant as it exceeded the world installed nuclear power capacity in the same year, for the first time [1–3]. In the period 1996–2006 the global capacity has increased more than 12 times and has reached 73,904 MW in 2006 (Fig. 5.1) [4, 5].

The undisputed global leader is the European Community EU-27 with a 65% share, followed by the USA and India (Fig. 5.2). Such a spectacular development knows no other global industry sector worldwide. In the years 2007–2010 an annual growth over 21% was expected; for 2010 the global installed capacity should have reached 160,000 MW. With the launch of the European Technology Platform on wind energy issues the EU Commissioner A. Piebalgs said [7]: *"Wind energy technology is certainly one of the fastest growing and plays an important role, contributing to create a sustainable and competitive energy policy in Europe"*. In 2005, the EU countries produced about $69.5 \cdot 10^6$ MWh of wind electrical energy or more than 26 times the electricity consumption of Moldova in 2005. The wind provides electricity to more than 35 million households in the EU, but very few know it – a symptom which indicates a lack of knowledge about this technology. Globally, by 2020 about 12% of the produced electricity will be of wind origin. Table 5.1 lists the most advanced five countries and five global companies in the field of the wind energy.

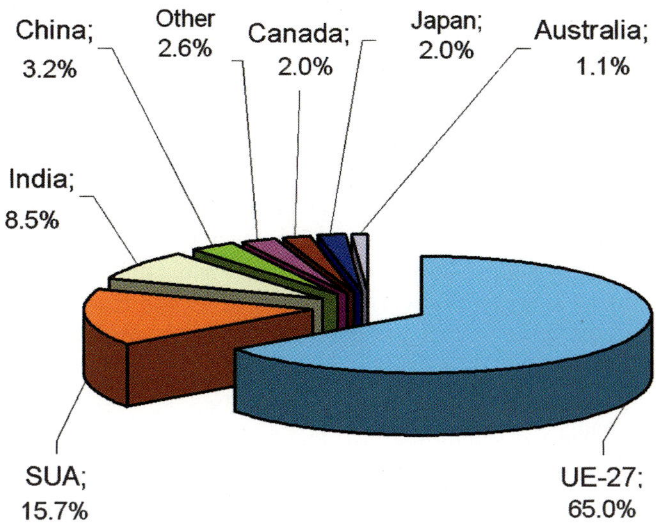

Fig. 5.2 Distribution of installed wind power at global level [4, 6]

Table 5.1 World most advanced countries and companies in the wind technology

Country	Installed wind power, MW	World share, %	Company	Wind market share, %
Germany	20,622	27.9	Vestas, Denmark	27.4
Spain	11,615	15.7	Gamesa, Spain	15.5
USA	11,603	15.7	GE Wind, USA	15.3
India	6,270	8.5	Enercon, Germany	14.5
Denmark	3,136	4.2	Suzlon, India	7.5

To know the whole technology of the wind energy conversion into electricity, knowledge in various fields is required, including meteorology, aerodynamics, electrical engineering, mechanical engineering and civil engineering. In order to make a correct decision for investment it is necessary to have knowledge of the economic analysis of projects.

The following chapter describes the principle of converting the kinetic energy of an air stream into mechanical energy, the formulas for calculating the airflow power and the technical limitations are presented, which reduce the conversion efficiency. Then, a review of the wind technology developments, the major construction schemes, modern trends in the construction of wind turbines and the use of materials for blades are presented. The technical characteristics of high power turbines are presented, including the turbines suitable for use in the wind conditions of Moldova and the steps needed to be done at the initial phase of a wind power plant construction. The last paragraph describes low-power turbines, including the elaborations drawn by the authors and examples of their rational application.

5.2 Estimation of the Wind Energy Resources

5.2.1 Characteristics and Parameters of the Wind Energy

Wind serves as *fuel* for wind power plants. Taking into account that the wind power density (see Sect. 5.4.1) is proportional to the cube of the wind speed it is very important to know the wind energy resources over the whole country, the wind sources of the region and the site where possibly the wind power plant will be built. Typically, the wind energy resources are explained by two main characteristics of the wind – the wind speed and its power density, which determines the wind energy potential of the location.

For policy makers at central level, it is important to know the wind energy resources for strategic planning; in this regard the following questions should be answered [8]:

- What are the wind energy resources and how are they spread across the regions?
- What is the share from the total electricity consumption that can be covered by the wind energy?
- How can we exploit this potential?

At the local level or for an investor in the wind energy at the initial stage of implementing a project, it is important to know the answers to the following questions:

- What is the wind energy potential on a certain site?
- How much electricity will be produced in 1 year by a turbine with specified technical characteristics?
- What will the cost price of wind power be?
- What is the period of return on investments?
- What is the annual and diurnal variation of the wind speed, and respectively, of the wind power density?

The correct answers can be obtained as a result of measurements of wind characteristics at the given site at the height of the turbine axis of rotation for a minimum of 1 year period. But this way is expensive and requires a long period of time.

Countries with a high degree of wind energy use have chosen a different way – computer modelling of the wind speed for large areas, using special software that would consider the orography and the land surface characteristics, the obstacles, etc. In these models, the so-called historical wind data is applied collected from the weather stations across the country or region. As a result, the Wind Atlas (WA) is developed, which includes information about the speed and wind power density in the form of a contour or graded map. The Wind Atlas can be produced at a global, country or region level, but it does not substitute the need for instrumental measurements; it only identifies the region where to focus investigations and to indicate the location where necessary to perform the measurements.

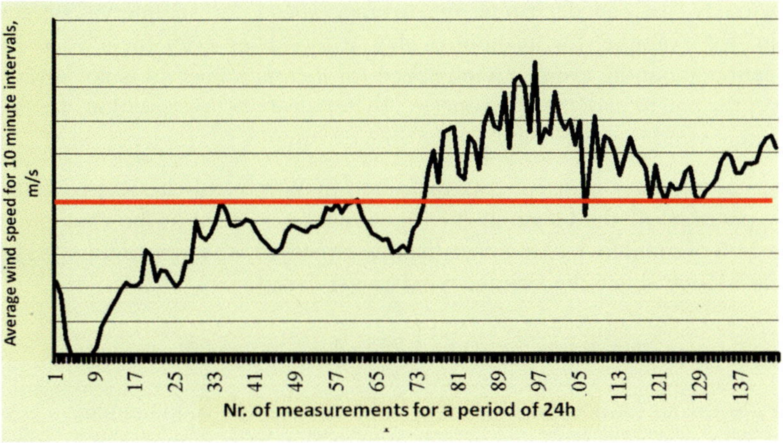

Fig. 5.3 An example of wind variation for 24 h

At the next phase of investigations, a virtual wind turbine with known technical characteristics may be located in a certain geographical point and, using the WA data, it can determine the amount of electricity that can be produced in a certain period – a month or a year, etc. Obviously, there are constraints that limit or make it difficult to use mathematical models to estimate the wind energy resources. First of all, we refer to the availability of reliable primary data on wind and digital topographic maps for the scale required. No less important is the availability of wind data measurement characteristics – speed and direction, carried out at least 50 m height above the ground in order to validate the results obtained by calculation.

The *wind speed and direction* are the main characteristics of wind for a certain location. At meteorological stations the wind speed is measured by cup anemometers, which are also fitted with wind vanes to determine the wind direction. According to the standards, the wind speeds have been obtained as a result of every 3 h records, respectively, at 0, 3, 6, 9, 12, 15, 18, 21 o'clock. The wind speed for each 3 h interval is considered the average fixed velocity for an interval of 10 min, i.e. between 0^{00}–0^{10}, 3^{00}–3^{10}, etc. The weather stations measure wind characteristics at a height of 10–12 m above the ground.

The wind is characterized by a pronounced change in both its speed and direction and, in order to obtain accurate information, the primary data for a minimum period of 10 years is necessary. Figure 5.3 shows an example of the variation in the wind speed over a period of 24 h, at a height of 50 m above the ground. The diagram shows the processed results of the wind speed collected every 3 s. As a result of the speed measurement, the average speed over a period of time equal to 10 min (the arithmetic average of 200 measurements) is considered. Thus, we will have 144 results in 24 h. We state that the average wind speed for 10 min intervals varies in 24 h from 0 to 8.71 m/s.

Obviously, we can determine the average speed for an interval bigger than 10 min, for example, for an hour, a day, a month or even for a year. But the information about the average wind speed for a certain interval is not sufficient to consider the potential of the wind energy. To demonstrate this assertion, we calculate the average wind power density for the above example, i.e. for a period of 24 h.

Average wind power density. It is measured in W/m^2 and characterizes the wind energy potential of the location. Average arithmetic speed in the example above, within 24 h is equal to 4.49 m/s (see the line *parallel* with x-axis in Fig. 5.3). Using formula 5.9, we obtain the average wind power density

$$p = 0.5\rho V^3 = 0.5 \cdot 1.225 \cdot 4.49^3 = 54.5\,\text{W/m}^2$$

where ρ is the air (atmospheric) density; V is the average wind velocity.

But the wind velocity is a variate and it should be characterised in terms of probability theory. Such a term is the probability density function of the wind velocity, $F(V)$, which is defined as the fraction of time for which the average wind velocity falls within a specified interval ΔV_i. In other words, the probability density function of the wind velocity characterizes the share of velocity in the range of V_{min} and V_{max} obtained during the measurements. To determine the probability density function of the wind velocity for the example above (Fig. 5.3), let us proceed as follows:

- The velocity variation range during the measurements is determined. In our case $V_{min} = 0.0$ and $V_{max} = 8.71$ m/s;
- The velocity variation range is divided into n equal intervals, usually between 0.1 and 1.0 m/s. It was chosen $\Delta V_i = 1.0$ m/s. The speed of calculation for each period is equal to the average velocity. For example, within the interval 6 velocities between 5 and 6 m/s are enclosed, the average speed of calculation is considered equal to 5.5 m/s;
- 144 measurements are scanned and the number of measurements n_i is determined falling in each interval;
- Function $F(V) = n_i/N \cdot \Delta V_i$ is determined.

The obtained results have been included in Fig. 5.4 (the histogram or the bar chart).

By considering the probabilistic nature of the wind velocity, the power density is calculated by formula

$$p = 0.5\rho \sum_{i=1}^{9} V_i^3 F(V_i) = 83.9\text{W/m}^2 \tag{5.1}$$

and is 54% higher than the power density calculated above, using only the average wind velocity for 24 h. Graphical interpretation of results is shown in Fig. 5.5. The largest rate of power density belongs to the velocity range between 6 and 7 m/s and is 35.5%. However, the highest rate of velocity belongs to the speed range between

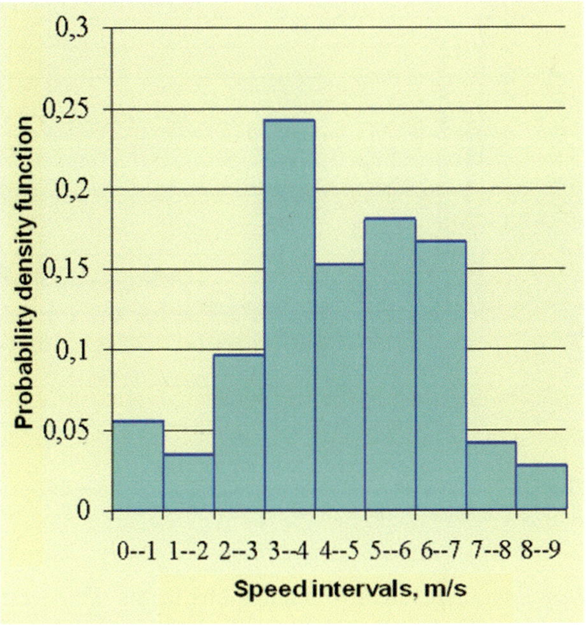

Fig. 5.4 Velocity probability density function

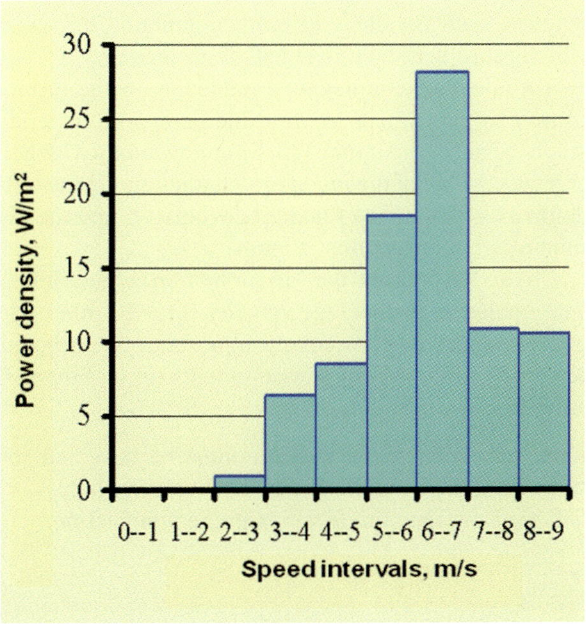

Fig. 5.5 Wind power density

3 and 4 m/s (see Fig. 5.4). Modern turbine start-up speed is equal to or bigger than 4 m/s. From Fig. 5.4 it results that, for the analyzed period (24 h), the lucrative velocities duration ($\geq$4 m/s) is about 60%.

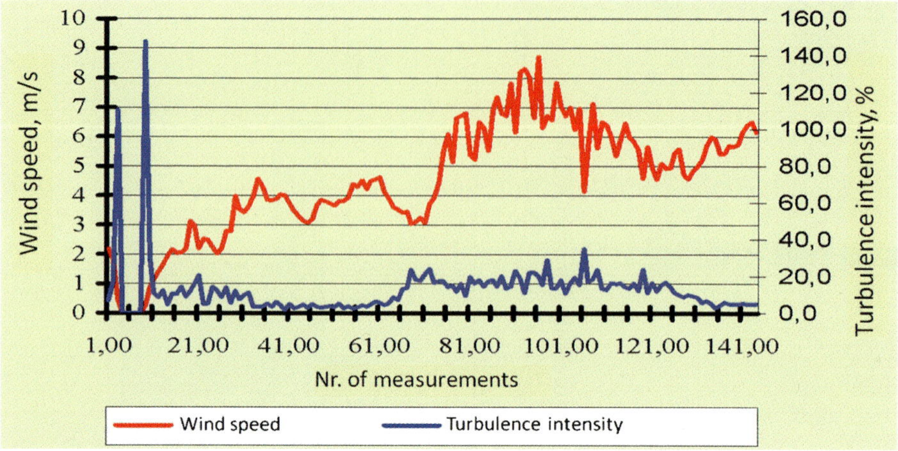

Fig. 5.6 Variation of wind velocity and turbulence intensity: period of observations – 24 h; number of measurements —144

Turbulence. It refers to fluctuations in the wind velocities over a short period of time, usually less than 10 min. Turbulence is caused by two phenomena: first, the friction between the airflow and the Earth surface, often magnified by topographical features, such as valleys, hills and mountains; the second is related to thermal effects which cause vertical movement of air masses.

Turbulence has a negative influence on the turbine rotor, as mechanical stress caused by short gusts of wind increases, the material from which the propeller is made exhausts and it may fail. Simultaneously with height increasing, the turbulence is reduced. One of the indicators characterizing the turbulence is turbulence intensity defined as the ratio of standard deviation σ and the average velocity for a period of time equal to or less than 10 min.

Figure 5.6 presents the variation of turbulence intensity for a period equal to 24 h. The smaller is the average velocity for a 10 min period, the bigger is the intensity of turbulence. To draw conclusions about the turbulence it is necessary to have the results of wind velocity measurements for less than 10 min for periods of 10 years, at least.

Extreme winds. Wind turbines must be designed so as to withstand the extreme winds or gusts of wind. If the wind velocity is bigger than 25 m/s the wind turbine is broken or taken out from under the wind action.

5.2.2 Methodology of Wind Power Potential Estimation

To calculate the average wind velocity, power density, wind rose and the probability density function of the wind speed, and, further on, to estimate the wind energy potential, two models are currently used: the model developed by the EU countries, known as WAsP (*Wind Atlas Analysis and Application Program*) [10, 11], on which basis the European Wind Atlas [6] for 15 EU countries was developed; this model is based on the theory of air currents, and the American model, developed by NASA and U.S. Air Force, based on the theory of climate dynamics [12].

The American model has been recently developed; it has many possibilities, including modelling of wind climatology in the mountainous territory. The MesoMap soft requires a huge capacity of the computer network – four Cray C90 supercomputers. For this reason, the soft is not sold, and the *AWS truewind* company offers only services in the field.

High efficiency of the WASP soft, the optimum ratio *price/quality* has induced several countries in the Central and Eastern Europe, including the ten new EU members, to use it for the assessment of the wind potential to draw their own atlases similar to the European approach.

In the framework of some projects studies on wind climatology statistics were carried out by authors, the Wind Atlases for ten meteorological stations (out of the existing 17) were developed, measurements of the wind characteristics have been performed at a height of 50 m above the ground at three sites located in the south and central regions of Moldova, and the wind energy potential in the south of the country was estimated, as well. The results of the study were published by authors in [13–18]. These studies were based on the methodology accepted in the EU countries and on the WAsP software.

WAsP soft allows two operating modes:

1. Analysis of primary data about the wind in order to obtain the Wind Atlas for each weather station (observation point) separately.
2. Using the Wind Atlas and the power curves of wind turbines in order to assess the wind energy potential at any point in a radius of less than 50 km from the point where measurements were made.

To get the Wind Atlas of the site where the meteorological stations are located, the following initial information is required:

- Primary data on the wind for a period of not less than 10 years;
- Description of the meteorological station site by pointing out the roughness of surroundings and existing obstacles in the immediate vicinity of the measuring device;
- Digital map of the region.

5.3 Statistics of Wind Climatology and the Wind Atlas

Statistics of wind climatology is presented graphically by interpreting the wind direction (the wind rose) and the distribution of the wind speed probability density function (histogram obtained using the methodology described in Sect. 5.2.1) and

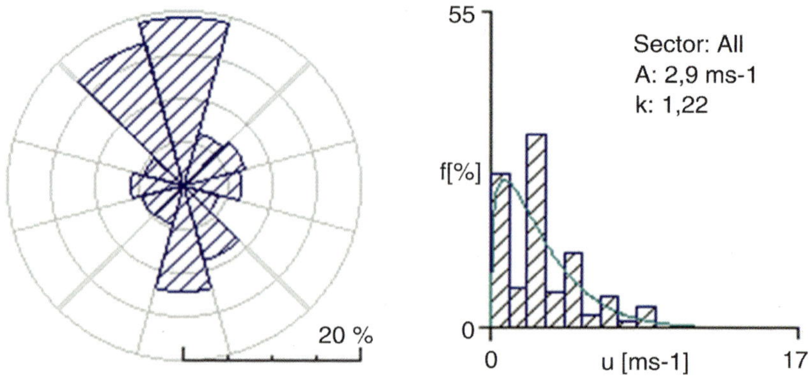

Wind rose		*Share of wind velocity for each sector, in %*											Weibull distribution	
		Wind velocity, m/s												
Sec. [°]	**%**	**1**	**2**	**3**	**4**	**5**	**6**	**7**	**8**	**9**	**10**	**11**	**A**	**k**
0	19.2	12	5	43	6	17	3	6	1	5	1	1	3.5	1.4
30	6.2	36	11	21	12	9	4	3	1	2	1	0	2.8	1.3
60	7.0	32	6	32	5	14	1	4	0	5	0	1	2.9	1.2
90	6.3	36	6	36	4	9	1	3	1	4	0	0	2.8	1.4
120	4.1	55	7	14	7	7	3	4	2	1	0	0	2.1	1.0
150	8.8	25	5	33	6	17	3	7	2	2	0	0	3.0	1.4
180	12.0	19	6	41	6	14	3	7	1	3	0	0	2.8	1.2
210	4.5	50	16	18	6	4	2	2	1	1	0	0	1.8	1.1
240	4.8	46	11	26	3	7	2	3	0	2	0	0	2.2	1.2
270	5.8	37	8	35	3	9	1	3	0	3	0	1	2.7	1.3
300	4.5	49	11	20	6	7	3	2	1	1	0	0	2.1	1.1
330	16.8	13	4	36	7	18	3	9	2	7	0	1	4.0	1.52
Total		**27**	**7**	**33**	**6**	**13**	**3**	**5**	**1**	**4**	**1**	**0**	**5.3**	**2.9**

Fig. 5.7 Example of wind climatology statistics at one of the meteorological stations

in tabular form – the rate of the wind velocity for each sector. Figure 5.7 shows an example of wind climatology statistics at one of the meteorological stations [6].

The distribution of wind speed probability density function, i.e. the histograms are approximated by the Weibull distribution function [6]

$$F(V) = \frac{k}{A}\left(\frac{V}{A}\right)^{k-1} \exp\left(-\left(\frac{V}{A}\right)^{k}\right) \tag{5.2}$$

where A and k are Weibull distribution parameters, A is the scale parameter, and k – is the shape parameter.

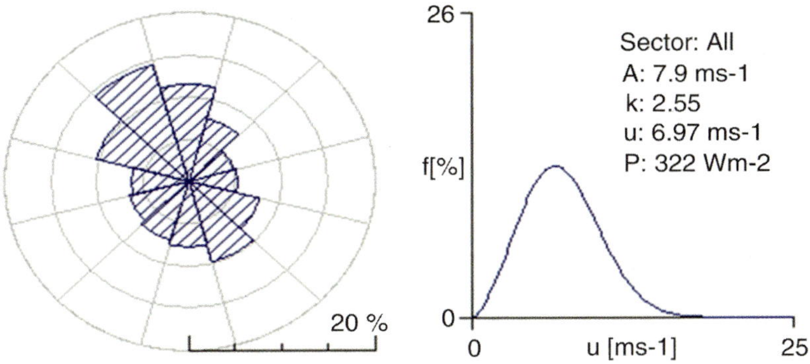

Height, m		Roughness class			
		R-0	R-1	R-2	R-3
10	m/s	7.0	4.8	4.2	3.4
	W/m^2	322	125	83	40
25	m/s	7.6	5.8	5.2	4.4
	W/m^2	414	201	147	88
50	m/s	8.2	6.7	6.1	5.3
	W/m^2	505	285	220	145
100	m/s	8.9	8.0	7.3	6.4
	W/m^2	656	458	352	235
200	m/s	9.8	9.9	9.0	7.9
	W/m^2	920	909	678	438

Fig. 5.8 An example of Wind Atlas for a site

By applying the Weibull parameters the average annual wind velocities at each meteorological station were estimated

$$\overline{V} = A \cdot G\left(1 + \frac{1}{k}\right) \tag{5.3}$$

where *G(1 + 1/k)* is the second Eulerian integral, which values are given in [6].

The Wind Atlas (WA). In the context of the wind energy sector, the concept of the Wind Atlas (Fig. 5.8) has a much broader content [8]. The Wind Atlas contains not only maps, graphs, images, which are characteristic for all common atlases, but, first of all, primarily numerical data in tabular form on the wind speed and power density in W/m^2. WA is developed both for presenting data on wind energy resources in a given area (weather station), and in order to provide these data to estimate the wind energy potential in the surrounding region. Its purpose is also to identify places where wind potential is stronger.

To compile the Wind Atlas the following is needed: statistics of wind climatology at meteorological stations, numerical description of the land around the anemometer's site with account of roughness and obstacles and the digital map. The table presents the annual average wind speeds and wind power densities for five predefined heights (10, 25, 50, 100 and 200 m above the ground level) and four roughness classes reported to standard conditions: 0.0; 0.03, 0.1 and 0.4 m. In the upper left side, the wind rose is given, and on the right – the average parameters (for all sectors) A and k of the Weibull distribution and probability density function of the wind speed. Both the wind and Weibull distribution can be obtained for each of the 12 sectors and the average for all sectors.

It should be noted that WA refers to a certain point – the station where measurements were made. Based on these data, we can calculate the same characteristics for any location in the neighbourhood, in the radius of 50 km. It is obvious that for a new site, where the weather data was not recorded, it is necessary to have information about the obstacles, roughness, and relief character. This information is obtained as a result of field investigations, which are processed and adapted to the WASP software requirements.

There is an essential difference between the information contained in the Wind Atlas and the statistics of the wind climatology presented above. The figures in the table mean the average wind speed and wind power density for various heights and different roughness classes. The figures in the first column are the speeds and power densities that would have occurred in the given location if the roughness of the terrain and obstacles would not overshadow the measuring device. In other words, these are data "*cleaned*" out of the negative influence of terrain peculiarities. For example, the R-0 roughness class corresponds to a stretch of calm water level. Moving us into another geographical point, for example, in the location of another wind plant, using the WA data calculated for the meteorological station nearby and the data describing the new location, the average wind speed and wind power density for the five heights and the four standard classes of roughness are calculated.

Methods used to validate the European Wind Atlas (EWA). Measuring the wind characteristics for a minimum of 1 year period is expensive. The historical data about the wind can be used to calculate the wind speed and wind direction at a point of interest where no measurements were made. Obviously, the question is: what is the certainty of the calculated results? To answer this question, various validation procedures have been used, taking into account both the methods used to validate EWA and the information available in the Republic of Moldova. EWA validation was performed using the following methods [6, 18]:

1. By mutual comparison of the results of measurements and calculations of the average wind speed and Weibull parameters. Mutual comparison procedures include the following: one of meteorological stations is the forecast weather station, e.g. for example the SM1 station in Fig. 5.9. The average speed and Weibull parameters for the forecast station are calculated using the historical data from other stations in the same area – SM2, SM3 and SM4, called forecasting stations. The calculations are repeated four times, pointing in succession as the forecast station another station;
2. The second method consists is the use of the measurements at heights of over 50 m and of wind characteristics for a minimum period of 12 months. In many

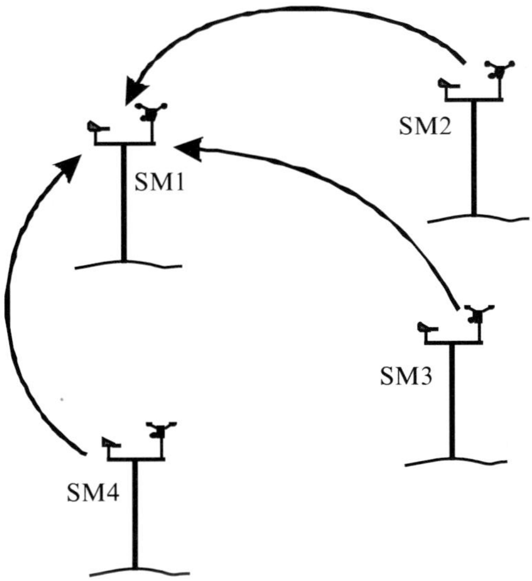

Fig. 5.9 Principle of validation by reciprocal comparison of meteo stations: SM1 – forecast meteorological station; SM2, SM3, SM4 – forecasting meteorological stations

Western European countries there are data on wind characteristics measured at heights of several tens and even hundreds of meters: the Netherlands – 200 m; Portugal – 100 m; Finland – 220 m; Sweden – 145 m. In this case, the validation of results estimated by calculation is performed as follows (see Fig. 5.10): for each tower where measurements were performed on a minimum of 12 months at different heights data are collected corresponding to heights *H1* equal to 10–12 m. These data are used as reference data for calculations. The average speed, Weibull parameters and density of wind power are calculated for heights *H2* and *H3* comprised between 50 and 200 m. The calculated results are compared with those measured.

The principle of validation (Fig. 5.11) consists in comparing the results of calculations of wind characteristics for the same point, using historical data as initial data from weather stations and recent data measurements. For a certain location, where no measurements of wind characteristics were made, the wind power density can be predicted with a relative error of 2–12% and the average annual speed – by 1.0–7.5%. The error is lower if the reference stations (forecasting) are located in open spaces and are not screened. If the forecasting stations are heavily shielded or are located in hilly areas, then the prediction error is larger.

The last validation procedure consists in comparing the annual average speed measured in several locations with the speeds calculated in the same locations at the same height, using the historical data from the nearest meteorological station.

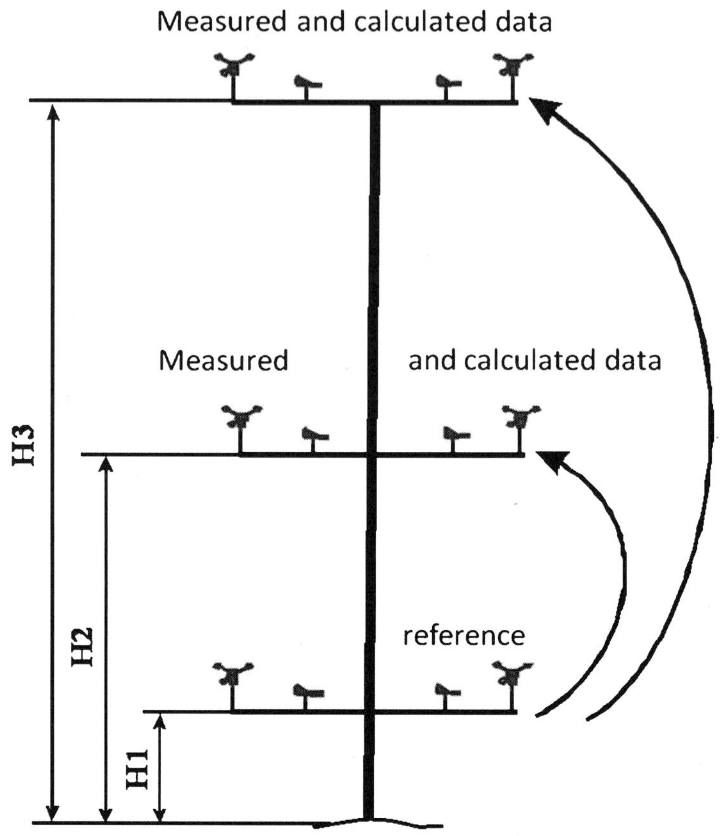

Fig. 5.10 Principle of validation by measuring at various heights

5.4 Conversion of the Air Flow Kinetic Energy into Mechanical Energy: Betz Limit

5.4.1 Wind Energy and Power

The energy of a stream of air moving at a linear velocity V is determined by the kinetic energy expression

$$E = m\frac{V^2}{2} \tag{5.4}$$

where m is the air mass in motion, determined by the air density ρ and the volume that crosses a certain surface S in a time unit

$$m = \rho S V \tag{5.5}$$

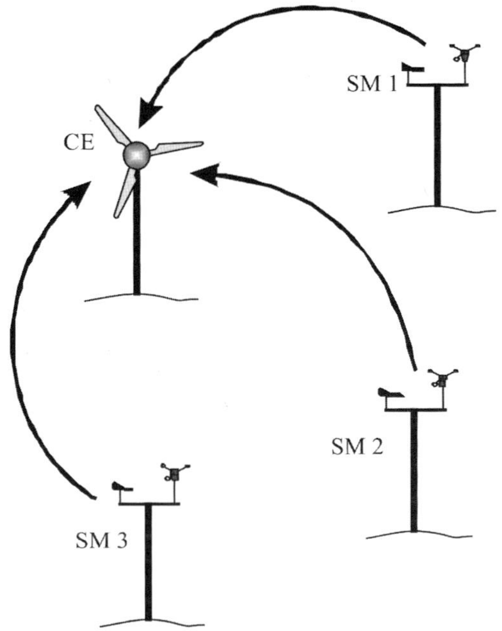

Fig. 5.11 Principle of validation by comparing the results for a wind plant located in a certain place: CE – wind plant; SM1, SM2, SM3 – historical or recent data at meteorological stations

The mass measure unit from expression (5.5) is kg/s and, substituting in Eq. 5.4, we will get the air flow power in watts

$$P = \frac{\rho}{2} S V^3 \tag{5.6}$$

The wind specific capacity or wind power density per 1 m^2 is

$$p = 0,5\rho V^3 \tag{5.7}$$

At normal atmospheric pressure and 15°C temperature, the air density is 1,225 kg/m^3. If the height above the sea level varies between 0 and 100 m (modern high-power turbine towers have heights of 60–120 m), the density variation does not exceed 5% and in the first approximation, we consider it constant. Figure 5.12 presents the specific capacity variation of a flow of air in terms of its speed.

The rated wind speed for modern high power turbines varies between 12.0 and 15.0 m/s (see the shaded area).

On the basis of expressions (5.6) and (5.7) the following conclusions can be drawn:

1. Formula (5.5) show the energy potential of a stream of air that crosses the surface S or a square meter of surface;
2. Double increase of the turbine rotor diameter will lead to an increase of four times of the air flow power that passes through the rotor surface;

Fig. 5.12 Variation of air flow specific capacity

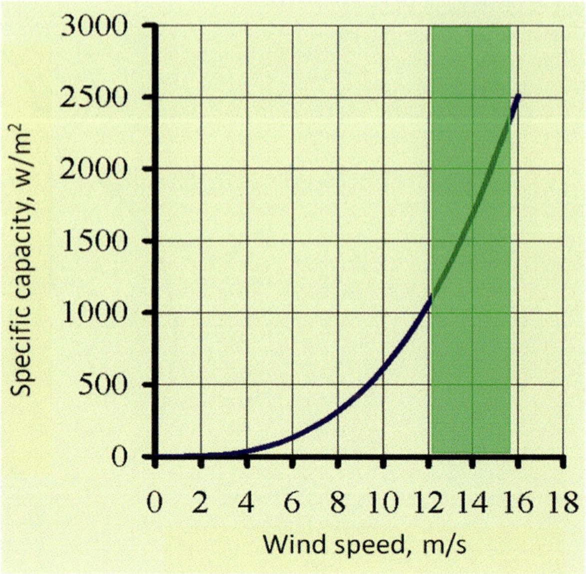

3. Double increase in the wind speed leads to eight times increase in the air flow power or specific power;
4. It is very important to know the value of the wind speed and how it varies over time in order to forecast exactly the wind energy potential in a certain locality.
5. It takes considerable effort to get the confidence that the wind station will be located in a locality with the highest wind speeds. In some countries, relatively high towers (more than 60–80 m) are exploited, to take advantage related to speed increase with increasing height.

It is obvious that not all air flow power expressed by the formula (5.6), will be transformed into mechanical energy and then into electricity by the wind turbine. A considerable part of energy will be preserved in the air flow leaving the area adjacent to the turbine; otherwise the last will not work.

5.4.2 Wind Turbine in the Air Flow

The wind turbine converts the kinetic energy of the air flow that passes through an area swept out by the rotor into mechanical energy, then using the generator – into electricity. The question arises: what happens by placing the turbine rotor in an air flow? It is obvious that the air flow gives up only part of the kinetic energy (see the next paragraph), the remaining energy is consumed so as the air leaves the area of flow – turbine interaction. Figure 5.13 shows an airflow with the initial speed

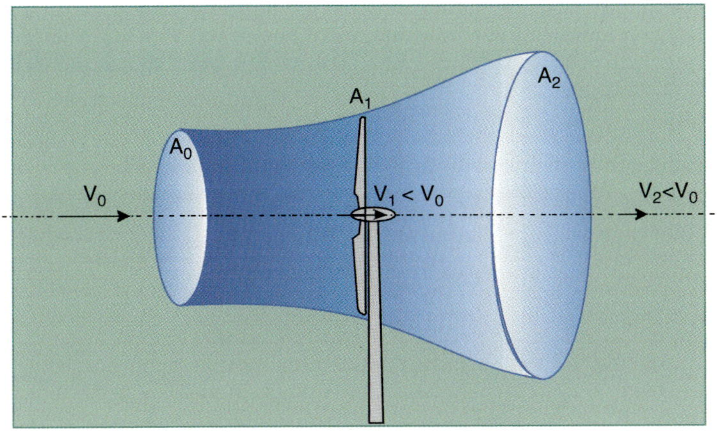

Fig. 5.13 The effect produced by the wind turbine on the air flow

V_0, which crosses the circular area A_0 and interacts with the turbine rotor with the scavenged area A_1. In section A_1, the air flow encounters resistance, the pressure increases, and the speed decreases to V_1. Giving up part of the energy the air flow leaves the turbine with the speed V_2 lower than V_1. As the air mass that passes through sections A_0, A_1 and A_2 remains constant and the speed drops, it follows that $A_2 > A_1 > A_0$, i.e., there is deflection of the air flow that passes through the turbine rotor, forming a funnel. The airflow formed immediately after the propeller is called the propeller slipstream, in which the static pressure is lower than in the free zone of the atmosphere. The static pressure is repressuring at longer distances from the propeller.

5.4.3 Betz Limit

In 1919 the German physicist Albert Betz formulated the law that answers the question: what part of the kinetic energy of the air flow can be converted into mechanical energy? Betz analysed a turbine with an ideal rotor: it is assumed that the rotor presents a disk with an infinite number of thin blades, energy losses are neglected, the flow of air runs through imaginary sections without turbulence in Fig. 5.13. Velocity V_0 is the speed of airflow till the rotor, V_2 is the speed of air flow leaving the rotor area, V_1 is the flow velocity in the A_1 section of the rotor. In accordance with the Newton's second law the variation of motion quantity is equal to the force acting on the body:

$$F = \frac{d(mV)}{dt} = m\frac{dV}{dt} \tag{5.8}$$

The variation of the air flow velocity for the above model during 1 s ($dt = 1\ s$) will be, $dV = V_0 - V_2$, so

$$F = m \cdot (V_0 - V_2) \tag{5.9}$$

We introduce the concept of air flow braking factor in the turbine $e = V_1/V_0$ and, assuming that the wind speed varies linearly, determine the air flow speed of the A_1 area of turbine:

$$V_1 = \frac{V_0 + V_2}{2}, \rightarrow V_2 = 2 \cdot V_0 \cdot e - V_0 \tag{5.10}$$

According to Eq. 5.5 the air mass crosses area A_1 in a second:

$$m = \rho \cdot A_1 \cdot V_1 = \rho \cdot A_1 \cdot V_0 \cdot e \tag{5.11}$$

Let substitute in Eq. 5.9 velocity V_2 and mass m according to Eqs. 5.10 and 5.11:

$$F = 2 \cdot \rho \cdot A_1 \cdot {V_0}^2 \cdot e(1 - e) \tag{5.12}$$

Power developed by turbine is the product of force and velocity:

$$P = F \cdot V_1 = 2 \cdot \rho \cdot A_1 \cdot {V_0}^3 \cdot e^2(1 - e) \tag{5.13}$$

According to Eq. 5.6, the air flow output with velocity V_0

$$P_0 = \frac{1}{2}\rho \cdot A_1 \cdot {V_0}^3 \text{ or } 2P_0 = \rho \cdot A_1 \cdot {V_0}^3 \tag{5.14}$$

Substituting in Eq. 5.13 we obtain

$$P = 4 \cdot P_0 \cdot e^2(1 - e) = P_0 \cdot C_p \tag{5.15}$$

where

$$C_p = 4 \cdot e^2(1 - e) \tag{5.16}$$

is called the power factor (efficiency factor) of Betz limit. Differentiating the expression (5.15) with respect to e, we determine its value for which power P will be the maximum. The result is $e = 2/3$, $Cp = 16/27 = 0{,}593$.

We can draw the following conclusion: the air flow will give up to an ideal turbine not more than 59.3% of its original power P_0 and this will be achieved when the braking factor $e = 2/3$ and air flow velocity after the turbine will be $V_2 = 1/3\ V_0$. In fact, the best three-bladed wind turbines have the Betz coefficient equal to 0.45–0.50.

5.4.4 Number of Blades and Rotor Diameter Effect

Betz limit states that an ideal wind turbine can extract from the wind a power not exceeding 59.3%, but the analysis made above does not indicate the operating system of the turbine or the construction that the rotor must have in order that the maximum power factor is achieved. Next, we will undertake a qualitative analysis of the turbine operating mode and of the effect of the number of blades or the solidity factor on the value of power factor. Also, the dependence of the rated power on the commercialized turbine rotor diameter is analysed.

The efficiency of air flow energy conversion into mechanical energy will be lower than the optimum value if:

1. The turbine rotor has a bigger number of blades (the solidity factor is big) or the rotor turns with a very high frequency and each blade moves in an air flow distorted (turbulent) by the front blade.
2. The turbine rotor has a small number of blades (the solidity factor is small) or the rotor turns with a very slow frequency and the air flow crosses rotor's surface without interacting with it.

In order to achieve maximum energy conversion efficiency it is necessary to correlate the rotor's speed of rotation with the wind speed. To characterize the wind turbines with different aerodynamic parameters, the dimensionless parameter is applied, called the tip *speed ratio* λ. The speed ratio links in a single formula three important variables of the turbine: the rotational speed ω, the rotor radius (or diameter) R and the wind speed V, and is defined as the ratio between *linear* turbine *blade* tip *speed* U to the wind *speed*

$$\lambda = \frac{U}{V} = \frac{\omega R}{V} \tag{5.17}$$

A turbine of certain design will operate in a large range of variations of the tip *speed ratio* λ, but will have maximum efficiency C_p only for an optimal value of the speed, or, if linear velocity U is equal to the wind speed multiplied to the optimal value of the tip *speed ratio*.

Figure 5.14 shows C_p–λ characteristics taken from [57] for turbines with different number of blades.

Analysis of these characteristics allows drawing the following conclusions:

1. The smaller the number of blades the greater the optimum *tip speed ratio* for which the power factor or energy conversion efficiency is maximized.
2. Two equal power turbines, but with a different number of blades are distinguished by the fact that the multi-blade turbine will develop a higher moment and will have lower rotation speed and vice versa – the turbine with few blades will develop a small moment but will have a higher rotational speed.
3. Three-bladed turbine has the biggest factor of efficiency. The differences between the factors of maximum efficiency of 2–5 blade turbines are not significant.

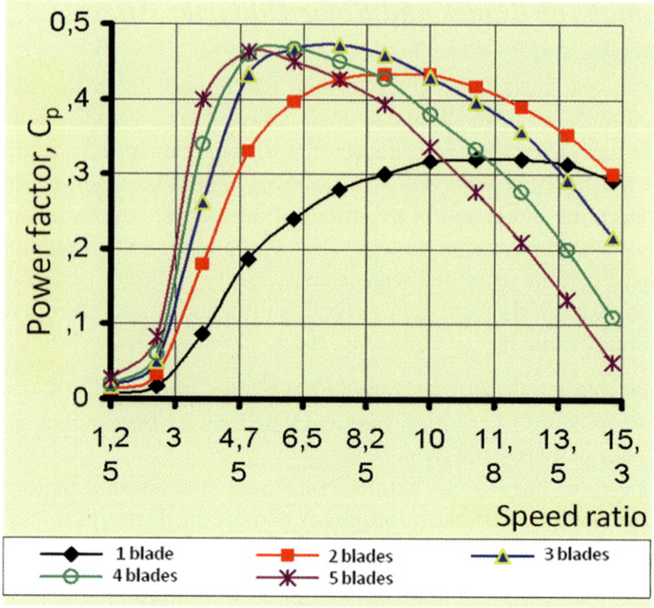

Fig. 5.14 Aerodynamic characteristics of different turbines

Advantages of one or two blade turbines consist in the possibility of operation in a wider range of *tip speed ratio* variation, in which the efficiency factor is maximum or close to maximum.

4. The maximum efficiency factor (Betz) of the turbine with 12–18 blades is smaller than that of the turbine with three blades and does not exceed 0.35.

Dependence of the turbine output power on the rotor diameter. Low power turbines have towers with relatively bigger heights than the high power turbines. This is explained by the necessity to exclude the negative influence of the ground surface layer and obstacles on the wind speed. For rotor diameters between 5 and 10 m the ratio of the tower height and rotor diameter is equal to 6–2. Starting with diameters equal to or bigger than 30 m, this ratio varies around the digit 1. Obviously, the specific costs of smaller turbines will be higher.

Mechanical power generated by the turbine is proportional to the rotor diameter square. The wind speed will increase simultaneously with the increase of diameter and height of the tower. Usually, the increase of the wind speed is considered proportional to the heights ratio of the exponent 1/7 [3, 5]. Thus the turbine power is proportional to the rotor diameter of the exponent $(2 + 3\ 1/7) = 2.42$. For turbines currently marketed a good approximation is given by the expression [3]:

$$P = 0.06 \cdot D^{2,42} \tag{5.18}$$

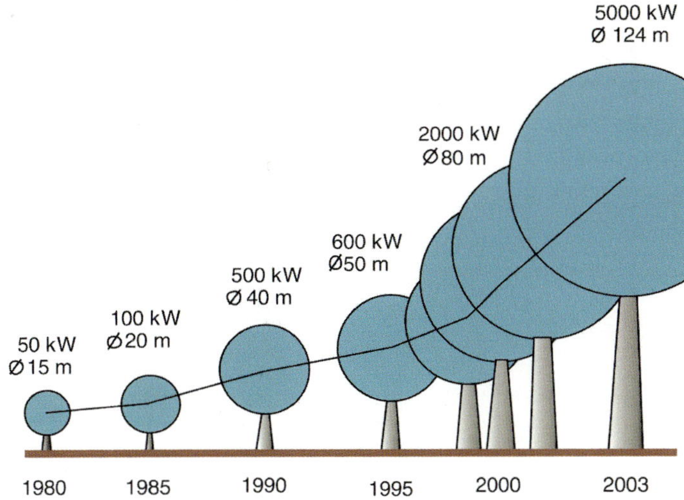

Fig. 5.15 Rotor diameter and capacity increase for sold turbines

where D is the rotor diameter, in m; P is rotor power, in kW.

Figures 5.15 and 5.16 show qualitative and quantitative evolutions of modern turbines capacities. The solid line (Fig. 5.16) corresponds to the analytical expression (5.18).

There is a tendency worldwide to increase the rotor diameter, even when the rated output remains the same. For example, the average diameter of the rotor of 1.5 MW turbines designed before 1997 was equal to 65.0 m in diameter; in 2000 the diameter reached 69.1 and 73.6 m in 2003. Increase in the rotor diameter leads to an increase of the extracted energy from the wind. If rated output remains the same the wind calculated speed can be decreased. Thus, the area of wind turbines use widens, including the new areas with average and small wind energy potential. This trend is also reflected by the empirical expression (5.19): for the turbines designed after 2003 the rated output can be calculated with the expression

$$P = 0.000195 D^{2.156} \tag{5.19}$$

Linear velocity of the blade tip is the product of the rotational speed and rotor radius. For turbines with rated power of 0.6–3.6 MW the range of linear velocity varies between 43.0 and 90 m/s (155–325 km/h). Such linear velocities require strict design of the aerodynamic profile, ensurance of surface good quality and excellent dynamic balancing of the rotor. All these measures lead to considerable reduction of noise and allow immediate placement of modern turbines near the villages and towns.

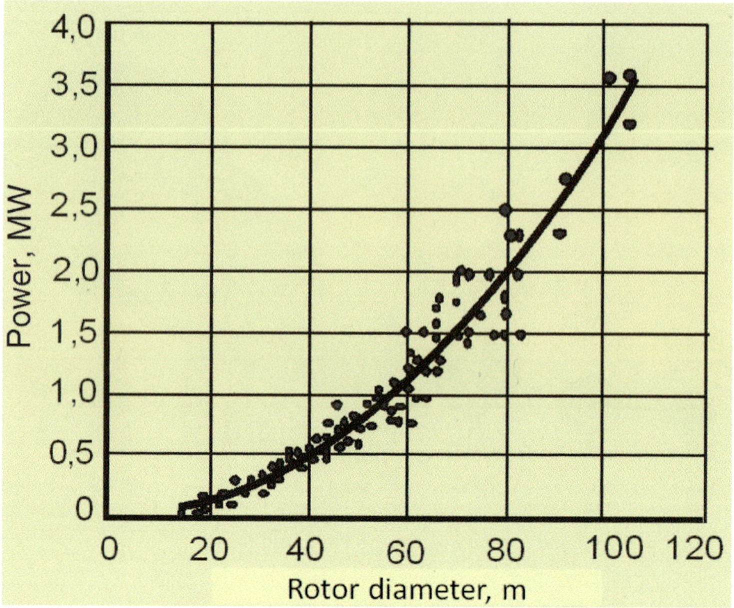

Fig. 5.16 Rated power of commercial turbines depending on rotor diameter

5.5 Evolution of Wind Technology Development

5.5.1 Commencement of Commercial Technologies

The period before 1970. The largest wind turbine for electricity production was built in the area of Grandpa Knob, Vermont, USA [3, 5]. The 1,250 kW turbines and 53 m rotor diameter was the outcome of collaboration between engineers Smith Putnam, von Karman and den Hartog. This was the first turbine which power was more than 1 MW and served as a platform for experimental studies of the fatigue effects of materials for blades and tower, EECS dynamics and verification of the data imposed by the project. However, the landmark year for launching the modern wind technology is considered 1957, when Johannes Juul, Danish engineer, designed the first wind 200 kW turbine, manufactured in northern Denmark, Gedser area (Fig. 5.17) [23]. He was the first who took the theory of airplane wing and transposed it in the construction of wind turbines. The turbine was installed on a 25 m tower and had three rotor blades. It was equipped with self-control system and self-stopping at the wind speed allowable limit exceeding, with electromechanical orientation drive and induction generator. It worked until 1967 with an average power factor of about 20%. Subsequently, the turbine is known as "*the Gedser turbine*" or the "*Danish concept*". Currently over 75% of the wind power (medium and large) turbines are based on the "*Danish concept*" [5], characterized by three

Fig. 5.17 The first turbine manufactured on the basis of the "*Danish concept*" by J. Juul in Gedser, Denmark [23]

thin-bladed rotor with aerodynamic profile oriented towards the wind and rotates at a relatively high speed – tens or hundreds of revolutions per minute depending on the diameter of the propeller. The innovative concept and characteristics of this model were soon recognized worldwide; Denmark became the main exporter of wind turbines and has over 33% of the global market.

The period 1970–1990. The results of experiments with different materials for blades have led to the abandonment of steel as a material too hard, as well as the aluminium that did not cope with dynamic stresses. Gougeon Brothers (USA) proposed wood and epoxide resin based material used in the construction of small and medium power turbines.

Due to a successful combination of federal and state laws concerning facilities granted to the wind energy sector, the first wind boom started in California: 1,700 MW wind power were installed (5.7 times more than the installed capacity of Thermo-Electrical Power Plants in the Republic of Moldova) in the period 1980–1995. The Californian boom has had a negative side – technical performances of the wind turbines were poor and have been placed irrationally which fact generated distrust in the wind energy. However, the U.S. market and facilities have enabled European producers to export and test more types of turbines and develop the foundation of modern technology.

Period between 1990–present. The wind energy development in California was not sustainable. After cancellation of some facilities a period of stagnation has started. Instead, the European markets have developed rapidly. In Germany, in the early 90s, the rate of wind power growth reaches 200 MW/year. New producers have emerged in Germany, Spain, and USA. New technological concepts have been developed: the outstanding innovative scheme of the wind generator with direct coupling (*direct drive generator*), variable rotational speed turbine, flow control systems for the power supplied to the grid, composite materials for blades, etc.

5.5.2 Wind Turbine Design

Wind turbines can be classified into four major groups, depending on the power developed at the wind rated speed, which is between 11 and 15 m/s. Micro turbines cover powers between 0.05 and 3.0 kW. Low-power wind turbines range from 3 to 30 kW and medium power turbines range between 30 and 1,000 kW. Both micro turbines and low power turbines are designed to operate in autonomous mode; these turbines supply electricity to territorially dispersed consumers that are not connected to the public power grid. To this end, the turbines are fitted with electrical energy batteries and power conditioning devices – controllers and frequency converters. The fourth group includes turbines with greater power – over 1,000 kW, called high-power turbines or multi-megawatt turbines. The current trend is power increase per unit, the absolute majority of the turbines operate in parallel with public electricity network, and priority is given to turbines with bigger power than 1 MW.

Over the years hundreds of wind turbine construction schemes have been proposed and patented, but several dozen have been tested only, of which only a few have penetrated the wind turbines market. Figure 5.18 presents the most significant structural schemes of the wind turbines.

Wind turbines: with horizontal or vertical axis of rotation? The answer to the question is in favour of the turbines with horizontal axis of rotation (Fig. 5.18). The absolute majority of the sold turbines are horizontal axis. The turbine axis of rotation coincides with the direction of the wind turbine and is parallel to the ground surface. In turbines with vertical axis the wind direction is perpendicular to the axis of rotation, respectively, perpendicular to the ground surface (Fig. 5.18). Although vertical axis turbines have lost the competition, engineers come back again and again to this design scheme, the main reason being the following two indisputable advantages:

- The generator, the multiplier and other functional components can be placed on the ground surface; the gondola and massive tower are not required;
- The turbine does not require a special device to track wind direction.

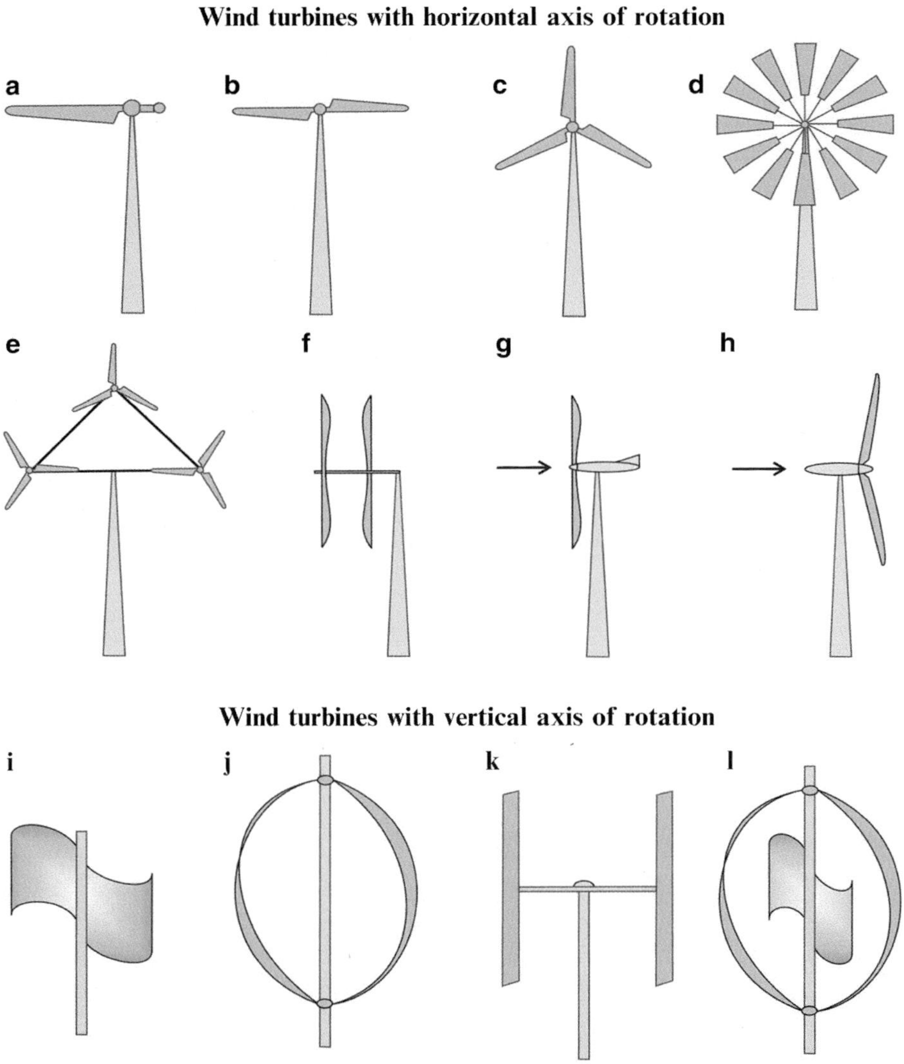

Fig. 5.18 Wind turbine design: (**a**, **b**, **c**) with 1-, 2- or 3-blades; (**d**) multi-blade; (**e**) with more rotors; (**f**) with two rotors turning in opposite directions; (**g**) with rotor at tower front and weathercock (up–wind); (**h**) with rotor at tower back with self-control (down – wind); (**i**) Savonius; (**j**) Darrieus; (**k**) Evence; (**l**) combined Darrieus – Savonius

Unfortunately, the disadvantages of these turbines prevail compared to the advantages:

- Wind speed in the adjacent to the surface layer is small. Thus, we save on tower construction, but lose the power developed by the turbine.
- Wind energy conversion factor into mechanical energy is lower.

Fig. 5.19 Modern horizontal

- Some types of turbines, such as the Darrieus or Evence turbines do not provide starting. An auxiliary motor is required to start the turbine or a small turbine of Savonius type.
- High power turbines need support cables, which considerably increase the occupied land area.
- Replacing the main thrust bearing requires complete disassembly of the turbine.

Below we briefly describe the construction schemes shown in Fig. 5.18. A key feature of the horizontal axis turbine is the number of blades. There may be one, two, three or more blades (see Fig. 5.18). If the turbine has more blades, the solid area of the rotor swept area is bigger. The theory of wind turbines states that the number of blades is taken into account with the solidity factor that shows the relationship between all blades' area and the area swept by the rotor. It is obvious that one- or three-blades turbine have a lower solidity factor than the turbines with 12 or 18 blades. The effect of the number of blades on the turbine performance is described in Sect. 5.4.4. The bigger the solidity factor (multi-bladed rotor), the smaller is the rotational speed of the turbine, and the developed moment is bigger too, and inversely. For this reason, the turbines with fewer blades are used to generate electricity, and the multi-blade turbines are applied for pumping water, driving saws, crushers, grinding rollers, etc. In other words, multi-blade turbines are used in machines that require low speed of rotation and big moments on start-up.

Other constructive schemes of horizontal axis turbines are shown in Fig. 5.18: e – multi-blade turbine; f – turbine with two rotors, that turn in opposite directions; g – with rotor at tower front (*up – wind*) and weathercock for orientation; with rotor at tower back with self-control (*down – wind*). Figure 5.19 shows a modern horizontal axis turbine, type V90 – 2. 2 MW power, rotor diameter is 90 m [23].

Fig. 5.20 Darrieus turbine [23]

Fig. 5.21 Turbine H, 420 kW, Crimean peninsula, Ucraine

Vertical axis turbines are shown in Fig. 5.18: Savonius turbine with S-type rotor (i), Darrieus turbine with rotor – ellipse (j), Evence turbine with H-type rotor (k), Turbine with combined rotor – Darrieus – Savonius (l). Examples of vertical axis turbines are presented in Figs. 5.20, 5.21.

5.5.3 *Principles of Control of Wind Turbine Power Output to the Grid*

The wind turbine will provide rated output in the network, if the wind speed is equal to the calculation speed, usually 11–15 m/s. For higher wind speeds it is necessary to limit the mechanical power, respectively, the overstressing of the rotor blades, multiplier, generator, tower etc. Thus, there is need to control the power turbine. The most common control methods are as follows [56, 57]:

- *Passive stall control*;
- *Active pitch control*;
- *Active stall control*;
- *Yaw control.*

Power control using the passive stall control method. It is the simplest method and can be used for turbines with constant rotation speed, i.e. the speed of rotation does not depend on the wind speed or varies slightly (1–2%). Constant rotation speed of the turbine can be obtained in EECS equipped with asynchronous or synchronous generators connected directly to the public power grid (Fig. 5.22). Rotor blades are fixed rigidly and have aerodynamic shape, providing a laminar character to the air flow for wind speeds between the start V_p and calculated V_c velocities (Fig. 5.22). For wind velocities bigger than V_c (Fig. 5.22), the air flow motion above the blade becomes turbulent, the lifting force decreases and the resistance force increases, and respectively, the mechanical power decreases.

The typical form of power characteristics $P = F(V)$ for a turbine with stall control is shown in Fig. 5.22. In the velocity zone $V_p < V \leq V_c$, power output to the network is proportional to the cube of the wind velocity. At the calculated velocity, the turbine generates rated output, and if the wind velocity continues to increase, aerodynamic braking intensifies and the increase power output is limited. In most cases, for velocities equal to or greater than 25 m/s, the turbine is disconnected and broken by the mechanical brake from the equipment. The main advantage of this control method is to achieve simplicity, the disadvantages consist in the fact that this method requires a rigorous calculation of the aerodynamic blade profile; the generator must withstand overstresses of 20–30%, for high wind velocities (turbulent movement intensifies) the generated power in the network becomes less than the rated output.

Power control using the active pitch control method. Control of power output is achieved by controlling the adjustment angle α (Fig. 5.23). For this purpose the blade is rotated by a special device around the longitudinal axis. Turbine rotational speed can be variable. To maintain constant frequency, the synchronous generator is connected to the network using frequency converter (Fig. 5.23). For small adjustment angles, ranging from 0° to 13–15°, the aerodynamic *lift force* increases linearly with increasing the angles of adjustment:

$$F_L = \frac{\rho}{2} C_L A_P V^2 \tag{5.20}$$

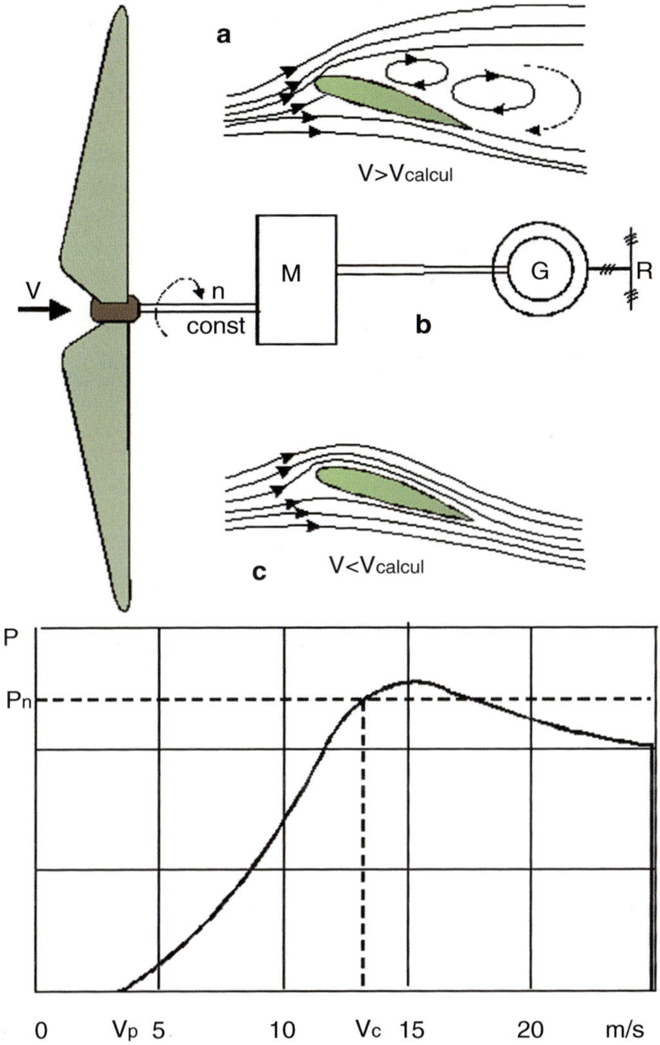

Fig. 5.22 Principle of control of rated output to the grid using the stall control method

where C_L is the lift coefficient; A_P is the blade area (so-called master-sectional area or area of the blade projection to the surface perpendicular on the wind direction); V is the wind velocity; ρ – air density.

Also, the *drag force FD* appears, which direction coincides with the wind direction:

$$F_D = \frac{\rho}{2} C_D A_P V^2 \tag{5.21}$$

where C_D is the drag coefficient.

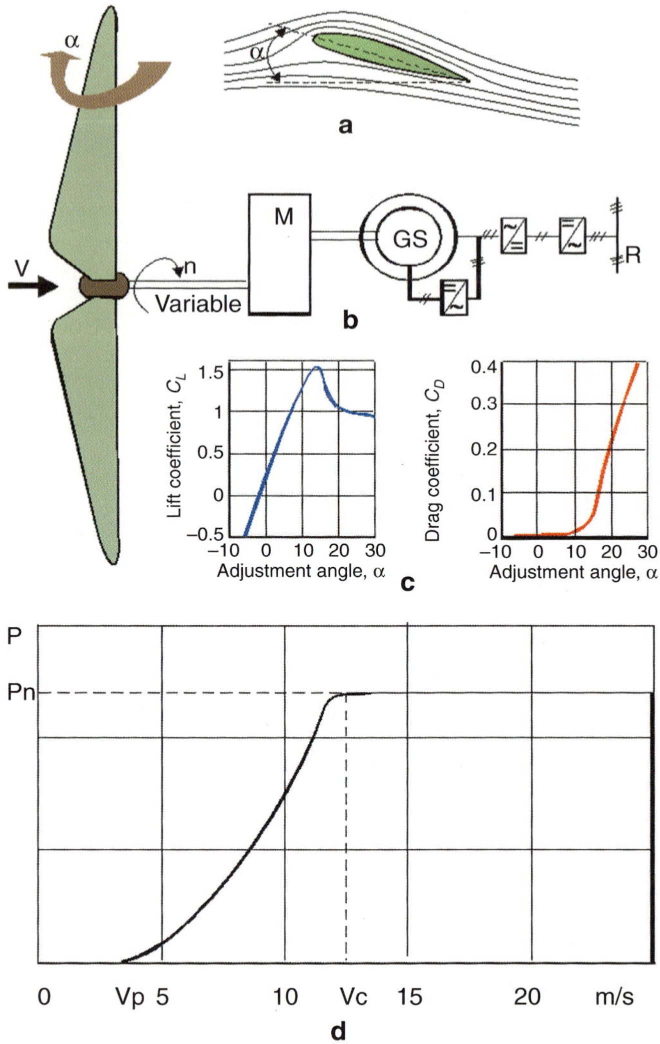

Fig. 5.23 Principle of control of power output to the grid by using active pitch control method

Typical variation of coefficients C_L and C_D, and respectively, of the forces F_L*and* F_D for aerodynamic blade with NACA632XX profile is shown in Fig. 5.23. For wind velocities bigger than the nominal velocity the mechanical power is kept constant by increasing the angle of adjustment α, coefficient C_L decreases sharply and C_D increases, power output remains constant (see Fig. 5.23). The main advantage of the active pitch control method consists in reducing the mechanical stress on the blades, rotor and tower; wind energy conversion efficiency increases with 2–4% at lower velocities than the rated one. Disadvantages – the complexity of implementation, the need for a system of rapid adjustment of the entering angle.

Active stall control method. It is a combination of the two methods – passive stall control and active pitch control. At wind speeds lower than the rated velocity, the angle of adjustment is controlled to obtain greater efficiency of the wind energy conversion into mechanical energy. For wind velocities higher than the nominal velocity, the angle of adjustment is controlled in the opposite direction to the original one for the pitch control, usually the setting range is $0 < \alpha < -5^{\circ}$. From Fig. 5.23 it shows that for these angles of adjustment the lift coefficient decreases, and respectively, the lifting force is reduced, so the mechanical power is constant.

Power limit by removing the turbine rotor from the wind direction. This method is recommended only for power turbines less than 30 kW. At high wind velocities the turbine rotor is removed from the direction of wind either by turning it around the tower axis (*yaw control*), or around the axis perpendicular to the tower axis (*tilt control*). In both cases, passive mechanical systems are applied that react to wind velocities higher than the nominal velocity.

The analysis of small power turbines projects depending on the power control method is shown in Fig. 5.24. Figure 5.25 shows turbines with power more than 1 MW [3, 5]. In 57% of the implemented projects the method of turbine rotor removal from the wind direction is applied, and about 17% of turbines are not equipped with power control systems. The turbine rotor without power control is calculated to withstand extreme winds in the area. In high-power turbines clear trend is found of increasingly wide use of control by adjusting the entering angle or by pitch control (Fig. 5.24).

5.5.4 Constructive Schemes for Generator Operating

The first marketed wind turbine, built by J. Juul, in the area of Gedser, that generated the "*Danish concept*" (see Sect. 5.5.1) was equipped with multiplier and asynchronous generator. This diagram (Fig. 5.26) prevails today in most turbines with rated power greater than 100 kW. To reduce weight and increase efficiency, asynchronous generator must be designed for rotational speed as big as possible – 3,000 min^{-1} at a frequency of 50 Hz and 3,600 min^{-1} at a frequency of 60 Hz. However, the 600 kW turbine speed of rotation is about 30 $\text{min}^{-1}$and that of the turbines with power bigger than 1,000 kW is even smaller. An optimum mass of the multiplier – generator train is obtained for generator rotation speeds of (1,000–1,500) min^{-1}. In this case the multiplier gear ratio should be about 1:50. The multiplier – asynchronous generator design diagram is not universal and can not be recommended for all wind turbines. The solutions are different depending on the turbine power and area of use.

Micro turbines, rated power equal to or less than 3 kW. The rated speed of rotation of micro turbines is relatively high (200–500 min^{-1}) and used in autonomous regime. Over 95% of turbines are equipped with low speed synchronous generators with permanent magnets (PMSG) and coupled directly with the rotor (Fig. 5.26).

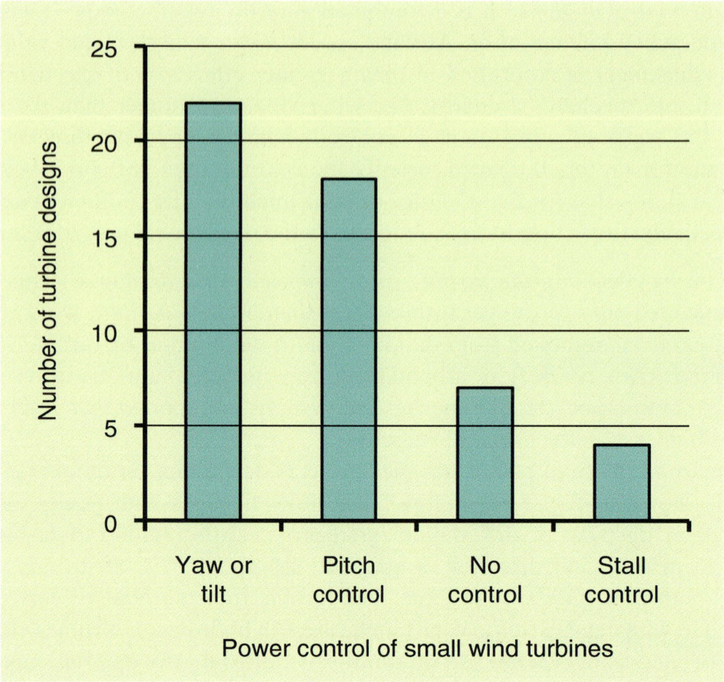

Fig. 5.24 Methods of power control used for small-power turbines

At low speeds of rotation, the technical performance of synchronous generators decreases essentially and in isolated electrical systems requires special excitation equipment and voltage stabilization. The reference literature has not identified any examples of wind micro turbines equipped with asynchronous generator operating autonomously or supplying a remote power network.

Small-power turbines – (3–30 kW). Absolutely all turbines with rated power up to 10 kW are equipped with PMSG coupled directly to the wind turbine. In the 10–30 kW power range there are some exceptions: the wind turbine produced by "*Atlantic Orient Corporation*" (USA) with nominal power of 20 kW is equipped with variable reluctance generator [3].

Medium-power turbines – (30–1,000 kW) and multi – megawatt turbines. This situation is uncertain. The multiplier-generator diagram prevails on the market; the generators can be asynchronous or synchronous with electromagnetic excitation (EMSG), or with permanent magnets. After the 90s of the last century several prototypes of wind turbines with direct coupling were brought to the market, following several objectives:

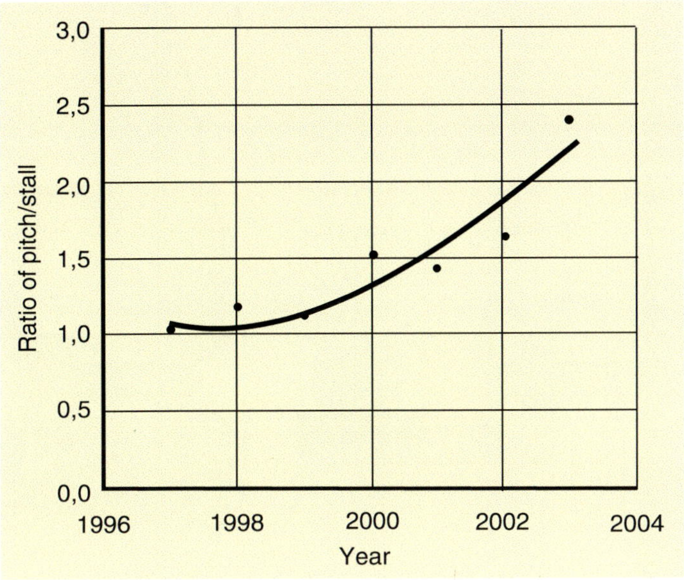

Fig. 5.25 Ratio of control methods for power used in turbines larger than 1,000 kW

- Reducing the operating and maintenance costs;
- Increasing the wind energy conversion efficiency, including areas with moderate winds;
- Decreasing the start speed (start) of the turbine, thus increasing the range of lucrative wind speeds;
- Reducing the length and weight of the gondola;
- Reducing vibration and noise;
- Increasing the availability and reliability of wind turbine.

WinWind Finnish Company launched a 1,100 kW power wind turbine on the market: the rotor diameter is 56 m, which is a hybrid, a compromise between the diagram with the multiplier and the diagram with direct coupling. The so-called "*Multibrid*" concept [5], which is put at the basis of the new turbines, consists in the use of one-step planetary multiplier with gear transmission ratio of 1:5,7 and PMSG rotation speed between 40 and 146 min^{-1}. Weight of the multiplier – PMSG train has remained the same as for traditional turbines, but the gondola is a more simple and compact construction.

5.5.5 *Blades Manufacturing Materials*

The ideal material for blades should combine the following structural properties: hardness- weight specific optimal ratio; longevity and flexibility to fatigue; low cost

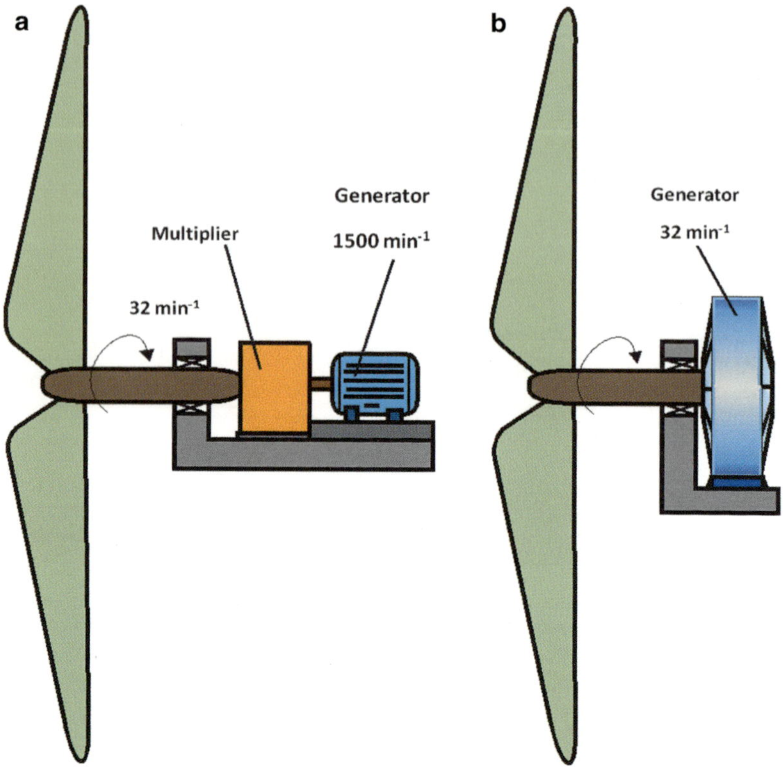

Fig. 5.26 Wind turbine with multiplier (**a**) and direct coupling (**b**)

and easy process to get the desired aerodynamic shape. After multiple experiences at the early stage of wind technology development, based on the use of steel, aluminium, etc., nowadays global manufacturing technology for turbine blades is entirely based on the following composite materials:

1. GRP – *Glass Reinforced Plastic*, producers: LM Glasfiber Denmark; Aerpac, Rotorline, Polymarin, The Netherlands.
2. Wood plus epoxide resin, producer: Taywood Aerolaminates (TAL), USA.
3. CFRP – *Carbon Fibre Reinforced Plastic*, producer: ATV Enterprise, France.

Share of the most common materials for blades and the main actors on the world market are shown in Fig. 5.27. The trend of decreasing the rotor mass, and increasing the blade elasticity require manufacturers to return to the so-called "*boat building*" technology. Known for decades, this technology is used in large scale construction of different crafts, motorboats, boats, etc. The material is based on polyester resin and contains less glass fibre.

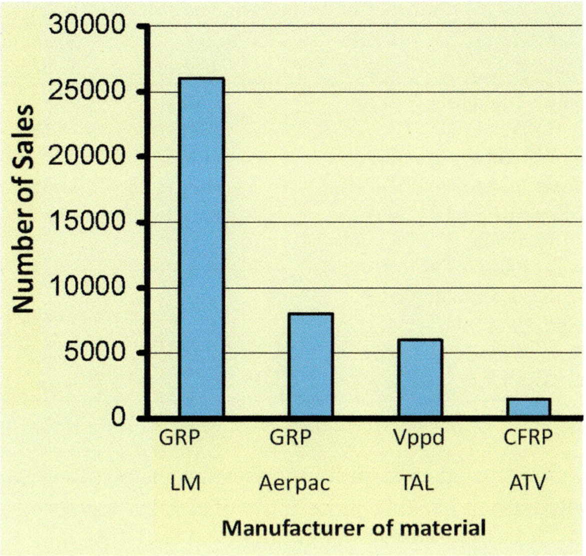

Fig. 5.27 Materials for blades and main producers on the world market

5.6 Large Wind Turbines and Farms

5.6.1 Large Wind Turbines: Trends and Objectives

Although the first 1,000 kW wind turbines prototypes have been tested long before the 80s of the last century, none of them have been sold. Even operated for short periods of time (Table 5.2) they served as experimental research platforms to gain new knowledge and create the technology for the next step made in the late twenty to early twenty-first century.

The *"Danish concept"* for wind energy conversion technology has dominated the world market about 25 years. Although different projects were carried out, they all had the following common characteristics: three blades, constant rotor rotation speed, aerodynamic braking (stall) for power control, asynchronous generator. On world market, turbines of up to 200 kW were mainly sold, rotor blades were made of PFS or wood plus or epoxide resin. And in the late twentieth century an increase in power per unit took place (without changing the "*Danish concept*") – 250, 400, 600 and 750 kW.

Since the year 2000, 1 MW power turbines are emphasized. Rotor diameter reaches 100 m and more. Although the constructive ideology retains some features of the "*Danish concept*" in the newly completed projects the following prevail: rotor variable speed, power control through changing the adjustment angle, more widespread direct coupling, and more frequent use of carbon fibre based material. As a result, the efficiency of converting wind energy into mechanical energy increased, the quality of the electricity network supply improved, noise and vibration

Table 5.2 Performance of the first wind turbines prototypes with power over 1.0 MW [9]

Turbine type, country	Rated output, MW	Rotor diameter, m	Operating hours	Generated power, GWh	Years of production
Smith-Putnam, USA	1.25	53	695	0.2	1941–1945
Mod-1, USA	2.0	60	–	–	1979–1983
Mod-2, USA	2.5	91	8,658	15.0	1982–1988
Growian, Denmark	3.0	100	420	–	1981–1987
WEG LS-1, Great Britain	3.0	60	8,441	6.0	1987–1992
Mod-5B, USA	3.2	98	20,561	27.0	1987–1992
Ecole, Canada	3.6	64	19,000	12.0	1987–1993
WTS-4, USA	4.0	78	7,200	16.0	1982–1994

were reduced, and multiplier-related problems were excluded. The wind turbine market has become more attractive, sales increased and specific costs decreased. To compare the two technologies, Figs. 5.28 and 5.29 show the composition of gondolas of two modern turbines: Z72, Zephyros, The Netherlands [21], rated power – 2,000 kW, rotor diameter – 72 m, starting wind speed – 3 m/s, rated speed – 15 m/s, permanent magnet synchronous generator, direct coupling, variable speed, pitch control, gondola weight – 61 t; and V80, Vestas, Denmark [23], rated power – 2,000 kW, rotor diameter – 80 m, starting wind speed – 4 m/s, nominal speed – 15 m/s, dual feed asynchronous generator with multiplier 1:60, variable speed, pitch control, gondola weight – 67 t. In the first variant, the gondola is much more robust and lighter by 6 t, the number of components and the length of rotor-generator train are significantly decreased.

The main objectives of wind technologies development can be formulated as follows:

1. Widespread use of intelligent monitoring and control of mechanical loads and vibrations;
2. New construction concepts adapted for rotor manufacturing;
3. Large-scale use of carbon fibre for the construction of high-power turbine blades;
4. Flexible rotors placed behind the tower with small thermal solidity coefficient;
5. Direct coupling, new topologies of permanent magnet generators;
6. Studies of 5 MW turbines design;
7. Variable speed rotor, high voltage generation;
8. Studies of the development of offshore turbines: high rapidity coefficient turbine, special foundations and structures, floating turbines, hydrogen production and transportation, special lifting systems, etc.;
9. Design and testing of wind turbines with rated power of 8–12 MW per unit [19], work started in 2005 and to be completed in 2009. Twelve megawatt turbine diameter shall be 190 m, tower height – 170 m.

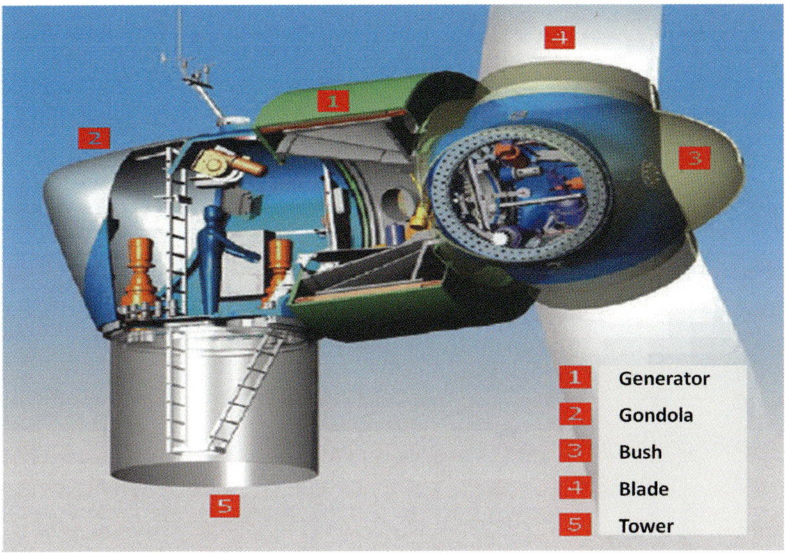

Fig. 5.28 Z72 turbine with direct coupling and permanent magnet synchronous generator

Table 5.3 presents the main characteristics of marketed wind turbines with power over 1.0 MW [22–29]. The symbols mean:

P_n – rated power; D_R – rotor diameter, H_T – tower height, n – nominal rotation speed, γ – range of rotational speed variation;
CIG – cage induction generator; WIG – wound induction generator; DFIG – doubly fed induction generator; PMG – permanent magnet generator;
U – multiplier gear ratio, V_P – starting wind speed, V_n – wind rated speed (calculation value); indicates lack of data.

The largest 5 MW wind turbine, developed by the German Repower Company [26], was put into operation in August 2006 in the coastal area of the North Sea. It is the first pilot wind farm structure located near the "*Beatrice*" oil platform (Fig. 5.30).

Based on Table 5.3 data the following conclusions can be drawn:

- There is a clear trend for the use of power adjustment through changing the pitch control, which allows a finer control of power supplied into the grid and reduction of mechanical loads (Sect. 5.5.3);
- Cage induction generators with constant rotation speed are being replaced with doubly fed induction generators or permanent magnets synchronous generators. In both cases the rotor speed is variable;
- For moderate climatic conditions turbines with lowest starting and rated speed are suitable, for example: D6, E82, FL2500, N100/2.5, designed for areas with moderate wind speeds.

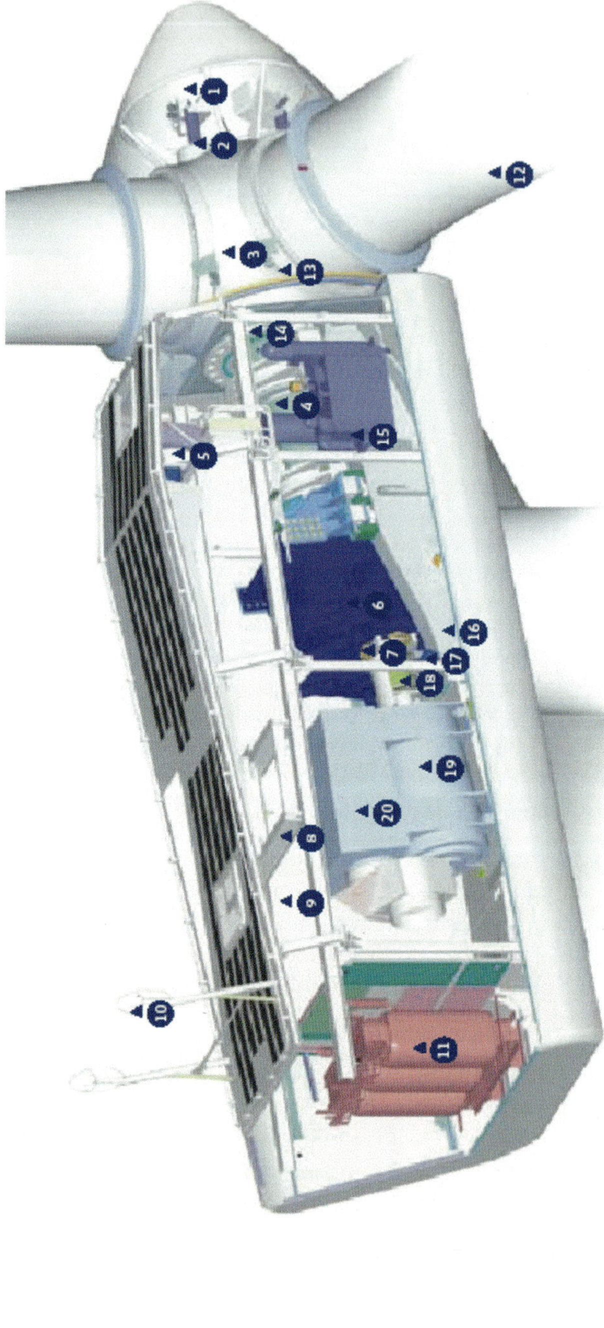

1- Hub controller; 2 - Pitch cylinders; 3 - Blade hub; 4 - Main shaft; 5 - Oil cooler; 6 - Gearbox; 7 - Mechanical disc brake;
8 - Service crane; 9 - VMP-Top controller with converter; 10 - Ultrasonic sensors; 11 - High voltage transformer;
12 - Blade; 13 - Blade bearing; 14 - Rotor lock system; 15 - Hydraulic unit; 16 - Machine foundation;
17 - Yaw gears; 18 - Composite disc coupling; 19 - Induction generator; 20 - Air cooler for generator

Fig. 5.29 V80 turbine with multiplier and asynchronous generator

Table 5.3 Main characteristics of modern turbines, over 1.0 MW power

	P_n	D_R/H_T	n/γ	Generator type	U	Power adjustment	V_P/V_n
Turbine type, country	MW	m	rot/min	–	–	–	m/s
D6, Germany	**1.25**	**64/60–91**	**21.1/13.2–24.5**	**DFIG**	**1:53.1**	**Pitch**	**2.8/12.5**
V82-1,65, Denmark	1.65	82/78	14.4/0.0	CIG	?	Stall	3.5/13.0
Z72, the Netherlands	2.0	72/86-80	23.5/?	PMG	1:1	Pitch	3.0/15.0
E-82, Germany	**2.0**	**82/70–138**	**?/6–19.5**	**PMG**	**1:1**	**Pitch**	**2.0/12.0**
V-90-2, Denmark	2.0	90/80–105	14.9/9.0–14.9	CIG	?	Pitch	2.5/13.0
D8, Germany	2.0	80/80–100	18.0/11.1–20.7	DFIG	1:94.4	Pitch	3.0/13.5
E-70, Germany	2.3	71/58–119	?/6.0–21.5	PMG	1:1	Pitch	2.0/15.0
FL2500, Germany	**2.5**	**100/65–160**	**?/9.4–17.1**	**WIG**	**1:79.6**	**Pitch**	**3.5/11.5**
N100/2,5, Germany	**2.5**	**100/100**	**?/9.6–14.9**	**DFIG**	**1:74.4**	**Pitch**	**3.0/12.5**
V90-3, Denmark	3.0	90/80–105	16.1/8.6–18.4	CIG	?	Pitch	4.0/15.0
E-112, Germany	4.5	114/120	?/8.0–13.0	PMG	1:1	Pitch	2.5/?
5 M, Germany	5.0	126/100–117	?/6.9–12.1	DFIG	1:97	Pitch	3.5/13.0

Fig. 5.30 Mounting of 5 MW wind turbine, Germany, North Sea coast

The need to adjust the rotation speed of the turbine results from Fig. 5.31, which depicts the $P_r(\Omega_r)$ characteristics for different wind speeds: wind velocity brings variation of the maximal (optimal) power produced by the rotor. To extract maximal power from the wind it is necessary to change the rotation speed Ω_r so as to maintain the operating point of the turbine in the optimal (see the red line) field. Below (Fig. 5.32) there are three of the most frequently used generator-public electricity grid connection schemes, which differ by the generator type:

- Cage induction generator;
- Doubly fed induction generator;
- Permanent magnet synchronous generator directly coupled with turbine rotor.

5.6.2 Wind Farms

Large scale wind energy production is done by tens or hundreds of wind turbines forming the so-called "wind farms" or wind plants, connected to the public

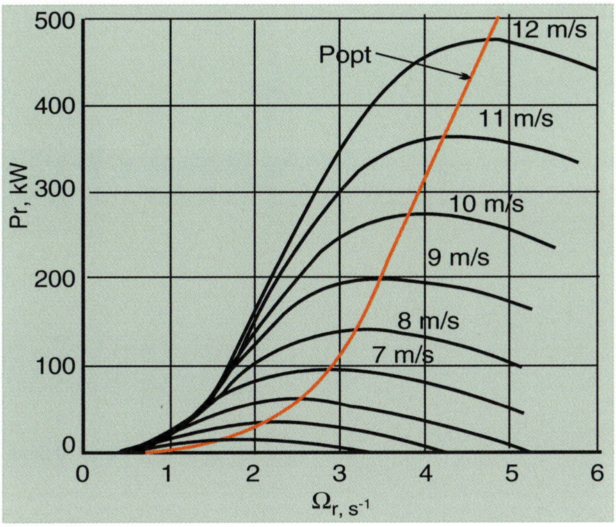

Fig. 5.31 Rotor power variation due to rotation speed and wind velocity

electric grids. Synchronous operation mode employing traditional generating units is beneficial for wind energy producer: all energy produced is supplied to consumers connected to the same public electric networks. At speeds below 3 m/s or when the wind is missing, the energy producer becomes a consumer. Medium and high power wind turbines are not operated in autonomous regime due to the high cost of electric batteries. In this case, wind turbine operation in parallel with an electric unit is rational. Such schemes are common in Africa, South America, China, India, where there are isolated areas not connected to the public electric grids. Figures 5.33 and 5.34 are examples of wind farms located on hills.

5.7 Small Power Wind Turbines

5.7.1 Global Market Overview, Manufacturing Companies and Incentive Policies

Small systems for wind energy conversion were among the first used by humans for their own or community energy needs. Further over-centralisation of power distribution system led to a substantial costs increase in the case of isolated consumers. From this point of view, the idea to de-centralize power supply systems for isolated consumers is appropriate (e.g. smart grids).

It is noticeable that most of the Third World countries do not have centralized electricity distribution networks. Therefore, the area of small power wind turbines

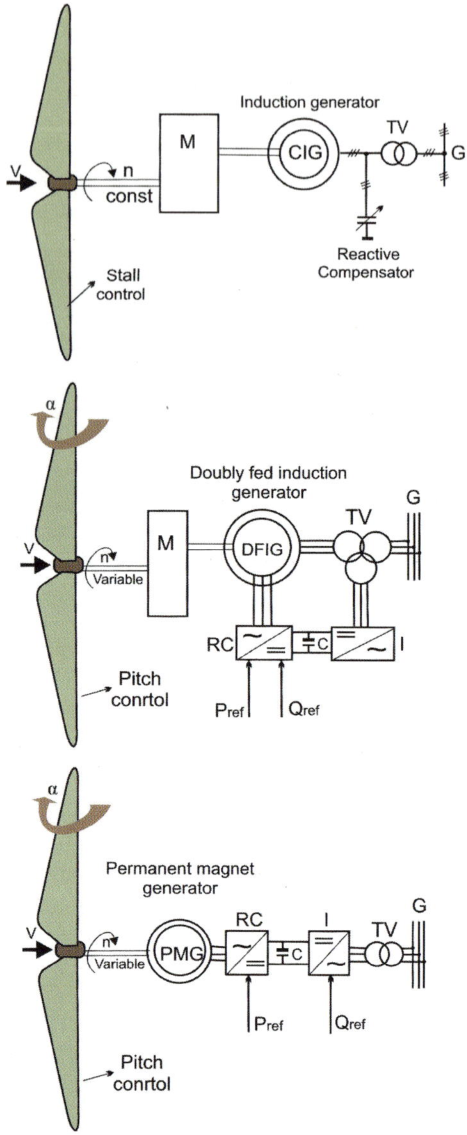

Fig. 5.32 Wind turbine with permanent magnet and direct coupling synchronous generator

tends to enlarge, especially with account of the inevitable fuel costs increase and environmental issues the mankind is challenging. In this case, the state policy promoting green energy production is of major importance. The USA, pioneer of small wind industry, adopted for this purpose in 1985 an incentive system for green wind energy production, which, since 1990 recorded an annual 14–25% increase. About 30% of federal investment tax credit will bring an annual 40% growth of small power wind turbines sales [30]. About half of small power wind turbines

Fig. 5.33 Wind farm located on the hills of Challicun, Australia [27]

Fig. 5.34 Wind farm on a hill, California, USA [28]

Table 5.4 Analysis of ways to install small wind turbines

Connection type	Units	Installed capacity, kW	Sales, USD
Network	706	5,158	18,197,600
Isolated	5,933	10,935	32,706,750
Total	**6,639**	**16,093**	**50,904,350**

manufactured in USA were exported, and this shows that the small wind turbine industry is one of the dominant renewable energy technologies in the US.

The term *"small power wind turbine"* is defined as an electricity generator with up to 100 kW capacity. A small wind system can include, if necessary, a turbine, a tower, an inverter, batteries, foundation, etc. To examine what the market for small power wind turbines in the US is about, one should consider the following. According to [31], the number of small wind turbines sold in the US in 2006 was 6,807 units, of which 6,639 (about 98%) produced in the USA, with a total installed capacity of 17,543 kW (including 16,093 kW of US produced units). In 2006 outside US, 9,502 turbines were sold having a total installed capacity of 19,483 kW [32]. The correlation analysis of turbines connected to the network and installed separately (Table 5.4) shows that the great majority (5,933 to only 706) of turbines are not connected to the network, thus they are employed by isolated consumers.

The cost of small wind turbines remains the only major factor affecting the market growth. For small wind turbines, the generation volume is not the only question, the increasing costs of construction materials, such as copper and steel, also has to be considered. Market growth is also a major function of state policies in the field, particularly regarding ways to exempt companies of various taxes [33], other financial incentives, etc.

Another impediment is the investment payback period. In the USA, for small wind turbines this period ranges from 6 to 30 years, depending on many factors, such as the quality of wind resources, installation site, turbines purchase costs, turbine performance and energy costs. A study conducted in 2006 at Lawrence Berkley National Laboratory estimated that 30% decrease of the federal investment tax credit will reduce turbines payback period to approximately 4.5 years, and, similarly, exemption of the state property tax can reduce this period to 4 years [33]. This case is similar to photovoltaic systems industry, whose market grew due to federal investment tax credit, thus helping consumers to purchase solar systems.

The demand for small wind turbines arises from consumers' interests with regard to climate change, unpredictable gas prices and energy security. Many consumers, representing nowadays a small percentage of the market, are driven by the decision or need to meet energy needs independent of the electric network.

The analysis of the impediment factors on small wind turbine market is presented in Table 5.5, based on a 72 respondents' survey. The answer to each of the ten questions was appreciated by an 8-degree scale (1 – no barriers, and 8 – greatest barriers).

Table 5.5 Small wind turbine market growth barriers

Indicator	1	2	3	4	5	6	7	8	Average
Economic/buyer's price									6.53
Restrictive areas and permission rules and/or costs									6.03
Lack of support incentives and/or of significant subsidies									5.35
Lack of financial incentives									5.73
Visual impact/community opposition									5.14
Lack of public information/awareness									5.00
Lack of turbine certification									4.87
Lack of installer certificate									4.42
Lack of access to information regarding wind resources and maps									4.19

Table 5.6 Small wind turbine application areas

Application	Area (%)
Rural or suburban houses	51
Farms	19
Small businesses	10
Schools or public facilities	10
Urban use	5
Other	5

The top export markets for US producers are Canada, UK, China, Europe (excluding UK) and India. Table 5.6 presents applications of turbines sold in 2006. Most small power wind turbines (80%) are used for lighting and heating rural and suburban houses, for farms irrigation and other works, for small businesses electricity supply (mainly for processing and storage of agricultural production) located in areas without electricity networks, etc.

The growth of wind turbine market is driven by cost reduction of 1 kWh of electricity produced, which decreased from 0.15–0.18 USD/kWh to 0.1–0.11 USD/kWh in order to achieve 0.07 USD/kWh over the next 5 years. For comparison, electricity currently produced by big wind turbines is 0.04–0.07 USD/kWh and by solar modules – 0.18 USD/kWh [34]. Estimates were made according to 2004 data, which include federal policies for commercial systems and federal investment tax credit. The cost of electricity for high power turbines includes the federal tax credit of 0.02 USD/kWh. According to AWEA 2005, after reducing the production costs by 20% until 2010, the cost of installed capacity kW will reach 1,700 USD. The US estimates for the potential market of wind systems indicate that the number of small wind turbines could reach between four and eight million units.

Conclusion: consumer prices and state policies are the most important components in supporting and developing the small power wind turbine market. Geopolitical, climate and economic forces shall further increase the market demands.

Table 5.7 USA companies producing small wind turbines

Producing company	Models (capacity)
Abundant Renewable Energy www.abundantre.com	AWP 3.6 (1 kW)
Bergey Windpower Co. www.bergey.com	BWC XL.1 (1 kW), BWC EXCEL (10 kW)
Distributed Energy Systems www.distributed-energy.com	NPS 100 (100 kW)
Energy Maintenance Service www.energyms.com	E15 (35 kW or 65 kW)
Entegrity Wind Systems www.entegritywind.com	EW15 (50 kW)
Lorax Energy www.lorax-energy.com	FL 25 (25 kW), FL 30 (30 kW), FL 100 (100 kW)
Solar Wind Works www.solarwindworks.com	Proven WT600 (600 W), WT2500, (2.5 kW) WT6000 (6 kW), WT15000 (15 kW)
Southwest Windpower Co. www.windenergy.com	AIRX (400 W), Whisper 100 (900 W), Whisper 200 (1 kW), Whisper 500 (3 kW)
Wind Turbine Industries Corp. www.windturbine.net	

Producers of small wind turbines. In the context of action taken to solve the imminent energy crisis, to stop the global environmental impact and the trend towards decentralization (especially for isolated consumers) of power supply and of other types of energy, the small wind turbine industry is continuously growing. Presently, in the world there are a number of companies producing a wide range of small wind turbines rated less than 1 kW and up to 100 kW. The USA became the largest producer of small wind turbines. The American Wind Energy Association (AWEA) has made a compilation of companies manufacturing and selling wind turbines for residential applications, industrial/commercial and farms use (Table 5.7).

Small wind turbines produced in the USA are currently employed in over 140 countries. For example, Fig. 5.35 presents the most popular product of the company Bergey Windpower Co. – Bergey Excel-S 10 kW turbine, with a cost of 21,450 USD [35]. Following the adoption of new incentives, the payback period is 6–7 years in California. From a financial standpoint it is net superior to purchase a wind turbine than to purchase electricity from an energy company. Important is that the company developed a set of new very high resolution wind maps for the entire state of California. Similar results have other US companies included in the above table. There are also many other companies in various countries that manufacture and operate small wind turbines. The most known are:

- Iskra Wind Turbine Manufacturers Ltd, Nottingham, United Kingdom;
- WestWind, J.A. Graham Renewable Energy Services, Northern Ireland;
- Gazelle Wind Turbines Ltd, UK;
- TairuiWindpower CO, China;
- Shenzhen Lemon Digitals Limited, China;
- HEFEI HUMMER DYNAMO CO, LTD, China.

Fig. 5.35 Bergey Excel-S, 10 kW turbine [35]

Further, a brief analysis of wind turbines manufactured by the mentioned companies comes.

Iskra Wind Turbine Manufacturers Ltd [36], Nottingham, UK. Iskra wind turbines are some of the most efficient and cost-effective small wind turbines on the market. The manufacturing company has become a reliable supplier of small wind turbines on the European market. The company has full support of Nottingham City Council, which promotes the "*green energy*" and environmental protection policy, implemented also in schools. Figure 5.36 presents an image of Iskra AT5-1 turbine installed in 2005 at a school in Redland, UK under a program of CO_2 emissions reduction. The turbine has a three blade rotor mounted on a 12 m tower. It is connected to the network. This is important both economically and environmentally,

Fig. 5.36 Iskra AT5-1 wind turbine installed near the Redland school, UK [36]

and also educationally, as part of an educational program to inform the community about opportunities for using alternative energy sources.

WestWind, J.A. Graham Renewable Energy Services [37] is located in Northern Ireland and since 1983 has produced thousands of wind turbines rated 3, 5, 10 and 20 kW, installed worldwide and operating in severe weather conditions. Ten kilowatt turbine is designed to supply small communities. It can be installed in one or more units in one place (in case of high demand for electricity). Figure 5.37 presents the picture of two wind turbines installed (with a solar installation) at Euda wind farm (Australia). The turbine has a three-blade rotor and 6.2 m diameter, starts at 2 m/s speed. The rotor speed varies within 110–600 min^{-1} and is equipped with an automatic protection against excessive wind speed. The cost of turbine (without batteries) is $53,820.

Gazelle Wind Turbines Ltd, UK [31] was founded in 1998 when the MKW group, responsible for research within North Energy Associates, pointed a market niche for small and medium wind turbines. The prototype turbine was developed with the assistance of the Department of Labour and Industry SMART, UK. Gazelle turbine represents an elegant energy solution for small and medium consumers, such as schools, small businesses, rural companies, water pumping and eco-centres. Its 11 m carbon and epoxide fibre rotor, through a two-stage planetary multiplier, drives the four-pole generator at 1,500 min^{-1}, generating 20 kW of electricity (Fig. 5.38). The turbine is connected to the network. At 6.5 m/s wind speed and 13 m tower it is designed to generate approximately 60 MW per year. Electricity is standard – three phases 400/415 V, nominal frequency – 50 Hz. Due to its new director Ken Chaplin, Gazelle company today is well positioned for growth in this vibrant sector.

Fig. 5.37 Two WestWind 10 kW turbines installed at Euda wind farm (Australia) [37]

Fig. 5.38 Gazelle 20 kW wind turbine, installed at Sunderland

Fig. 5.39 Ten kilowatt Tairui wind turbine

TairuiWindpower CO [38] is a leader in the production of small wind turbines in China. Wind turbines have a power ranging 200–20,000 W and are exported to many countries around the world, partly solving the problem of carbon emissions mitigation. The company offers quality wind turbines at relatively low prices. The wind turbine shown in Fig. 5.39 includes a three blade rotor made of rubber-based composite material. Starting speed is 2.5 m/s, calculated speed – 11 m/s. Number of rotations – 160 min^{-1}. The rotor is connected directly to a permanent magnet 10 kW generator, current's voltage – 240 V. Turbine weight is 540 kg. Turbine cost (without batteries) is $9,708.

Hefei Hummer Dynamo CO, Ltd is another company manufacturing small wind power turbines in China [39], located in Hefei, the National Modern Technologies Development Zone. The company specializes in research and development, production and promotion of small wind turbines. The steppes located in northeast China are rich in wind energy. Wind power density is usually 200–300 W/m^2. A three blade turbine with 5 m diameter (Fig. 5.40) can cover electricity needs of an isolated consumer, including a refrigerator or a pump.

5.7.2 Small Power Wind Turbines Designed at the Technical University of Moldova

Small wind turbines should be mostly robust and simple, have maximal resistance and little maintenance, and optimal wind energy conversion efficiency. Given the

Fig. 5.40 Grassland Well model wind turbine, China

topical interest and relatively high costs of imported wind turbines, a team of authors developed two types of small power wind turbines. The wind turbines with servo motors have the ability to track wind direction and remove the bladed rotor out of the wind action at wind speeds exceeding 15–25 m/s. The advantages of these turbines compared to vane wind turbines are:

- Angular positioning stability of the bladed rotor at dynamic fluctuations of air currents direction;
- Bladed rotor protection from overloads, caused by wind speeds exceeding the highest allowed values.

5.7.2.1 Wind Turbine with Tail Vane

Based on the study of wind energy potential and specific orographic terrain surface of Moldova, characterized mainly by gorges oriented along North–south direction, the authors of the present book developed the concept of a three-bladed rotor with asymmetric aerodynamic profile [40–48]. Theoretical research on the developed rotor was carried out using modern software ANSYS CFX5.7 and Autodesk MotionInventor. As a result, the basic parameters of the aerodynamic

Fig. 5.41 3D model of rotor and vane wind turbine (Designed in Technical University of Moldova)

profile characterizing the efficiency of wind energy conversion by the rotor blades were determined.

Taking into account that across the gorges the wind prevails on North–south direction with minor fluctuations, the authors have designed a prototype wind turbine facing the wind by means of a vane. This turbine has a simple construction and requires neither cinematic wind guidance devices, nor devices for the protection of turbine rotor from the excessive wind action. Construction simplicity of the wind turbine with vane leads to about 20–30% cost reduction compared to turbines with cinematic guidance devices. Figure 5.41 presents a 3D model of rotor and vane wind turbine having certain parameters (Table 5.8).

The choice of three blade rotor scheme provides a greater dynamic stability, minimizing related vibrations and sonic background, thus resulting a longer life

Table 5.8 Basic parameters of the vane wind turbine

Parameters	
Bladed rotor diameter	8.6 m
Rotor swept area	58 m^2
Number of blades	3
Blades profile	Aerodynamic asymmetric
Rated power at wind speed of 10 m/s	10 kW
Wind guidance	Vane
Blades positioning	Fixed
Voltage	240 V c.c.
Starting wind speed	2 m/s
Calculated wind speed	10 m/s
Generator	Three-phase, permanent magnet
Generator drive	Direct
Rotation speed, rot/$\min^{-1}$	160
Blade material	Composite based on glass fibre-reinforced resin
Turbine weight	335 kg
Variable height modular telescopic tower	7–18 m
Modular tower weight	650 kg
Accumulator battery	12 V, 200 Ah × 20

period of all components. Direct connection of the rotor to the generator ensures rotor start up at lower wind speeds, production of a larger amount of energy, requires less demanding maintenance compared to turbine multiplier case. Specially designed permanent magnet generator combines efficiency with the simplicity of construction. The outer coverage of blades featuring asymmetric aerodynamic profile, also the gondola cone and the vane are manufactured from composite materials reinforced with glass fibre employing modern technologies in the CESCER laboratory, TUM:

- MEKP solidifier (methyl-ethyl-ketone-peroxide);
- Luperox K1 Standard, ATOFINA, France;
- CRYSTIC polyester resin;
- Gelcoot (white and black colours);
- Glass fibre.
    - Scrint Gobain Vetrotex;
    - Woven Rowind;
- Spray Up Rowind;
- Aluminium Hydroxide ATH;
- Polyurethane glue IMFI (France).

The manufacturing technology for blade covering, vane and gondola cone is similar to that of the blades for multi-blade rotor of micro hydro power plants described in Chap. 4, Sect. 4.5.2.11. Figure 5.42 shows the gondola cone and blades fabrication. Composite materials components resistance is comparable to that of the

Fig. 5.42 Blades (**a**, **b**) and gondola parts (**c**, **d**) manufacturing from composite material in the CESCER Laboratory, TUM

metal structures and has competitive properties advantages such as small weight, corrosion resistance, fatigue resistance, low starting torque, and relatively low costs of small series production.

The CESCER Laboratory at TUM is endowed with modern equipment, fully fitted for the production cycle of composite materials machine parts using modern technologies. IT-supported technological and infrastructural equipment provide mobility and diversity in terms of timely implementation of various technical and technological solutions, as well as design and research solutions in the field of machine building.

5.7.2.2 Wind Turbine Equipped with Servo Motor

The wind turbines with servo motors have the ability to track wind direction and remove the bladed rotor out of the wind action at wind speeds exceeding 15–25 m/s. The advantages of these turbines compared to vane wind turbines are:

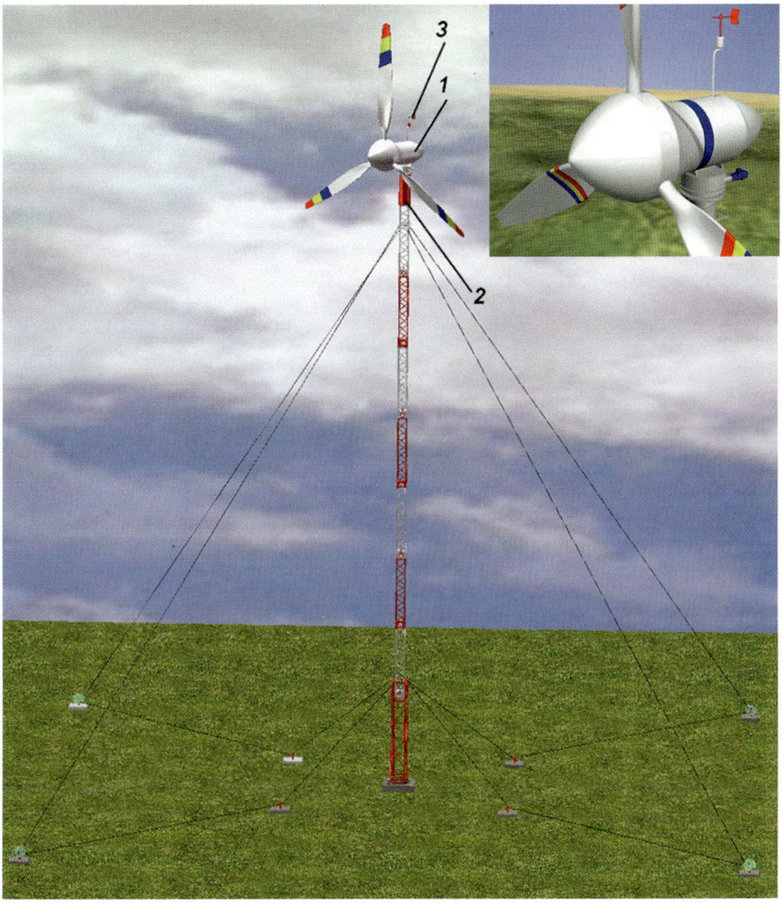

Fig. 5.43 Servo motor wind turbine 3D model

- Angular positioning stability of the bladed rotor at dynamic fluctuations of air currents direction;
- Bladed rotor protection from overloads, caused by wind speeds exceeding the highest allowed values.

Figure 5.43 shows a 3D model of a rotor and an overview of the servo motor wind turbine, developed by a team of authors. Both wind direction rotor orientation and its removal out of the action of air currents is done by means of a device, called servo motor, which performs the kinematical liaison of gondola 1 with tower 2 and which is controlled by a wind vane electronic transducer 3. When the wind direction is changed the vane 3 performs an angular re-positioning, a deviation signal occurs and the control system starts the servo motor, that rotates gondola with rotor in one direction or another until the rotor axis coincides with the direction of air currents.

Fig. 5.44 Assembling stage of the component of the wind turbine

Angular positioning stability is ensured by a certain time delay of the servo motor switch depending on the wind flow action in one direction or another. Repositioning period of bladed rotor perpendicular to airflow velocity vector depends on the kinematical characteristics of the driving mechanism (the servo motor), and actually determines time repositioning stability of the gondola. Kinematical characteristics of the servo motor were determined by the dynamics of the airflow velocity vector variation.

The wind turbine project, developed by authors, was implemented at the Scientific Technical Centre for Implementation of Advanced Technologies at the Technical University of Moldova in cooperation with companies INCOMAS SA, Chişinău, ELECTROMAŞ SA, Tiraspol, Reupies SRL and SA Topaz from Chişinău etc. Figure 5.44 presents the assembling stage of the component of the wind turbine. The team developed the manufacturing technology for blades and gondola parts from composite material reinforced with glass fibre. The rotor blades and gondola cone were made of composite materials in the Laboratory of New Technologies within the Centre for Renewable Energy Conversion Systems Development (CRECSD), Technical University of Moldova. The wind turbines with servo motor, shown in Fig. 5.45, are installed at the Râşcani campus of the Technical University of Moldova and is designed for lighting and irrigation system of the adjacent dendrologic park.

Fig. 5.45 Wind turbine with telescopic tower (**a**) and tubular (**b**) mounted at Râşcani campus

By installing a wind turbine in the park of the Technical University of Moldova, the designers have pursued a specific educational purpose for the student community: the opportunities of using the "*green energy*" without negative impact on the environment.

The wind turbine, developed by authors, was demonstrated at the various international exhibitions [49–55], where was appreciated by gold and silver medals.

Structural and functional parameters of the servo motor turbine are presented in Table 5.9.

The electricity generation depend on the turbine power characteristics (Fig. 5.46) and wind speed probability density function. Figure 5.47 shows the connection schemes of the wind turbine for electricity supply to the dendrologic park lighting system of the Technical University of Moldova. The tower has modular construction with telescopic height advancing, or tubular with special hydraulic device, due to which lifting cranes are not required when changing the turbine installation height.

The chosen construction of the tower ensures the reduction of turbulence induced by the wind passing through the tower. Constructive peculiarities of the turbine correspond to local industry technological possibilities, which fact allowed manufacturing of 75–80% of components at domestic industrial enterprises.

Table 5.9 Basic parameters of the servo motor wind turbine

Parameters	
Bladed rotor diameter	8,6 m
Rotor swept area	58 m^2
Number of blades	3
Blades profile	Aerodynamic asymmetric
Rated capacity at 10 m/s wind speed	10 kW
Wind guidance	Actuator
Blades positioning	Fixed
Voltage	240 V c.c.
Starting wind speed	2 m/s
Calculated wind speed	10 m/s
Generator	Three-phase, permanent magnet
Generator driving	Direct
Rotation frequency	160 min^{-1}
Blade material	Resin-based composite material reinforced with glass fibre
Turbine weight	392 kg
Variable height modular telescopic tower	7–18 m
Modular tower weight	708 kg
Accumulator battery	12 V, 200 Ah × 20

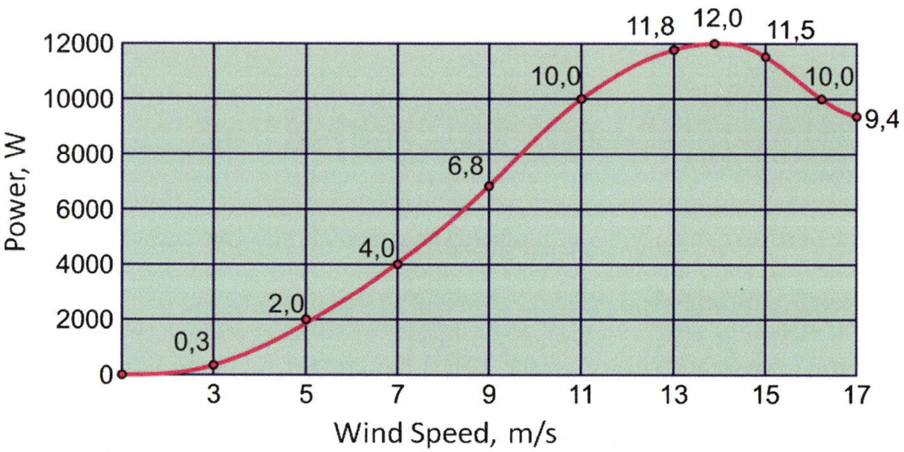

Fig. 5.46 Wind turbine power characteristic

5.7.2.3 Wind Energy Supply of the Dripping Irrigation System

The electricity supply of agricultural land irrigation systems from the public power grid becomes inefficient, that is why various autonomous sources of energy are becoming more widespread [8, 46–48]. Figure 5.48 shows a drip irrigation system powered by electricity from a wind turbine 1 designed by the authors and described

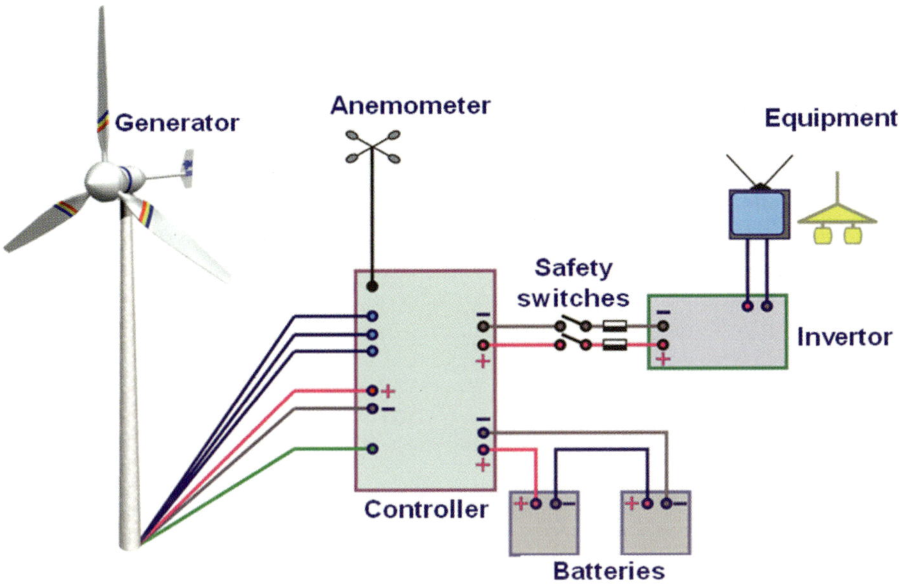

Fig. 5.47 Wind turbine connection schemes for lighting system supply

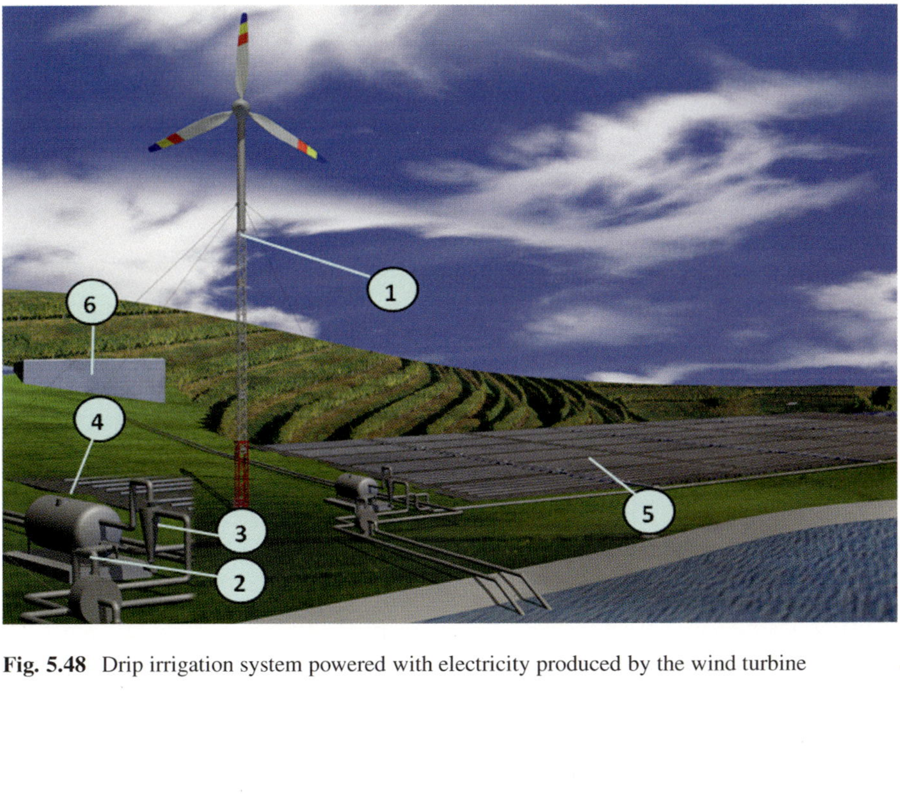

Fig. 5.48 Drip irrigation system powered with electricity produced by the wind turbine

in Sects. 5.7.2.1 and 5.7.2.2. Centrifugal pump 2 with productivity parameters Q (m^3/h) and pumping height H corresponding to the needs for irrigation is supplied with electricity from a wind turbine generator 1. Centrifugal pump 2 sucks water from the lake (or river) and pumps it into the system through the fertilizing dispenser 3 and filtering device 4 connected consecutively in the pump discharge pipe. Fertilized and filtered water under pressure is pumped into the pipe network 5. The irrigation system must include a water storage tank 6 located at an altitude higher than the irrigated ground. Water in the tank may be used during the periods when the wind speed is insufficient to produce the demanded electricity. Subject to the launch of a new generation of electric batteries on the market, more efficient and cheaper, the irrigation systems equipped with batteries could be an alternative that will load at times when irrigation is not appropriate.

References

1. Le baromètre de l'énergie éolienne. Systèmes solaires. Le journal des énergies renouvelables, nr. 141, Janvier – Février 2001, pp 21–29
2. Le baromètre de l'éolien. Systèmes solaires – Le jourmal des énergies renouvelables, nr. 135, Janvier–Février 2000, pp 29–36
3. European Wind Energy Association (1999) Wind energy. The facts. European Wind Energy Association, Luxembourg
4. www.wwindea.org. Accessed 19 July 2007
5. Wind Energy: the facts. An analysis of wind energy in the EU-25. http://www.ewea.org/index.php?id=91. Accessed 30 Apr 2007
6. Troen I, Petersen EL (1989) European wind atlas. Directorate General for Science, Research and Development, Brussels
7. Piebalgs A (2007) Wind energy for future. Mesagerul Energetic J 66:14–16
8. Bostan I, Dulgheru V, Sobor I, Bostan V, Sochirean A (2007) Renewable energy conversion systems. Ch.: Tehnica-Info,2007, 592 pp
9. Gipe P (1995) Wind energy comes of age. Wiley, New York
10. Mortensen NG, Landberg L, Troen I, Petersen EL (1998) Wind atlas analysis and application program (WAsP), vol 1, Getting started. RISO, Roskilde
11. Mortensen N, Landberg L, Troen I, Petersen EL (1993) Wind atlas analysis and application program (WAsP), vol 2, User's guide. RISO, Roskilde
12. www.awstruewind.com. Accessed 30 Apr 2007
13. Todos P, Sobor I, Chiciuc A (2001) Study of statistic meteorological data on the wind characteristics on the territory of the Republic of Moldova. In: Proceedings of the international conference SIELMEN 2001, vol II. Chişinău, 4–6 Oct 2001, pp 23–26
14. Todos P, Sobor I, Chiciuc A, Grosu M (2002) Processing the results of wind raw data on the territory of the Republic of Moldova. Bull Polytech Inst Iaşi XLVIII (LII), Fasc. 5C. Electrotechnics, Energetics, Electronics, Iaşi, pp 301–306
15. Todos P, Sobor I, Chiciuc A, Grosu M (2003) About wind energy potential of the Republic of Moldova. In: 70th anniversary of the State Agrarian University of Moldova, international symposium, 7–8 Oct 2003. Ch.: UASM Publishing House, 2003. Agrarian engineering, pp 155–157
16. Todos P, Sobor I, Chiciuc A (2004) Renewable energy sources in RM: current situation and perspectives. Energetics 1:14–18, ISSN: 1220–5133

17. Sobor I, Caragheaur D, Nosadze Ş (2006) Renewable energy sources: course of lectures//Ministry of Education and Youth. Technical University of Moldova. Ch.: UTM, 2006, 380 p. ISBN 978-9975-45-020-1
18. Sobor I (2007) Wind energy potential in the Republic of Moldova: models, estimations, measurements and validations. Meridian Ingineresc J 2:59–66, ISNN 1683-853X
19. www.risoe-staged.risoe.dk. Accessed 24 July 2007
20. www.zephiros.com. Accessed 27 July 2007
21. www.compositetechcorp.com. Accessed 25 Nov 2007
22. www.vestas.com. Accessed 25 Nov 2007
23. www.friendly-energz.de. Accessed 25 Nov 2007
24. www.enercon.de. Accessed 25 Nov 2007
25. www.repower.de. Accessed 26 Nov 2007
26. www.theage.com.au. Accessed 26 Nov 2007
27. www.science.howstufffworks.com. Accessed 26 Nov 2007
28. www.ewea.org. Accessed 19 July 2007
29. Edwards JL et al (2004) Evaluating state markets for residential wind systems: results from an economic and policy analysis tool. Lawrence Berkeley National Laboratory, p 55. http://repositories.cdlib.org-ibnl/LBNL-56344
30. www.mkw.co.uk
31. AWEA Small Wind Turbine Global Market Study (2007). Published by the American Wind Energy Association
32. Edwards JL et al (2004) Evaluating state markets for residential wind systems: results from an economic and policy analysis tool. Lawrence Berkeley National Laboratory, p 39. http://repositories.cdlib.org-ibnl/LBNL-56344
33. Solar Energy Industries Association road map http://www.seia.org/roadmap.pdf
34. www.bergey.com
35. www.iskrawind.com
36. www.westwind.com.au/turbines.htm
37. www.nb-tairui.com
38. http://www.chinahummer.cn/english
39. Bostan I, Dulgheru V, Bostan V, Ciupercă R (2009) Anthology of inventions: renewable energy conversion systems, vol 4. Ch.: Bons Offices SRL, Chişinău, 458 pp
40. Bostan I, Dulgheru V, Bostan V, Sobor I (2008) Industrial prototype development and manufacture of small wind turbine power. In: Proceedings of the international conference on new and renewable sources of energy CNSNRE 2008, IXth edn. Bucureşti, 23–25 Oct 2008
41. Dulgheru V, Bostan V, Crudu R (2009) Some aspects of increasing the efficiency of converting wind turbine rotor with three blades. In: Proceedings of the technical-scientific conference of the collaborators and doctoral students. Technical University of Moldova, Chişinău, 11 Dec 2009
42. Bostan I, Dulgheru V, Guţu M (2009) Some aspects of designing three-bladed wind micro turbines. In: Proceedings of the scientific conference of the collaborators and doctoral students. Technical University of Moldova, Chişinău, 11 Dec 2009
43. Bostan I, Dulgheru V, Ciupercă R (2006) Theoretical aerodynamic analyses of airfoils for use on small wind turbines and selection of optimal profiles for the wing of helical wind turbine functional model (Part I). Buletinul Institutului Politehnic din Iaşi, Tomul LII(LVI) Fasc. 5D. Section Machine Building, Iaşi, p 1280–1284. ISSN: 1011–285
44. Bostan I, Dulgheru V, Ciupercă R (2006) Theoretical aerodynamic analyses of airfoils for use on small wind turbines and selection of optimal profiles for the wing of helical wind turbine functional model (Part II). Buletinul Institutului Politehnic din Iaşi, Tomul LII(LVI) Fasc. 5D. Section Machine Building, Iaşi, p 1285–1288. ISSN: 1011–285
45. Bostan I, Dulgheru V, Bostan V, Sobor I (2008) Utilisation of the renewable energy sources – wind, solar and hydro in the Republic of Moldova. In: Proceedings of the 1st international scientific and technological conference on radio electronics, informatics, technology, TOPAZ, ASM, UTM. Chişinău, 15–16 Oct 2008, pp 297–304. ISBN 978-9975-45-092-8

46. Bostan I, Dulgheru V, Sobor I (2008) Some aspects of renewable energy utilisation in Moldova. In: Proceedings of the international conference on new and renewable sources of energy CNSNRE 2008, IXth edn. Bucureşti, 23–25 Oct 2008
47. Dulgheru V (2009) Utilisation of the renewable energy sources – wind, solar and hydro in the Republic of Moldova. Scientific Journal "Eng Meridian", no 3. Edited by Technical University of Moldova, p 63–69. ISSN 1683-853X
48. Bostan I, Dulgheru V, Sobor I, Ciupercă R (2008) Low power Aeolian turbine. Cycle of inventions. Salon International des Inventions, Bruxelles, 12–15 Nov 2008 (silver medal)
49. Bostan I, Dulgheru V, Sobor I, Sochireanu A (2009) Small power wind turbine with horizontal axle. Expoziţia Internaţională a Creativităţii şi Inovării EUROINVENT, Iaşi, România, 7–9 mai 2009 (gold medal)
50. Bostan I, Dulgheru V, Bostan V, Sobor I, Sochireanu A (2009) Turbină eoliană cu ax orizontal de putere mică. Universitatea Tehnica din Cluj-Napoca. SC expo Transilivania SA Cluj-Napoca. International Salon of Research and Innovations PROINVENT, VIIth edn. Cluj-Napoca (gold medal)
51. Bostan I, Dulgheru V, Bostan V, Sobor I, Sochireanu A, Ciupercă R, Ciobanu O, Ciobanu R, Bodnariuc I, Dicusară I, Malcoci Iu, Odainâi V, Trifan N, Rusu E, Crudu R, Guţu M (2010) Low power Aeolian plant. World Exhibition and Innovation, Research and New Technology "EUREKA 2010". 18, 20 Nov 2010 (gold medal)
52. Bostan I, Dulgheru V, Sobor I, Bostan V, Sochireanu A, Ciupercă R, Ciobanu O, Ciobanu R, Dicusară I, Trifan N, Cibotari V, Malcoci Iu, Rusu E, Crudu R, Guţu M (2010) Low power aeolian plant with horizontal axis. Patent no. 2,431 MD. International Inventions Fair, INVENT-INVEST SIR20! Iasi, România, 22, 27 Nov 2010
53. Bostan I, Dulgheru V, Sobor I, Bostan V, Sochireanu A, Crudu R, Guţu M, Ciobanu O, Ciobanu R, Trifan N (2010) Renewable energy conversion systems. National exhibition "Made in Moldova", IXth edn. Chamber of Commerce and Industry of the Republic of Moldova (MOLDEXPO – International Exhibition Center). 27, 30 Jan 2010
54. Bostan I, Dulgheru V, Sobor I, Bostan V, Sochireanu A, Ciupercă R, Ciobanu O, Ciobanu R, Dicusară I, Trifan N, Cibotari V, Malcoci Iu, Rusu E, Crudu R, Guţu M (2011) Horizontal axis wind turbines with an output of 10 kW. International Salon of Research and Innovations PROINVENT IXth edn. Cluj Napoca, 14–18 Mar 2011 (gold medal)
55. Boyle G (2004) Renewable energy: power for a sustainable future. Edited by Oxford University Press. Oxford University Press, Oxford, 453 p. ISBN 0-19-926178-4
56. Burton T et al (2001) Wind energy handbook. Wiley, New York, 643 p. ISBN ISBN 0 471 48997 2

Chapter 6
Permanent Magnet Generators (PMG) for Wind Turbines and Micro Hydro Turbines

6.1 Technical Requirements Imposed to Electric Generators for Hydraulic and Wind Energy Conversion

There are essential differences in the design of systems for water/air flow kinetic energy conversion into mechanical energy. First of all, speeds of rotation of the turbine rotors are different: the speed of those which operate in open current does not exceed a few rev/min, for example, flow turbines (water current turbines); hydraulic turbines of low pressure rotate at speeds of tens to hundreds rev/min; small power wind turbines (up to 50 kW) rotate at speeds between 100 and 500 min^{-1}; high-power wind turbines – 20 to 40 min^{-1}. However, both systems are classified as low-speed systems having important similarities: the physical processes that occur at the interaction of the turbine rotor with an open water or air current are similar; due to variations in water or air current speed, the systems function under random; the power cubic dependence of water or air current speed, etc. Therefore, studies on electrical generators, which follow below, refer to both conversion systems, hydraulic and wind as well.

New terminology appeared in the literature with the exploitation of renewable energy sources: "solar pump," "solar battery", "solar generator" "wind generator", "hydro generator", etc. The use of the adjective nearby traditional designation does not necessarily mean that the principle of operation of a certain component of energy conversion system differs considerably from that known. The purpose is to emphasize some technical and design features of this component, arising from the specific technical requirements imposed.

One of the most important technical requirements imposed on all components of the renewable energy conversion systems, in particular, on electric generators is their efficiency. Electricity produced from renewable sources is still expensive because of high initial investment. Efficiency increase of energy conversion is a real way of reducing the investment (to obtain a unit of electricity will require a smaller area of hydro turbine blades or a wind turbine with a smaller diameter, a more robust and easy hub, a lower tower height, etc.).

I. Bostan et al., *Resilient Energy Systems*, Topics in Safety, Risk, Reliability and Quality 19, DOI 10.1007/978-94-007-4189-8_6,

Table 6.1 Technical parameters of wind turbines of various powers: estimated wind speed – 12 m/s; tip speed ratio – 6

Rated output (kW)	Turbine diameter (m)	Turbine rotational speed (rev/min)	Generator rotational speed (rev/min)	Number of generator's poles
2	2.8	487	500	12
3	3.5	390	400	15
5	4.5	303	300	20
10	6.4	213	200	30
20	9.0	151	150	40
50	14.3	95	100	60
100	20.2	68	75	80

The second requirement follows from the fundamental property of turbine rotors used to convert the kinetic energy of water or air flow into mechanical energy – low speed of rotation. For example, for water current speeds between 1.0 and 1.6 m/s, calculated rotation speed of a 5-blade rotor with a 4 m diameter will vary between 1.8 and 2.5 min^{-1}. Rotational speeds of the wind turbines are bigger, but not exceeding a few hundred revolutions per minute for low power turbines and dozens of speed revolutions – for high power turbines. The empirical formulas given in [1] can calculate approximate values of the diameters and speeds of rotation of wind turbines (see Table 6.1 and Fig. 6.1):

$$D = 2,02\sqrt{P} \tag{6.1}$$

$$n = 1364/D \tag{6.2}$$

where
D is the wind turbine rotor diameter, m; P – turbine power, kW;
n – rotational speed, min^{-1}.

Traditional generators with electrical excitation designed for general use do not correspond to the above mentioned two requirements. For powers up to 15 kW their efficiency is about 80%, rotational speed – 1,500 min^{-1}. The design of these generators for speeds of 500 min^{-1} and smaller will lead inevitably to higher decrease of efficiency, which cannot be accepted in the case of renewable energy sources use. Coupling of hydro turbine rotor with the generator will be done with a gear with multiplication ratio between 833 and 600. Such a solution will lead to lower efficiency of the whole system, increase of weight, dimensions, vibration and noise. The alternative solution is to optimize the linkage "turbine rotor – gear – generator" by reducing the rotational speed of the generator, and respectively – the multiplication factor.

The third requirement relates only to wind turbines and corresponds to the global trend of abandoning the scheme "wind turbine – gear – generator" in favor of the scheme "turbine rotor – generator", i.e., using generators with low rotational speed directly coupled to the turbine directly.

We must emphasize that the above requirements are contradictory. From the theory of electrical machines it is known that decrease of the rotating speed leads to

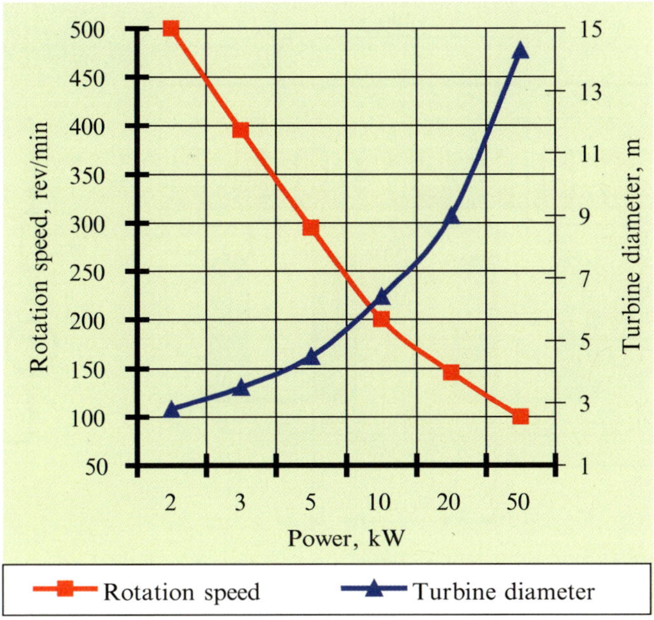

Fig. 6.1 Rotational speed and turbine diameter as function of power

lower technical performance of the machine, including efficiency. A second negative impact of small speeds consists in the excessive increase of the generator diameter and the number of magnetic poles (see Table 6.1). Thus, technological difficulties occur in mounting a large number of magnetic poles on the rotor. Contradiction can be eliminated significantly if traditional electric excitation is substituted by excitation of rare earths permanent magnets, for example, neodymium – iron – boron (NeFeB). Production technology of rare earth permanent magnets was set up in the 1980s of last century, but due to high prices of raw materials, areas of use have been limited. With the reduction of costs, there is a real boom in manufacturing technology of synchronous generators – electric excitation is replaced by permanent magnets, resulting in increased efficiency, lower costs of active materials (copper and electrical steel) and mass [2, 3]. Over 90% of wind turbines with power up to 30 kW are equipped with permanent magnet generators.

6.2 Synchronous Generators with Electric Excitation or Permanent Magnets

The main advantages of synchronous generators with electric excitation (EEG) consist in the possibility of voltage and power factor regulation. Most high power hydro- and turbo-generators are of EEG type, and operate in parallel with electric

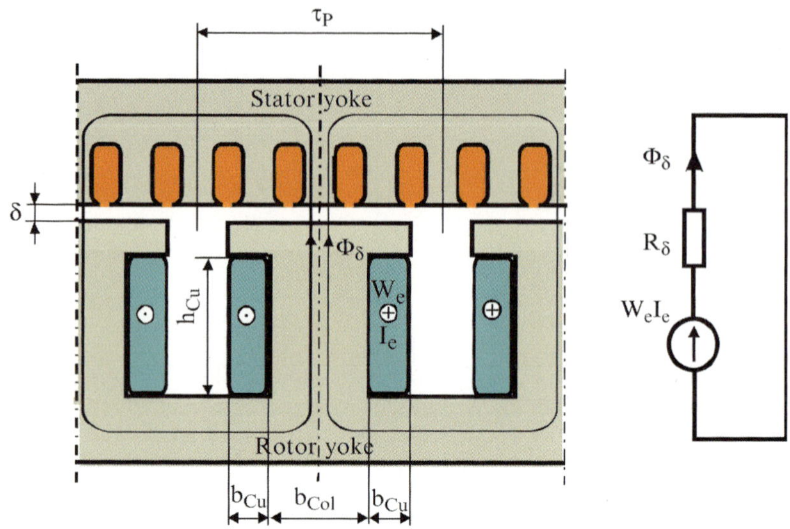

Fig. 6.2 View of the magnetic circuit of EEG and equivalent diagram

grid with a constant rotation speed. Connection to the grid is direct without converters; voltage is adjusted by varying the excitation current. In the case of water current turbines or wind turbines, frequency and voltage varies simultaneously with the speed of water or wind. To stabilize the voltage or frequency, electronic converters are used and the need to adjust the excitation current is not necessary.

EEG disadvantages are notable with increasing the number of poles to be placed on the periphery of the rotor. From Table 6.1 you see that the number of poles is inversely proportional to the speed of rotation. If the goal is reducing the rotational speed to optimize the kinematics chain or direct coupling of the generator with the turbine rotor, you will need to use an EMG with a rotation speed as low as possible. Placing a large number of poles on the rotor leads to the reduction of pole pitch and the surface of the window reserved for the excitation winding. Consequently, we have to increase considerably the column height to achieve the same surface of the window, and respectively, the entire generator. Further, it will be demonstrated that the problem can be solved if electric excitation, which requires an additional source of direct current, will be replaced with permanent magnets. Figure 6.2 shows a sequence of EEG magnetic circuit, and Fig. 6.3 – a permanent magnet synchronous generator (PMG). For both cases, on the right, equivalent diagrams of magnetic circuits are presented. We will use the analytical method of analysis and, for simplicity, we accept the following:

1. Leakage flow is neglected;
2. Magnetic voltage drops on the iron parts shall be deemed null;
3. The nonlinear character of the magnetization curve is not taken into account;
4. Analysis refers to a single air gap.

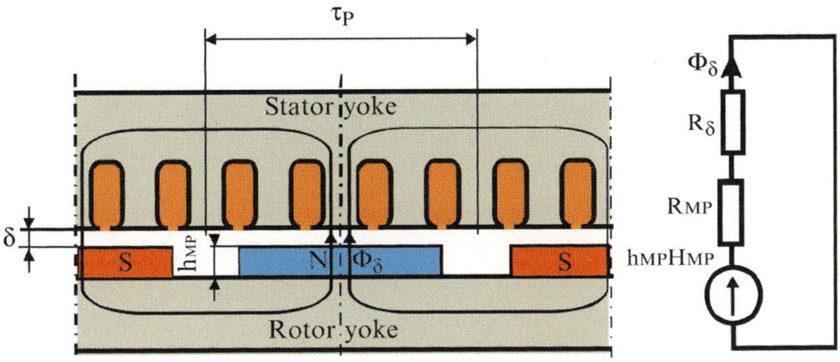

Fig. 6.3 View of the magnetic circuit of PMG and equivalent diagram

Magnetic flux in the air gap Φ_δ is determined by Ohm's law for magnetic circuit

$$\Phi_\delta = \frac{I_e W_e}{R_\delta} = S I_e W_e \frac{\mu_0}{\delta} \tag{6.3}$$

and, taking into account that $\Phi_\delta = SB_\delta$, we obtain flux density in the air gap

$$B_\delta = I_e W_e \frac{\mu_0}{\delta}, \tag{6.4}$$

where $I_e W_e$ is the magnetomotive force of the excitation winding; R_δ – magnetic reluctance of the gap; δ – gap length; μ_0 – magnetic permeability of vacuum; S – pole area.

When using permanent magnets, the same air gap flux Φ_δ will be created by a magnet of h_{PM} thickness, which provides a H_{PM} field strength. The h_{PM} H_{PM} product is analogue to the magnetomotive force $I_e W_e$ and the magnetic field flux in the air gap is determined by the formula:

$$\Phi_\delta = \frac{h_{PM} H_{PM}}{R_{PM} + R_\delta} = h_{PM} H_{PM} \frac{\mu_0 S}{h_{PM} + \delta} \tag{6.5}$$

Taking into account that $H_{PM} = B_r/\mu_0$, we obtain magnetic induction in the gap for the second case

$$B_\delta = B_r \frac{h_{PM}}{h_{PM} + \delta} \tag{6.6}$$

where B_r is the remanence flux density of magnet.

We solve Eq. 6.6 compared with h_{PM} and obtain

$$h_{PM} = \delta \frac{B_\delta}{B_r - B_\delta} \tag{6.7}$$

It results from Eq. 6.4 that flux density in the air gap B_δ is determined by the magnetomotive force of operating winding $I_e W_e$, or otherwise said, by the surface of the window reserved for winding. The bigger is the length of air gap δ, the smaller is the induction B_δ. When using permanent magnets the magnetic flux density in the air gap B_δ depends on the remanence flux density B_r and magnet thickness h_{PM}. The bigger is the magnet thickness h_{PM}, compared to air gap length δ, the smaller is the difference between B_δ and B_r. Magnets made of NeFeB possess remanent induction between 1.1 and 1.3 T, which fact allows fulfilling values of B_δ equal to 0.85–1.0 T. Such values cannot be obtained by small power EEG, except those with rated output bigger than 100 kW and rotor diameter bigger than 1.5 m.

Next, we will determine the ratio between the pole column height and permanent magnet thickness h_{Cu}/h_{PM} (see Figs. 6.2 and 6.3) and how it varies depending on the magnetic flux density in the air gap B_δ and the polar pitch τ_P, i.e. on the number of poles $2p$. The magnetic flux in the pole column is determined by formula:

$$\Phi_{Col} = b_{Col} l_\delta B_{Sat} \tag{6.8}$$

where b_{Col} is the pole column width; l_δ – is the rotor packet length; B_{Sat} – saturation induction of the electrical steel, the column is made of. Taking into account that $\Phi_\delta = \Phi_{Col}$ and $b_{Col} = \tau_P - 2b_{Cu}$, where b_{Cu} is the width of window occupied by the winding, we obtain

$$b_{Cu} = \frac{\tau_P \left(B_{Sat} - B_\delta\right)}{2 B_{Sat}} \tag{6.9}$$

The magnetomotive force $(I_e W_e)$ can be expressed as function of the window dimensions: height h_{Cu} and length b_{Cu}

$$I_e W_e = W_e J A_{Cu} = J\left(h_{Cu} b_{Cu} K_u\right) \tag{6.10}$$

where J is current density; A_{Cu} – is conductor area; K_u – window copper fill factor (window utilization factor).

We substitute Eq. 6.4 $I_e W_e$ according to Eq. 6.10, and b_{Cu} according to Eq. 6.9 and resolve against to h_{Cu}

$$h_{Cu} = \frac{2\delta B_\delta B_{Sat}}{\mu_0 J K_u \tau_P \left(B_{Sat} - B_\delta\right)} \tag{6.11}$$

We divide Eqs. 6.11 to 6.7 and obtain the report h_{Cu}/h_{MP}

$$\frac{h_{Cu}}{h_{MP}} = \frac{2 B_{Sat}\left(B_r - B_\delta\right)}{\mu_0 J K_u \tau_P \left(B_{Sat} - B_\delta\right)}. \tag{6.12}$$

Formula (6.12) determines the ratio of column height necessary to place the excitation winding of EEG inductor, and, respectively, the magnet height depending

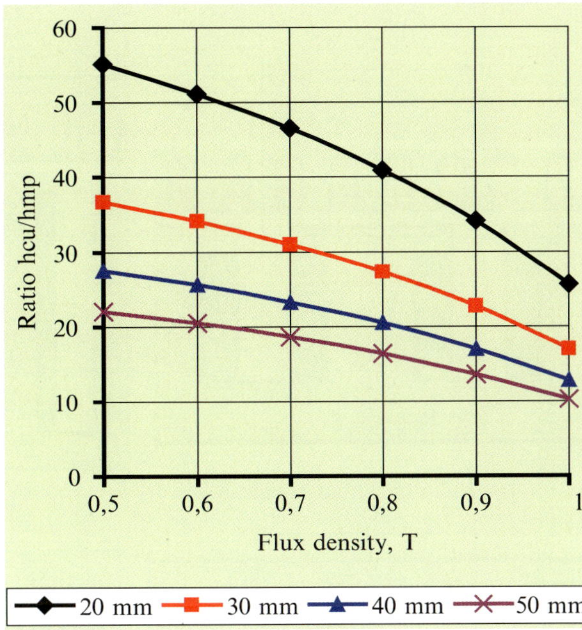

Fig. 6.4 Ratio $h_{Cu}/h_{MP} = F(B_\delta, \tau_P)$

on induction B_δ in the air gap flux and length of polar pitch τ_P. For small power generators (1–10 kW), the length of polar pitch τ_P varies between 20 and 50 mm, flux density $B_r = 1.2$ T, saturation induction $B_{Sat} = 1.8$ T, fill factor $K_u = 0.7$, current density in excitation winding $J = 2$ A/mm^2. This data was used to estimate the family of characteristics $h_{Cu}/h_{MP} = F(B_\delta,\ \tau_P)$, presented in Fig. 6.4. The smaller is the polar pitch τ_P and the magnetic induction in the air gap, the bigger is the ratio h_{Cu}/h_{MP}. For instance, for a 3.0 kW generator, velocity speed 375 rev/min, re-designed on the basis of an induction motor, the polar pitch does not exceed 30 mm. If induction in the air gap is equal to 0.9 T, the ratio h_{Cu}/h_{MP} will be equal to 22.7. From Eq. 6.7 for $\delta = 1.0$ mm, and $B_r = 1.2$ T, we obtain $h_{MP} = 3.0$ mm, and $h_{Cu} = 22.7{\cdot}3 = 78.1$ mm. In the case of EEG it will be necessary to leave more space to place the inductor excitation winding (see Fig. 6.5).

6.3 Constructive Schemes of PMG

The following two criteria are the basis of classification of PMG constructive schemes:

1. Orientation of vector perpendicular to the surface air gap relative to the axis of rotation. The vector can be perpendicular to the axis of rotation or parallel to the axis of rotation

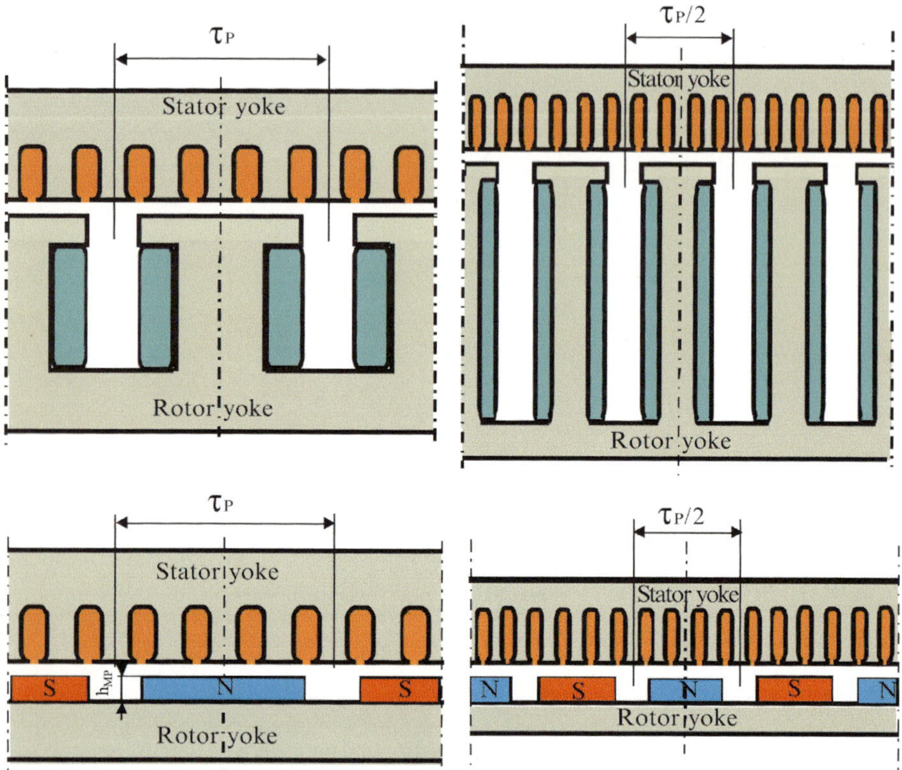

Fig. 6.5 Explanation on how to raise the height of EEG winding (*above*); as for PMG the magnet thickness remains the same (*below*)

2. Orientation of magnetic field lines in the stator (rotor) yoke relative to the direction of rotation. Magnetic field lines can be parallel or perpendicular to the direction of rotation

The combination of these two general principles has generated four different construction schemes, which brief description follows below.

Permanent magnet generator with radial air gap and longitudinal flux. Figure 6.6 shows the simplified diagram of the magnetic circuit. The permanent magnets 2 are mounted on the rotor 1 surface. The stator 3 with winding 4 is separated from the rotor by a portion of air of 1–2 mm air gap. The magnetic flux lines 6 close by: the gap – the stator tooth 7 – the stator yoke 5 – the stator tooth 7 – the rotor yoke 8 – the gap. The direction of flow lines in the stator yoke 5, respectively, the rotor yoke 8, are parallel to the direction of rotation. The vector, perpendicular to the air gap surface, has a radial direction and is perpendicular to the axis of rotation too. Stator 3 and rotor 1 may be interchanged, we get an inverted design, which combines better with the wind rotor. PMG manufacturing technology under this diagram has a high

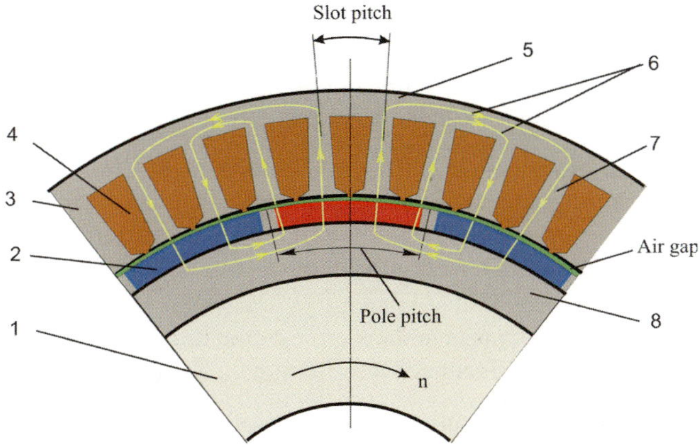

Fig. 6.6 PMG with radial gap and longitudinal flux: *1* – rotor; *2* – permanent magnets; *3* – stator; *4* – winding; *5* – stator yoke; *6* – line of flux; *7* – tooth; *8* – rotor yoke

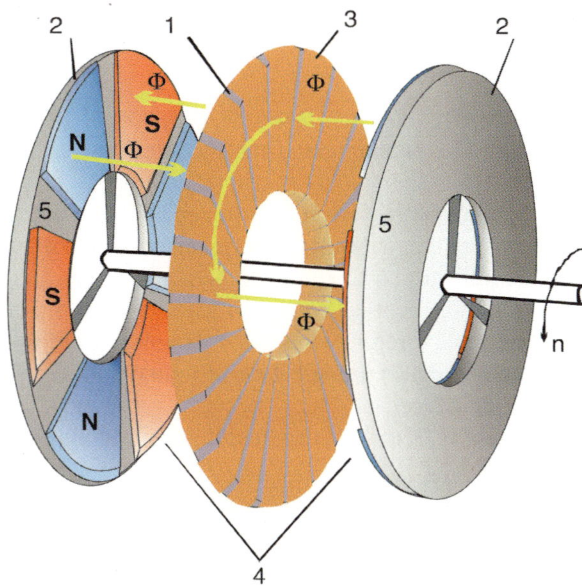

Fig. 6.7 PMG with axial gap and longitudinal flux: *1* – stator; *2* – permanent magnet rotor; *3* – toroidal winding; *4* – air gaps; *5* – rotor yoke

degree of maturity, does not differ from that used in the manufacture of traditional electrical machines and does not require purchase of new production equipment. Most low speed PMGs are produced in conformity with this diagram.

PMG with axial air gap and longitudinal flux. The constructive diagram of generator is shown in Fig. 6.7. The stator 1 and the rotors 2, placed on both sides of stator, take the form of a ring. The rotors are made of a massive steel plate, and the stator is made of electrical sheet steel strip with thickness 0.35–0.50 mm which can be with or without slots (Fig. 6.7 shows the version without slots). The winding

3 is wrapped around the stator ring and is called toroidal or "*torus*". The air gap 4 forms between the two rotors and the stator. The vector perpendicular to the air gap surface is parallel to the axis of rotation, and the magnetic flux in the yokes of rotors and stator is parallel to the direction of rotation.

The main advantages of this design are:

- Small length and can have modular construction;
- In the design without slots the cogging torque is equal to zero, but in the design with slots it can be diminished considerably, and it is easy to fix the stator winding;
- The use of permanent magnets in the simplest constructive form – the plane one;
- End winding (inactive) are reduced to minimum.

As disadvantages the following can be mentioned: it is difficult to make two air gaps with identical lengths and as result, unbalanced electromagnetic forces appear; in the construction without slots (the winding wires are placed in air) the length of the air gap increases scientifically, and respectively, the thickness and mass of permanent magnets increase; although several prototypes have been built [4–6], today we cannot say that the technology of construction of these generators is mature.

Permanent magnet generators with axial air gap and transverse flux. An important peculiarity of this diagram is the modular design of both the stator and the rotor [6]. Figure 6.8 shows only two modules: of the stator and the rotor, respectively. Each modulus of the rotor 1 is stiffened by disc 5 and consists of two pieces 4 made of ferromagnetic material; the permanent magnet 6 is placed between these pieces. Each modulus forms a pair of magnetic poles. The stator 2 of mono phase generator consists of two identical parts: U-form laminated magnetic circuit, made of electric sheet steel, and toroidal windings 3. Magnetic field lines Φ in the U elements yokes are perpendicular to the direction of rotation, and in the air gap they are parallel to the axis of rotation.

The authors of this concept mention the following advantages of this diagram:

- Modular construction allows separate fabrication of each phase and due to this separate reparation;
- Stator windings construction is simple, the process of manufacture can be automated with low cost;
- The concentration of magnetic field lines allows obtaining bigger (than 1.0 T) values of flux density in the air gap;

Also, we would like to note that in the mono phase design of the generator the cogging torque is great and there are no possibilities to be reduced without significantly complicating the construction. Lack of information on cost efficiency and specific expenditures of active materials. Manufacturing technology has not been proven at large-scale.

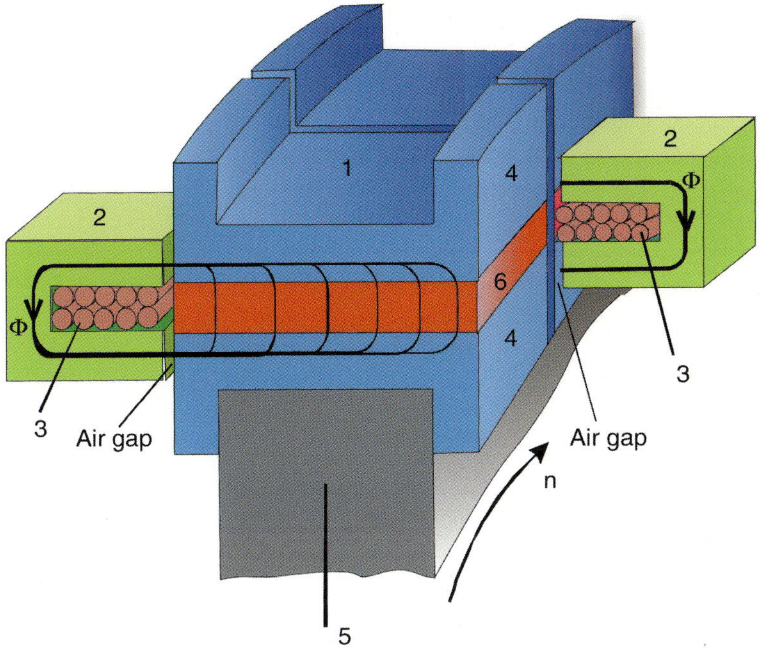

Fig. 6.8 PMG with axial gap and transverse flux: *1* – permanent magnet rotor; *2* – stator; *3* – toroidal winding; *4* – air gaps; *5* – non-ferromagnetic disc

Permanent magnet generators with radial air gap and transverse flux. The design was proposed in 1988 [7]. On peripheral surface of the disc rotor 1, made of non-ferromagnetic material (Fig. 6.9), I-shape piece 2, made of ferromagnetic material, and permanent magnets 3 are mounted. Toroidal winding 4 and U-shape ferromagnetic pieces 5 form the mono phase stator. Number of permanent magnets is two times higher than the number of U-shape pieces. The space between the permanent magnets and U-form armature form the air gap 6.

PMG with transverse flux differs essentially from the longitudinal flow, primarily through the stator winding construction: a phase consists of one or two sections (see Figs. 6.8 and 6.9) of circular shape. Second, the section or phase is linked with the flux of all poles, but not with the flux of a single-pole as it occurs in PMG with longitudinal flux. Double increase in the number of poles leads to twice decrease of the magnetic flux of a PMG pole with longitudinal flux (see Figs. 6.6 and 6.7). In the case of PMG with transverse flux the magnetic flux remains constant (provided the leakage flow neglect). Theoretically, specific power developed by generator (kW/kg) will increase twice. Thus, a low speed generator with transverse flow will weigh two times less than the induction generator with gear.

It is necessary to mention that the real design of the transverse flux generators are more complicated, the leakage fluxes are bigger, and the manufacture technology is in the initial phase. The literature describes only the experimental prototypes.

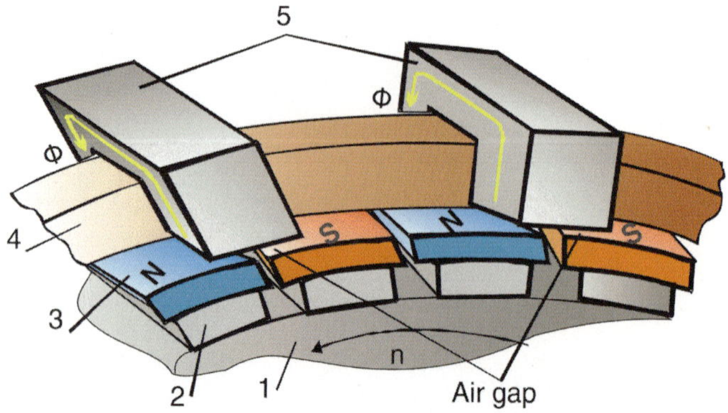

Fig. 6.9 PMG with axial gap and transverse flux: *1* – permanent magnet rotor; *2* – stator; *3* – toroid winding; *4* – air gaps; *5* – non-ferromagnetic magnet disc

The wide-spread diagram is the one with the radial gap and longitudinal flux. Most PMG are produced according to this diagram as well as synchronous and induction motors and the electric excited generators.

6.4 Example of Wind Turbine PMG Design

6.4.1 Main Dimensions of PMG

To design an electrical machine it is necessary to specify some basic features forming the input data:

1. Type of electrical machine: permanent magnet generator.
2. Topology: with radial air gap and longitudinal flow, outer-rotor design, permanent magnets are mounted on the internal cylindrical surface of the rotor cylinder (Fig. 6.10).
3. Rated output, $P_n = 12.5$ kW.
4. Rated rotational speed, $n_n = 150$ tur/min, respectively, angular frequency, $\Omega_n = 15.7\ \mathrm{s}^{-1}$.
5. Number of poles, $2p = 40$.
6. Rated frequency, $f_n = 50$ Hz.
7. Rated phase voltage, $U_{nf} = 165$ V.
8. Number of phases, $m = 3$.
9. Power factor, $Cos\varphi = 0.9$.

At the first stage of the calculation it is necessary to determine the main dimensions of PMG: diameter D_a and length L_a of the armature core (Fig. 6.10). The main dimensions D_a and L_a are related to apparent power S and angular velocity Ω by the so-called electrical machine constant [8].

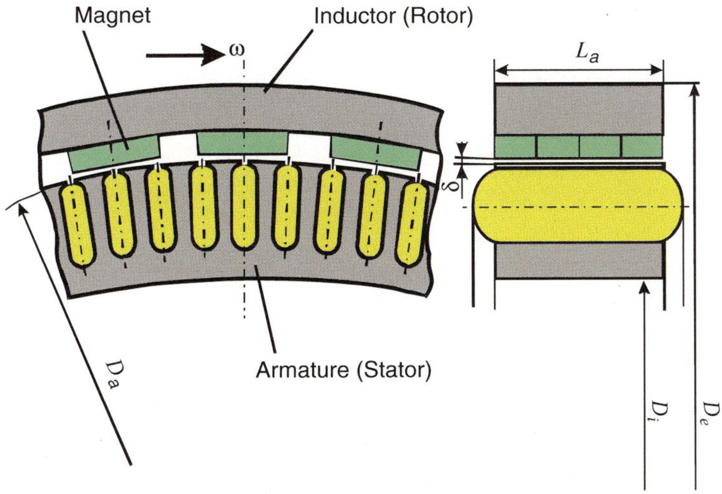

Fig. 6.10 PMG main dimensions

$$C_A = \frac{D_a^2 L_a \Omega}{S} \tag{6.13}$$

which is measured in m^3/VAs and defines the volume of active materials necessary to produce a unit of electricity. Inverse value, $K_A = 1/C_A$ [9, 10] expressed in VAs/m^3 determines energy density

$$K_A = \frac{S}{D_a^2 L_a \Omega} \tag{6.14}$$

At the same time, constant C_A, implicitly, main dimensions D_a and L_a depend on the electromagnetic loadings – air gap flux density B_δ and linear current density A

$$C_A = \frac{D_a^2 L_a \Omega}{S} = \frac{2}{\pi \alpha_\delta K_B K_W A B_\delta} \tag{6.15}$$

where α_δ is the ratio between magnet width b_m and pole pitch $\tau_p = \pi D_a/2p$; K_B – air gap magnetic field shape factor; K_W – winding factor. Initial numerical values of coefficients α_δ, K_B, and K_W can be selected from the literature on traditional electrical machines that recommend the following values [8]: $\alpha_\delta = 0.65$–0.8; $K_B = 1.05$–1.22; $K_W = 0.95$–1.0. For small power electrical machines the allowable values of B_δ and A are, respectively, 0.85–1.05 T and 20–27 kA/m [10].

Design of PMG is an iterative process rather complicated in which the main dimensions are chosen in advance, and then the electromagnetic and optimization calculations follow. Optimization calculation is not the subject of this book, the main dimensions will be determined from Eq. 6.15 with consideration of low power

PMG optimizing results obtained [3]. Thus, ratio L_a/D_a for 5–10 kW PMG is 0.45–0.75. The smaller is this ratio the better are cooling conditions. Taking into account the design topology of PMG (Fig. 6.10) in which the source of loss (armature) is inside the machine, we select the ratio $L_a/D_a = 0.45$, in such a way we increase the diameter and compensate the worsening of cooling conditions.

By substituting the ratio L_a/D_a in Eq. 6.15 we determine the length L_a and diameter D_a:

$$L_a = \sqrt[3]{\frac{2 \cdot 0.202 P_n}{\pi \alpha_\delta K_B K_W A B_\delta \Omega Cos\varphi}} = \sqrt[3]{\frac{2 \cdot 0.202 \cdot 12500}{\pi \cdot 0.66 \cdot 1.1 \cdot 0.954 \cdot 20000 \cdot 0.85 \cdot 15.7 \cdot 0.9}}$$

$= 0.213$ m. Let us accept $L_a = 0.21$ m or 210 mm.

Armature diameter $D_a = L_a/0.45 = 0.21/0.45 = 0.467$ m. Let us accept $D_a = 0.47$ m or 470 mm.

6.4.2 Design of Teeth and Slots Zone and Armature Winding

Number of armature slots

$$Z = 2pmq = 40 \cdot 3 \cdot 1 = 120$$

where $q = 1$ is the number of slots per pole and phase.

Armature slot pitch

$$\tau_a = \frac{\pi D_a}{Z} = \frac{\pi \cdot 470}{120} = 0.0123 \text{ m or } 12.3 \text{ mm.}$$

Slot pitch magnetic flux

$$\Phi_S = B_\delta \tau_a L_a = 0.85 \cdot 12.3 \cdot 210 \cdot 10^{-6} = 21.95 \cdot 10^{-4} \text{ Wb.}$$

The teeth width is determined from the condition that flux density will not exceed allowable values that are equal to 1.5–2.0 T [10]. The teeth has trapezoidal shape and for the value of induction in maximal section equals to 1.7 T we calculate the teeth width

$$b_{t\,max} = \frac{\Phi_S}{B_Z L_a K_{Fe}} = \frac{21.95 \cdot 10^{-4}}{1.7 \cdot 210 \cdot 0.97 \cdot 10^{-3}} = 0.00633 \text{ m, or } 6.3 \text{ mm,}$$

where $K_{Fe} = 0.97$ is the iron lamination space factor.

Admit oval shape of the slot and, respectively, trapezoidal shape of teeth (Fig. 6.11).

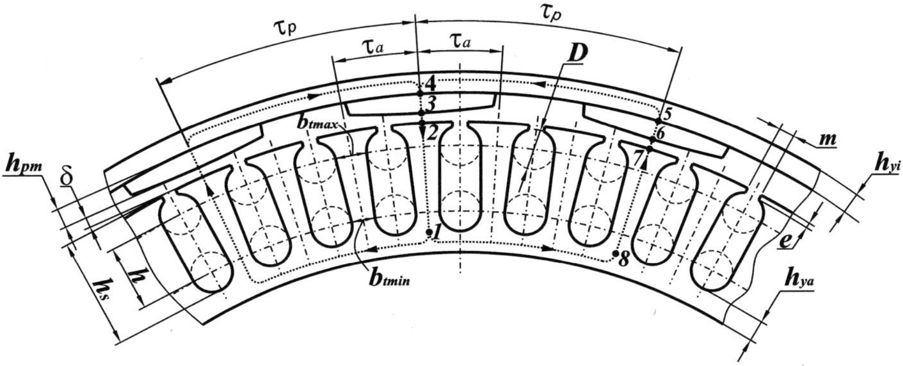

Fig. 6.11 Teeth and slot zone

Slot diameter D is determined from the equation below

$$\pi\,(D_a - D - 2\,(e - 0.4)) = Z\,(b_{t\,\max} + D)$$

or

$$D = \frac{\pi D_a - Zb_{t\,\max} - 2\pi\,(e - 0.4)}{Z + \pi} = \frac{\pi \cdot 470 - 120 \cdot 6.3 - \pi}{120 + \pi} = 5.8\,\text{mm}.$$

Due to technological considerations dimension e is accepted equal to 0.9 mm.
Rated current for an active load

$$I_n = \frac{P_n}{3U_n} = \frac{12500}{3 \cdot 165} = 25.3\,\text{A}.$$

Effective surface of copper conductor

$$S_{Cu} = \frac{I_n}{J} = \frac{25.3}{5.8} = 4.36\,\text{mm}^2,$$

where J is current density in A/mm^2. Allowable value for PMG is equals to 4.0–6.5 [10].

Number of effective conductors in the slot

$$U_s = \frac{A\tau_a}{I_n} = \frac{20000 \cdot 12.3 \cdot 10^{-3}}{25.3} = 9.7.$$

Admit ten active conductors.

Slot surface occupied by the copper

$$S_{S,Cu} = S_{Cu}U_S = 4.36 \cdot 10 = 43.6\,\text{mm}^2.$$

We select the copper conductor with non-isolated diameter $d = 1.4$ mm and with isolated diameter $d_{is} = 1.485$ mm.

Number of elementary conductors in an active conductor $n_{W,el} = \frac{4S_{Cu}}{\pi d^2} = \frac{4 \cdot 4.36}{\pi \cdot 1.4^2} = 2.83$. Admit three elementary conductors in a real one.

Corrected surface of real conductor

$$S_{Cu,cor} = n_{W,el} \cdot \frac{\pi d^2}{4} = 3 \cdot \frac{\pi \cdot 1.4^2}{4} = 4.62\,\text{mm}^2.$$

Corrected surface of slot occupied by copper

$$S_{S,Cucor} = S_{Cu,cor} U_S = 4.62 \cdot 10 = 46.2\,\text{mm}^2.$$

Slot available surface to mount the conductors and the insulation

$$S_S \geq \frac{S_{S,Cucor}}{K_{S,Cu}} \geq \frac{46.2}{0.35} = 132\,\text{mm}^2.$$

where $K_{S,Cu}$ slot copper space factor. For low voltage electrical machines and round conductor winding value $K_{S,Cu}$ varies between 0.4 and 0.45 [10]

Height of rectangular part of slot

$$h = \frac{S_S - \pi D^2/4}{D} = \frac{132 - \pi \cdot 5.8^2/4}{5.8} = 18.2\,\text{mm}^2.$$

Let's verify if condition in above point is respected

$$S_S = hD + \frac{\pi D^2}{4} = 18.2 \cdot 5.8 + \frac{\pi \cdot 5.8^2}{4} = 131.98\,\text{mm}^2.$$

Teeth width in minimal section is determined by equality

$$\pi \left(D_a - 2h - D - 2\left(e - 0,4\right)\right) = ZD + Zb_{t\,\min}$$

or

$$b_{t\,\min} = \frac{\pi D_a - 2\pi h - \pi D - 2\pi\left(e - 0.4\right) - ZD}{Z} = \frac{\pi \cdot 470 - 2\pi \cdot 18.2 - \pi \cdot 5.8 - 2\pi \cdot 0.5 - 120 \cdot 5.8}{120} = 5.37\,\text{mm}.$$

Accept $b_{tmin} = 5.4$ mm.

6.4.3 Calculation of Magnetic Circuit

The magnetic circuit of an electrical machine consists of ferromagnetic materials and air portions. Assuming that the design of electrical machine is symmetric, we can say that the magnetic fluxes of each pole pairs are equal. From Fig. 6.11 we can identify the magnetic circuit of generator, flux lines of a pair of poles and geometrical dimensions. A flux line contains the following parts of magnetic circuit:

1–2 and 7–8 – height of armature tooth;
2–3 and 6–7 – air gap length;
3–4 and 5–6 – height of permanent magnet;
4–5 – inductor (rotor) yoke length;
1–8 – armature (stator) yoke length.

The purpose of this paragraph is to calculate the size of the magnetic circuit, including a permanent magnet, taking into account the specific values of admissible magnetic loading as recommended in [8, 10].

Air gap length and Carter's coefficient. Special literature recommends that air gap length δ to be equal or bigger than $0.001 \cdot D_a$ [3, 10]. In our case, $\delta \geq 0.5$ mm. At the same time, to decrease the influence of armature reaction, permanent magnet losses and their excessive heating, air gap length should be increased. Thermal extension, mechanical stresses and need for decrease of cogging torque, lead to the increase of air gap length, as well. For 5–10 kW generators and rotational speed between 150 and 50 rev/min, the air gap length is 1.5–3.0 mm [11, 12]. For our generator the minimal air gap length was selected (see Fig. 6.11) equal to $\delta = 1.5$ mm. To mention that the air gap length increase will lead to permanent magnet thickness growth and respectively the cost will increase too. To take into account the influence of slots, the air gap equivalent length is calculated

$$\delta_e = K_C \cdot \delta = 1.066 \cdot 1.5 = 1.6\,\text{mm},$$

where K_C is Carter's coefficient and is calculated by formula

$$K_C = \frac{\tau_a}{\tau_a - \gamma\delta} = \frac{12.3}{12.3 - 0.51 \cdot 1.5} = 1.066$$

where factor γ is calculated depending on the slot opening m which is equal to 2.8 mm (selected from technological point of view) and the air gap δ (Fig. 6.11) by formula [8]

$$\gamma = \frac{(m/\delta)^2}{5 + m/\delta} = \frac{(2.8/1.5)^2}{5 + 2.8/1.5} = 0.51$$

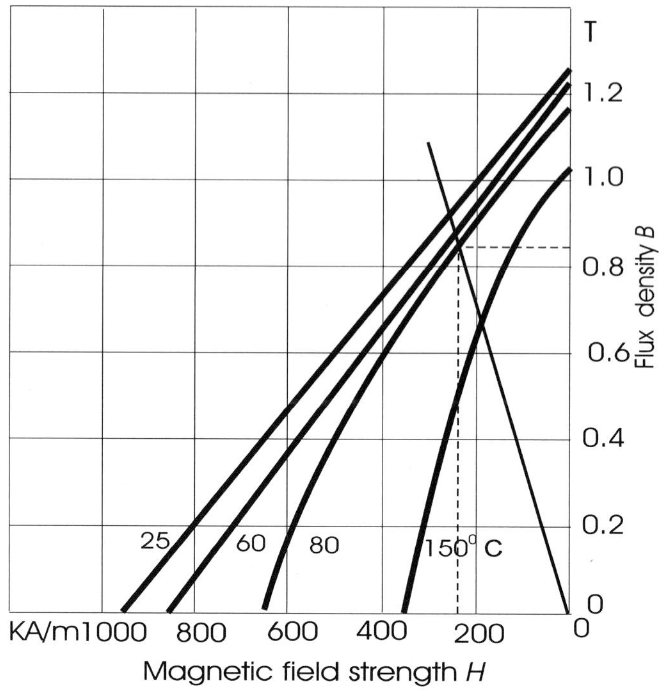

Fig. 6.12 Demagnetizing characteristics of NdFeB magnet

Permanent magnet height. The magneto motive force produced by the magnet is equal to the product between magnetic field strength H_m and the magnet height (thickness) h_{pm}. This force must be equal to the sum of magnetic voltage drops on all parts of magnetic voltages. For the magnetic circuit in Fig. 6.11 results

$$H_m h_{pm} = H_\delta \delta_e + H_t h_s + H_{ya} L_{ya} + H_{yi} L_{yi}$$

Where H_δ, H_t, H_{ya}, *şi* H_{yi} are magnetic field strengths of air gap, tooth, armature yoke and inductor (rotor) yoke. The sum of the last three terms constitutes magnetic voltage drops on the iron portions. At this phase their numerical values are unknown. For permanent magnet generators their sum does not exceed 15–20% from of air gap magnetic voltage. The above equality can be written in the following way

$$H_m h_{pm} = K_U H_\delta \delta_e \quad \text{or} \quad h_{pm} = K_U \frac{B_\delta \delta_e}{\mu_0 H_m}$$

where K_U is the factor that takes into account the magnet voltage on the iron portions.

From the characteristics of demagnetization of magnet N38SH (Fig. 6.12) for $B_\delta = 0.85$ T and operating temperature of about 80°C, the magnetic field strength H_m is equal to 250,000 A/m. Thus, the demanded thickness of magnet is

$$h_{pm} = K_U \frac{B_\delta \delta_e}{\mu_0 H_m} = 1.15 \frac{0.85 \cdot 1.6 \cdot 10^{-3}}{4\pi \cdot 10^{-7} \cdot 250000} = 0.00498\,\text{m}, \quad \text{or} \quad h_{pm} = 5\,\text{mm}.$$

Permanent magnet width. To decrease the variation of magnetic reluctance between the magnet and the armature teeth when the inductor rotates and implicitly of the cogging torque it is necessary that magnet width is equal to a natural number of teeth pitch, i.e. $W_{pm} = n\tau_a$. As well, for permanent magnet generators the relative magnet width $\alpha_{pm} = W_{pm}/\tau_p$ should be about 2/3 [2, 3], where the pole pitch is equal to

$$\tau_p = \frac{\pi\,(D_a + 2\delta)}{2p} = \frac{\pi\,(470 + 3) \cdot 10^{-3}}{40} = 0.03715\,\text{m or } 37.15\,\text{mm}.$$

Compliance with these two conditions can occur if the magnet width is equal to

$$W_{pm} = 2\tau_a = 2 \cdot 12.3 \cdot 10^{-3} = 0.0246\,\text{m or } 24.6\,\text{mm}$$

and

$$\alpha_{pm} = W_{pm}/\tau_p = 24.6/37.15 = 0.66.$$

Inductor yoke width (thickness)

$$h_{yi} = \frac{\Phi_\delta K_\sigma}{2B_{yi} L_a K_{Fe}} = \frac{44 \cdot 10^{-4} \cdot 1.24}{2 \cdot 1.3 \cdot 210 \cdot 10^{-3} \cdot 0.97} = 0.01\,\text{m or } 10\,\text{mm},$$

where $B_{yi} = (1.1\text{–}1.5)$ T – admissible flux density in rotor yoke [10]; $K_\sigma = 1.24$ – leakage factor. The report between the magnet flow Φ_{pm} and the air gap flow was calculated by using the finite element method.

Armature yoke height (thickness)

$$h_{ya} = \frac{\Phi_\delta}{2B_{ya} L_a K_{Fe}} = \frac{44 \cdot 10^{-4}}{2 \cdot 1.1 \cdot 210 \cdot 10^{-3} \cdot 0.97} = 0.0098\ \text{m}, \quad \text{accept } 10\,\text{mm}.$$

Magnetic induction in tooth. As the tooth has V-shape (see Fig. 6.11) we calculate the magnetic flux density in three sections: minimal, average and maximal.

$$B_{t\,\max} = \frac{\Phi_s}{b_{t\,\min} L_a K_{Fe}} = \frac{21.95 \cdot 10^{-4}}{5.4 \cdot 210 \cdot 0.97 \cdot 10^{-6}} = 1.99\,\text{T}.$$

$$B_{tavr} = \frac{\Phi_s}{b_{tavr} L_a \cdot K_{Fe}} = \frac{21.95 \cdot 10^{-4}}{5.85 \cdot 210 \cdot 0.97 \cdot 10^{-6}} = 1.84\,\text{T}.$$

$$B_{t\,\min} = \frac{\Phi_s}{b_{t\,\max} L_a \cdot K_{Fe}} = \frac{21.95 \cdot 10^{-4}}{6.3 \cdot 210 \cdot 0.97 \cdot 10^{-6}} = 1.71\,\text{T},$$

where b_{tavr} is tooth average width

$$b_{tavr} = \frac{b_{t\,\max} + b_{t\,\min}}{2} = \frac{6.3 + 5.4}{2} \cdot 10^{-3} = 0.00585\,\text{m or } 5.85\,\text{mm}.$$

We state that the maximal value of induction does not exceed the admissible value equal to 2.0 T [10].

Magnetic field strength and magnetic voltage on parts of magnetic circuit:

- Segment 1–2 or armature teeth;

Field strength is determined from the magnetization curve of electric sheet steel 2,212 [8, 13] for certain values of magnetic flux density

$$B_{tmax} = 1.99\text{ T},\ H_{tmax} = 6{,}420\text{ A/m}.$$

$$B_{tevr} = 1.84\text{ T},\ H_{tavr} = 3{,}220\text{ A/m}.$$

$$B_{tmin} = 1.71\text{ T},\ H_{tmin} = 1{,}940\text{ A/m}.$$

Magnetic voltage on segment 1–2

$$U_{1-2} = \frac{1}{6}\left(H_{t\,\max} + 4H_{tavr} + H_{t\,\min}\right) h_s = \frac{1}{6}(6420 + 4 \cdot 3220 + 1940) \cdot 24.5 \cdot 10^{-3}$$
$$= 86.7\text{ A}.$$

- Segment 2–3 or air gap;

$$U_{2-3} = H_\delta \delta_e = \frac{B_\delta}{\mu_0}\delta_e = \frac{0.85}{4\pi \cdot 10^{-7}} \cdot 1.6 \cdot 10^{-3} = 1082\text{ A}.$$

- Segment 4–5 or inductor (rotor) yoke;

Inductor cylinder is made of massive steel. From the magnetization curve [8] for $B_{yi} = 1.3$ T magnetic field strength is $H_{yi} = 1{,}590$ A/m. Magnetic voltage on segment 4–5 is calculated by formula

$$U_{4-5} = C_{yi} H_{yi} L_{yi} = 0.4 \cdot 1590 \cdot 0.0387 = 24.6\text{ A},$$

where $C_{yi} = 0.4$ – factor which takes into account non-uniform distribution of induction in the yoke [10]; L_{yi} – yoke length

$$L_{yi} = \frac{\pi\left(D_a + 2\delta + 2h_{pm} + h_{yi}\right)}{2p} = \frac{\pi\,(470 + 3 + 10 + 10) \cdot 10^{-3}}{40}$$
$$= 0.0387\,\text{m} \quad \text{or} \quad 38.7\,\text{mm}.$$

- Segment 1–8 or armature yoke;

Armature is made of electric sheet steel 2,212 [10] plates with 0.5 mm thickness. For $B_{ya} = 1.1$ T magnetic field strength $H_{ya} = 332$ A/m.

Magnetic voltage on segment 1–8

$$U_{1-8} = C_{ya} H_{ya} L_{ya} = 0.5 \cdot 332 \cdot 0.0323 = 5.4\,\text{A},$$

where $C_{ya} = 0.5$ – factor which takes into account non-uniform distribution of induction in the armature yoke [10]; L_{ya} – armature yoke length

$$L_{ya} = \frac{\pi\left(D_a - 2h_s + h_{ya}\right)}{2p} = \frac{\pi\,(470 - 2 \cdot 24.5 - 10) \cdot 10^{-3}}{40}$$
$$= 0.0323\,\text{m or } 32,3\,\text{mm}.$$

Coefficient K_U

$$K_U = \frac{U_{2-3} + U_{1-2} + U_{4-5} + U_{1-8}}{U_{2-3}} = \frac{1082 + 86.7 + 24.6 + 5.4}{1082} = 1.11$$

and does not exceed preliminary accepted value.

6.4.4 Calculation of Parameters

Permeance factor of the slot. For an oval slot (see Fig. 6.11) the permeance factor is calculated by formula [10]

$$\lambda_S = \frac{h}{3D} + \frac{e}{m} + 0.685 = \frac{18.2}{3 \cdot 5.8} + \frac{0.9}{2.8} + 0.685 = 2.05.$$

Tooth tip permeance factor, according to [10]

$$\lambda_t = 0.875 \frac{5\left(\frac{\delta_{ecv}}{m}\right)}{5 + 4\left(\frac{\delta_{ecv}}{m}\right)} = 0.875 \frac{5\left(\frac{6.26}{2.8}\right)}{5 + 4\left(\frac{6.26}{2.8}\right)} = 0.7$$

where δ_{ecv} is the equivalent length of the slot calculated with account of magnet space permeability, which is approximately equal to the air one

$$\delta_{ecv} = \delta + \frac{h_{PM}}{\mu_{PMr}} = 1.5 \cdot 10^{-3} + \frac{5 \cdot 10^{-3}}{1.05} = 6.26 \cdot 10^{-3}\,\text{m}.$$

Harmonic permeance factor according to [8]

$$\lambda_d = 0.03\frac{\tau \cdot \alpha_\delta}{\delta \cdot K_C \cdot q} = 0.03\frac{36.9 \cdot 0.66}{1.5 \cdot 1.066 \cdot 1} = 0.45$$

where $\tau = \pi D_a/2p = \pi \cdot 0.47/40 = 36.9 \cdot 10^{-3}$ m.

Permeance factor of winding lateral sides. For different types of electrical machines this coefficient was determined experimentally. According to [10, 14] for synchronous machines with armature winding in a single layer the permeance factor of the sides $\lambda_W = 0.32$.

Leakage reactance of winding

$$X_\sigma = 4\pi f \mu_0 \frac{W^2}{pq} L_a \sum \lambda = 4\pi \cdot 50 \cdot 4\pi \cdot 10^{-7} \cdot \frac{200^2}{20 \cdot 1} \cdot 0.21 \cdot 3.53 = 1.17\,\Omega$$

where $\Sigma\lambda = \lambda_S + \lambda_S + \lambda_S + \lambda_W = 2.05 + 0.7 + 0.46 + 0.32 = 3.53$; $W = U_S Z/2$ m $= 10 \cdot 120/2 \cdot 3 = 200$ – number of turns of one phase.

Magnetization or main reactance of armature winding

$$X_a = \frac{4\mu_0 m f (W K_W)^2 \tau L_a}{\pi p \delta_{ecv}} = \frac{4 \cdot 4\pi \cdot 10^{-7} \cdot 3 \cdot 50 \cdot (0.955 \cdot 200)^2 \cdot 0.0369 \cdot 0.21}{\pi \cdot 20 \cdot 6.26 \cdot 10^{-3}}$$
$$= 0.54\Omega$$

Synchronous reactance

$$X_S = X_\sigma + X_a = 1.17 + 0.54 = 1.71\Omega$$

DC resistance. For low voltage machines with windings of enamelled copper the average length of a coil turn is calculated by formula [10]

$$l_{av} = 2L_a + 2.4\frac{\pi (D_a - h_s)}{2p} + 0.1 = 2 \cdot 0.21 + 2.4\frac{\pi (0.47 - 0.0245)}{40} + 0.1$$
$$= 0.55\,\text{m}.$$

Total length of conductor in a coil

$$L_C = W l_{av} = 200 \cdot 0.55 = 110\ \text{m}.$$

DC resistance at room temperature (+20°C)

$$R_{20} = \frac{\rho_{Cu} l_C}{S_{Cu}} = \frac{0.0175 \cdot 10^{-6} \cdot 110}{4.36 \cdot 10^{-6}} = 0.44\Omega$$

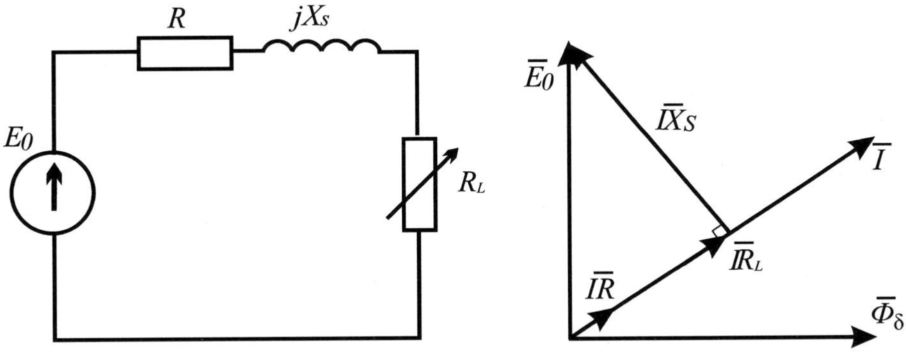

Fig. 6.13 PMG equivalent scheme and phasor diagram

And at temperature +80°C

$$R_{80} = R_{20}\,(1 + \alpha_{Cu}\Delta t) = 0.44(1 + 3.81 \cdot 10^{-3} \cdot 60) = 0.53\ \Omega$$

where $\alpha_{Cu} = 3.81 \cdot 10^{-3}$ is the temperature coefficient of resistivity for copper.

6.4.5 External Characteristic of Generator

According to definition, external characteristic presents graphic interpretation of voltage on load as function on the current or the power output obtained at a constant rotational speed. If the generator operates in autonomous regime, change of the wind speed will cause changes in rotational speed, generator's voltage and frequency. Further, the external characteristics are calculated for three rotational speeds of 80, 150 (rated value) and 180 rev/min. This is the range in which the speed of rotation of a 3-blade turbine with a diameter of 8 m varies, when operating in autonomous mode. Calculated external characteristics are compared with measured ones obtained at an active load ($Cos\varphi = 1$).

Electro motive voltage generated

$$E_0 = 4.44 K_W W f B_\delta \alpha_\delta \tau L_a = 4.44 \cdot 0.955 \cdot 200 \cdot 50 \cdot 0.85 \cdot 0.66 \cdot 0.0369 \cdot 0.21$$
$$= 184.3\ \mathrm{V}$$

where $K_W = 0.955$ is the winding ratio of the diameter pitch and slot inclination of the tooth pitch.

Equivalent diagram of the generator and phasor diagram. Equivalent diagram and phasor diagram for the live loading are shown in Fig. 6.13. The current of rated speed or any other rotational speed n results from the phasor diagram.

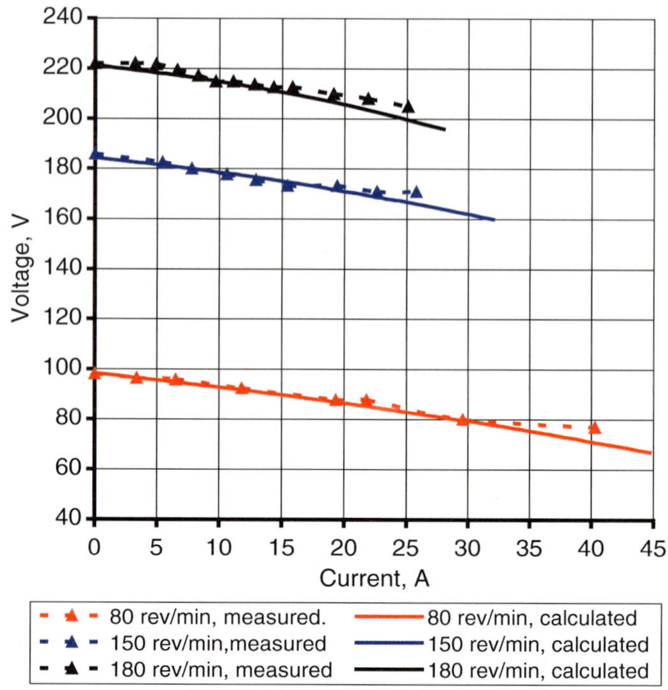

Fig. 6.14 PMG external characteristics

$$I = \frac{E_0}{\sqrt{(R+R_L)^2 + X_S^2}} \text{ or } I = \frac{E_0 \frac{n}{150}}{\sqrt{(R+R_L)^2 + \left(X_S \frac{n}{150}\right)^2}}.$$

For various values of load R_L, we calculate the current I by using one of the above formulas and the voltage $U = IR_L$. Figure 6.14 shows external characteristics $U(I)$ calculated and measured. In the high currents zone magnetic circuit saturation is more pronounced and the voltage measured is higher than that calculated because the reactance X_S decreases. The variation of reactance X_S is not taken into account.

6.4.6 *Mass of Active Materials*

Mass of inductor rim steel

$$M_{ri} = \rho_{Fe}\pi\left(D_{ex} - h_{yi}\right)h_{yi}L_a = 7700 \cdot \pi\,(0.503 - 0.01) \cdot 0.01 \cdot 0.21 = 25.0\,\text{kg},$$

where $\rho_{Fe} = 7{,}700\,\text{kg/m}^3$ is density of solid steel; D_{ex} – external diameter of the rim (inductor cylinder)

$$D_{ex} = D + 2\left(\delta + h_{PM} + h_{yi}\right) = 0.47 + 2\,(1.5 + 5 + 10) \cdot 10^{-3} = 0.503\,\text{m}.$$

Armature rim steel mass

$$M_{ra} = \rho_{Efe}\pi\left(D - 2h_s - h_{ya}\right)h_{ya}L_a K_{Fe}$$
$$= 7650 \cdot \pi\,(0.47 - 2 \cdot 0.0245 - 0.01) \cdot 0.01 \cdot 0.21 \cdot 0.97 = 20.1\,\text{kg}$$

where $\rho_{Efe} = 7{,}650$ kg/m^3 – density of electrical sheet steel.

Teeth steel mass

$$M_Z = \rho_{Efe}h_s b_{tevr}L_a Z K_{Fe} = 7650 \cdot 0.0245 \cdot 0.00585 \cdot 0.21 \cdot 120 \cdot 0.97 = 26.7\,\text{kg}.$$

Copper mass

$$M_{Cu} = \rho_{Cu} m L_C S_{Cu} = 9800 \cdot 3 \cdot 110 \cdot 4.36 \cdot 10^{-6} = 14.1\,\text{kg},$$

where ρ_{Cu} – copper density.

Magnets mass

$$M_{PM} = 2\rho_{PM}pL_a W_{PMavr}h_{PM} = 7500 \cdot 2 \cdot 20 \cdot 0.21 \cdot 0.0258 \cdot 0.005 = 8.1\,\text{kg}.$$

where ρ_{PM} – NdFeB magnet density.

Total mass of active materials

$$M_\Sigma = M_{ri} + M_{ra} + M_Z + M_{Cu} + M_{PM} = 25 + 20.1 + 26.8 + 14.1 + 8.1$$
$$= 94.1\,\text{kg}$$

Specific mass of active materials $M_S = M_\Sigma/P_n = 94.1/12.5 = 7.5$ kg/kW. Compared to other prototypes of permanent magnet generators: 7.0 kg/kW, $P_n = 30$ kW, $n = 60$ rev/min [2]; 6.6 kg/kW, $P_n = 20$ kW, $n = 211$ rev/min [15].

6.4.6.1 Power Losses and Efficiency

Resistive losses in conductors (armature winding)

$$P_{Cu} = mI_n^2 R = 3 \cdot 25.3^2 \cdot 0.53 = 1017\,\text{W}.$$

Losses in teeth steel

$$P_{Fe,s} = K_{Fe,s}P_{15}\left(\frac{B_{Zevr}}{1.5}\right)^2 M_Z = 2 \cdot 5.5 \cdot \left(\frac{1.84}{1.5}\right)^2 \cdot 26.8 = 444\,\text{W},$$

where $K_{Fe,s} = 2$ – empirical correction coefficient for teeth [10]; $P_{15} = 5.5$ W/kg – specific losses of steel at field induction equal to 1.5 T [13].

Armature yoke steel losses

$$P_{Fe,ra} = K_{Fe,y} P_{15} \left(\frac{B_{ay}}{1.5} \right)^2 M_{ra} = 1.5 \cdot 5.5 \cdot \left(\frac{1.08}{1.5} \right)^2 \cdot 20.1 = 86\,\mathrm{W},$$

where $K_{Fe,y} = 1.5$ – empirical correction coefficient for yoke [10].

Losses in permanent magnets. Electrical resistivity of NdFeB permanent magnet material is comparable with the common steel, the average value is equal to $144 \cdot 10^{-8}$ Ωm [16]. Slot harmonics will induce eddy currents in magnets that will provoke losses in the same mode as in usual steel. Calculation of losses in magnets are done according to [10]

$$P_{PM} = a_\nu \left(1 + \frac{\tau_a}{2L_a} \right) \frac{K_\nu^3}{\beta_\nu^2} \left(\frac{B_0}{\mu_0 \mu_{rpm}} \right)^2 \rho_{PM} S_{PM} = 17.4 \left(1 + \frac{0.0123}{2 \cdot 0.21} \right) \frac{29.4^3}{511^2} X$$

$$X \left(\frac{0.034}{4\pi \cdot 10^{-7} \cdot 1.05} \right)^2 \cdot 144 \cdot 10^{-8} \cdot 0.21 = 350\,W$$

Where the terms from the certain formula are calculated as follows

$$a_\nu = \frac{1}{\sqrt{2}} \sqrt{\sqrt{4 + \left(\frac{\beta_\nu}{K_\nu} \right)^4} + \left(\frac{\beta_\nu}{K_\nu} \right)^2} = \frac{1}{\sqrt{2}} \sqrt{\sqrt{4 + \left(\frac{510.6}{29.4} \right)^4} + \left(\frac{510.6}{29.4} \right)^2}$$

$$= 17.4$$

$$\beta_\nu = \frac{2Z}{D} = \frac{2 \cdot 120}{0.47} = 510.6$$

$$K_\nu = \sqrt{\frac{\omega_\nu \mu_{PM}}{2\rho_{PM}}} = \sqrt{\frac{1885 \cdot 1.05 \cdot 4\pi \cdot 10^{-7}}{2 \cdot 144 \cdot 10^{-8}}} = 29.4\,\mathrm{m}^{-1}.$$

Angular frequency of magnetic field teeth harmonics

$$\omega_\nu = \frac{\pi n Z}{30} = \frac{\pi \cdot 150 \cdot 120}{30} = 1885\,\mathrm{s}^{-1}.$$

Magnet flux density generated by teeth harmonics

$$B_0 = \beta B_\delta = 0.04 \cdot 0.85 = 0.034\,\mathrm{T},$$

where

$$\beta = \frac{1 + u^2 - 2u}{2\left(1 + u^2\right)} = \frac{1 + 1.52^2 - 2 \cdot 1.52}{2\left(1 + 1.52^2\right)} = 0.04$$

$$u = \frac{m}{2\delta_f} + \sqrt{1 + \frac{m}{2\delta_f}} = \frac{2.8}{2 \cdot 3.9} + \sqrt{1 + \frac{2.8}{2 \cdot 3.9}} = 1.52$$

The fictive air gap is

$$\delta_f = \delta + \frac{h_{PM}}{2\mu_{rpm}} = 1.5 + \frac{5}{2 \cdot 1.05} = 3.9\,\text{mm}.$$

Total losses

$$P_\Sigma = P_{Cu} + P_{Fe,s} + P_{Fe.ra} + P_{PM} = 1017 + 444 + 86 + 350 = 1897\,\text{W}.$$

Calculated efficiency

$$\eta_C = 1 - \frac{P_\Sigma}{P_n + P_\Sigma} = 1 - \frac{1897}{12500 + 1897} = 0.868 \text{ or } 86.8\%.$$

Efficiency determined on the basis of testing in factory conditions

$$\eta_M = 1 - \frac{P_{\Sigma M}}{P_n + P_{\Sigma M}} = 1 - \frac{1600}{12500 + 1600} = 0.886 \text{ or } 88.6\%.$$

6.5 Permanent Magnet Generator as a Component Part of a Wind Turbine

Based on the calculations made in Sect. 6.4 there has been elaborated the outer-rotor design of a permanent magnet generator (PMG) presented in Fig. 6.15a. Compared with a traditional electric machine, here the rotor with permanent magnets (inductor) is placed outside (outer-rotor) and the stator (induced) – inside. Several advantages of this construction can be easily identified from the figures presented below:

1. Wind turbine blades can be conveniently assembled in front of the rotor cylinder, thus directly coupling the wind turbine with PMG.
2. For the same outer diameter of PMG, the outer-rotor design provides more space for mounting a larger number of magnetic poles.
3. Permanent magnet can have the simplest flat shape. Also, a variable air gap is obtained – in the centre of the magnet the air gap length is minimal, and at the periphery – is maximum. Thus, the distribution of magnetic flux per polar step is closer to the sinusoidal distribution.
4. Centrifugal force imposes a pressure on the magnets and a greater reliability of the magnet- glue medium is ensured.

The generator includes the following basic components. A stator 1 with winding 2 is connected via rigid flange 3 with gondola 3 of the wind turbine (Fig. 6.18). Rotor 4 with permanent magnets 5 is fixed by rigid flange 6 with axle 7, mounted

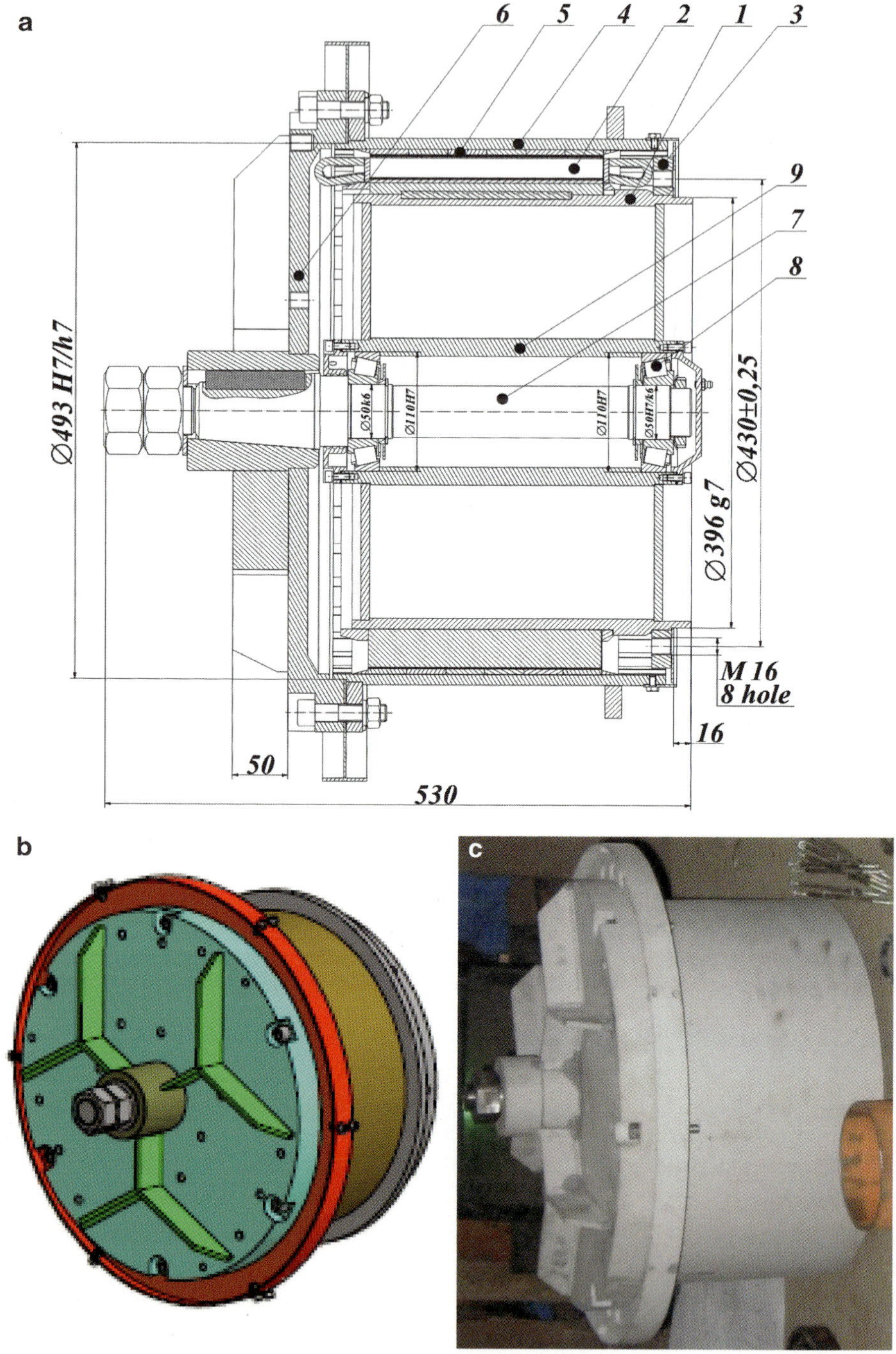

Fig. 6.15 Reverse design of a permanent magnet generator (PMG) Generator (**a**), 3D design (**b**) and industrial prototzpe (**c**)

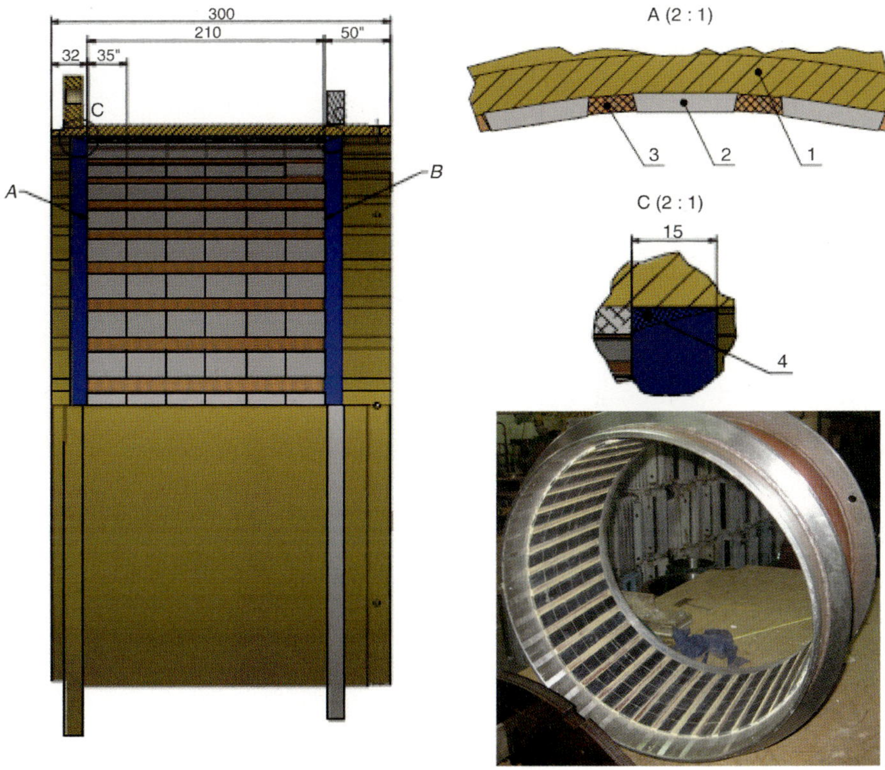

Fig. 6.16 Rotor (inductor) GMP

on radial-axial bearings with rollers 8 inside the bushing 9 of the stator 1. The space between the permanent magnets and fittings form the air gap. On the flange 6 there are fixed the rotor blades 1 of a wind turbine (Fig. 6.18).

Rotor (inductor) GMP (Fig. 6.16) includes a cylinder 1 made form ferromagnetic massive steel on the inner surface of which there are glued magnets 2. The space 3 between magnets 2 and the ring spaces A and B at the ends of magnets are filled with glue in order to exclude the magnets damage during mounting operations.

Another important node of the PMG is the stator or inducer. On the cylindrical surface of the frame 1 (Fig. 6.17a) it is mounted the laminated core 2 with slots. For diminishing the cogging torque the slots are skewed (Fig. 6.17b). The skewed angle is of the same magnitude as one slot pitch. The three phase single layer windings 3 are placed on the slots. Positions 4, 5 and 6 show the insulating pieces for coil ends 3.

In Fig. 6.18 there is presented the 3D model of the wind turbine. Blades 1 are mounted on a flange 2 of the PMG 3 and together with cylinder 4 form the wind turbine rotor. GMP 3 through flange 5 is assembled with gondola 6. The orientation to the wind direction mechanism of the wind turbine rotor includes the servomotor

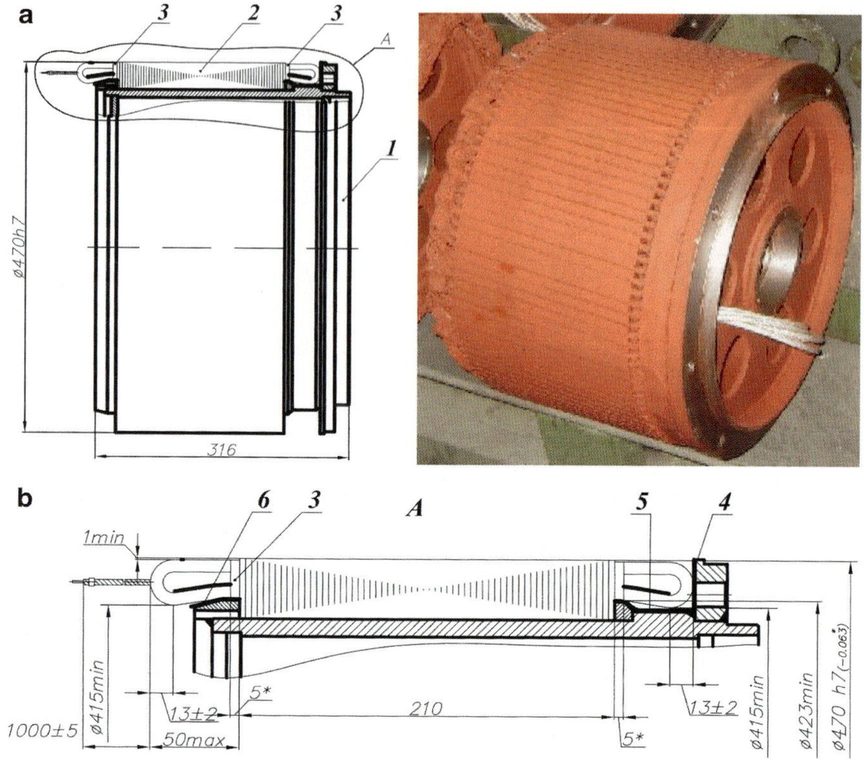

Fig. 6.17 PMG stator

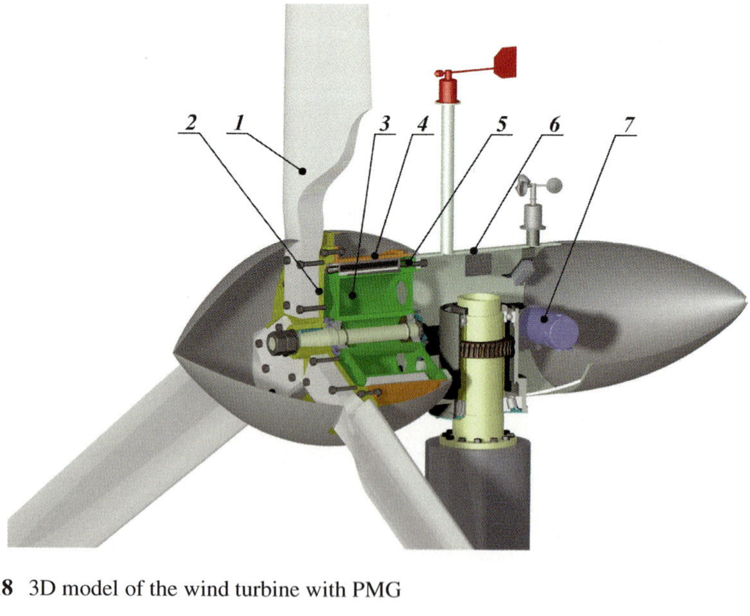

Fig. 6.18 3D model of the wind turbine with PMG

7 connected to the bushing with flange 5 rigidly fixed to the top of the tower 6. The turbine is also equipped with anemometer 8 used to determine the wind speed and weather cock 9 to determine the wind direction.

References

1. Twidel JW, Weir AD (2006) Renewable energy resources, 2nd edn. Taylor & Francis, New York, 625 p. ISBN ISBN 0–419–25320–3
2. Grauers A (1996) Design of direct-diven permanent-magnet generators for wind turbines. Technical Report no. 292, Chalmers University of Technology, Goteborg. ISBN 91-7197-373-7. ISSN 0346-718X
3. Lampola P (2000) Directly driven, low-speed permanent-magnet generators for wind power applications. Acta Polytechnica Scandinavica, Electrical Engineering Series, no. 101, Espoo. ISBN 951-666-539-X. ISSN 0001–6845
4. Dubois MRJ (2004) Optimized permanent magnet generator topologies for direct – drive wind turbines. Les Imprimeries ABC Inc, Lévis, 245 p. ISBN ISBN 0–9734585–0–X
5. Chalmers BJ, Wu W, Spooner E (1999) An axial-flux permanent-magnet generator for a gearless wind energy system. IEEE Trans Energy Convers 14(2):251–257
6. Muljadi E, Butterfield CP, WanY-H Axial flux, modular, permanent-magnet generator with a toroidal winding for wind turbine applications. In: IEEE industry applications conference, National Wind Technology Center, National Renewable Energy Laboratory
7. Weh H, Hoffmann H, Landrath L (1988) Directly driven permanent magnet excited synchronous generator for variable speed operation. In: Proceedings of the European community wind energy conference EWEC'88, Herning, pp 566–572
8. Proektirovanie e'lectricheskih mashin: Uceb. dlya vuzov,- V 2-h kn.:/I. P. opy'lov, B.K. Klokov, V. P. Morozkin, B. F. Tokarev; Pod red. I. P. opy'lov.- 2-e izd. – Moskva.: E'nergoatomizdat, 1993. ISBN 5-283-00724-3
9. Upadhyay KG (2008) Design of electric machine. New Age International, Daryaganj Delhi, p I. ISBN ISBN (13): 978-81-224-2952-7
10. Pyrhonen J, Jokinen T, Hrabovcova V (2008) Design of rotating electrical machines. Wiley, Chichester. ISBN 978-0-470-69516-6
11. Bala C, Ignat M, Stoica V, Ionescu DM (2008) The design and technological aspects on an eolian 5 kW syncron generator with permanent magnets. In: Proceedings of the XVIII international conference on electrical machines, ICEM 08, Vilamoura, 6–9 Sept 2008, pp 1–4
12. Bequé S, Dils J, Van Dessel M (2003) Optimisation of a direct drive low speed permanent magnet wind generator. In: Proceedings of the Nordic Matlab conference 2003, Copenhagen, 21/22 Oct 2003, pp 215–219
13. Isotropic Electrical Sheet Steel under EN 10106–96, GOST 21427.2-83, ASTM A677/A677M-89
14. Shuiskij VP (1968) Raschet e'lectricheskih mashin. E'nergia, Leningradskoe otdelenie
15. Wu W, Ramsden VS, Crawford T, Hill G (2000) A low-speed, high-torque, direct-drive permanent magnet generator for wind turbines. In: Industry applications conference IEEE, Rome, 08–12 Oct 2000, pp 147–154
16. www.x-magnet.net/hard-magnet/Sintered-NdFeB-Magnet. Accessed 12 July 2011

Chapter 7
Sea Waves Energy

7.1 Generalities

If at the end of the nineteenth century the most widespread energy used – the electricity – had an auxiliary and insignificant role in the global energy balance, then in 1930 about 300 billion kWh of electricity were produced in the world, and in 2004 this figure reached 21,000 billion kWh [1]. The material and the spiritual level of mankind are directly dependent on the amount of the available energy. The stringent laws of the nature state that useful energy can be obtained only by converting it from other forms. The World energy structure analysis shows that 4 out of 5 kW are obtained, in principle, using the same method by which the primitive man heated himself, that is by burning fuel, or by using its chemical energy converted into electricity at power plants. Of course, fuel combustion methods have become much more perfect.

But the largest energy reserves are stored in the oceans – a large area of water currents continuously moving and covering about 71% of the planet's surface. The Planet Ocean has a huge energy potential that can be employed to produce electricity. The main sources of the ocean energy considered, at least in the current technical level, refer to: tides, currents, waves, temperature differences of the seawater layers. The first mathematically documented explanation of the tidal forces was done in 1687 by Isaac Newton in his work *"Philosophiae Naturalis Principia Mathematica"*. Tides occur regularly in certain coastal areas of the planet at amplitudes reaching sometimes 14–18 m, resulting in slow oscillations of the sea water level (Fig. 7.1). The principle of using the tidal energy at tidal power plants, by the way the only source presently used from all above mentioned, envisages the arrangement of dammed pools in order to make possible the capture of water energy, triggered by these oscillations, both at tide rise (during flow) and at tide fall (during ebb). The use of the World's sea and ocean energy is still behind the use of the wind and other renewable energy sources. Certain opinions identify a gap of about two decades. Perhaps, the coming years will witness the emergence

I. Bostan et al., *Resilient Energy Systems*, Topics in Safety, Risk, Reliability and Quality 19, DOI 10.1007/978-94-007-4189-8_7,

Fig. 7.1 Shore line after tide withdrawal

Table 7.1 Comparative analysis of energy resources

Energy source	Renewable source	Low capital costs	Low current costs	Minimal environ-mental impact	Predictability	Minimal visual impact	Modular structure
Fossil	No	Yes	No	No	Yes	No	No
Nuclear	No	Yes	No	No	Yes	No	No
Wind	Yes	No	Yes	Yes	No	No	Yes
Solar	Yes	No	Yes	Yes	No	No	Yes
Hydro	Yes	Yes	Yes	No	Yes	No	No
Waves	Yes	No	Yes	Yes	No	Yes	Yes
Sea currents	Yes	No	Yes	Yes	Yes	Yes	Yes

of new technological concepts that will allow to employ the energy stored in the terrestrial hydrosphere. Table 7.1 [2] demonstrates a comparative analysis of the energy resources on the basis of various parameters. This short presentation shows that the sea and ocean energy can become a serious resource capable to satisfy the energy needs of the mankind. Probably in the coming years we shall witness dramatic developments and emergence of new ideas. In order to outline long-term strategies for the development of alternative energy resources the political factor is also expected to intervene. The European Union has agreed on the price of electricity pumped into the grid – 5 €c/kWh (5 cents per kWh) [3]. The protocol is valid for sea wave energy conversion systems with buoys or for systems established by the second generation of wave energy converters (WEC), including a prototype or a pilot station.

Important issues for wave energy exploitation. Several important events have taken place in the past 10 years:

- *The Kyoto Protocol* is a treaty between the governments to establish the tasks related to the increase of renewable energy use in the first decade of the new millennium. Unfortunately, China, the former USSR countries and India have not adopted significant decisions to implement these tasks; as well, in 2001 the USA has announced that it does not intend to implement these decisions;
- *The UK study on renewable sources (1999)* – as part of it, R&D financing in the field of wave energy development has been reintroduced for Great Britain. A certain influence had the Report "Sea Energy – until 2020";
- *The increasing focus on climate change issues* has led to increasing the consensus within the scientific community, highlighting the effects of environmental degradation due to so-called "greenhouse effect". The so-called pole phenomenon related to the increase of icebergs, glacier melting, sea levels raise is the product of this greenhouse effect.
- *The alarming increase in the oil prices (in December 2007 it reached the psychological level of US$100 per oil barrel) from the lowest level since 1998* was the reason of reassessment of conventional energy projects and imposed the attractivity of renewable energy technologies, including conversion of wave and tidal energy. Modern technologies for converting tidal and wave energy are already economically competitive, for example, for isolated communities that use electric generators powered by diesel engines.

In 2005 the Ocean Renewable Energy Coalition (OREC) was established [4] – an association for the promotion and advanced commercial application of the ocean energy. On January 14, 2005, the Electric Power Research Institute (EPRI) presented its final report on wave energy conversion in the U.S. The report described the analysis of the energy potential in various places of the U.S. coast, the viability and economics of various technologies for capturing oceanic energy. In the same year, EPRI launched the second phase of its ocean programme – evaluation of the tidal energy conversion technologies and of energy potential in various coastal areas of the U.S. and Canada.

In April 2005, the Federal Energy Regulatory Commission (FERC) approved the promotion of limited pilot projects, having the right to issue licenses for ocean and hydrokinetic technologies. FERC and the U.S. Department of Energy attempted to explore ways to capture the kinetical energy of rivers, sea currents and waves. The Ocean Power Technologies (OPT) company has a significant role in the U.S.; this company has funded a 2.8 million dollars submarine project at a naval base in Hawaii. In May 2005, the Ocean Power Delivery (OPD) Company announced a contract with an electricity company from Portugal on the construction of the world first commercial farm for wave energy conversion with a capacity of 2.25 MW. More and more investors express their interest in the ocean energy conversion projects, being motivated by low technology costs and increasing taxes for environmental protection.

7.2 Energy potential

7.2.1 Tidal Energy Potential

Tidal energy is a major source of ocean energy. For effective utilization of tidal energy certain natural conditions are required. Firstly, it is necessary that the tidal amplitude is at least 8 m, and, secondly, a natural basin (usually an estuary) is necessary to communicate with the ocean through a narrow opening. These natural conditions exist only in 20 regions of the globe (such as: the Atlantic coasts of France, Great Britain, USA, Canada, in northern Australia, in eastern China, etc.) (Fig. 7.2) [5]. The amount of energy available from this source, if fully employed by tidal power plants, would produce about 100,000 times more energy than all hydro plants currently in operation worldwide. So far, tidal power plants produce electricity at a price that is twice bigger than the price for electricity produced at hydro power plants.

Tidal waves are caused by the gravitational attraction of the moon and the sun, exercised on oceans during the Earth's rotation. Relative movements of these space bodies cause periodic rise and fall of the sea level, according to the number of interaction cycles. These include:

- The semidiurnal cycle that is predominant, due to the rotation of the Earth relative to the gravitational field of the Moon and the Sun (approx. 12 h 24 min);
- The diurnal cycle, due to the rotation of the Earth relative to the gravitational field of the Moon (approx. 24 h 48 min);

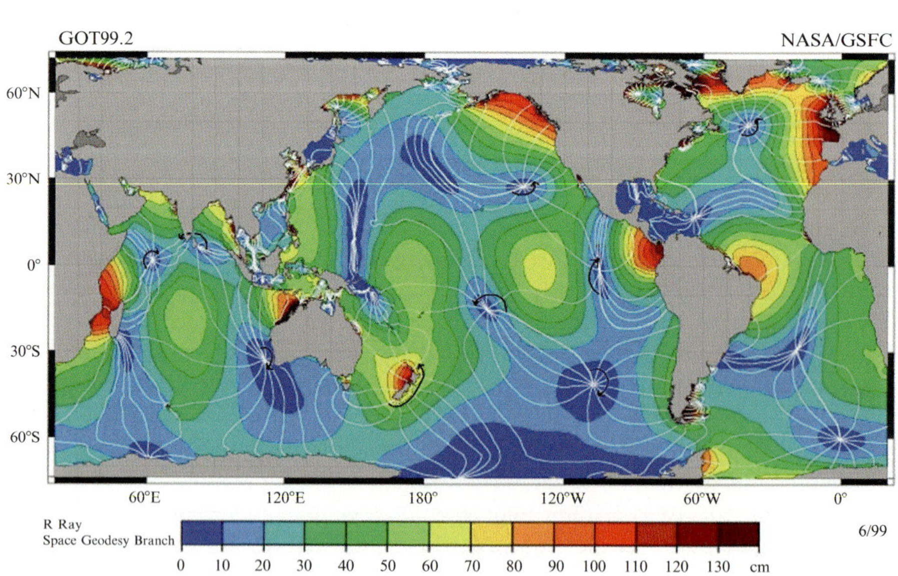

Fig. 7.2 Global distribution of tidal energy resources (height of tides in cm)

- The 14-day cycle, which results from the gravitational field of the Moon combined with that of the Sun, generating the highest and lowest tides;
- The half-year cycle, due to the inclination of the Moon orbit to the Earth orbit, which provides the highest tidal current growth in March and September;
- Other cycles, such as the 19-year and 1,600-year cycles, generated by other complex gravitational interactions.

The magnitude of the tide generating force is about 68% for the Moon and 32% for the Sun, due to their mass and distance from the Earth [5]. High tides are twice stronger than low tides, with some small disturbances over a long cycles' period. In the open ocean the highest tidal amplitude is approximately 1 m. The tides rise substantially closer to the shore. In some cases, tides can be amplified by the tidal waves reflection along the coastline or by resonance. This effect occurs especially in large river deltas and in marine bays. For example, these effects cause up to 11 m high tides in the Severn Bay (UK). Under the influence of these various factors, the size of tidal waves can vary substantially between different segments of the coastline.

The amount of energy that can be obtained from tidal energy varies depending on the location and the period of year. Available energy is roughly proportional to the tidal wave size square. The extraction of energy from tides is considered suitable only in areas where energy is concentrated in large waves or in geographical locations for facilities' construction. Such places have been identified in Britain, France, Eastern Canada, Northern and Eastern Coasts of Russia, Korea, China, Mexico and Chile. Other places have been identified on the Patagonian coast of Argentina, Western Australia and Western India (Table 7.2) [6].

Tidal energy can be exploited directly from the sea currents generated by the combined gravitational forces of the Sun and the Moon. These forces generate semidiurnal movements in the shallow sea water, particularly in places where the coastal morphology creates natural systems, for example around the sea foreland and among the islands. These phenomena produce strong currents and tidal waves, which are prevalent around the British Isles and in many other regions of the world with similar conditions. The kinetic energy of these currents can be converted into electricity by placing turbines and other generating equipment within the coast area.

Technical concepts of tidal energy exploitation. The majority of countries that have investigated the exploitation potential of tidal energy focused on the use of dams to create spaces that can be used to control the natural tidal currents. British and other specialists have concluded that the construction of dams in creeks is most advisable in terms of their costs for small water depths. This technology was used in the Netherlands to close Schelde.

Tidal dams include a sluice that is opened to allow the tide to flow into the basin and turbo-generators. During the flow phase the sluice is opened and water flows in, ensuring the maximum water level in the basin. Upon reaching the maximum water level the sluice is closed. At this stage of the cycle, the turbines can be used in reverse mode – as pumps – to increase the water level in the basin. This leads to an increase of the amount of produced energy by 10%. When the sufficient water

Table 7.2 Prospected zones for tidal energy projects

Countries	Locations	Height of tidal waves (m)	Water basin surface (km^2)	Installed capacity (MW)	Rough annual production (TWh/year)	Annual plant load factor (%)
ArgentinaS	San José	5.8	778	5,040	9.4	21
	Santa Cruz	7.5	222	2,420	6.1	29
	Rio Gallegos	7.5	177	1,900	4.8	29
Australia	Walcott Inlet	7.0	260	2,800	5.4	22
Canada	Cobequid	12.4	240	5,338	14.0	30
	Shepody	10.0	115	1,800	4.8	30
India	Khambat Bay	7.0	1,970	7,000	15.0	24
South Korea	Garolim	4.7	100	400	0.836	24
Mexico	Rio Colorado	6–7			5.4	
Great Britain	Severn	7.0	520	8,640	17.0	23
	Mersey	6.5	61	700	1.4	23
	Duddon	5.6	20	100	0.212	22
USA	Knik Arm	7.5		2,900	7.4	29
	Turnagain Arm	7.5		6,500	16.6	29
Russian Federation	Mezen	6.7	2,640	15,000	45	34
	Tugur	6.8	1,080	7,800	16.2	24
	Penzhinsk	11.4	20,530	87,400	190	25

level is reached, the water flow is directed through the turbines to generate electrical energy; the process is similar to hydro power plants with dams. Double energy generation is technically possible during the tide rise and fall, but in this case less energy will be produced, because the water height is lower compared to the previous case.

Besides, the Kaplan turbines, in horizontal configuration, are optimized for the generation with currents in one direction. The construction of such a tidal power plant is preceded by a detailed feasibility study considering both risks –technical and commercial ones, with account of the environmental impact forecasts, study of hydraulic currents on both sides of the dam, hydraulic modelling to determine the energy produced during each tidal cycle.

Other concepts, based on the development of artificial storage systems, are also investigated and promoted. One of these concepts is based on the creation of two or more pools that can increase water motion control and can allow turbines to operate longer than in a single basin scheme. Secondary basins have been proposed for the Severn system (Scotland), but this project was rejected due to higher cost of the produced energy. It is necessary to consider several factors to assess the feasibility of tidal energy conversion systems use.

Economic considerations. Tidal energy projects employing dams have high installed kilowatt costs (1,500 $/kW). Long erecting period and reduced load factor result in high energy unit cost, not competitive with the alternatives based on conventional fossil fuel. The economic prospects for alternative forms of tidal energy remain uncertain, because there is little information regarding the cost and performance of tidal current generators or basin schemes. However, without detailed information (for investors) and thorough analysis of the environmental effects, the tidal energy shall not be developed under any form. Still, there are certain non-power benefits that will stimulate the development of tidal energy schemes.

Environmental issues. The dams for capturing tidal energy will alter the existing gulf ecosystems in various ways. Firstly, the inter-tidal before-dam space will be permanently flooded, thus creating problems to the gulf ecosystem. The change of hydraulic regime will alter the natural sedimentation processes, more sediments accumulating in front of the dam. The amount depends on the position of the dam. The reduction of tidal after-dam currents will help to decrease the turbulence effects, to assure a better penetration of the light and, simultaneously, to increase the plankton productivity. Gulfs are of key importance both for the migratory fish species, many of which are used for commercial purposes, and also for migratory bird populations.

Following an extensive analysis of papers in this field (national, international, governmental/web sites), made by James Craig from AEA Technology company, UK, the most important achievements in the field from different countries have been selected [7].

Argentina. Tidal wave heights up to 7.5 m have been recorded at the southern coast between Tierra del Fuego and Golfo San Matías. In terms of practical application of the estimated tidal energy potential, five locations with an estimated overall potential of 37 TWh/year were identified. A tidal power plant construction began in the San José Bay, which will have a 780 km^2 basin and will be connected to the sea by a large natural 7 km long channel. The dam in this location will be approximately 13.4 km long, will have an installed capacity of 5,040 MW and will generate about 9.4 TWh/year.

Australia. Tidal energy potential is concentrated particularly on the Northwest coast of Australia, where the world's highest wave heights are observed. This coastline has many bays and inlets suitable for dam construction, such as Walcott Inlet, Secure Bay, St. George Basin and the largest of them – King Sound. In the late 1990s, the Tidal Energy Australia Company, Western Australia, proposed a joint project with double basin/double current for the King Sound bay near Kimberley, Derby district. The advantage of this scheme is continuous electricity production. One basin maintains a high level of water and the other one – a lower level. The 48 MW capacity plant will be the second largest tidal power plant in the world and the only one that will continuously produce electricity.

A special interest represented the Western Australian tidal energy potential actively promoted in the area of Derby town, located at the height of two adjacent bays near King Sound. The project envisaged the construction of an artificial channel that was supposed to connect the two bays. After analysing the technical and financial aspects compared to the scheme of the gas burning plant, in July 2000, the community decided not to promote this project taking into consideration the environmental impact.

Canada. Featuring a huge tidal energy potential Canada has decided to develop substantially its conventional hydropower reserves. The heights of the seashore provinces of New Brunswick and Nova Scotia register some of the biggest tides in the world. The most promising prospects of tidal energy potential have been made for two locations in this area: Cumberland and Minas Basins. Meantime, the only tidal power plant that was sold is located in the town of Annapolis Royal, another one – below the Bay of Fundy. 20 MW power plant was put into operation in 1984. The Annapolis plant employed the world's largest Straflo-type turbine to produce more than 30 millions kWh/year.

China. The areas of the south-eastern coast in Zhejiang, Fujian and Guangdong provinces are considered to have substantial tidal energy potential. The use of tidal power began in 1956: a few small tidal plants were built to pump water for irrigation. Since 1958, a small power 12 kW plant for electricity generation was built. Around 1970 it was supplemented with other plants, among which the largest are: at Jiangxi – 3 MW and at Baishakou – 960 kW. Currently, in China there are seven tidal power plants with a total capacity of 11 MW.

France. Several tidal power plants were built in the modern period. The first and the largest plant of 240 MW power is the dam in La Rance Gulf in northern Bretagne. The 0.8 km dam also serves as a highway between Saint-Malo and Dinard. The dam was built in the period 1961–1966 and currently has 34 years of successful commercial operation. Annual energy production is about 640 million kWh. At first, the dam was designed to use both tidal currents: the flow and the ebb, but this kind of operation was a partial success. Now the dam is operated exclusively for ebb streams; at the same time, the two ways generation is regularly used during high tides. The 10-year program of the 24 turbines renovation began in 1996 at the 30th anniversary of the plant.

India. Locations with the highest tidal energy potential are the Gulf of Kutch and Khambat (Cambay) Gulf, situated in the west of Gujarat state, and, the Ganges delta in the Sunderbans area of West Bengal, eastern India, respectively. The size of tides in the two bays is 5 and 7 m, theoretical capacity – 900 and 7,000 MW, annual production – estimated at about 1.6 and 15.0 TWh, respectively. After a feasibility study the construction of a 3 MW tidal energy plant started in the Durgaduani area, Sunderbans region.

Russian Federation. Design studies for tidal energy applications were carried out in Russia since 1930. As part of this work, at Kislogubsk, near Murmansk, a 400 kW small pilot plant was built and put into operation in 1968. The success of this facility helped to design bigger tidal power plants in the northern and eastern areas of the country: Lumbov (67 MW) and Mezen Bay (15,000 MW) in the White Sea, Penzhinsk gulf (87,400 MW) and Tugur gulf (6,800 MW) in the Okhotsk Sea. The Tugur gulf plant has the most feasible scheme. The preliminary design work began in 1972, but the finishing period remains uncertain.

Great Britain. High tides on the western coasts of England and Wales provide for the most favourable conditions in the world for tidal energy use. If all reasonably exploitable bays would be used, annual tidal electricity production would reach 50 TWh, equivalent to about 15% of the current electricity consumption in Great Britain. Six places with significant tide potential of 5.2–7.0 m have been identified, feasibility studies have been entirely carried out for two major projects – the Severn bays (8,640 MW), where tidal wave height reaches 12 m, and Mersey (700 MW), where tidal wave height reaches 8 m, and for smaller projects in Duddon (100 MW), Wyre (64 MW), Conwy (33 MW) and Loughor (5 MW) bays. Great Britain invested around £20 million in tidal energy R&D. Most efforts (during 1980–1992) were concentrated in co-financing feasibility and development studies in partnership with industrial consortiums. Based on the estimates carried out within the Tidal Energy Government Program (1978–1994) it was concluded that these projects are not attractive because of high capital costs, long construction periods and relatively low load factor (21–24%).

7.2.2 Ocean Currents Energy Potential

One of the primary purposes of physical oceanography is to know the general movement of ocean water around the globe. It is also important to know how currents change throughout the day, year, 10 years and more. Currents are essential components of biochemical and hydrological cycles. Knowledge of the ocean currents is also highly important for maritime operations, such as navigation, sea research and protection, pollutants dispersing, etc. It is known from the observations of ocean currents that the wind moves water and the wind is the main force that moves the ocean currents. In the first part of the twentieth century Fridtjof Nansen, Norwegian scientist, noted that the North Atlantic icebergs are moving because of the wind. His student, V. Walfrid Ekman has shown, however, that this effect is mainly due to the Earth's rotation, and the Coriolis forces are responsible for it. This can be confirmed by the example of the Gulf Stream that moves eastward when the winds blow mainly from West to East. Other factors that influence the direction of the ocean currents are water density and temperature [8]. Figure 7.3 [9] shows a map of the ocean currents, drawn at the beginning of the twentieth century; it illustrates very well the formation of currents in different parts of the globe. Large water currents, resulting from water deepening in the North Sea and Greenland, and also around Antarctica, are called the conveyor belt (Fig. 7.4) [8].

Sea currents are bearers of very large kinetic energies. Thus, it was calculated that an ocean current with about 100 m width, 10 m depth and 1 m/s speed, would provide about 2 million kWh in a year. Intensive studies have been performed regarding the use of energy from the Gulf Stream, whose dimensions are truly impressive: 500 km initial width, 10 km/h speed, 20–25°C temperature, stirring up the ocean waters at 2,500–3,000 m depths and carrying a water volume of 80 millions m^3/s [10]. The Gulf Stream is an Atlantic Ocean current, which takes its origin in the Gulf of Mexico, leaves the Gulf through the Florida Bay and follows

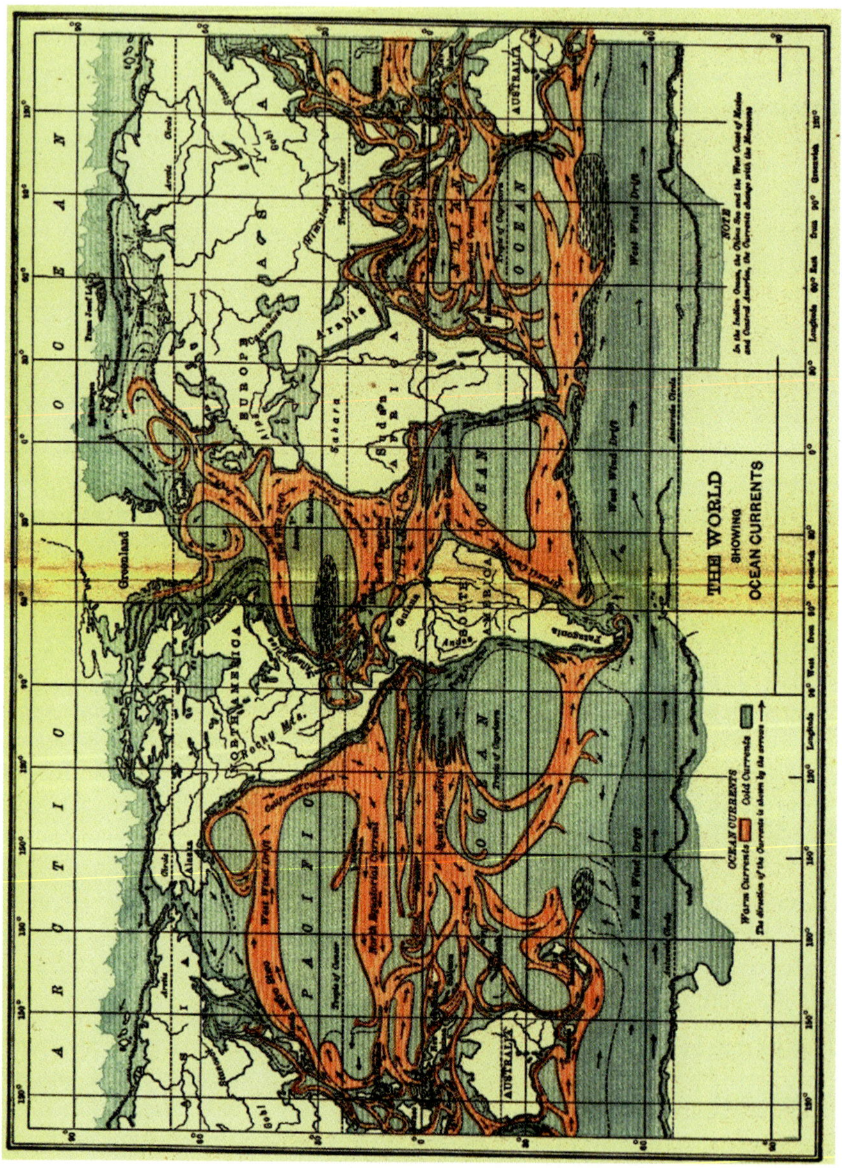

Fig. 7.3 Global scenery of the sea currents, year 1911

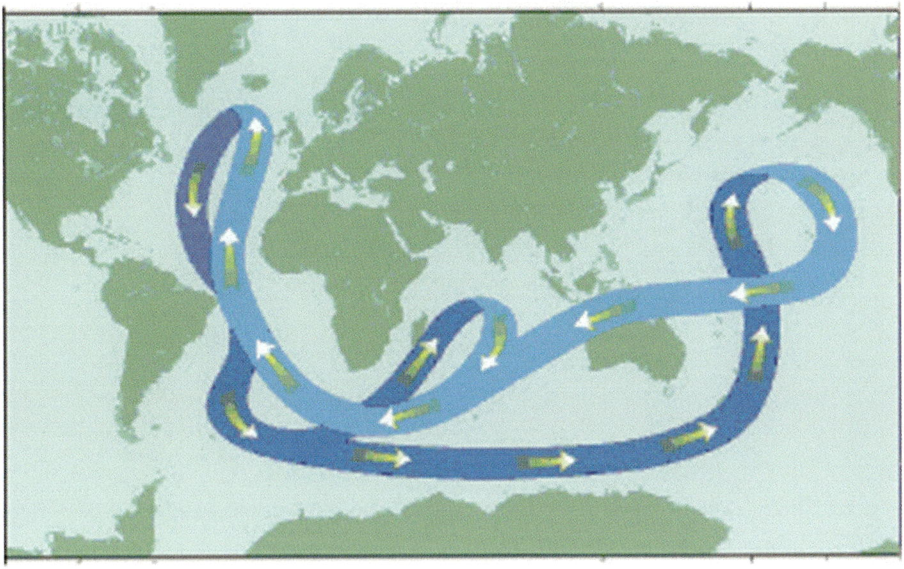

Fig. 7.4 Direction of ocean currents movement

the U.S. eastern coastline before it crosses the Atlantic Ocean. At approximately 30° West and 40° North it splits in two – with the northern stream crossing to northern Europe and the southern stream recirculating off West Africa.

7.2.3 Ocean Waves Energy Potential

Sea waves represent another form of ocean energy. It is a form of energy storage, passed by the wind to the oceans and seas; this energy is measurable and worthy of consideration. Calculations show that 1 m high-, 40 m long- and 5 s period waves have a 5 kW available power per 1 m width front. Despite the climate change phenomenon, the world energy resources of ocean waves remain very big. The biggest wave energy is concentrated at the Western Coasts at 40–60° latitude in the northern and southern hemispheres. The energy contained in the wave crests varies between 30 and 70 kW/m; it reaches about 70 kW/m in the Atlantic Ocean, west of Ireland, and in the extreme south of Latin America, and 100 kW/m – in the New Zealand (Fig. 7.5) [11]. Using only 10% of this energy would be sufficient to meet the current global energy needs. The annual sea wave energy resources reach the astronomical figure of 320 GW for Europe (Fig. 7.6) [11]. This is an engineering problem and it deals with the development of the most efficient systems to capture these enormous amounts of energy. The analysis of Fig. 7.6 shows that

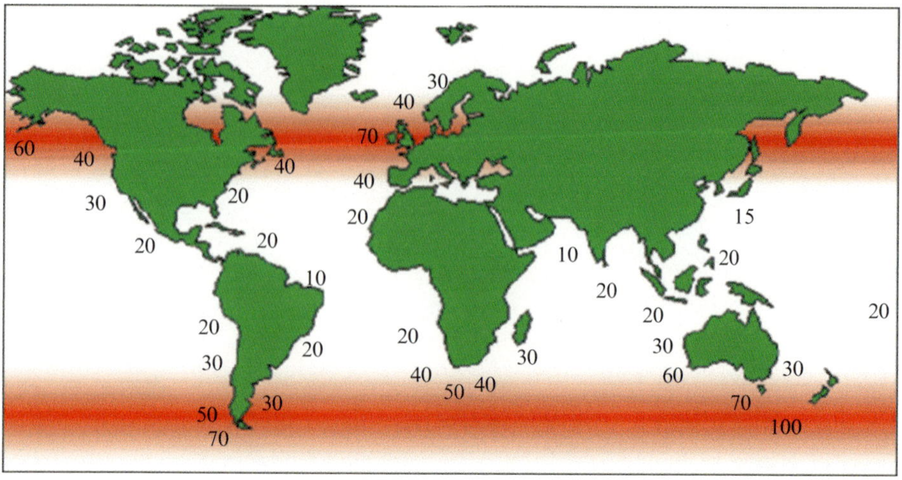

Fig. 7.5 Global distribution of wave energy sources, in kW/m on the wave crest (Numerical scenery)

Fig. 7.6 Energy potential of European water waves, kW/m

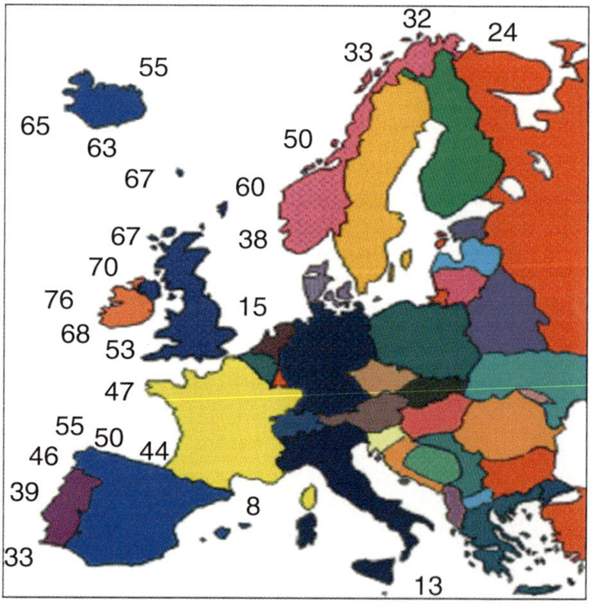

western Atlantic waters near Ireland and Scotland have the highest energy potential (exceeding 70 kW/m), more than twice higher than the average for Europe. A further analysis of the chart (Fig. 7.7), drawn according to the British Met Office data of 24.12.1999, shows that the western coasts of Scotland and Ireland are the richest in the available wave energy, the waves reaching up to 8 m on the south-western coast of Ireland.

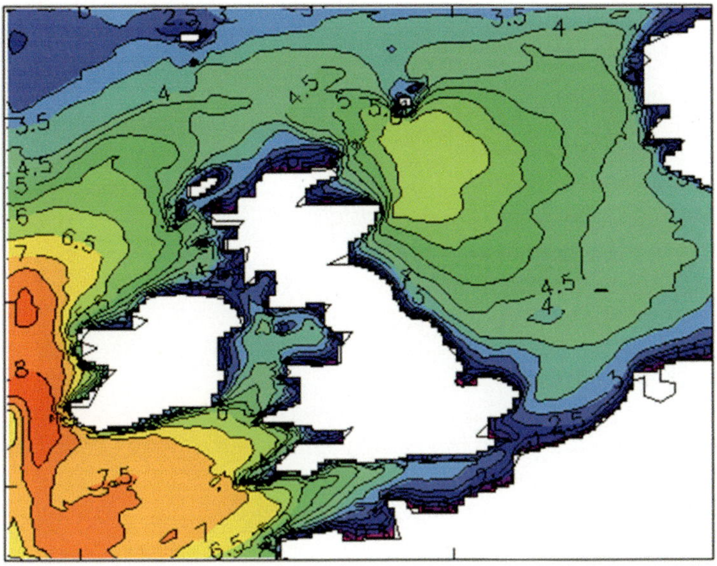

Fig. 7.7 Wave height (m) according to water wave model developed by UK met office

7.3 Tidal Energy: Theoretical Considerations

7.3.1 General Notions

The maximum level of water is called "high tide" or "high water" and the minimal – "low tide" or "shallow water". In any point of the ocean, normally two large tides and two low tides occur each day. In general, high tides last for 12 h and 24 min. Twelve hours are due to the Earth's rotation and 24 min – to the Moon's orbit. This is the main semi-diurnal monthly period and it usually represents half the time that separates one monthly zenith from another. The height of high and low tides also varies. In the period of "the new moon" and "the full moon" when the Sun, the Moon and the Earth form a line (Fig. 7.8) [5], the solar tidal forces add to the moon forces. The tide size in this case is the highest. When the Moon is in the first or third quadrant, the Sun and the Moon are in the phase quadrature of 90° (seen from the Earth) and the tidal forces generated by the Sun and the Moon are partly compensated. In these coordinates, the size of the waves is minimal during the monthly cycle. The relative distance between the Moon and the Earth also affects the tide height. When the moon is in the perigee, the tides rise, and when the moon is in the apogee the tide height decreases. Every 7.5 "lunations", the perigee and alternatively, other positions of the new moon and the full moon coincide, and in these periods the tidal height value is bigger than others; and if, at the same time, the waves move toward the shore, the consequences can be particularly severe. Tides are

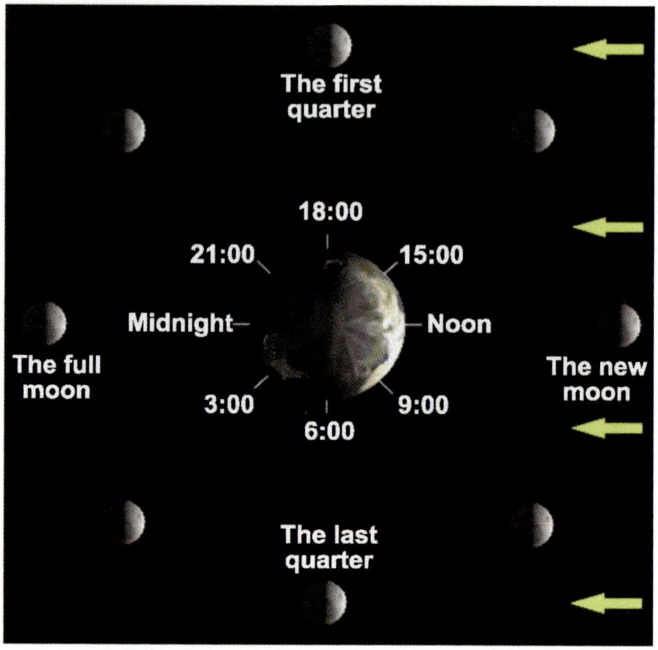

Fig. 7.8 The Earth and the Moon seen from the North Pole

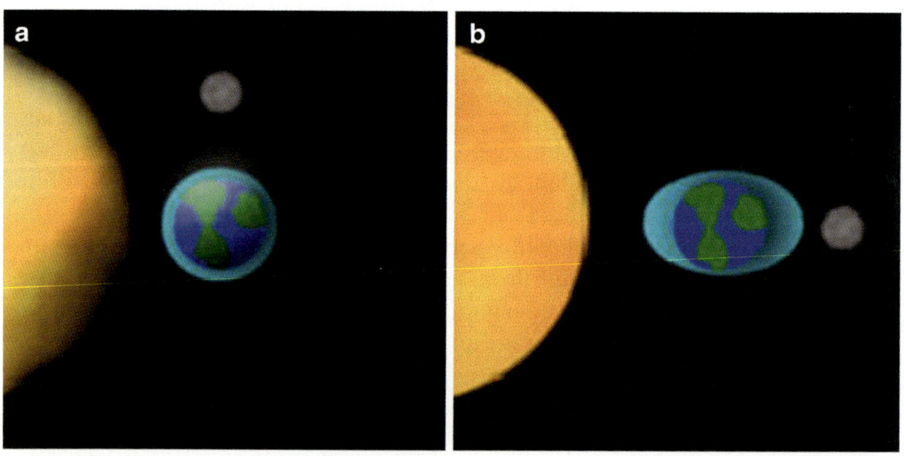

Fig. 7.9 Concepts of minimal (**a**) and maximal (**b**) tides

minimal during the first and the third quarters of the Moon (Fig. 7.9a) [5]. Maximal tides (called syzygy) occur at the new and the full moon, when the Earth and the Moon are very close (Fig. 7.9b).

In many places a delay between the phases of the moon and its effects on tides occurs. For example, in the North Sea the maximal and minimal tides are noticed at

2-day delay towards the phases of the new/full moon, and towards the first and the third quadrant, respectively. This is explained by the fact that the origin of tides is in the southern oceans. The resulting effect of the tide amplitude or height travels through the oceans. It is known that their journey as a single broad wave is pushed from the north through the Atlantic Ocean. This causes relatively low tides in some locations (nodes) and high tides elsewhere. It must not be confused with tidal levels caused by local geography, which can be found in Nova Scotia, in the Bristol Channel, in the Iceland Channel and in the English Channel. In these locations tides may be higher than 10 m. The tidal wave from the Atlantic Ocean reaches the English Channel in a day and needs another day to travel around the British Isles in order to produce an effect in the North Sea. The exact timing and height of tides are largely influenced by the local bathymetry (depth) in different coastal points. There are some extreme cases: in the Bay of Fundy, on the Eastern coast of Canada, the highest tides values have been recorded – 16 m. Northern Quebec, north-eastern Canada, is considered by many experts to have the highest tidal value (about 17 m), however it is glacier-free only 4 months a year, while the Bay of Fundy rarely freezes in winter. Very low tides are recorded in the Mediterranean Sea and the Baltic Sea because of the narrow connection with the Atlantic Ocean. Due to the same causes, extremely low tides are recorded in the Gulf of Mexico and in the Sea of Japan.

7.3.2 Tidal Physics

Although gravitational forces exerted by the Sun on the Earth are 200 times bigger than those exerted by the Moon, the moon-generated tidal force is approximately two times bigger than the force generated by the Sun. This is explained by the fact that tidal forces are not enforced by the gravitational field, but by its gradient. With the increase of distance from the source, the field gradient decreases more rapidly than the gravitational field, because the Sun is about 400 times farther from the Earth than the Moon. Ignoring the complications caused by the marine currents, the ocean surface can be accurately approximated to an equipotential surface, which usually is the surface of the Earth. Because the gravity is equal to the potential gradient, there are no tangential forces on such a surface and thus the ocean surface experiences gravity equilibrium. Now let's consider the effect of massive external bodies such as the Sun and the Moon. These celestial bodies have strong gravitational fields that weaken as the distance in space increases, and that react to change the shape of the Earth's equipotential surface. This deformation of the Earth's shape has a fixed orientation in space compared to the influential bodies. It is the Earth's rotation toward this shape that causes diurnal tidal cycles.

Gravitational forces follow the inverse square law (the force is inverse proportional to the distance square), and the tidal forces are inverse proportional to the distance cube. Since the Sun's gravitational attraction on the Earth is 179 times

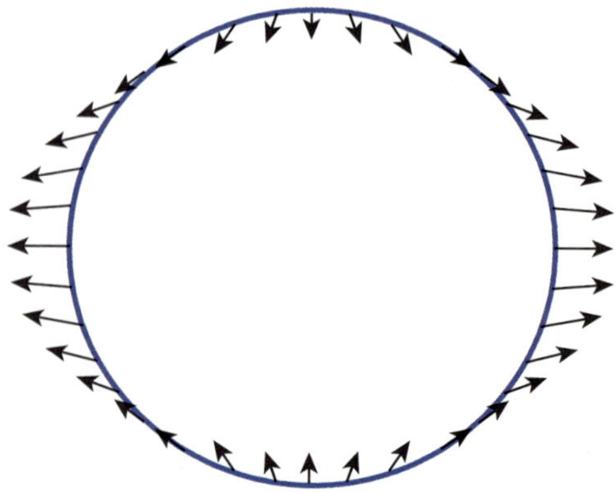

Fig. 7.10 Lunar gravity differential action on the Earth

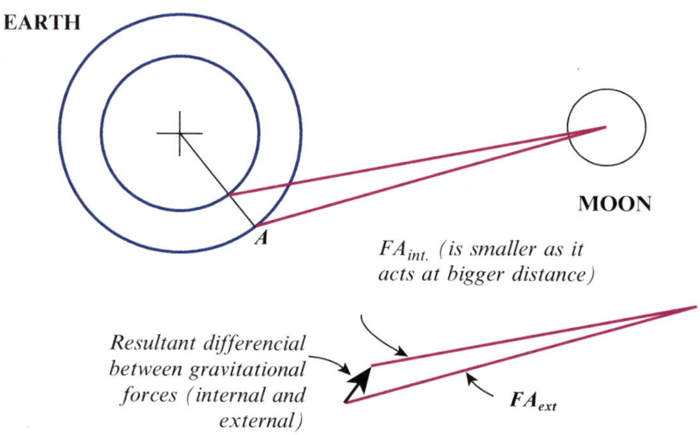

Fig. 7.11 Calculation of resultant differential between gravitational forces

bigger than that of the Moon, and the Sun is 389 times farther from the Earth than the Moon, the Sun's tidal effect is not more than 46% of the Moon effect [5].

The Moon exerts different gravitational attraction in different parts of the Earth. Removal of the Moon positioning reduces its attraction exerted on the Earth. Figure 7.10 [5] explains better the differential gravitational action of the Moon on the Earth's coating layer. Differential gravitational field of the Moon on the Earth is known as the Tide Generating Force. It is the primary mechanism that generates tidal action and explains the rates of the two diurnal tides. Other forces, such as the solar gravity force also supports the tidal action. For simplicity, we consider the Moon-caused tidal influence (Fig. 7.11) only. Description of the Sun tidal action may be the same. Water in the sublunar points is more compact than the solid earth, so water is strongly attracted in its growth too. In the antipode areas of the Earth situated far from the Moon (antipode point), the water is farther from the Moon than the solid earth is. Thus, the Earth (its solid part) moves around the Moon more than the water, thus increasing the water's lag behind. At lateral sides the water is drawn insignificantly in different directions than toward the centre. It can be shown that forces in sublunar and antipode points are approximately equal and that internal forces in lateral points are twice lower. The point in which the tidal forces are parallel on the Earth's surface is approximately at 55° from the orbital plane. These parallel components contribute most of all to the formation of tides because water particles do not follow these forces. The force acting on the water particle is only about 1/10,000,000 part of the force caused by the Earth attraction.

7.3.3 Amplitude, Cycle Period and Tidal Braking

At relative rotation of the Earth towards the Moon during one lunar day (24 h 48 min.) each of the two tidal heights rotates with relative velocity that is observed on the Earth, reaching the tidal height every 12 h and 24 min. The theoretical amplitude of ocean tides generated by the Moon is about 54 cm in the highest point, which corresponds to the amplitude that could be achieved if the ocean would have uniform depth and the Earth would not rotate.

Similarly, the Sun generates tides with theoretical amplitude of about 25 cm (46% of the Moon amplitude) with a 12-h cycle. Maximal tides occur through the overlapping of the two effects, theoretically reaching 79 cm, when the minimal tide is reduced to 29 cm. Because the orbit of the Earth's rotation around the Sun and the orbit of the Moon rotation around the Earth are elliptical, the tidal amplitudes vary depending on the variation of the Earth–Sun and the Earth–Moon distances. This causes the variation of tidal forces and of the theoretical amplitude by about ±18% for the Moon and ±5% for the Sun [5]. If the Sun and the Moon are in maximal proximity and aligned to the new Moon positions, the theoretical amplitude shall reach 93 cm. The real amplitude differs significantly not only due to the variations in the oceanic depth and flow obstacles caused by continents, but also because the natural period of wave propagation is the same as the rotation period of the magnitude: about 30 h. If there were no underwater mountain areas, high wavelength wave propagation at half-way around the Earth would take 30 h (for comparison, the Earth's lithosphere natural period is about 57 min). If the Moon would not "brake" suddenly and would not be a solid body, the ocean's level would oscillate with a 30 h period at amplitude slowly decreasing due to energy dissipation. This value of 30 h is a simple function of terrestrial gravity, of the average depth of the oceans and of the Earth's circumference. Since the Moon tidal forces drive the ocean with a period of approximately 12.42 h, which is considerably less than the ocean natural period, a complex resonance phenomenon occurs. This phenomenon, due to the friction effects, causes increased braking of 12 min between the high tide and the monthly zenith. This tidal braking corresponds to a 3° angle between the position of the Moon, the Earth's centre and the location with an average high tide. This tidal braking brings an increase in the gravity torque moment of the

Moon, resulting in the gradual transfer of the angular momentum on its orbit and in the gradual increase of the Earth–Moon separation. According to the principle of angular momentum conservation, the Earth's rotation speed is reduced respectively. Thus, in the geological time period, the Moon remains behind the Earth and the length of the Earth day increases.

7.4 Theoretical Aspects of Wave Energy

7.4.1 Some Aspects of Wave Formation

Ocean waves behave rather unpredictably. They say that the first wave rule, especially in the open ocean is that there is no rule at all. It is hard to see it, but a "normal" wave, described as a simple physical process, is seriously distorted by a number of factors generated by the forces of nature. However, for an engineer working in the development of sea wave energy conversion systems it is strictly necessary to have a description, even a summary, of the wave actions. What follows are some examples of the wave formation laboratory. In fact, the alternative models of water, the variable water depths, the reverse currents, the natural obstacles, and many other factors can change substantially the way and form of waves.

Ocean surface waves propagate horizontally along the *"air-sea"* border. They are described as progressive orbital waves. They are orbital because when the wave passes through several points, the particles of water beneath the waves move in an orbital path (Fig. 7.12) [12]. They are progressive, as the wave form moves (progressive) horizontally from one location to another. In Fig. 7.12 the wave moves from the left to the right and causes this rotational motion of the water particles in a complete rotation cycle at full wave flow.

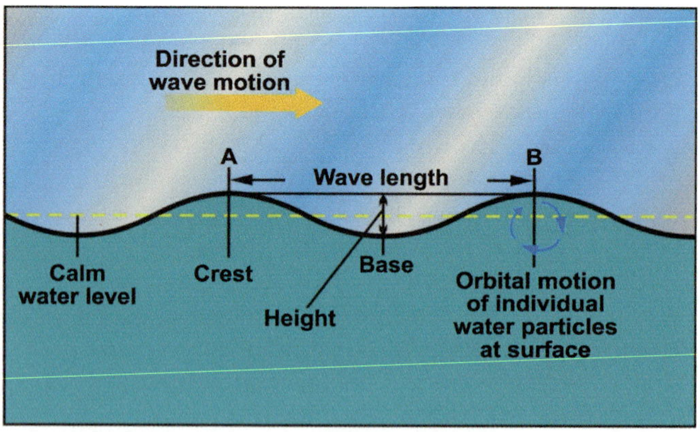

Fig. 7.12 Waves trajectory

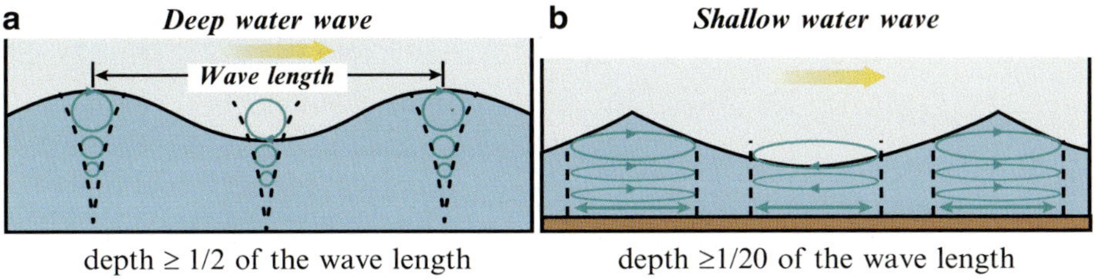

Fig. 7.13 Deep water waves (**a**) and shallow water waves (**b**)

When the wind blows above the water, it changes its surface, causing some undulations in the beginning, which then turn into waves. Storms occur when the wind blows in the same direction for a long time. In reality, the waves are not a perfect and harmonic process, but an irregular one.

Usually, they consist of several overlapping waves with different frequency and speed [13]. Waves behave differently in deep and shallow waters. Figure 7.13 [13] shows two cases of wave propagation: in deep and shallow waters. In deep waters, the wave action decreases rapidly with the increasing depth (Fig. 7.13b). In shallow waters, the wave action does not decrease with the increasing depth (Fig. 7.13b). Depending on their size there are long (with distances of 100 m between the wave crests) and short waves.

7.4.2 Types and Basic Characteristics of Waves

Waves need time to develop; they do not erupt spontaneously from the ocean. It takes a certain wind speed to blow on a certain distance for a considerable length of time to finally form the waves. There are three types of waves:

- Capillary waves or ripples (small);
- Sea waves;
- Swells (Intense waves).

Capillary waves appear on smooth water when the wind blows. Under the action of wind, they transform into sea waves or swells (intense waves) (Fig. 7.14). Sea waves are generated by the wind blowing a certain period with a certain speed. Then they tend to become longer and straighter after the wind ceases to beat. Swells (intense waves) are waves that move in any direction from their origin and do not relate to local wind conditions, in other words, they are sea waves that have long been left without wind. It is a confusing term, because water does not act together with the waves, but only together with water currents – there are two exclusive features in water animation. As shown in Fig. 7.15, the waves can be classified according to the wave period. According to the cumulative distributed energy in the ocean waves, the principle of force generation and the principle of damping forces

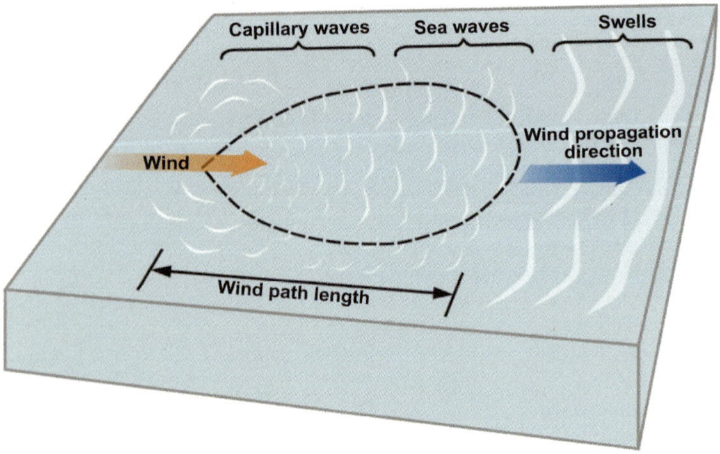

Fig. 7.14 Transformation of waves under wind action

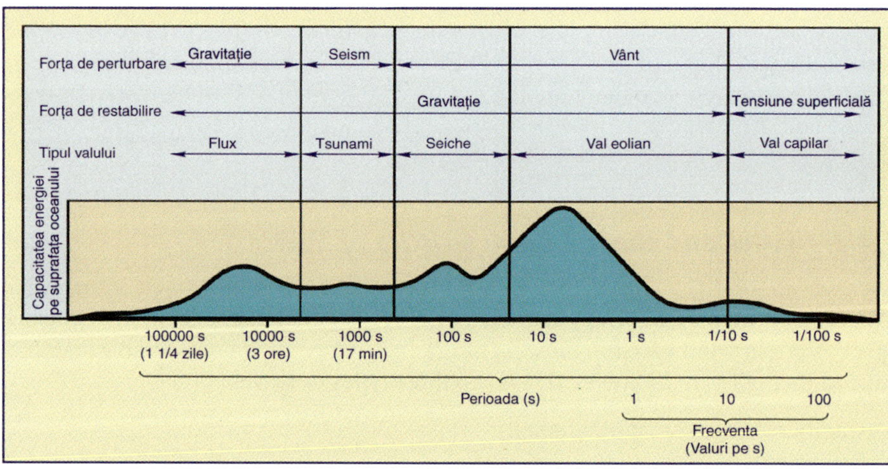

Fig. 7.15 Classification of waves depending on their period

change when the wave period increases. The smallest waves (capillary waves) have periods <0.1 s and are generated by the small wind breeze. Because they are so small, they are damped by the surface tension of water (see Fig. 7.15).

The most common gravitational waves are waves that have periods between 1 and 30 s. They are generated by wind and hurricanes and are damped by gravitational forces. Waves with periods longer than 5 min are long waves, generated by intense hurricanes and earthquakes, quieted by the forces of gravity and Coriolis forces. The longest wave periods of 12 and 24 h are tides generated by the Sun and the Moon and damped by bottom friction and Coriolis forces.

Fig. 7.16 Mechanism of orbital motion of water particles

Detaching ourselves from reality, let us try to take into account the basic characteristics of waves. There are three basic factors to be taken into account when analysing waves:

- The wind speed;
- The time of wind blowing in one direction;
- The distance of open water over which the wind blows.

All these factors work together to generate waves. The bigger the ratio of each of these variables in the equation, the higher the waves are.

An ocean surface wave, as shown in Fig. 7.12, has sinusoidal form (with crests and troughs, which have identical shapes and a fixed wave length) and is orbital progressive with the water-particles in the wave moving on an orbital trajectory with a completed cycle at the passage of a complete wave. The waves are characterized by the following features (Figs. 7.12, 7.13). The distance between two succeeding wave crests is defined as the *wave length* (**L**). The vertical distance from the crest top to the lower point of the succeeding trough is defined as the *wave height* (**H**). The time required by the crest to get from point A to point B is defined as the *wave period* (**T**). The number of crests that pass through point A or B every second is defined as the *wave frequency* (**f**) (*frequency* is a term reverse to *period*). And finally, the speed with which the wave crest moves horizontally on the ocean surface is defined as the *wave speed or wave phase speed* (**c**).

Water particles in a wave have different models of motion. In a normal wave it is an orbital motion of water particles. This is best demonstrated by a cork floating on water. When the wave increases, the cork swirls in a point (it is attracted by the orbital motion). This is a very passive motion, as linear motion of the powerless wave particle is very aggressive, therefore it is very destructive. The images in Fig. 7.16 show the orbital motion of the cork that floats on water when the wave passes from the left to the right. Cork position, in fact, is unchanged, making only a slight rotation.

7.4.3 *Waves and Wind*

The relationship between the wind and waves is very important, so a new classification system was developed that incorporates the relationship between the waves and the wind speed. This system, called the Beaufort wind scale was developed

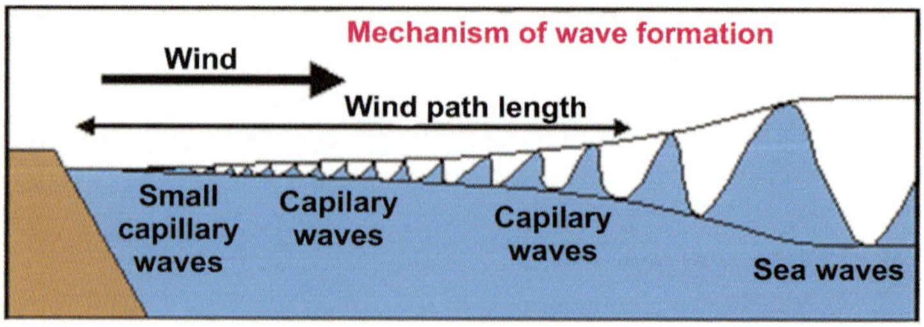

Fig. 7.17 Mechanism of wave formation

in 1805 by Sir Francis Beaufort, British admiral and hydrographer to the Royal Navy. The observations of Sir Francis Beaufort on the wind speeds, based mainly on the observed sea conditions, are listed in Table 7.3 [14]. Since the wind causes the formation of water waves, it is easier to understand the mechanism of their generation. For a perfectly calm sea, there is no wind, practically. At a certain wind speed, surface layers begin to move, forming tiny undulations. It is interesting that these waves do not move exactly in the wind direction, but under a 70–80° angle to the wind direction. Undulations at minimum speed of 0.23 m/s increase up to small waves and begin to move in the wind direction. Figure 7.17 shows the evolution of the wave generated by the wind. At the wind speed of 7–11 km/h, these waves move at an angle of 30° to the wind direction. The wind, with some degree of turbulence, reaches the surface of the water and starts the transfer of energy to the waves. Depending on the duration and distance, the wave size increases quadratic to the maximum size. When the maximum wave size is reached, they are moving more slowly than the wind.

When the wind speed increases, the amount of energy transferred to the water rises more rapidly in proportion to the fourth power of the wind speed. The diagram in Fig. 7.18 shows the energy spectrum of power for wind speeds of 20, 30, 40 knots. The area under the curve represents the total energy of the sea state. The vertical scale is the measured amplitude wave (A + A) in m^2. A major influence on the wave formation parameter is the so-called "*distance of wave development*". Table 7.4 presents the wave parameters for different wind speeds and the distance of wave development (Beaufort scale).

Also, the change of the wind blowing direction over the existing waves can cause confusion in the wave motion and as a result the new waves are formed that overlap the existing ones, creating huge waves. If the wave system meets a current in its path, then two situations might occur: if the wind and water currents have the same direction, the waves become calm, creating long ocean waves. Contradictory motion of the water currents and wind leads to the formation of larger and more aggressive waves.

In addition to the wind speed, temperature is also a factor in forming the waves. Warm air moving over the water has an angle of incidence towards the water surface smaller than the cold air. Cold air moves over open water and creates waves faster than warm air masses.

Table 7.3 Sea conditions at various wind speeds (Beaufort scale)

Wind intensity, degrees	Description of the wind	Wind speed (m/s)	Wave height (m)	Effects generated by the wind on the surface of aquatic basins
0	Calm	0–0.2	0	Sea like a mirror
1	Light air	0.3–1.5	0.083–0.165	Ripples without crests
2	Light breeze	1.6–3.3	0.083–0.165	Small wavelets. Crests with glassy appearance, but do not break
3	Gentle breeze	3.4–5.4	0.66	Large wavelets, crests begin to break. Glassy looking foam, occasional white horses
4	Moderate breeze	5.5–7.9	1.32	Small waves with breaking crests. Fairly frequent white horses
5	Fresh breeze	8.0–10.7	1.98	Moderate waves of pronounced long form. Many white horses, some spray
6	Strong breeze	10.8–13.8	3.30	Long waves begin to form. White foam crests are very frequent. Some airborne spray is present
7	Near gale	13.9–17.1	4.62	Sea heaps up, white foam from breaking waves blowing in streaks with the wind
8	Gale	17.2–20.7	5.94	Moderately high and long waves. Crests break into spin drift, blowing foam in well marked streak
9	Severe gale	20.8–24.4	7.59	High waves, dense foam streaks in wind, wave crests topple, tumble and roll over. Spray reduces visibility
10	Storm	24.5–28.4	9.57	Very high waves with long overhanging crests. Dense blowing foam, sea surface appears white. Heavy tumbling of sea, shock-like, poor visibility
11	Severe storm	28.5–32.6	12.21	Sea completely covered with long white patches of foam. Poor visibility
12	Hurricane	Over 33	14.85	Sea is completely white with foam and spray. Air is filled with driving spray, greatly reducing visibility

Note: the rounded average values of the wind speed are presented in parenthesis

Table 7.4 Relationship of the wind speed with the wave parameters

Wind speed (km/h)	Distance of wave development (km)	Wave duration (h)	Height (m)	Length (m)	Period (s)
20	24	2.75	0.33	10.6	3.2
40	176	11.50	1.80	39.7	6.2
60	660	27.50	5.10	89.2	9.1
80	1,682	50.00	10.30	158.6	12.4

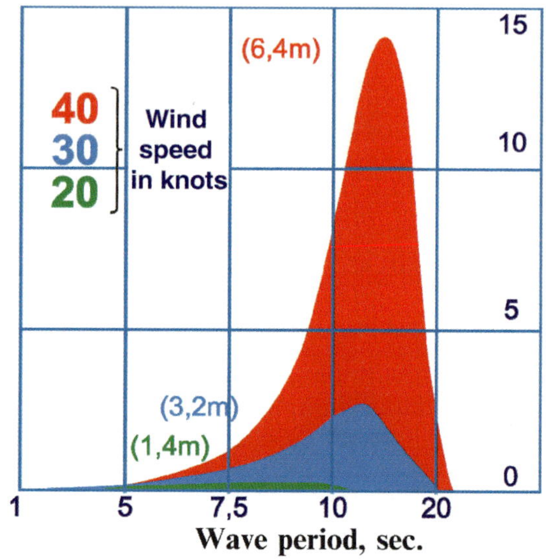

Fig. 7.18 Spectrum of total energy developed by the waves for various wind speeds

7.4.4 *Some Aspects of the Wave Theory*

The ocean surface waves are mechanical waves that propagate along the interface between *water and air*. The restoring force is provided by gravity and so they are often referred to as surface gravity waves. As the wind blows, pressure and friction forces perturb the equilibrium of the water surface. These forces transfer energy from the air to the water, forming waves. In the case of monochromatic linear plane waves in deep water, particles near the surface move in circular paths, making wind waves on the ocean surface as a combination of longitudinal (back and forth) and transverse (up and down) wave motions. When waves propagate in shallow water (where the depth is less than half the wavelength) the particle trajectories are compressed into ellipses (Fig. 7.19b) [14]. As the wave amplitude (height) increases, the particle paths no longer form closed orbits; rather, after the passage of each crest, particles are displaced slightly from their previous positions, a phenomenon known as the Stokes drift. The physical concept of waves is presented in Fig. 7.20.

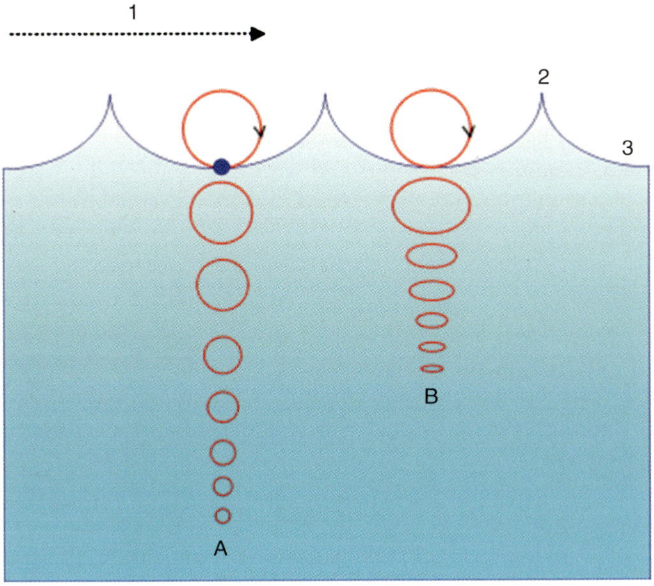

Fig. 7.19 Diagram of water particles in deep water waves propagation direction

Fig. 7.20 Physical concept of waves

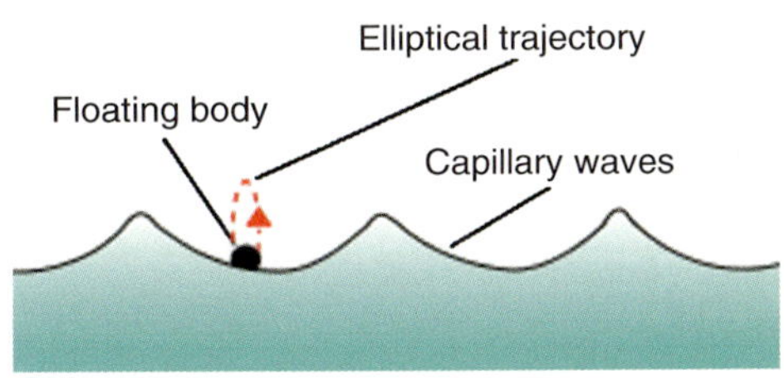

As the depth below the free surface increases, the radius of the circular motion decreases. At a depth equal to half the wave length λ, the orbital movement is reduced to zero. The phase speed of the surface wave (also called the celerity) is well approximated by the relation:

$$c = \sqrt{\frac{g\lambda}{2\pi} + \tanh\left(\frac{2\pi d}{\lambda}\right)}, \tag{7.1}$$

where c is the phase speed; λ is the wavelength; d is the wave depth; g is gravity acceleration.

In deep water, where $d \geq \frac{1}{2}\lambda$, so $\frac{2\pi d}{\lambda} \geq \pi$, and the hyperbolic tangent approaches 1, the speed c, in m/s, approximates $1,25\sqrt{\lambda}$, when λ is measured in meters.

This expression tells us that waves of different wavelengths travel at different speeds. The fastest waves in a storm are the ones with the longest wavelength. As a result, after a storm, the first waves to arrive on the coast are the long–wavelength swells.

The movement of ocean waves can be captured by wave energy devices. Energy density (per unit area) of regular sinusoidal waves depends on the water for density ρ, gravity acceleration g and the wave length h, which is equal to two amplitudes a and is determined by the relationship:

$$E = \frac{1}{8}\rho g h^2 = \frac{1}{2}\rho g a^2. \tag{7.2}$$

The velocity of propagation of this energy is the group velocity. An important practical application of the science of waves is the solution of 3D equation of the wave crest in time [14].

Equations describing shallow water (also called Saint Venant equations) are a set of equations describing the flow under the action of horizontal surface force in the fluid. These equations can be used in the atmospheric and ocean simulations, but are simpler than primitive equations. Models of shallow water equations have only one vertical level, so they can not compensate for any factors which vary with height. In general, large waves contain more energy. Wave energy is determined by the wave height, wave speed, wavelength and water density. Wave size is determined by the wind speed, by the distance at which the wind excites the waves, by the depth and topography of the location. Wave motion is the highest at the surface and reduces exponentially with the increasing depth, but wave energy is present as pressure waves in deep tide waters.

The potential energy of a set of waves is proportional to the height and the quadrate of the wave period. Waves with longer periods have lengths that are relatively larger and faster. The potential energy equals kinetic energy (when it can be converted). Wave energy is expressed in kW/m (in locations of the "shoreline type") and is calculated from the relation:

$$P = 0.5H^2T, \tag{7.3}$$

where H is the wave height, in meters; T – wave period, in seconds.

The relationship below allows the calculation of energy generated by the wave. Excluding the waves generated by major hurricanes, the biggest waves have a height of 15 m and a period of 15 s. Under this formula, such waves contain approximately 1,700 kW of potential energy per meter of the wavelength. Locations with the biggest potential of wave energy will have less power flow than the above – about 50 kW/m.

How much energy can be obtained from the wave? Linear wave theory assumes that the motion of water in a point is sinusoidal. The period of time T for a wave passing through this point can be expressed as:

$$T = \sqrt{\frac{2\pi\lambda}{g}}, \tag{7.4}$$

where λ is the wavelength (m); g is the gravity (kg/m^2).

Energy contained in a wave can be expressed by the equation:

$$P = \frac{\rho g^2 a^2 T}{8\pi}, \tag{7.5}$$

where a is the wave amplitude (height).

According to the existing technologies there are three types of wave energy collection:

- Use of buoy systems;
- Use of fixed contoured systems;
- By oscillating water columns.

Wave equation. Wave equation is a differential equation that describes the evolution of harmonic wave in time. The equation has slightly different forms depending on how and in which circumstances the wave is transmitted. Considering that a one-dimensional wave moves along an axis x with velocity v and amplitude a (which generally depends on x and t), the equation is:

$$\frac{1}{v^2}\frac{\partial^2 a}{\partial t^2} = \frac{\partial^2 a}{\partial x^2}. \tag{7.6}$$

The 3D equation is:

$$\frac{1}{v^2}\frac{\partial^2 a}{\partial t^2} = \Delta a, \tag{7.7}$$

where Δ is the Laplacian.

The general solution for one-dimensional wave equation was obtained by d'Alembert:

$$a(x,t) = F(x - vt) + G(x + vt) \tag{7.8}$$

The equation can be viewed as two frequencies crossing the length x in different directions: F in the direction $+x$ and G – in the direction $-x$. The three-dimensional wave propagation can be described by substituting x with directions x, y, z.

Migratory waves. Moving waves are called migratory waves or progressive waves. Their disturbance varies depending on time t and distance:

$$y(z,t) = A(z,t)\sin(kz - \omega t + \phi), \tag{7.9}$$

where $A(z,t)$ is the wave surface amplitude; k is the wave number and ϕ is the wave phase.

Phase velocity of the wave is:

$$v_p = \frac{\omega}{k} = \lambda f, \tag{7.10}$$

where λ is the wavelength.

As with other types of waves, wave energy is proportional to the quadrate of their height. For example, a 3 m height wave has nine times more energy than a 1 m height wave. The waves have direction and speed. Tidal waves caused by the underwater volcanic eruptions in Chile, May 1960, had a pathway of 6,000 sea miles (11,000 km) and reached New Zealand in about 12 h, with the speed of about 900 km/h.

The relationship between the speed of long surface waves and their depth in shallow water is described by the formula:

$$c^2 = gd\frac{p_2 - p_1}{p_2} \tag{7.11}$$

where c is wave velocity; g is free fall acceleration (9.8066 m/s^2); d – wave depth, m; p_2 – water density (=1); p_1 – air density.

The relationship shows that the wave velocity increases with the increase of wave depth and the relative difference of the environments' density. For ocean depth of 4,000 m the speed is:

$$v \approx \sqrt{10x40000} = 200\,km/s = 720\,km/h. \tag{7.12}$$

In deep water the relationship between the velocity of long surface waves and their depth is presented by the formula:

$$l = \frac{gt^2}{2\pi}, \tag{7.13}$$

where $l = tc$ for all wave cases. By substituting in the above equation we obtain:

$$c = \frac{gt}{2\pi} \text{ or } t = 0.641c(\text{s}), \tag{7.14}$$

where t is the wave period (s), f is wave frequency; l is wavelength (m). It becomes possible to calculate c and l by wave periods t (s):

$$c = 1.56t(m/s) = 5.62(km/h), \qquad l = 1.56t^2(m).$$

The waves with a wave period of 10 s that travel with the speed of 56 km/h with the wave length of approximately 156 m can produce waves with periods of 17 s and lengths of 450 m in 24 h.

7.5 Conceptual Systems of Ocean Energy Conversion

7.5.1 Classification of Conversion Systems

The energy of the sea currents is at an early stage of development, therefore there are very few tested prototypes and demonstrations at present. The first patent for a wave energy system is granted in Paris to a father and son named Girard in 1799. Their installation used direct mechanical action to drive pumps, wood cutting saws and other machinery. In 1970 a major interest in this type of energy was recorded in Europe, particularly in Great Britain. In November 2000, the world's first commercial plant was installed on a small island in Scotland.

Due to the placement areas, wave energy conversion systems fall into three main categories:

- Systems that can be placed in the area close to the shore;
- Systems that can be located on the shore;
- Systems that can be located on the coast.

The area close to the shore is considered an area of up to 20 km. Operation in such areas must take into account the aesthetic and landscape aspects, and also the impact on marine life should be well-defined. The construction cannot be large as in the case of coastal installations. Usually shoreline power plants are incorporated into the existing structures to minimize the wave effects over the shore, for example, in the sea walls. If water depth exceeds 50 m, the power plants are considered coastal.

Ocean energy conversion technology is still very young and therefore there is no very clear information on energy efficient solutions from the technological point of view. All modern competitive technological methods transform the action of waves into electrical or mechanical energy, using different ways. There are four basic ways of wave energy conversion [11]:

- Submersible turbines that capture the motion of ocean currents and use their energy to propel the blades. The rotation of blades acts on the generator like a submerged wind turbine. An example: Blue Energy's Davis Hydro;
- Digging a system of channels that directs the water flow in the reservoirs, increasing the water level above the sea level. Then this water drives the turbines similarly to the case of hydroelectric dams;
- Floating systems containing buoys installed on the ocean surface. When the sea level increases or decreases the relative motion of the floating body drives the pistons and hydraulic pumps, and this kinetic energy is used to drive the turbine;
- Systems based on the oscillation of water column, mounted in generation plants; the water enters the space especially reserved and forces the air to move the turbine. When the wave withdraws, the air pressure decreases, causing the return of the turbine. Example: LIMPET facility in Scotland.

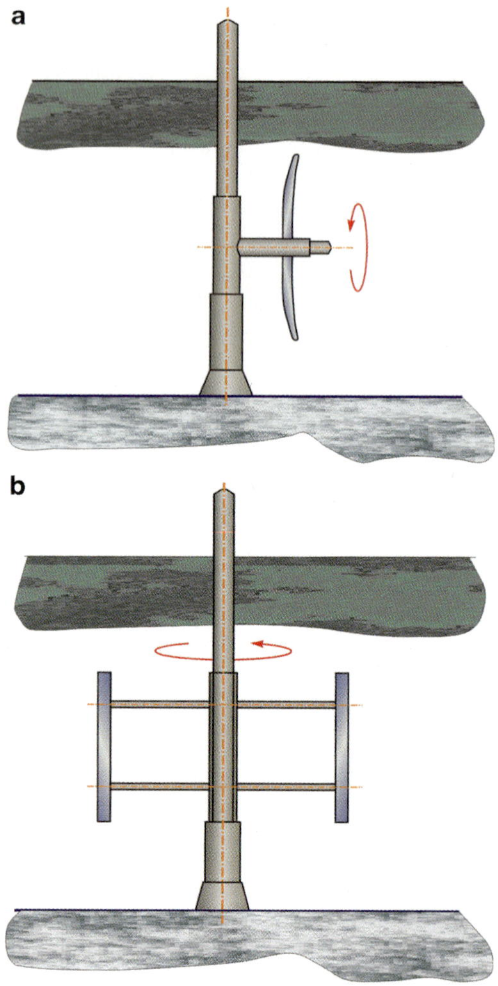

Fig. 7.21 Water turbine: (**a**) with horizontal axis; (**b**) with vertical axis

7.5.2 Water Turbines with Submersible Blades

Useful energy can be generated by sea currents using completely submerged turbines. Water turbines with horizontal axis (Fig. 7.21a) and, respectively, with vertical axis (Fig. 7.21b) [15] operate like the wind turbines, using the kinetic energy of the moving fluid, transforming it into a rotational motion and electricity. The speed of water currents is lower than the wind speed, but with a higher density (835 times bigger than the air density); water turbines are smaller than the wind turbines for the same installed capacity. The power generated by tides depends on the velocity of the water streams, the explorable area and the water turbine efficiency. It can be calculated with the equation:

$$P_{t.s.} = \frac{1}{2}\rho A v^3 C_p, \tag{7.15}$$

where δ is the sea water density (1,025 kg/m^3);

A is the area of multi-blade rotor (m^2);
v is velocity of water currents;
C_p is power coefficient, measurement of turbine efficiency.

The prototypes of turbines with horizontal axis and 10 kW power, produced on the basis of conceptual schemes, (Fig. 7.21a) have been tested. It is envisaged to install and demonstrate a 300 kW installation on the Southern Coast of Great Britain (MCT, 2000). Flux concentrators can be used to increase the power generated.

In turbines with vertical axis (Fig. 7.21a), both cases based on the use of frontal resistance and buoyancy were investigated. The best known example is the Darrieus turbine with three or four thin blades or with aerodynamic section. Several prototypes of this kind have been tested, including 5-kW Darrieus turbines, installed in Kurushima area, Japan. It is planned to install such turbines with a total capacity of 30 MW in Canada [15].

7.5.3 Systems Based on the Oscillating Water Column

The Government of Scotland has funded three wave energy projects. Only one of them was implemented and generates electricity. This is the LIMPET 500 plant (Land Installed Marine Powered Energy Transformer) project (Applied Research & Technology – ART, Scotland) [11] put into operation on a small island of Islay on the west coast of Scotland. Plants of this type (OWC) are the most important ones from the class of shore facilities. The operating principle is shown in Fig. 7.22. The motion of water column in the structure, due to changes in the sea level, compresses (at the wave motion in one direction) and decompresses (return water) alternately the air inside the structure, which drives the turbine (e.g. Wells type) and the generator. Another similar OWC 400 kW plant is built on Pico Island in Portugal.

Figure 7.23 shows OSPREY (Ocean Swell Powered Renewable Energy) device for wave energy conversion, operating under the same principle (Applied Research and Technology – ART). The prototype is designed for seashore areas with water depth up to 20 m. The prototype with installed capacity of 2 MW was developed by *Wavegen* LTD, Great Britain. The plant operates at a depth of 15 m at a distance of about 1 km from the shore.

7.5.4 System with Floatable Elements

An example of plants floating on the water surface is the **Salter's Duck facility**, developed by Professor Stephen Salter, University of Edinburgh, Scotland, as

Fig. 7.22 Operation principle of LIMPET plant, Scotland

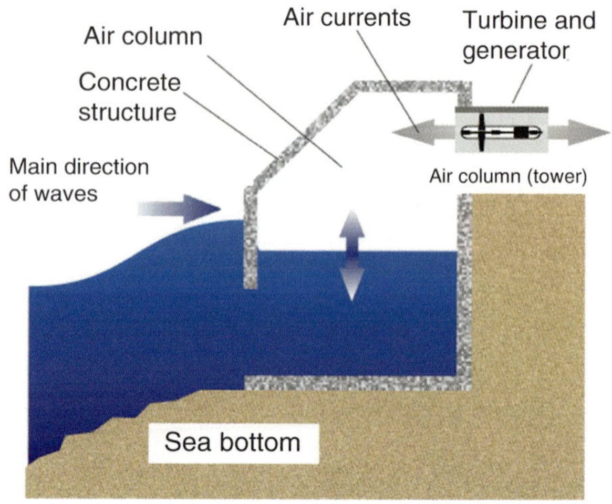

Fig. 7.23 OSPREY device for wave energy conversion

response to the oil crisis of 1970. The conceptual diagram of the facility is shown in Fig. 7.24 [16]. Duck elements possess alternating rotational motion under the alternative action of waves above their crests. The installation requires at least 80 m depth and is anchored to the seabed with a flexible (adjustable) link.

Small scale tests have shown that the Duck system can stop up to 90% of wave motion which is converted into electrical energy. There are several projects aimed at reducing as much as possible the ratio weight/power.

Installation with oscillating configuration. This plant tracks the wave movement. It generates energy, using alternative rotational motion generated by the junctions of cylindrical or flat sections. An example is the Pelamis WEC facility, which was

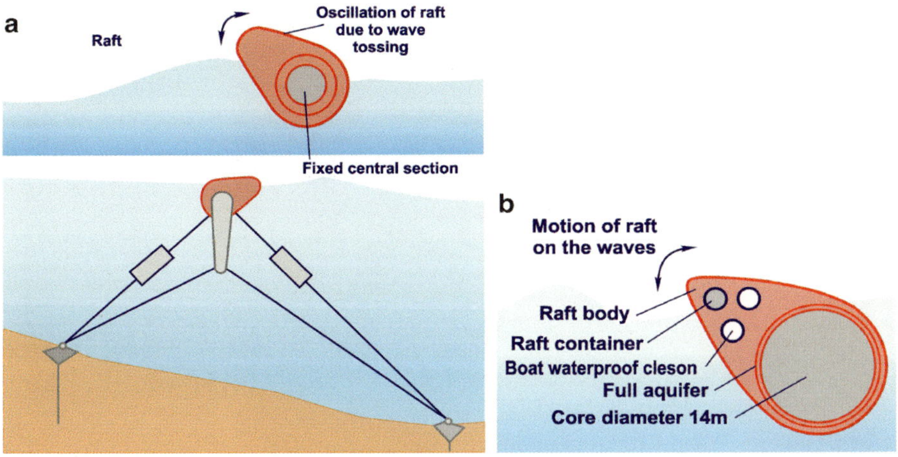

Fig. 7.24 The conceptual diagram of the Salter's Duck facility

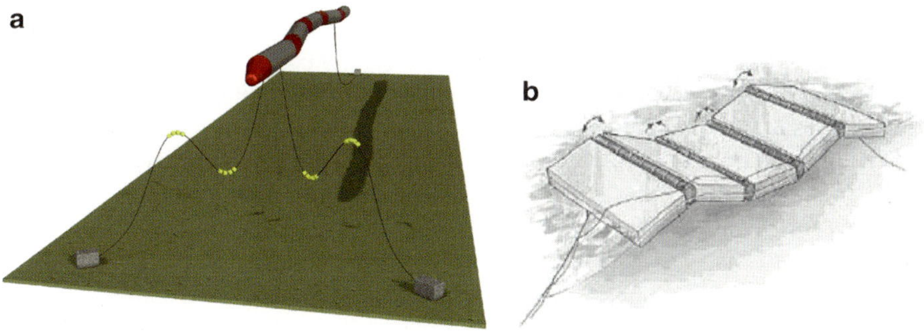

Fig. 7.25 Pelamis wave energy converter (*Ocean Power Delivery LTD*)

developed by the Ocean Power Delivery company (Fig. 7.25) [16]. When the plant moves with the waves, the motion is transmitted to the joints via hydraulic frames, which pump oil under high pressure to the hydraulic power engines through the batteries. These engines are used to drive the power generators. A plant with the installed capacity of 750 kW is 130 m length, its diameter is 3,5 m and includes five sections.

7.5.5 *Cost Analysis of a Number of Coastal Systems*

The most important achievements in the development of wave energy conversion systems belong to the following companies: *Land Installed Marine Powered Energy Transformer* (LIMPET) (Applied Research and Technology – ART, Scotland) is

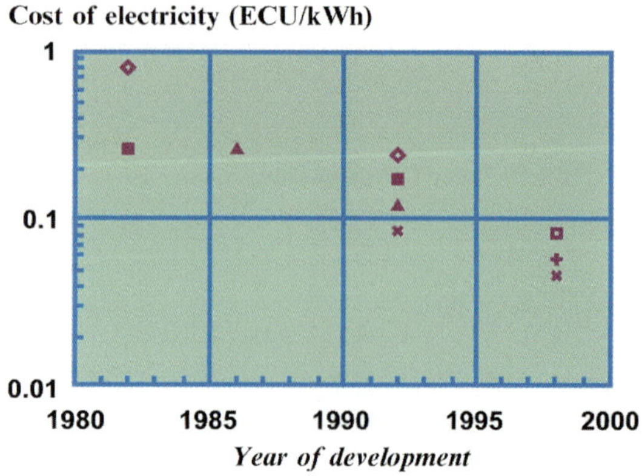

Fig. 7.26 Evolution of coastal systems cost generation: ◇ *Edinburgh Duck* ■ *Bristol cylinder* ▲ *SEA clam* × *PS Frog Mc* □ *Cabe wave pump* + *sloped IPS Buoy*

leader in the field of systems located in areas close to the seashore [17]; Ocean Swell Powered Renewable Energy (Osprey) is leading in the area of systems located on the shore (Applied Research and Technology – ART); *Duck systems* developed at the University of Edinburgh and *PS Frog* systems developed at the University of Lancaster are the most interesting ones located on the coast. Today, these systems are the most representative (but incomplete) from a wide range of wave energy technologies.

The analysis of cost evolution of coastal systems presented in the diagram in Fig. 7.26 shows cost savings of more than two times in just 20 years. The trend remains valid for the next years due to the new advanced technologies for wave energy capture. At the same time, with some notable exceptions, the development of wind, wave and tidal energy conversion systems usually is an isolated activity, with some cooperation between different research teams. This fact leads to losses of resources and talent, especially in areas of common interest for the three technologies, including: the geology of foundations; the impact on the environment; the selection of materials; sustainability standards; design software and research; legal requirements; capacity restrictions; electricity transmission; energy storage; hydrodynamic simulations, etc.

7.6 Wave Energy Capture Systems

Rapid depletion of fossil fuel reserves influenced researchers to consider new energy sources. As the oceans and the seas have always been close to the man and occupy 71% of the Earth's surface, the engineers and experts in the field of renewable energy resources have built special devices to capture the wave energy. Many research institutes on energy and hydraulics (U.S., France, Great Britain and Japan) have planned to design devices to capture the wave energy.

Following the regulations of the Scottish Executive, the energy generated by the waves and tides may represent up to 10% of the electricity generated in the country

Fig. 7.27 The first tidal power plant, La Rance, France, 1966

in 2010. To this end, major changes are made in the legislation on renewable energy, in order to facilitate significant investments in the renewable marine energy. This is equivalent to complete replacement of some fossil fuel power plants. Scotland has access to one of the richest marine energy resources in the world. In 2001, Scotland's Renewable Resource report stated that Scotland can generate a capacity of up to 21.5 GW (79.2 TWh/year) with the help of the wave and tidal energy. According to the report "*Using marine energy potential of Scotland*", developed by *Marine Energy Group* in 2004, capacities of 1,300 MW can be installed in Scottish waters till 2020, adding 100 MW each year. In order to support the green energy producers by *Renewables Obligation (Scotland),* electricity distributors are required to supply electricity to customers with a certain proportion coming from the renewable sources. The Scottish Executive target is to raise the percentage of renewable energy to 18% in 2010 and to 40% by 2020. Recently the Scottish Government announced funding for the largest *"wave farm"* in the world. The Scottish Government announced funding of 13 million pounds for testing other marine power plants too [18].

7.6.1 Tidal Energy Capture Systems

Between 1980 and 1990 two notable successes in the development of ocean energy conversion systems have been announced. One is a 240 MW tidal power plant in La Rance, France (Fig. 7.27). The second success is the large tidal power plant in Nova Scotia with an installed capacity of 20 MW. Currently, there are several development

Fig. 7.28 "*SeaGen*" project, proposed by the *Marine Current Turbine* company

projects that seek commercial exploitation of tidal and wave energy: Rhode Island, 500 kW in 2006; Northern Ireland, 1 MW in 2006; Cantabria, Spain, 1,25 MW in 2007; Cornwall, United Kingdom, 5 MW in 2008; in northern Portugal, 24 MW in 2007; Northern Devon, United Kingdom, 10 MW since 2010, etc. There is a real explosion of projects that seek to exploit the wave and tidal energy. Perhaps, the 20 years gap, mentioned earlier, will be quickly recovered.

Nowadays, the most ambitious project seems to be the one of the Severn Estuary, UK. In 2006 the British Government has approved a feasibility study for a 16 km dam along the estuary. The Severn barrage will cost about $25 billion and will produce 8.6 GW of electricity. A tremendous amount of energy, but its price will be too high, as the environmentalists say. It will ruthlessly disturb the estuary ecosystem. Therefore, the future seems more interesting to other systems, such as the "*SeaGen*" proposed by the Marine Current Turbines Company (Fig. 7.28). Basically, this is a submersible wind turbine which blades represent submarine propellers that rotate under the action of water during the tidal currents. The main mechanical components of the turbine are presented in Fig. 7.28. If such plants are installed on the British coast, they could generate as much electricity as in the Severn Dam, but without environmental impact on the coast. "SeaGen" will consist of a set of pairs of submarine propellers with a 15–20 m diameter, which will engage the electrical generators, as in the case of the wind propellers.

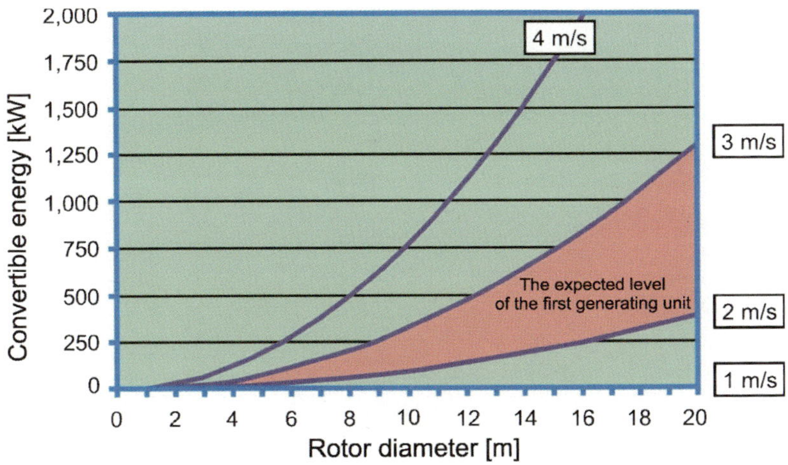

Fig. 7.29 Converted energy of marine currents at indicated velocities with approx. 30% efficiency

Each pair of propellers will be mounted on several 3 m diameter steel columns, which base will be cemented on the foundation dug on the seabed. Each pair of submarine propellers will produce from 750 to 1,500 kW, depending on the velocity of the underwater currents. Figure 7.29 shows the graphs of energy converted from the marine currents for various velocities, reaching approximately 30% efficiency. It is envisaged to develop similar farms of systems, which will be located in places where the speed has a maximum displacement of currents.

This principle is applied in the wind power plants too, but in the case of marine plants the density can be higher. Since the rotational speed of the submerged propellers is relatively small (10–20 min^{-1}), the environmental impact will be minimal. Figure 7.30 presents a computational modelling technology developed by the company "*Blue Energy*" (Canada) [19], which is at the stage of commercial implementation. This technology is based on the use of Davis wind turbine for high efficiency conversion of the sea currents, which fact allows the reduction of electricity cost to $0.02 per kWh. The use of tidal energy has the advantage to be completely predictable (unlike the wind energy, which depends on meteorological factors), but has the disadvantage that there are few places on the continental coasts where it reaches economically exploitable value.

7.6.2 Wave Energy Capture Systems

A study conducted by Reger Bedard, director of the "*Institute of Electric Power Research*", shows that by designing plants to exploit (with an efficiency of 50%) only 20% of the wave energy along the USA coast, the produced energy would exceed

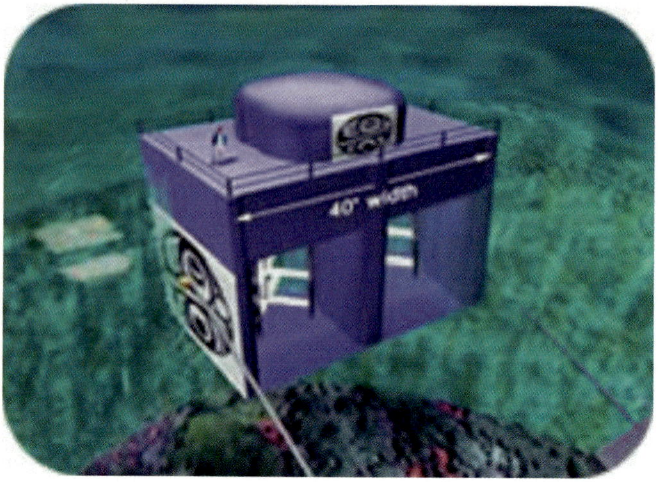

Fig. 7.30 "*Blue Energy*" tidal turbine, Canada

the generated energy of all hydropower plants of the United States! Unfortunately, unlike the tides, waves are whimsical, but nowadays there are projects aiming at the use of wave energy.

A special interest is shown for *the Pelamis* device [20], manufactured by the *Ocean Power Delivery* company, which soon will produce 2.55 MW of electricity on the northern coast of Portugal. The *Pelamis* device resembles a huge snake made up of several rigid segments. The entire system is anchored with a single cable, allowing free oscillation of the segments. The *Pelamis* system in northern Portugal (Fig. 7.31) has a length of 120 m and can provide electricity for 1,500 homes, a complete system that would produce 30 MW, and would cover 1 km^2 of ocean. These systems have been tested at the European Marine Energy Centre (EMEC) in Orkney by the *Ocean Power Delivery* company. *Ocean Power Delivery* has already exported the Pelamis systems to be used in commercial wave farms. The operating principle of the existing *Pelamis devices* is presented in Fig. 7.32. Each *Pelamis* system includes three conversion energy modules that, together, generate 750 kW of electricity (Fig. 7.32a). The movement of the sea wave forces the *Pelamis* segments to move up and down, doing a serpentine-like oscillatory motion. These oscillations are converted into hydraulic energy, after which, through an ingenious system, it drives some hydraulic engines, which, in their turn, engage the electric generators.

The mounting of three *Pelamis* devices of 750 MW each has already started in Aguçadora Wave Park in Portugal. It is envisaged to install wave farms with a total power of 3 MW in Orkney, Northern Scotland; these farms will include four 750 kW *Pelamis* units. There are plans to develop 20 MW ‚*Wave Hub*" systems on the North Coast of England, in Cornwall.

The Wave Hub Initiative, proposed by the USA to UK, will facilitate economically future actions, believes Colin Palmer, Director of Ocean Prospect: "Without

Fig. 7.31 *Pelamis* facility installed in the coastal waters of Portugal

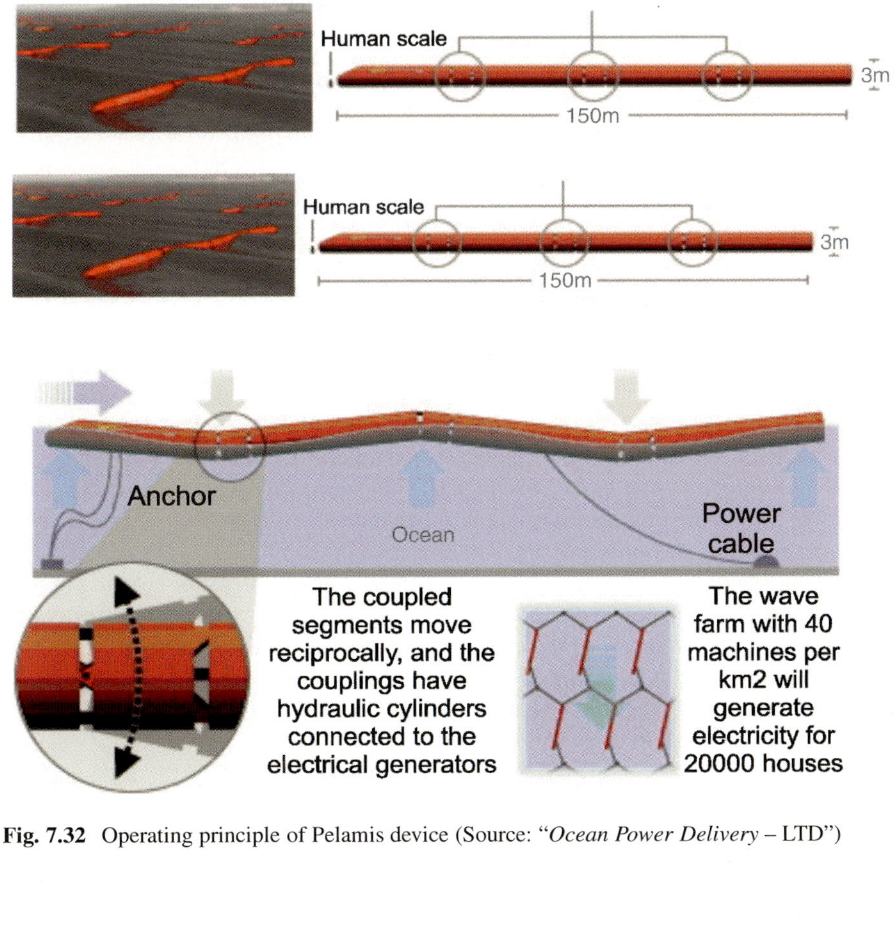

Fig. 7.32 Operating principle of Pelamis device (Source: "*Ocean Power Delivery* – LTD")

Fig. 7.33 PowerBuoy systems designed by OPT company

it, we would have to wait a long time until market conditions become suitable for a solitary and large-scale project. The "Wave Hub" can be a demonstration of what wave energy means to the world". Three British companies have been selected for ocean energy exploitation within the *"Wave Hub"* project; the financial means available for this project amount 15 million pounds starting 2007. *"Ocean Power Technologies Inc"*. (OPT), *"Fred Olsen & Co."* and *"Wind Prospect Ltd"* will build a *"farm"* of marine energy, lying on the seabed at a distance of 15 km from the beaches of Cornwall County. The generated electricity will be transmitted to the national grid via an underwater cable. It is based on *"OPT PowerBuoy"* system. The project is financed by the West of England Agency of Regional Development. *Ocean Power Technologies* Company (ORT) is the leader specializing in cost effective technologies for sea wave energy conversion, which are advanced and environmentally friendly technologies. The extraction of natural energy from the waves is based on the integration of patented technologies in the field of hydrodynamics, electronics, energy conversion and computer control systems. OPT power plant has a capacity of 5 MW and contains 30 *PowerBuoy* systems of 150 and 250 kW [18, 20] (Fig. 7.33). *PowerBuoy* technology consists of an ocean energy converter, submerged at a depth of more than one meter. Inside it, a piston is operated by up and down motion of waves. This motion is feeding a generator placed on the seabed that, in its turn, transmits electricity to the shore. This technology is environmentally benign, does not affect marine life and does not produce electric shocks or strong electromagnetic fields due to good insulation of the transmission cables.

Fig. 7.34 "*AquaBuoy Finavera*" coastal floating buoy structure, Oregon (USA)

Pacific Gas and Electric (PG & E) Company is the first American company to use the ocean energy. In 2007 the above company announced support for the projects envisaging construction of the first U.S. commercial plant for wave energy conversion on the coasts of North Carolina [21]. The facility will contain eight buoys installed over a length of about 4 km along the shoreline. The rising and falling of waves move the *buoy*-like structure generating electricity. It is scheduled that the plant will start operation in 2012, generating about 2 MW of electricity. Together with *Finavera Renewables,* Canadian company, they build a *"wave farm"* nearby the Eureka coast of Humboldt area (Fig. 7.34).

Sea water currents drive the turbine which, in its turn, generates electricity. The wave farm links several "AquaBuoys Finavera" converters. Figure 7.35 presents such a wave farm connected to a central storage system for the produced electrical energy, which is transported through an underwater cable to the shore. According to the experts' estimates, the energy potential of about 1,200 km California coast is about 37,000 MW. Using modern technologies, maximum 20% of wave energy can be converted into electricity.

7.7 Wave Energy Capture Systems, Authors' Elaborations

Among the clean and non-pollutant energy sources, the wave energy of sea and ocean is one of the most abundant to take advantage of. The wave energy potential is estimated at 219 G-watts along the coast of the European Union, or more than

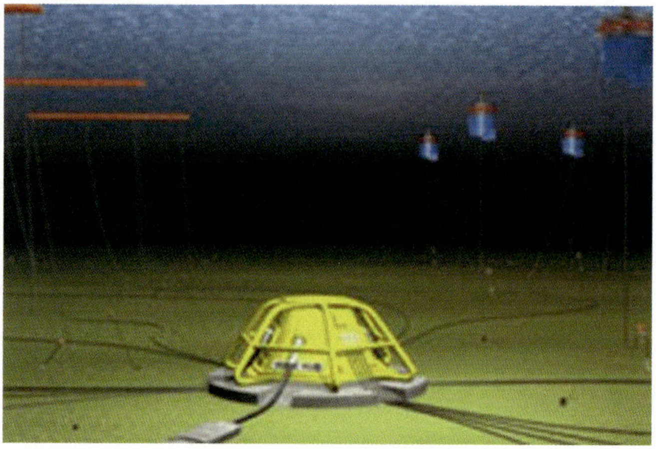

Fig. 7.35 Farm of wave energy converters

180 TW/h per year. Over the years, various devices have been designed to extract the wave energy, exploiting the lifting force of the waves with floating bodies.

Although Moldova does not border a sea or an ocean, the topic is of interest primarily in terms of international cooperation. Taking into account the importance of the topic, research is conducted at the Centre for Renewable Energy Conversion Systems Design (CESCER), at the Technical University of Moldova. Research is at an early stage. In this respect, some technical solutions have been developed for wave energy capture devices at conceptual level. These solutions are protected by six patents. Future development is expected based on these inventions, namely, laboratory samples and their testing in laboratory conditions.

7.7.1 Wave Energy Conversion Plants

7.7.1.1 Fixed Tower Plants and Floating Bodies

Such devices are part of the coastal power plants based on the use of a fixed tower and some floating bodies that are linked to the tower via a swivel connection. Further on, two patented technical solutions will be considered. The installation (Fig. 7.36a, b) [22] includes a floating body 1, bound with the arm 2 to the tower 3 through the shaft 4, forming together with the gear segment 5 (Fig. 7.37) a lever. The segment 5 is linked kinematically to the gear wheels of the transmission 6. Alternate rotational motion of the segment 5 is transformed into one-way rotational motion of the output shaft through transmission 6. The output shaft of the motion conversion device 6 is connected via a multiplier 7, to an electric generator 8. In order to standardize the rotational motion, a flywheel is fixed to the multiplier 7 output shaft.

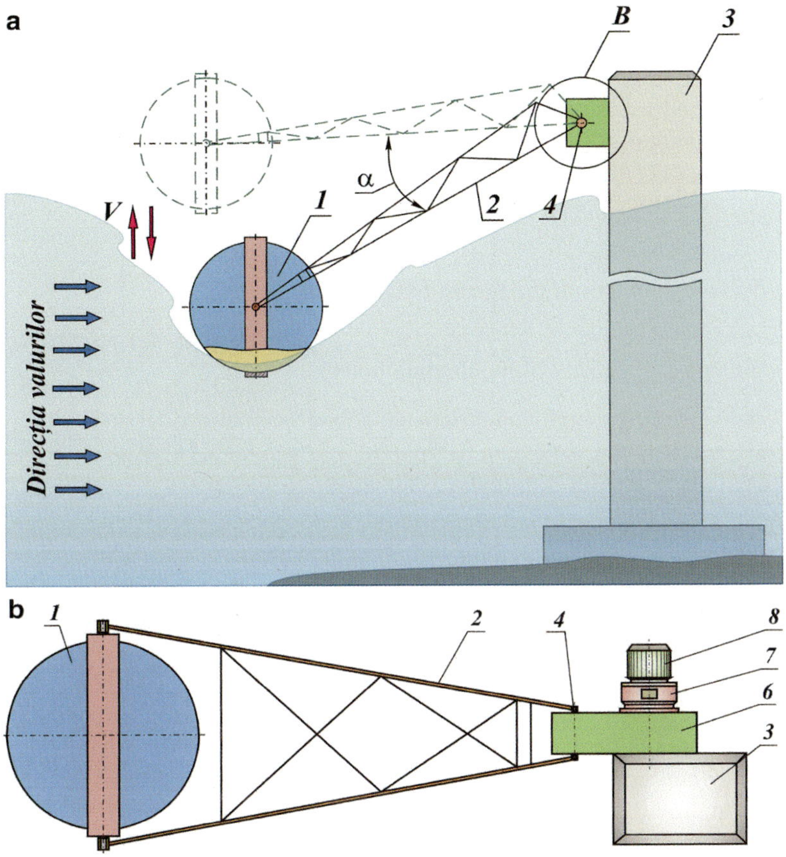

Fig. 7.36 Wave energy conversion plant

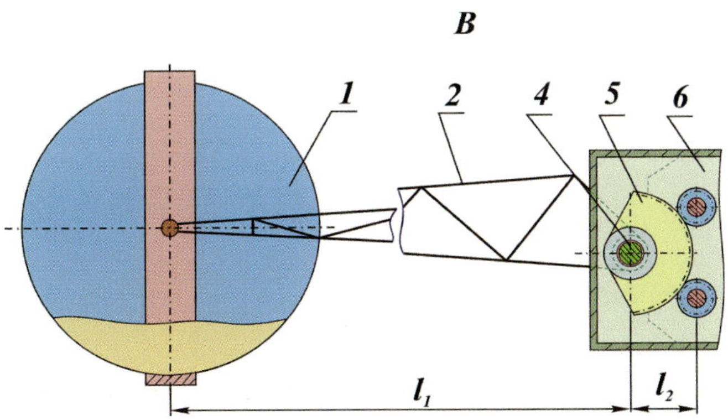

Fig. 7.37 Wave energy conversion plant with adder of alternate motions

At wave travelling with a certain period λ in the direction shown in Fig. 7.37, the floating body 1 will perform a motion in vertical plane. At a certain wave period, the body will make a motion equal to the height of the wave (amplitude). The lifting force developed by the floating body at its vertical motion is determined by the Morison's equation:

$$dF = \pi\rho\frac{D}{4}C_M v^2 + \frac{\rho}{2}DC_D\,|v|\,v, \tag{7.16}$$

where D is the diameter of the floating body,

ρ is the fluid density;
v is the upward velocity of the floating body;

C_M and C_D– are the coefficients that depend on the geometrical shape of the body, Reynolds number and Keulegan–Carpenter number (can be defined empirically, experimentally or by CFD computer simulation).

Next, the summary force $\boldsymbol{F_1}$, developed by the floating body *1*, will be transmitted to the gear segment *5* via arms $\boldsymbol{l_1}$ and $\boldsymbol{l_2}$:

$$F_5 = F_1\frac{l_1}{l_2}. \tag{7.17}$$

The moment of torsion developed by this force will be transmitted to the electric generator 8 through the gearing 6 and the multiplier 7. Alternate rotational motion of the rods 7 and, respectively, the gear segment 5 is transformed and multiplied by gearing 6 and multiplier 7 into continuous rotational motion, which is transmitted to the electrical generator 8. The wave energy is pulsating in rather wide limits and its frequency is high – from 1–2 s to 10–20 s. In this case, the normal operation of the plant needs a flywheel. In general, the wave energy is constant for a long period of time (several hours). In order to reduce rotational non-uniformity of the electrical generator rotor 8 and to improve the quality of the produced power, the flywheel 10 is mounted on the multiplier 7 output shaft. The proposed solution is advantageous as its construction is simple, easy to maintain and very efficient. Due to the provision of an arm driving the wave lifting force, a multiplied moment is transmitted to the input shaft that ensures an increase of the transformed energy.

Figure 7.38 [23] shows **a plant**, in which spherical or cylindrical floating bodies 1 and 2 are placed on both sides of the fixed tower 3 and are jointed to the tower by means of two arms 4 and 5. Since arms 4 and 5 are driven by the floating bodies 1 and 2 into alternate rotational motion with different directions, a motion adder, with bevel-gear wheels 6 and 7, was designed to sum up these motions; bevel-gear wheels are linked rigidly to the arms 4 and 5 (Fig. 7.39).

The summary motion is multiplied and transmitted to the electrical generator of the plant similarly to the unit from Fig. 7.36. Spherical or cylindrical floating bodies

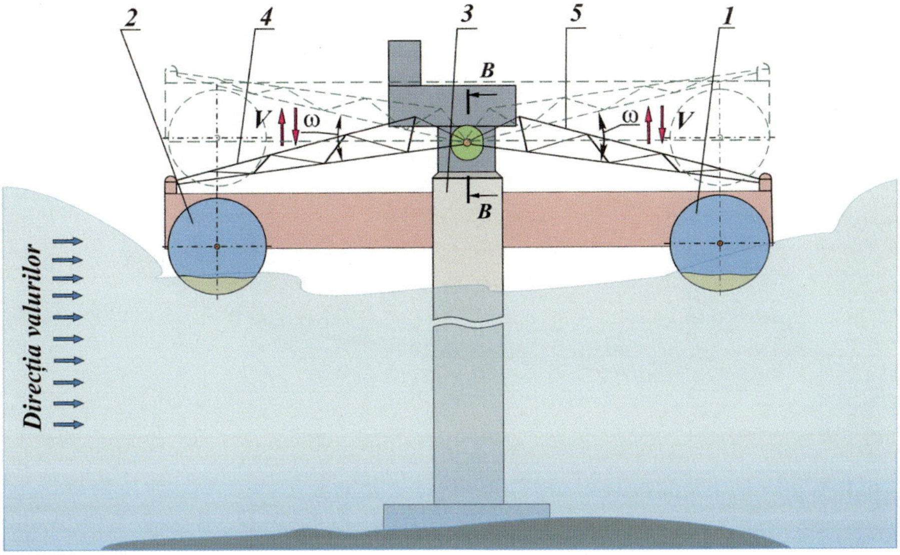

Fig. 7.38 Unit for wave energy conversion

Fig. 7.39 Adder of alternate rotational motions and their conversion to one-way motion

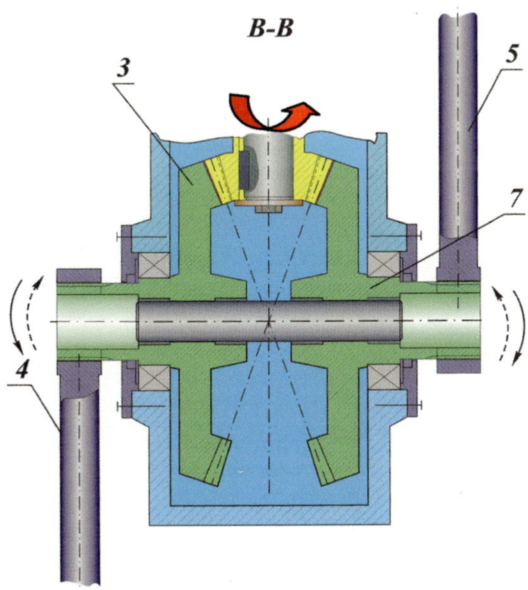

1 and 2 can be manufactured as a single toroidal floating body that has a bigger contact surface with water, ensuring an increased rate of wave energy conversion. Also, the symmetric placement of the arms ensures stability to the fixed tower facility.

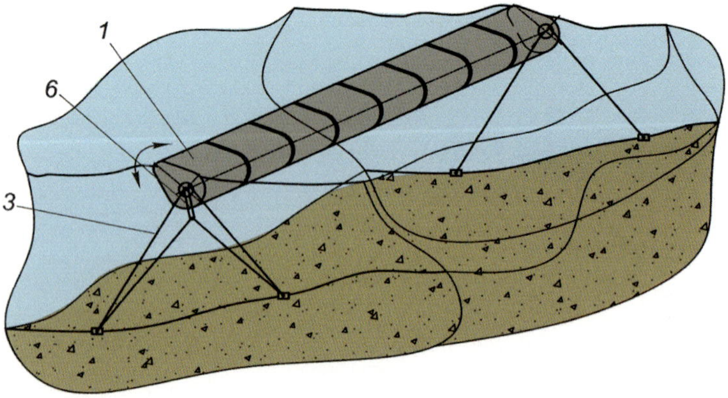

Fig. 7.40 Floating unit for wave energy conversion (DUCK-type)

Fig. 7.41 DUCK type floating body (with counterweight to bring it back into initial position)

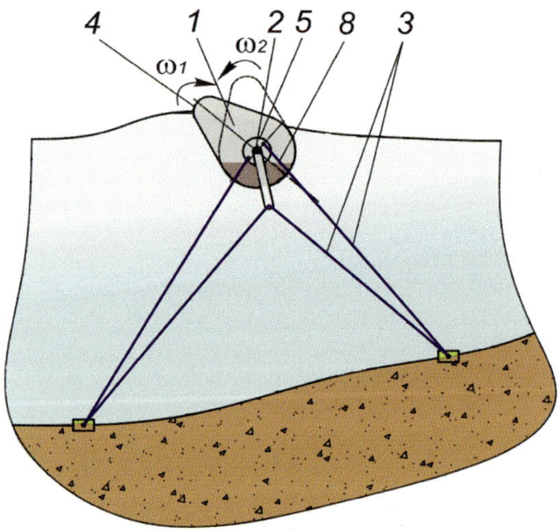

7.7.1.2 DUCK-Type Wave Energy Device

This is another floating wave energy device based on the use of several floating bodies linked by hinged joints. The operating principle of this wave energy device was described in Sect. 7.4 of this chapter.

The device shown in Fig. 7.40 [24] is a version of the *DUCK* wave energy device; it includes the floating bodies 1, mounted on the fixed shaft 2 (Fig. 7.41), which is anchored by anchoring cable 3. The floating body 1 includes an oval casing 4, made of plastic; its submerged part is bigger than the outside part. The casing 4 is located on a fixed shaft 2 with limited possibility of alternative rotation towards it. On the shaft end 2 a disk 5 and a bar with two holes executed at a certain distance

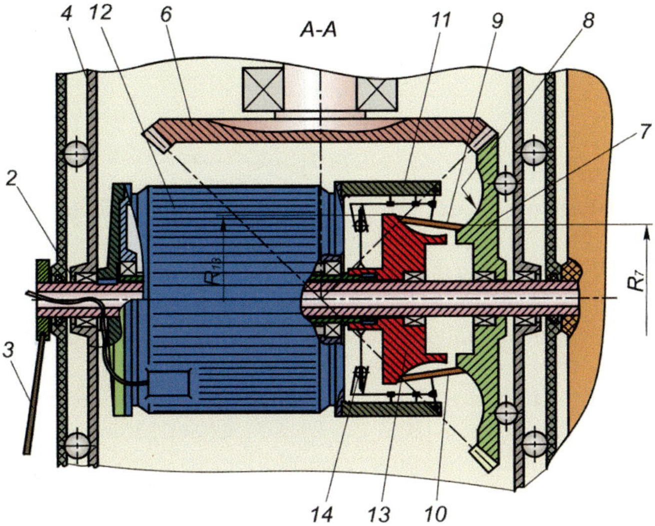

Fig. 7.42 Adder of alternative rotary motion, based on the use of bevel-gear transmission and friction toroidal speed variator

are installed through which the anchor cable 3 is passed. The centre of buoyancy of the "DUCK" floating body is shifted to its bottom by filling it with solid material. A summary mechanism of alternative rotational motions, called adder, is placed in the casing 4.

The adder (Fig. 7.42) includes a bevel wheel 6 geared with the bevel pinion 7; on the inner flange of the pinion the toroidal surface 8 is made. Disks 9 and 10 enter in contact with the toroidal surface; their axles are linked to the sleeve 11, fixed rigidly on the casing of the electrical generator 12. Disks 9 and 10 contact with their diametrically opposed sides with the toroidal surface of the friction wheel 13. The inertial mechanism 14, linked kinematically via a system of levers with the disk axles 9 and 10, is fixed on the wheel hub 13 linked rigidly to electrical generator rotor 12. This inert mechanism provides homogenisation of rotational speed of the friction wheel.

The wave linear theory shows that the motion of a point on the wave surface is sinusoidal. At wave travelling to a certain wave length $\boldsymbol{\lambda}$ in the indicated direction (Fig. 7.40, 7.41), the floating body 1 performs alternate rotational motion around the fixed shaft axis 2. The wave period $\boldsymbol{T}$ that moves a point can be expressed by:

$$T = \sqrt{\frac{2\pi\lambda}{g}}, \tag{7.18}$$

where $\boldsymbol{\lambda}$ is the wavelength (m).

The power contained in a wave can be expressed by the following expression depending on the wavelength (kW/m):

$$P = \frac{\rho g^2 a^2 T}{8\pi}, \tag{7.19}$$

where a is the wave amplitude (m).

This potential energy of the waves can be converted into electricity by using floating bodies (Fig. 7.41). Rotational motion $\boldsymbol{\omega}_1$ of a floating body 1 generated by the wave and the rotational motion $\boldsymbol{\omega}_2$ at its coming back under the action of the counterweight is transmitted to the rotating casing 4 of the adder (Fig. 7.42), where these two alternate rotational motions are summed up in a one-way rotational motion.

In the adder shown in Fig. 7.4, the rotational motion of the conical pinion 6 is transmitted via disks 9 and 10 to the friction wheel 13 and, further on, to the electrical generator rotor,12. Summary gear ratio is equal to:

$$i_{\sum} = i_{\Sigma 1} \cdot i_{2\text{var}}, \tag{7.20}$$

where

$$i_{2\text{var}} = \frac{R_{13\text{var}}}{R_{7\text{var}}}, \tag{7.21}$$

but $\boldsymbol{R_{13var}}$ and $\boldsymbol{R_{7var}}$ are contact radiuses of disks 9 and 10 with toroidal surfaces of the friction wheels 7 and 13.

In the result, the gear ratio will change and the electric generator rotor speed will become uniform. Adder kinematics allows partial equalization of summary angular velocity components values of the electrical generator rotor composed of $\boldsymbol{\omega}_1$ and $\boldsymbol{\omega}_2$ components, and thus improve the quality of generated electricity.

In the wave energy capture plant [25], unlike the previous facility, the returning of Duck bodies 1 to the original position is done through a system of curved elements 2 placed between the flange of the rotating disc 3 and the fixed part 4 (Fig. 7.43). A portion of energy stored in these curved elements at their decompression brings back the DUCK elements into the initial position (previous to the wave action). Summing up of alternative rotational motions is performed by an adder based on the use of coaxial planetary precessional transmission 18 in several steps and of a one-way clutch system (Fig. 7.44). This system leads to partial equalization as value of the summary angular speed components of the crank shaft and to the improvement of the produced electrical current quality, as well.

7.7.2 *Alternative Rotational Motion Summing Mechanisms for Wave Energy Conversion Systems*

In the majority of wave energy conversion plants the wave amplitude is converted into alternative rotational motion. However, electrical generators need one-way and

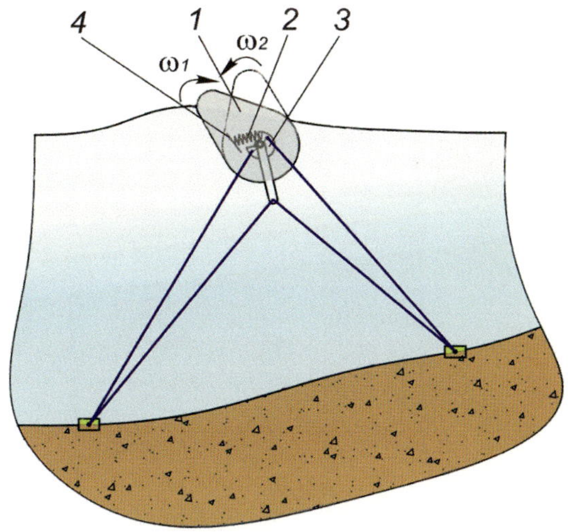

Fig. 7.43 DUCK floating body (with curved elements to bring it back to the original position

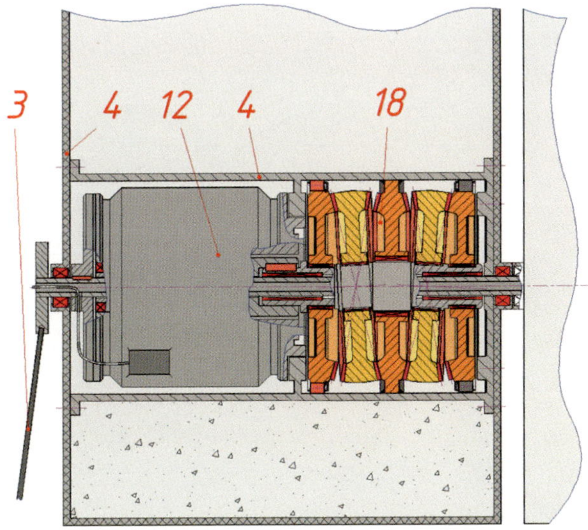

Fig. 7.44 Adder of alternative rotary motion based on the use of planetary precessional transmission

multiplied rotating motion. For this purpose a special mechanism is used to convert the motion.

The mechanism for motion conversion (Fig. 7.45b) [26] includes an input shaft 1, on which the gear segment 2 is installed, and two-step gear drives. The first step includes the gear wheel 3 mounted on the shaft 4, the gear wheel 5; the one-way clutch 6 is mounted on the hub of gear wheel 5. The gear wheel 7 is fixed on the shaft 8, on which the gear wheel 9 is fixed. The gear wheel 10 is mounted on the

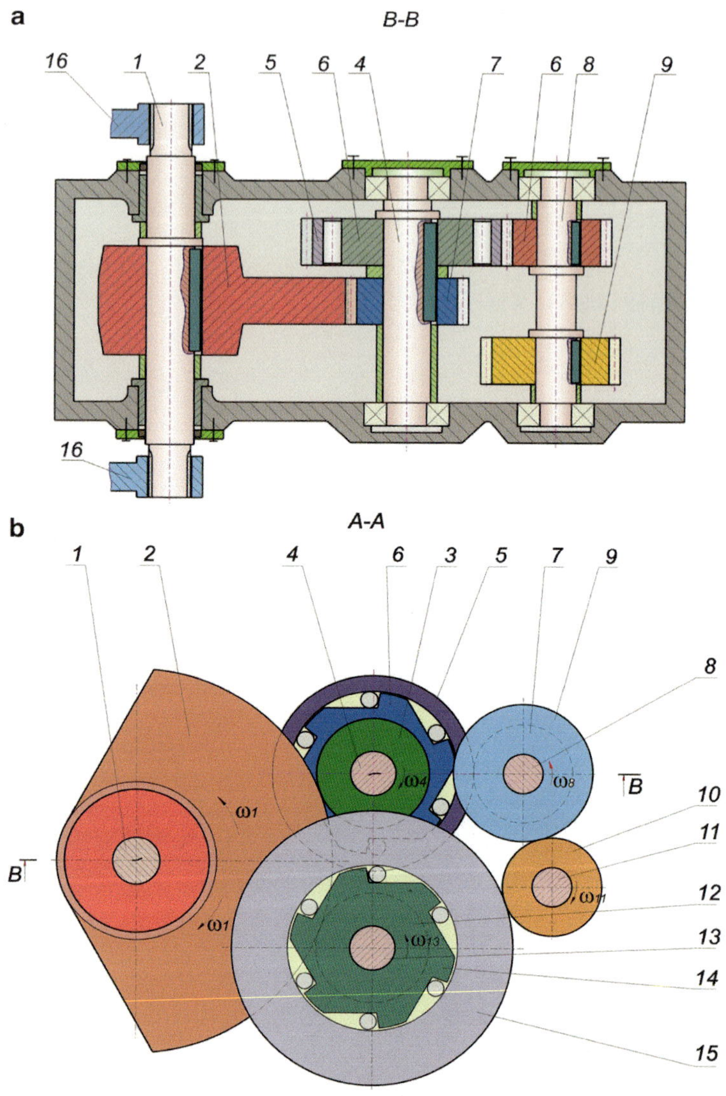

Fig. 7.45 Mechanism for conversion of alternative rotary motion into oneway rotary motion

output shaft 11. The second step includes the gear wheel 12, rigidly fixed on the shaft 13, on which the gear wheel 15 is mounted through the one-way clutch 14. One-way clutch 14 is installed on the shaft operating in the opposite direction compared to the one-way clutch 6. An input shaft is driven by two flows of the crank 16. Alternative rotational motion with angular velocity ω_1 is transmitted from the crank 16 to the gear segment 2 by an input shaft 1. From the gear segment 2 the alternative rotational motion is transmitted simultaneously to the two-step gear drives. In the first step, alternative rotational motion from the gear segment 2 is taken by gear wheel 3 and is transmitted via shaft 4 – one-way clutch 6 to the gear wheel 5. From the gear wheel 5, rotational motion in only one-way direction with angular velocity ω_4 (the direction of rotational motion corresponds to the direction of rotational motion, in which the one-way clutch 6 joins the shaft 4 with the gear wheel 5) is transmitted by gearing with the gear wheel 7 via shaft 8 to the gear wheel 9. From the gear wheel 9, the rotational motion with angular velocity ω_8 is transmitted by gearing with the gear wheel 10 to the output shaft 11. In the second step, alternative rotational motion from the segment 2 is taken via gearing by gear wheel 12 and transmitted through the shaft 13 – one-way clutch 14 to the gear wheel 15. From the gear wheel 15, one-way rotational motion with angular velocity ω_{13} (the direction of rotational motion corresponds to the direction of rotational motion, in which the one-way clutch 14 gears shaft 13 with the gear wheel 15) is transmitted by gearing with gear wheel 10 at the output shaft 11. One-way clutches 6, 14 are mounted so that an alternative rotational motion in one-way direction is transmitted from the shaft 4 to the gear wheel 5, and in the opposite direction it is transmitted from the shaft 13 to the gear wheel 15. Thus a continuous one-way rotary motion with angular velocity ω_{11} is obtained at the output shaft 11. For angular velocity ω_{11} transmitted by steps one and two to the output shaft 11 to be equal it is necessary to respect the equality:

$$i_{2,3} \cdot i_{5,7} \cdot i_{9,10} = i_{2,12} \cdot i_{15,10} \qquad (7.22)$$

where $i_{2,3}$ is the gear ratio of the gear wheels 2 and 3 step;

$i_{5,7}$ – is the gear ratio of the gear wheel 5 and 7 step;
$i_{9,10}$ – is the gear ratio of the gear wheel 9 and 10 step;
$i_{2,12}$ – is the gear ratio of the gear wheel 2 and 12 step;
$i_{15,10}$ – is the gear ratio of the gear wheel 15 and 10 step.

The structure of the designed device ensures a more uniform motion, reduced dynamicity and increased reliability at outlet.

The mechanism for motion conversion shown in Fig. 7.46 [26] is similar to the one shown in Fig. 7.45. This mechanism differs by its constructive design. It provides wave energy transmission through two separate flows of the two floating bodies. This possibility is provided by levers 1 and 2, mounted on shafts 3 and 4 and engaged with each other through the gear segments 5 and 6. The summing of alternative rotational motions and their conversion into oneway rotational motion is analogical to the device shown in Fig. 7.45.

Transmission of wave energy is also carried out by two levers engaged with each other through a system of bevel gears **in the mechanism for motion conversion** [27]. The summing of alternative rotational motions and their conversion into oneway rotational motion is analogical to the device shown in Fig. 7.45.

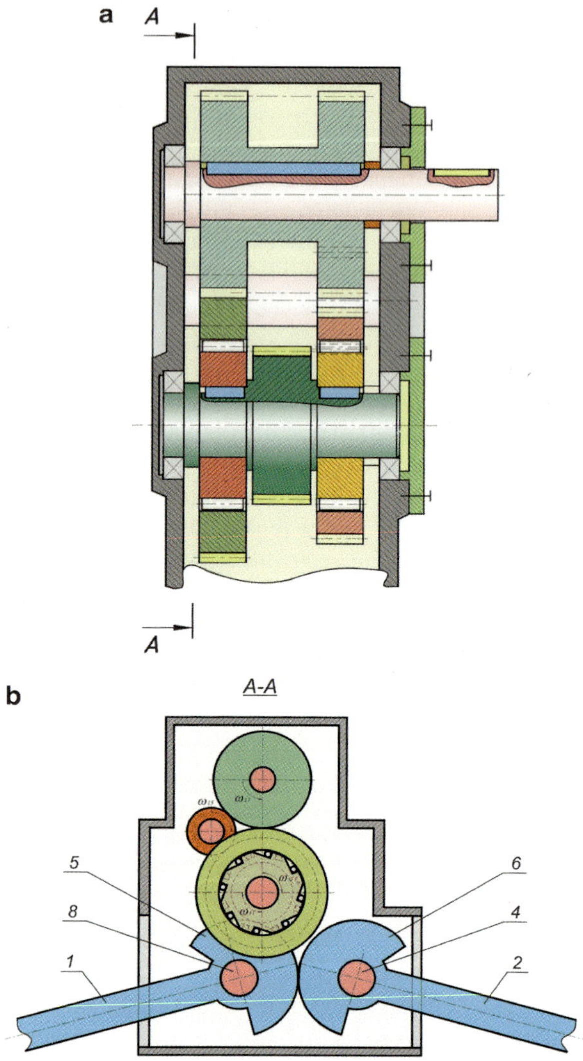

Fig. 7.46 Mechanism for conversion of alternative rotary motion into oneway rotary motion (with two input flows)

References

1. http://www.renewableenergyaccess.com/rea/news. Accessed 17.06.2007
2. Pat McCULLEN, ESBI; Bob MEIR, UK; Aodhan FITZGERALD, Marine Institute; Paul BREWSTER, ClearPower Technology (2003) Ocean energy: development & evaluation protocol. Always work to plan but be prepared to improvise. Part 1: wave power, HMRC
3. Rudkin EJ, Loughnan GL (2001) Vortec – the marine energy solution. In: Proceedings of the marine renewable energy conference 2001, Newcastle
4. Elefant C, O'Neill S (2005) Ocean energy report for 2005, Ocean Renewable Energy Coalition
5. http://en.wikipedia.org/wiki/Tide. Retrieved 20.04.2012
6. Open University Waves (1989) Tides and shallow water processes. Pergamon Press, Oxford

7. http://peswiki.com/index.php/PowerPedia:Fossil_fuel. Accessed 21.04.2012
8. Arey AL, Bryant FL, Clendenin WW, Morrey WT (1911) Physiography for high schools. D.C. Heath, USA
9. http://oceancurrents.rsmas.miami.edu/ocean-gyres.html. Retrieved 21.04.2012
10. Gorlov A, Rogers K (1997) Helical turbine as undersea power source. Sea Technology, United States
11. Centre for Renewable Energy Sources (CRES) (ed) (2002) Wave energy utilization in Europe: current status and perspectives. European thematic network on wave energy. EESD, ISBN 960-86907-1-4
12. Garrison I Ocean waves and standing waves, chapter 10. Reading assignment. In: Introductory oceanography. http://www4.ncsu.edu/eos/users/c/ceknowle/public/chapter10/part1.html. Retrieved 21.04.2012
13. Anthoni JF (2000) Oceanography: waves. www.seafriends.org.nz/oceano/waves.htm. Retrieved 21.04.2012
14. http://www.boatsafe.com/nauticalknowhow/waves.htm. Accessed 21.04.2012
15. MCT (Marine Current Turbines Ltd) (2000) http://www.globalmarinerenewable.com/images/stories/2011Presentations/IndustryDevelopmentDay/MCT_short_presentation_9_Mar11.pdf. Accessed 21.04.2012
16. Wave energy: technology transfer and generic R + D recommendations, DTI Pub/URN 01/799
17. http://www.wavegen.com/pdf/WavegenBrochure_may06_final.pdf. Accessed 21.04.2012
18. Lloyd E, Ramsson T (2007) Sector focus: marin power. Venture Business Research Ltd. Macnab R, Varma H (2008) Bathymetry from space
19. http://www.bluenergy.com/. Accessed 21.04.2012
20. http://news.bbc.co.uk/2/hi/uk_news/scotland/6377423.stm#graphic. Accessed 21.04.2012
21. http://www.myspace.com/video/aeo/aquabuoy/41273903. Accessed 21.04.2012
22. Bostan I, Dulgheru V, Bostan V, Ciobanu O, Ciobanu R, Sochireanu A (2006) Unit for wave energy conversion into electrical energy. Patent MD no. 2989. Publ.BOPI – No.2/2006
23. Bostan I, Ionescu F, Dulgheru V, Bostan V, Cozma T, Sochireanu A (2006) Unit for wave energy conversion into electrical energy. Patent MD no. 2990. Publ. BOPI –No.2/2006
24. Bostan I, Dulgheru V, Bostan V, Cozma T, Dicusară I (2008) Unit for wave energy conversion. Patent MD no. 3542. Publ. BOPI –No.3/2008
25. Bostan I, Dulgheru V, Bostan V, Cozma T, Ciobanu O, Ciobanu R (2008) Unit for wave energy conversion. Patent MD no. 3543. BOPI no. 3/2008
26. Bostan I, Dulgheru V,Bostan V, Ciobanu R, Sochireanu A (2006) Device for the transformation of alternating rotational motion into rotational motion. Patent MD no. 3017. BOPI no. 3/2006
27. Bostan I, Dulgheru V, Bostan V, Cozma T, Sochireanu A (2006) Device for the transformation of alternating rotational motion into unidirectional (one-way) rotational motion. Patent MD no. 2996. Publ. BOPI.- No.2/2006

Printed by Printforce, the Netherlands